A Life in Mathematical Physics: Conference in Honour of Tekin Dereli (DERELI-FS-2021)

Journal of Physics: Conference Series Volume 2191

Istanbul, Turkey
26 - 28 November 2021

ISBN: 978-1-7138-4527-0
ISSN: 1742-6588

Printed from e-media with permission by:

Curran Associates, Inc.
57 Morehouse Lane
Red Hook, NY 12571

Some format issues inherent in the e-media version may also appear in this print version.

Institute of Physics
Dirac House, Temple Back
Bristol BS1 6BE UK

Phone: 44 1 17 929 7481
Fax: 44 1 17 920 0979

techtracking@iop.org

Additional copies of this publication are available from:

Curran Associates, Inc.
57 Morehouse Lane
Red Hook, NY 12571 USA
Phone: 845-758-0400
Fax: 845-758-2633
Email: curran@proceedings.com
Web: www.proceedings.com

TABLE OF CONTENTS

A Life in Mathematical Physics: Conference in Honour of Tekin Dereli

1. Introduction

The friends and colleagues of Prof. Dr. Tekin Dereli came together to celebrate his 72nd birthday and the 50th year of his mathematical physics career. The three-days event took place online over the Zoom on 26-28 November 2021. His scientific and scholarly achievements, his kindness as a colleague, his thoughtful guidance of students and his service to the physics community in Turkey as a whole and last but not the least his open and easy accessibility to anyone needing his expert counseling have been appreciated by all.

We are grateful to the Turkish Academy of Sciences (TÜBA) for sponsoring the online, open-access publication of the proceedings as a special issue in The Journal of Physics: Conference Series.

2. Biography of Tekin Dereli

Tekin Dereli was born in Ankara in 1949 where he completed his early education at Ankara Science High School in 1967. He earned his Bachelor's degree in physics with high honors in 1971 at the Middle East Technical University (METU), where he also completed his PhD in 1976, with a thesis entitled "Fermions in Yang-Mills Potentials". He gave his habilitation in 1981 with a thesis on "Dynamics of Neutrinos in Gravitational Fields" (in Turkish). His academic career began at METU as an instructor in 1973, where he later became an associate professor. He was promoted to professorship in 1987 at Ankara University and then returned to METU in 1993. He was invited to Koç University as Professor of Physics in 2001 where he worked till 2021.

Tekin Dereli held various research positions abroad, being always on-leave from METU. These include a visiting scientist position at Yale University during 1974-1975 under the hosting of Feza Gürsey and a research associate position at Brandeis University during 1976-1977 under the hosting of Stanley Deser. He collaborated with Peter C. Aichelburg as Einstein-Memorial Foundation Fellow at the Institute for Theoretical Physics, University of Vienna during 1977-1978. He was SRC Research Associate at Lancaster University between 1979-1981 where a long and fruitful collaboration with Robin W. Tucker has begun. He had been an Associate Member of ICTP between 1986-1991 and Alexander von Humboldt Foundation Fellow at Karlsruhe University during 1988-1989 under the hosting of Julius Wess.

Tekin Dereli's research interests cover a broad range of subjects. He published papers on magnetic monopoles in Yang-Mills gauge field theories, supersymmetry and supergravity, the role of quaternions and octonions in physics. He studied Clifford algebras and spin structures associated with the space-time. He worked on generalised theories of gravity, both in higher and lower dimensions than four, emphasizing their differential geometric gauge structure. In particular he investigated various aspects of black hole and gravitational wave solutions.

Furthermore he constructed higher dimensional cosmological models. Tekin Dereli also worked on dualities in physics and mathematics in general and deformation quantisation in the context of the Landau problem.

Tekin Dereli's research output to this day amounts to 162 published articles. He has also written 47 papers in Turkish that are published in conference proceedings or popular science journals in Turkey. He authored a textbook on Quantum Mechanics in Turkish. Tekin Dereli has given courses for undergraduates at all levels both in physics and mathematics. He also offered a whole series of advanced theoretical physics courses over the years for graduate students in physics. Tekin Dereli supervised many young physicists whose work led to 26 Ph.D. theses, 31 M.S. theses and 75 graduation projects. Furthermore he collaborated with six post-doctoral research fellows.

Tekin Dereli received all the distinguished science awards in Turkey. They are, in chronological order, The Scientific and Technical Research Council of Turkey (TÜBİTAK) Science Encouragement Award (1982), Sedat Simavi Foundation Science Prize (1989), Prof. Dr. Mustafa Parlar Foundation Science Prize (1993), TÜBİTAK Science Prize (1996), Koç University College of Sciences Outstanding Teaching Award (2016), Turkish Physical Society Honor Award (2018). Tekin Dereli was an honorary member of the UNESCO Turkish National Commission between 1993-1996. He was elected a life-long member of the Turkish Academy of Sciences (TÜBA) in 1994 and is also a member of the Science Academy since 2012.

Tekin Dereli performed executive duties at the Turkish Physics Foundation and TÜBA. He served on the editorial boards of TÜBİTAK's popular Science and Technology Journal and the Koç University Press. He was the Director of the Graduate School of Sciences and Engineering at Koç University for two terms during the period 2009-2015.

Tekin Dereli is married since 1972 to Gülay Dereli who is also a physicist working in the field of computational condensed matter physics. They have a son, Cem Dereli and a daughter, Nazlı Dereli.

3. Scientific Bibliography of Tekin Dereli (1975-2021)

1. T. Dereli, J. H. Swank, L. J. Swank, "Fermions in Yang-Mills Electric and Magnetic Pole Potentials", Phys. Rev. D 11, 3541 (1975).
2. J. H. Swank, L. J. Swank, T. Dereli, "Fermions in Yang-Mills Potentials", Phys. Rev. D 12, 1096 (1975).
3. T. Dereli, L. J. Swank, "$SU(3)/SO(3)$ Magnetic Monopoles", COO-3075-137 Yale University Preprint (1976).
4. T. Dereli, S. Deser, "Supergravity Coupling to Nonlinear Realizations in Two-Dimensions", J. Phys. A 10, L149 (1977).
5. T. Dereli, S. Deser, "Addendum to Supergravity Coupling to Nonlinear Realizations in Two-Dimensions", J. Phys. A 10, 2189 (1977).
6. T. Dereli, S. Deser, "Fermionic Goldstone-Higgs Effect in $(2+1)$-Dimensional Supergravity", J. Phys. A 11, L27 (1978).
7. P. C. Aichelburg, T. Dereli, "Exact Plane Wave Solutions of Supergravity Field Equations", Phys. Rev. D 18, 1754 (1978).
8. P. C. Aichelburg, T. Dereli, "First Nontrivial Exact Solution of Supergravity", Czech. J. Phys. B 29, 252 (1979).

9. P. C. Aichelburg, T. Dereli, "Exact Plane Wave Solutions of $O(2)$ Extended Supergravity", Phys. Lett. B 80, 357 (1979).

10. T. Dereli, R. Güven, "Exact Solutions of Supergauge Invariant Yang-Mills Equations", Phys. Lett. B 84, 201 (1979).

11. I. M. Benn, T. Dereli, R. W. Tucker, "Gravitational Monopoles with Classical Torsion", J. Phys. A 13, L359 (1980).

12. I. M. Benn, T. Dereli, R. W. Tucker, "Gauge Field Interactions in Spaces with Arbitrary Torsion", Phys. Lett. B 96, 100 (1980).

13. T. Dereli, R. W. Tucker, "A Class of Exact Supergravity Solutions", Phys. Lett. B 97, 396 (1980).

14. I. M. Benn, T. Dereli, R. W. Tucker, "Double Dual Solutions of Generalized Theories of Gravitation", Gen. Rel. Grav. 13, 581 (1981).

15. I. M. Benn, T. Dereli, R. W. Tucker, "Solutions of Generalised Theories of Gravitation Derived from a Modified Double Duality Condition", Prepared for the Dynamical Systems and Microphysics: Geometry and Mechanics Conference, Udine, Italy, June 1981.

16. T. Dereli, R. W. Tucker, "A Gravitational Model for a Matter Free Torsion Ball", J. Phys. A 14, 2957 (1981).

17. T. Dereli, R. W. Tucker, "Exact Neutrino Solutions in the Presence of Torsion", Phys. Lett. A 82, 229 (1981).

18. T. Dereli, P. D. Morley, R. W. Tucker, "Electroweak Effects in a Superconductor", Phys. Lett. B 109, 497 (1982).

19. I. M. Benn, T. Dereli, R. W. Tucker, "A Critical Analysis of Some Fundamental Differences in Gauge Approaches to Gravitation", J. Phys. A 15, 849 (1982).

20. T. Dereli, R. W. Tucker, "An Intrinsic Analysis of Neutrino Couplings to Gravity", J. Phys. A 15, 1625 (1982).

21. T. Dereli, R. W. Tucker, "A Note on a Generalization of Weyl's Theory of Gravitation", J. Phys. A 15, L7 (1982).

22. T. Dereli, R. W. Tucker, "Weyl Scalings and Spinor Matter Interactions in Scalar-Tensor Theories of Gravitation", Phys. Lett. B 110, 206 (1982).

23. T. Dereli, M. Panahimoghaddam, R. W. Tucker, "Scalar Fields and Dynamics of the Early Universe", J. Phys. A 15, 3167 (1982).

24. T. Dereli, R. W. Tucker, "A Spin (1, 4) Gauge Theory Model with Kaluza-Klein Symmetry", Nucl. Phys. B 209, 217 (1982).

25. T. Dereli, R. W. Tucker, "Aspects Of A Spin (1, 4) Gauge Theory Model with Kaluza-Klein Symmetry", Prepared for the Conference on Group Theoretical Methods in Physics, İstanbul, Turkey, August 1982.

26. T. Dereli, "Dimensional Reduction of Einstein-Stephenson-Yang Theory of Gravity", Prepared for the 10th International Conference on General Relativity and Gravitation, Padova, Italy, June 1983.

27. T. Dereli, R. W. Tucker, "Dynamical Reduction of Internal Dimensions in the Early Universe", Phys. Lett. B 125, 133 (1983).

28. T. Dereli, M. Panahimoghaddam, A. Sudbery, R. W. Tucker, "Octonionic Geometry and Simple Supergravity in Eleven-Dimensions", Phys. Lett. B 126, 33 (1983).

29. T. Dereli, M. Önder, R. W. Tucker, "Einstein-Kahler Solutions in Some Cosmological Space-times", Class. Q. Grav. 1, L67 (1984).

30. A. A. Beler, T. Dereli, "Plane Waves in Supergravity", Class. Quant. Grav. 2, 147 (1985).

31. A. A. Beler, T. Dereli, "Plane Waves in $N = 2$ Extended Supergravity", Class. Quant. Grav. 2, 823 (1985).

32. A. A. Beler, T. Dereli, "Plane Waves in Supergravity Theories", Prepared for the First Torino Meeting on Supergravity and Extra Dimensions, Torino, Italy, August 1985.

33. T. Dereli, A. Eriş, "Conformal Scale Invariance and Scalar-Tensor Theories of Gravitation in Higher Dimensions", Turk. J. Phys. 9, 30 (1985).

34. T. Dereli, "Comments on Windows to Extra Dimensions", Phys. Lett. B 161, 307 (1985).

35. T. Dereli, M. Gürses, "The Generalized Kerr-Schild Transform in Eleven-Dimensional Supergravity", Phys. Lett. B 171, 209 (1986).

36. T. Dereli, A. Eriş, A. Karasu, "Harmonic Maps and Magnetostatic, Axially Symmetric Solutions of the Kaluza-Klein Theory", Nuovo Cim. B 93, 102 (1986).

37. T. Dereli, A. Verçin, "A Gauge Model of Amorphous Solids Containing Defects", Phil. Mag. B 56, 625 (1986).

38. M. Arık, T. Dereli, "Euler-Poincare Lagrangians and Kaluza-Klein Theory", Phys. Lett. B 189, 96 (1987).

39. T. Dereli, R. W. Tucker, "Variational Methods and Effective Actions in String Models", Class. Quant. Grav. 4, 791 (1987).

40. T. Dereli, R. W. Tucker, "Gravitational Interactions in $(2+1)$-Dimensions", Class. Quant. Grav. 5, 951 (1988).

41. T. Dereli, M. Mukherjee, R. W. Tucker, "Weyl Invariant Tensors in Odd Dimensions", Class. Quant. Grav. 5, L21 (1988).

42. M. Arık, T. Dereli, "Euler Form Actions and the Vanishing of the Cosmological Constant", Phys. Rev. Lett. 62, 5 (1989).

43. T. Dereli, G. Üçoluk, "Kaluza-Klein Reduction of Generalized Theories of Gravity and Nonminimal Gauge Couplings", Class. Quant. Grav. 7, 1109 (1990).

44. T. Dereli, G. Üçoluk, "Direct Curvature Yang-Mills Field Couplings Induced by the Kaluza-Klein Reduction of Euler Form Actions in Seven-Dimensions", Class. Quant. Grav. 7, 533 (1990).

45. T. Dereli, D. H. Hartley, M. Önder, R. W. Tucker, "Relativistic Elastica", Phys. Lett. B 252, 601 (1990).

46. T. Dereli, A. Verçin, "A Gauge Model of Amorphous Solids Containing Defects II. Chern-Simons Free Energy", Phil. Mag. B 64, 509 (1991).

47. T. Dereli, Ş. Özkurt, "Dynamical Reduction of First Order Effective String Field Equations in the Early Universe", Class. Quant. Grav. 8, 889 (1991).

48. T. Dereli, G. Üçoluk, "A Study of Stationary, Axially Symmetric Space-Time Geometries Satisfying Modified Double Duality Equations Using the Exterior Calculus Package XTR for Reduce", J. Comput. Phys. 99, 169 (1992).

49. T. Dereli, A. Verçin, "A Physical Realization of the Super-Sine Algebra", Phys. Lett. B 288, 109 (1992).

50. T. Dereli, A. Verçin, "Lie Algebra Deformation of $DiffT^2$ and Bloch Electrons in a Constant Uniform Magnetic Field", J. Phys. A 29, 6961 (1993).

51. T. Dereli, R. W. Tucker, "Signature Dynamics in General Relativity", Class. Quant. Grav. 10, 365 (1993).

52. T. Dereli, M. Önder, R. W. Tucker, "Signature Transitions in Quantum Cosmology", Class. Quant. Grav. 10, 1425 (1993).

53. S. Başkal, T. Dereli, "Higher Derivative Effective Yang-Mills Theory and Static Spherically Symmetric Field Configurations", J. Phys. G 19, 477 (1993).

54. T. Dereli, M. Önder, R. W. Tucker, "A Spinor Model for Quantum Cosmology", Phys. Lett. B 324, 134 (1994) [arXiv:gr-qc/9403005].

55. T. Dereli, R. W. Tucker, "Non-Metricity Induced by Dilaton Gravity in Two-Dimensions", Class. Quant. Grav. 11, 2575 (1994).

56. T. Dereli, M. Önder, R. W. Tucker, "The Classical Lorentz-Dirac Equation and Extrinsic Curvatures", Turk. J. Phys. 18, 77 (1994).

57. T. Dereli, M. Önder, R. W. Tucker, "On the Relation Between Classical and Quantum Cosmologies", Turk. J. Phys. 19, 2575 (1995).

58. T. Dereli, R. W. Tucker, "An Einstein-Hilbert Action for Axi-Dilaton Gravity in Four Dimensions", Class. Quant. Grav. 12, L31 (1995) [arXiv:gr-qc/9502018].

59. T. Dereli, M. Önder, R. W. Tucker, "Solutions for Neutral Axi-Dilaton Gravity in Four Dimensions", Class. Quant. Grav. 12, L25 (1995) [arXiv:gr-qc/9502019].

60. A. H. Bilge, T. Dereli, Ş. Koçak, "Self-Dual Yang-Mills Fields in Eight Dimensions", Lett. Math. Phys. 36, 301 (1996) [arXiv:hep-th/9603001].

61. T. Dereli, M. Önder, J. Schray, R. W. Tucker, C. Wang, "Non-Riemannian Gravity and the Einstein-Proca System", Class. Quant. Grav. 13, L103 (1996) [arXiv:gr-qc/9604039].

62. T. Dereli, J. Schray, R. W. Tucker, "Finite Action Yang-Mills Solutions on the Group Manifold", J. Phys. A 29, 5001 (1996) [arXiv:gr-qc/9605036].

63. A. H. Bilge, T. Dereli, Ş. Koçak, "The Geometry of Self-Dual Gauge Fields", Presented at the IX. Max Born Symposium, Karpacz, Poland, 25-27 September 1996 [arXiv:hep-th/9612230].

64. A. H. Bilge, T. Dereli , Ş. Koçak, "The Geometry of Self-Dual 2-forms", J. Math. Phys. 38, 4804 (1997) [arXiv:hep-th/9605060].

65. A. H. Bilge, T. Dereli, Ş. Koçak, "Self-Dual Gauge Fields in Eight Dimensions", Turk. J. Phys. 21, 361 (1997).

66. T. Dereli, A. Verçin, "W_∞-Covariance of the Weyl-Wigner-Groenewold-Moyal Quantization", J. Math. Phys. 38, 5515 (1997) [arXiv:quant-ph/9707040].

67. A. H. Bilge, T. Dereli, Ş. Koçak, "Seiberg-Witten Equations on R^8", Turk. J. Math. 21, 87 (1997) [arXiv:hep-th/9707178].

68. E. Atasoy, T. Dereli, M. Önder, "Constrained Euler-Poincare Supergravity in Five Dimensions", Turk. J. Phys. 21, 1181 (1997) [arXiv:gr-qc/9710091].

69. T. Dereli, M. Tekmen, "Wu-Yang Monopoles and Non-Abelian Seiberg-Witten Equations", Mod. Phys. Lett. A 13, 1803 (1998) [arXiv:hep-th/9806031].

70. A. H. Bilge, T. Dereli, Ş. Koçak, "Monopole Equations on 8-manifolds with Spin(7) Holonomy", Commun. Math. Phys. 203, 21 (1999) [arXiv:hep-th/9811161].

71. T. Dereli, R. W. Tucker, "String Fields and the Standard Model", Phys. Rev. Lett. 82, 4184 (1999) [arXiv:hep-th/9808059].

72. T. Dereli, R. W. Tucker, "Self-Gravitating Line Sources of Weak Hypercharge", Class. Quant. Grav. 16, L45 (1999) [arXiv:gr-qc/9902057].

73. A. H. Bilge, T. Dereli, Ş. Koçak, "Monopole Equations on R^8: The Energy Integral", Turk. J. Phys. 24, 253 (2000).

74. T. Dereli, Y. N. Obukhov, "On the Universality of Low-Energy String Model", Class. Quant. Grav. 17, 219 (2000) [arXiv:gr-qc/9909045].

75. T. Dereli, Y. N. Obukhov, "Massless Scalar Fields and Topological Black Holes", Phys. Rev. D 61, 084015 (2000) [arXiv:gr-qc/9912078].

76. T. Dereli, Y. N. Obukhov, "General Analysis of Self-Dual Solutions for the Einstein-Maxwell-Chern-Simons Theory in $(1+2)$ Dimensions", Phys. Rev. D 62, 024013 (2000) [arXiv:gr-qc/0001017].

77. T. Dereli, M. Önder, J. Schray, R. W. Tucker, C. W. Wang, "Non-Riemannian Gravitational Interactions: An Overview", Prepared for the Proceedings of the 9th Marcel Grossmann Meeting on General Relativity, Rome, Italy, July 2000.

78. T. Dereli, Ö. Sarıoğlu, "'Self-Dual' Solutions of Topologically Massive Gravity Coupled with the Maxwell-Chern-Simons Theory", Phys. Lett. B 492, 339 (2000) [arXiv:gr-qc/0009090].

79. T. Dereli, Ö. Sarıoğlu, "Supersymmetric Solutions to Topologically Massive Gravity and Black Holes in Three Dimensions", Phys. Rev. D 64, 027501 (2001) [arXiv:gr-qc/0009082].

80. A. Çatal-Özer, T. Dereli, "Non-Commutative Geometry and The Higgs Masses", Phys. Rev. D 63, 075006 (2001) [arXiv:hep-th/0011084].

81. A. Çatal-Özer, T. Dereli, "Electroweak Theory and Non-Commutative Geometry", Class. Quant. Grav. 18, 3251 (2001) Special Issue: Gürsey Memorial Conference II: M-Theory and Dualities, İstanbul, Turkey, 19-23 June 2000 [arXiv:hep-th/0102048].

82. M. Adak, T. Dereli, L. H. Ryder, "Neutrino Oscillations Induced by Space-Time Torsion", Class. Quant. Grav. 18, 1503 (2001) [arXiv:gr-qc/0103046].

83. H. Cebeci, T. Dereli, "Black Brane Solutions of Axi-Dilaton Gravity in D-dimensions", Phys. Rev. D 65, 047501 (2002) [arXiv:gr-qc/0009069].

84. D. Burton, T. Dereli, J. Gratus, W. Johnson, R. W. Tucker, C. W. Wang, "Towards the Control of Matter With Gravity", Prepared for the American Institute of Aeronautics and Astronautics Meeting, Utah, U. S.A., July 2001.

85. T. Dereli, R. W. Tucker, "On the Detection of Scalar Fields Induced by Space-Time Torsion", Mod. Phys. Lett. A 17, 421 (2002) [arXiv:gr-qc/0104050].

86. T. Dereli, R. W. Tucker, "A Broken Gauge Approach to Gravitational Mass And Charge", JHEP 0203, 041 (2002) [arXiv:hep-th/0112123].

87. M. Adak, T. Dereli, L. H. Ryder, "Dirac Equation in Space-Times with Torsion and Non-Metricity", Int. J. Mod. Phys. D 12, 145 (2003) [arXiv:gr-qc/0208042].

88. A. H. Bilge, T. Dereli, Ş. Koçak, "Seiberg-Witten Type Monopole Equations on 8-manifolds with Spin(7) Holonomy as Minimizers of a Quadratic Action", JHEP 0304, 003 (2003) [arXiv:hep-th/0303098].

89. D. Burton, T. Dereli, R. W. Tucker, "On the Motion of Matter in Gravitational Fields", in Symmetries in Gravity and Field Theory, V. Aldaya, J. M. Cervero, Y. P. Garcia (Editors) (Ediciones Universidad Salamanca, 2004) pp. 237-249 [arXiv:gr-qc/0107017].

90. A. H. Bilge, T. Dereli, Ş. Koçak, "Monopole Equations on 8-manifolds with Spin(7) Holonomy: An Overview of the Energy Integral", Proceedings of 5th International Conference on Symmetry in Nonlinear Mathematical Physics p.1266. Edited by A. G. Nikitin, V. M. Boyko, R. O. Popovych, I. A. Kehurchenkon (Kyiv, Institute of Mathematics of NAS of Ukraine, 2004).

91. M. Adak, T. Dereli, L. H. Ryder, "On the Possible Effects of Space-Time Non-Metricity on Neutrino Oscillations", Phys. Rev. D 69, 123002 (2004) [arXiv:gr-qc/0303080].

92. H. Cebeci, T. Dereli, R. W. Tucker, "Autoparallel Orbits in Kerr Brans-Dicke Space-Times", Int. J. Mod. Phys. D 13, 137 (2004) [arXiv:gr-qc/0311080].

93. T. Dereli, R. W. Tucker, "On the Energy-Momentum Density of Gravitational Plane Waves", Class. Q. Grav. 21, 1459 (2004) [arXiv:hep-th/0401130].

94. T. Dereli, Ş. Koçak, M. Limoncu, "Newton-Cartan Connections with Torsion", Bull. İstanbul Tech. Univ. 54, 1 (2004) [arXiv:gr-qc/0402116].

95. M. Adak, T. Dereli, "Spinor Couplings to Dilaton Gravity Induced by the Dimensional Reduction of Topologically Massive Gravity", Class. Q. Grav. 21, 2275 (2004) [arXiv:gr-qc/0403016].

96. A. B. Balantekin, T. Dereli, Y. Pehlivan, "An Exactly Solvable Model of Interacting Bosons", J. Phys. G 30, 1225 (2004) [arXiv:nuc-th/0407006].

97. T. Dereli, N. T. Yılmaz, "Dualisation of the Salam-Sezgin D=8 Supergravity", Nucl. Phys. B 691, 223 (2004)[arXiv:hep-th/0407004].

98. H. Cebeci, T. Dereli, "Axi-Dilaton Gravity in $D > 4$ Dimensional Space-Times with Torsion", Phys. Rev. D 71, 024016 (2005) [arXiv:gr-qc/0406119].

99. T. Dereli, N. T. Yılmaz, "Dualisation of the Symmetric Space Sigma Model with Couplings", Nucl. Phys. B 705, 60 (2005) [arXiv:hep-th/0507007].

100. T. Dereli, R. W. Tucker, "Gravitational Waves and Energy-Momentum Quanta", Proceedings of 3rd Workshop on Mathematical and Physical Aspects of Quantum Gravity, 28 July-1 August 2005, Blaubeuren, Germany.

101. A. B. Balantekin, T. Dereli, Y. Pehlivan, "Exactly Solvable Pairing Model Using an Extension of Richardson-Gaudin Approach", Int. J. Mod. Phys. E 14, 47 (2005) [arXiv:nucl-th/0505023].

102. A. B. Balantekin, T. Dereli, Y. Pehlivan, "Solutions of the Gaudin Equations and Gaudin Algebras", J. Phys. A 38, 5697 (2005) [arXiv:math-ph/0505071].

103. A. T. Bölükbaşı, T. Dereli, "On the $SU(3)$ Parametrization of Qutrits", J. Phys. Conf. Proc. 36, 28 (2006) [arXiv:quant-ph/0511111].

104. A. N. Aliev, H. Cebeci, T. Dereli, "Gravitational Field Equations in a Braneworld with Euler-Poincare Term", Class. Q. Grav. 23, 591 (2006) [arXiv:hep-th/0507121].

105. T. Dereli, Ş. Koçak, M. Limoncu, "Degenerate Spin Structures and the Levy-Leblond Equation", Turk. J. Phys. 30, 411 (2006) [arXiv:math-ph/0611028].

106. A. N. Aliev, H. Cebeci, T. Dereli, "Plane Symmetric Solutions of Gravitational Field Equations in Five Dimensions", Turk. J. Phys. 30, 401 (2006) [arXiv:gr-qc/0611065].
107. T. Dereli, J. Gratus, R. W. Tucker, "The Covariant Description of Electromagnetically Polarizable Media", Phys. Lett. A361, 190 (2007), [arXiv:math-ph/0610078] Nature News and Views, Vol 444/14 December 2006, 823.
108. T. Dereli, J. Gratus, R. W. Tucker, "New Perspectives on the Relevance of Gravitation for the Covariant Description of Electromagnetically Polarizable Media", J. Phys. A40, 5695 (2007) [arXiv:math-ph/0610082].
109. T. Birol, T. Dereli, Ö. Müstecaplıoğlu, L. You, "Coherence Lifetimes of Excitations in an Atomic Condensate Due to the Thin Spectrum", Phys. Rev. A76, 043616 (2007) [arXiv:quant-ph/0709.4268] Virtual Journal of Quantum Information, November 2007.
110. A. N. Aliev, H. Cebeci, T. Dereli, "Exact Solutions in Five Dimensional Axi-Dilaton Gravity with Euler-Poincare Term", Class. Q. Grav. 24, 3425 (2007) [arXiv:gr-qc/0703011].
111. A. B. Balantekin, T. Dereli, "An Exact Cosmological Solution of the Coupled Einstein-Majorana Fermion-Scalar Field Equations", Phys. Rev. D75, 024039 (2007) [arXiv:gr-qc/0701025].
112. T. Dereli, Ş. Koçak, M. Limoncu, "Linear Connections on Light-like Manifolds", Turk. J. Math. 32, 41 (2008) [arXiv:math.DG/0701934].
113. M. Adak, T. Dereli, "The Quadratic Symmetric Teleparallel Gravity in Two Dimensions", Europhys. Lett. 82, 30008 (2008) [arXiv:hep-th/0607058].
114. A. N. Aliev, H. Cebeci, T. Dereli, "Kerr-Taub-NUT Spacetime with Maxwell and Dilaton Fields", Phys. Rev. D78, 124022 (2008) [arXiv:hep-th/0803.2518].
115. T. Dereli, (Book Review) Giorgio Ferrarese, Donato Bini: Introduction to Relativistic Continuum Mechanics (Lecture Notes in Physics, Vol. 727, Springer, 2008) Gen. Rel. Grav. 41, 449-450 (2009).
116. T. Dereli, A. Teğmen, T. Hakioğlu, "Canonical Transformations in Three-Dimensional Phase Space", Int. J. Mod. Phys. A24, 4769 (2009).
117. T. Dereli, A. Teğmen, T. Hakioğlu, "Quantum Canonical Transformations in Weyl-Wigner-Groenewold-Moyal Formalism", Int. J. Mod. Phys. A24, 4573 (2009).
118. A. H. Bilge, T. Dereli, Ş. Koçak, "Maximal Linear Subspaces of Strong Self-Dual 2-Forms and the Bonan 4-Form", Lin. Alg. Appl. 434, 1200 (2010).
119. T. Dereli, Ş. Koçak, M. Limoncu, "Degenerate Spin Groups as Semi-Direct Products", Adv. App. Clifford Alg. 20, 565 (2010) [arXiv:gr-qc/0402116].
120. T. Dereli, R. W. Tucker, "Charged Relativistic Fluids and Non-Linear Electrodynamics", Europhys. Lett. 89, 20009 (2010); Erratum: ibid, 90, 49901 (2010) [arXiv:math-ph/1001.1282].
121. D. A. Burton, T. Dereli, R. W. Tucker, "Born-Infeld Axion-Dilaton Electrodynamics and Electromagnetic Confinement", Phys. Lett. B703, 530 (2011) [arXiv:hep-th/1012.4995].
122. T. Dereli, Ö. Sert, "Non-Minimally Coupled Gravitational and Electromagnetic Fields: pp-Wave Solutions", Phys. Rev. D83, 065005 (2011) [arXiv:gr-qc/1101.1177].
123. T. Dereli, Ö. Sert, "Non-Minimal $ln(R)F^2$ Couplings of Electromagnetic Fields to Gravity: Static, Spherically Symmetric Solutions", Euro. Phys. J. C71 1589 (2011) [arXiv:gr-qc/1102.3863].
124. T. Dereli, Ö. Sert, "Non-Minimal $R^{\beta}F^2$-Coupled Electromagnetic Fields to Gravity and Static, Spherically Symmetric Solutions", Mod. Phys. Lett. A26, 1487 (2011) [arXiv:gr-qc/1105.4579].

125. M. Adak, T. Dereli, "String-Inspired Chern-Simons Modified Gravity in 4-Dimensions", Euro. Phys. J. C72, 1979 (2012) [arXiv:gr-qc/0807.1832].

126. T. Dereli, Y. Gül, P. Forn-Diaz, Ö. Müstecaplıoğlu, "Two-Frequency Jahn-Teller Systems in Circuit QED", Phys. Rev. A85, 053841 (2012) [arXiv:quant-ph/1109.1199].

127. Ö. Akarsu, T. Dereli, "Cosmological Models with Linearly Varying Deceleration Parameter", Int. J. Theor. Phys. 51, 612 (2012) [arXiv:gr-qc/1102.0915].

128. Ö. Akarsu, T. Dereli, "A Comparison of the LVDP and ΛCDM Cosmological Models", Int. J. Theor. Phys. 51, 2995 (2012) [arXiv:gr-qc/1202.0495].

129. T. Dereli, N. Özdemir, Ö. Sert, "Einstein-Cartan-Dirac Theory in $(1+2)$-Dimensions", Euro. Phys. J. C73, 2279 (2013) [arXiv:gr-qc/1002.0958].

130. T. Dereli, T. Hakioğlu, A. Teğmen, "Quantum Canonical Transformations in Star-Product Formalism", J. Phys.: Conf. Series 462, 012054 (2013).

131. Ö. Akarsu, T. Dereli, "A Four-Dimensional ΛCDM-Type Cosmological Model Induced from Higher Dimensions Using a Kinematical Constraint", Gen. Rel. Grav. 45, 1211 (2013) [arXiv:gr-qc/1201.4545].

132. Ö. Akarsu, T. Dereli, "The Dynamical Evolution of 3-Space in a Higher Dimensional Steady State Universe", Gen. Rel. Grav. 45, 959 (2013) [arXiv:gr-qc/1210.1155].

133. Ö. Akarsu, T. Dereli, "Late Time Acceleration of the 3-Space in a Higher Dimensional Steady State Universe in Dilaton Gravity", J. Cosm. Astropart. Phys. (JCAP) 1302, 050 (2013) [arXiv:gr-qc/1210.8106].

134. Ö. Akarsu, T. Dereli, S. Kumar, L. Xu, "Probing Kinematics and Fate of the Universe with Linearly Time-Varying Deceleration Parameter", Europ. Phys. J. Plus 129, 22 (2014) [arXiv:gr-qc/1305.5190].

135. Ö. Akarsu, T. Dereli, N. Oflaz, "Accelerating Anisotropic Cosmologies in Brans-Dicke Gravity Coupled to a Mass-Varying Vector Field", Class. Q. Grav. 31, 045020 (2014) [arXiv:gr-qc/1311.2573].

136. A. K. Ardabili, T. Dereli, Ö. Müstecaplıoğlu, "Z_2 Topological Insulator of Ultracold Atoms in Bichromatic Optical Lattices" , Physica, B459, 1 (2015) [arXiv:cond-mat/1302.5524].

137. Ö. Akarsu, T. Dereli, N. Katırcıoğlu, M. Sheftel, "General Solution of a Cosmological Model Induced from Higher Dimensions Using a Kinematical Constraint", Gen. Rel. Grav. 47, 61 (2015) [arXiv:gr-qc/1407.2231].

138. Ö. Akarsu, T. Dereli, J. A. Vasquez, "A Divergence-Free Parametrization of Dynamical Dark Energy", JCAP 06 (2015) 049 [arXiv:gr-qc/1501.07598].

139. Ö. Akarsu, T. Dereli, N. Oflaz, "Accelerated Expansion of the Universe in a Higher Dimensional Modified Gravity with Euler-Poincaré Terms", Class. Q. Grav. 32, 215009 (2015) [arXiv:gr-qc/1507.08225].

140. A. Baykal, T. Dereli, "A Non-Minimally Coupled Einstein-Maxwell Model in a Non-Riemann Space-Time Wwith Torsion", Phys. Rev. D92, 065018 (2015) [arXiv:gr-qc/1508.01630].

141. T. Dereli, C. Yetişmişoğlu, "New Improved Massive Gravity", Europhys. Lett. 114, 600004 (2016).

142. T. Dereli, C. Yetişmişoğlu, "New Improved Massive Gravity and Three Dimensional Spacetimes of Constant Curvature and Constant Torsion" Phys. Rev. D94, 064067 (2016) [arXiv:gr-qc/1604.00463].

143. A. H. Bilge, T. Dereli, Ş. Koçak, "Self-duality in Higher Dimensions", J. Phys. Conf. Series 804, (2017) 012005.

144. M. Adak, Ö. Akarsu, T. Dereli, Ö. Sert, "Anisotropic Inflation with a Non-minimally Coupled Electromagnetic Field to Gravity", JCAP, 11 (2017) 026 [arXiv:gr-qc/1611.03393].

145. A. Baykal, T. Dereli, "Linearised Gravity in Terms of Differential Forms", Europ. Phys. J. Plus, 132 (2017) 52 [arXiv:gr-qc/1612.05510].

146. A. Baykal, T. Dereli, "Impulsive Gravitational Waves in General Massive $3D$ Gravity", Phys. Rev. D96, 084028 (2017) [arXiv:gr-qc/1706.04521].

147. S. Benli, T. Dereli, "Masses and Mixing of Neutral Leptons in a Grand Unified E_6 Model with Intermediate Pati-Salam Symmetry", Int. J. Theo. Phys. 57 (2018) 2343 [arXiv:hep-ph/1707.03144].

148. A. Baykal, T. Dereli, "Abbott-Deser Energy in Connection with Thirring's Superpotentials", Europhys. Lett. 122 (2018) 50002 [arXiv:gr-qc/1801.04219].

149. A. K. Ardabili, T. Dereli, Ö. Müstecaplıoğlu, "A Tight Binding Model for Quantum Spin Hall Effect on Optical Lattices", Aurum J. Engin. Syst. Arch. 3, 23 (2019) [arXiv:cond-mat/1401.3179].

150. M. Adak, T. Dereli, Y. Şenikoğlu, "Non-Riemannian Description of Robinson-Trautman Spacetimes in Brans-Dicke Gravity", Int. J. Mod. Phys. D28 (2019) 1950070 [arXiv:gr-qc/1807.07981].

151. B. Demirkaya, T. Dereli, K. Güven, "Analog Black Holes and Energy Extraction by Superradiance from Bose-Einstein Condensates (BEC) with Constant Density", Heliyon 5 (2019) e02497 [arXiv:cond-mat.other/1806.02139].

152. T. Dereli, K. İ. Ünlütürk, "Hamilton-Jacobi Formulation of The Thermodynamics of Einstein-Born-Infeld-AdS Black Holes", Eurohys. Lett. 125 (2019) 10005 [arXiv:gr-qc/1810.10505].

153. T. Dereli, C. Yetişmişoğlu, "Weyl Covariant Theories of Gravity in 3-Dimensional Riemann-Cartan-Weyl Spacetimes", Class. Q. Grav. 36 (2019) 215005.

154. T. Dereli, C. Yetişmişoğlu, "Weyl Covariant Quadratic Curvature Gravity in Riemann-Cartan-Weyl Spacetimes", Phys. Rev. D100 (2019) 044010 [arXiv:gr-qc/1906.11552].

155. B. Demirkaya, T. Dereli, K. Güven, "Acoustic Superradiance from a Bose-Einstein Condensate Vortex with a Self-Consistent Background Density", Phys. Scr. 95 (2020) 055001 [arXiv:cond-mat.quant-gas/1904.08113].

156. T. Dereli, Y. Şenikoğlu, "Gravitational Plane Waves in a Non-Riemannian Description of Brans-Dicke Gravity", Phys. Scr. 95 (2020) 045223 [arXiv:gr-qc/1904.11209].

157. T. Dereli, Y. Şenikoğlu, "A Non-Minimally Coupled, Conformally Extended Einstein-Maxwell Theory of pp-Waves", Turk. J. Phys. 44 (2020) 273 [arXiv:gr-qc/2002.09208].

158. T. Dereli, Y. Şenikoğlu, "Electric Charge Screening by Non-Minimal Couplings of Electromagnetic Fields to Gravity", Europhys. Lett. 131 (2020) 10006 [arXiv:gr-qc/2004.075051].

159. T. Dereli, K. Doğan, "Metric-Connection Geometries on Pre-Leibniz Algebroids: A Search for Geometrical Structure in String Models", J. Math. Phys. 62 (2021) 032301 [arXiv:hep-th/2006.05957].

160. T. Dereli, C. Yetişmişoğlu, "Variational Field Equations of Cosmological Topologically Massive Supergravity", Fort. d. Phys. 69 (2021) 2100009 [arXiv:gr-qc/2007.09706].

161. T. Dereli, Y. Şenikoğlu, "Non-Minimally Coupled Einstein-Yang-Mills Field Equations and Wu-Yang Monopoles in Bertotti-Robinson Spacetimes", Europhys. Lett. 135 (2021) 30001 [arXiv:gr-qc/2105.03055].

Journal of Physics: Conference Series **2191** (2022) 011001 doi:10.1088/1742-6596/2191/1/011001

162. T. Dereli, T. Popov, "Landau Problem and Noncommutative Tori of Magnetic Translations", J. Math. Phys. 62 (2021) 101701 [arXiv:math-ph/2106.11093].

4. Theses Supervised by Tekin Dereli

M.S. Theses:

1. *Solutions of Yang-Mills Equations in Euclidean Space* (Ahmet Satır, METU, 1980).
2. *Geometric Methods in Physics of Defects* (Tahir Çağın, METU, 1983).
3. *Quantum Adiabatic Theorem, Berry Phase and Its Applications* (In Turkish) (Adnan Teğmen, Ankara University, 1992).
4. *Coherent States in Quantum Mechanics* (Celal Harabati, METU, 1995).
5. *Symmetries and Algebras in String Models* (Vakıf Kemal Önemli, METU, 1997).
6. *The Mixmaster Cosmological Models as Chaotic Systems* (Wael al-Sawai, METU, 1997).
7. *Effective String Field Theories in Terms of Non-Riemannian Geometries* (In Turkish) (Muzaffer Adak, Pamukkale University, 1997).
8. *Weyl-Wigner-Groenewold-Moyal Quantization* (In Turkish) (İlhami Buğdaycı, Ankara University, 1998).
9. *Classical Solutions of Seiberg-Witten Monopole Equations* (Mohaç Tekmen, METU, 1998).
10. *Classical Dynamics of p-Branes* (Aziz Kolkıran, METU, 1999).
11. *Magnetic Black Branes in M-Theory* (Hakan Cebeci, METU, 1999).
12. *Geometry of Instantons* (Nejat Tevfik Yılmaz, METU, 1999).
13. *Electroweak Theory and Non-Commutative Geometry* (Aybike Çatal-Özer, METU, 2000).
14. *Oscillations of Disc Dynamos and Planetary Magnetism* (Yamaç Pehlivan, METU, 2000).
15. *Differential Geometry of Newton-Cartan Gravity* (In Turkish) (Murat Limoncu, Anadolu University, 2001).
16. *Classical Dynamics of 1-Dimensional Extended Objects* (Cem Yüce, METU, 2002).
17. *A Molecular Dynamics Simulation Study of Torsional Deformations of Single-Walled Carbon Nanotubes* (Ufuk Paralı, Koç University, 2005).
18. *Bloch Vector Formalism for Quantum Computation* (Ahmet Tuna Bölükbaşı, Koç University, 2006).
19. *Thin Spectrum and Excitation Lifetimes in Atomic Bose-Einstein Condensates* (Turan Birol, Koç University, 2007).
20. *The Schrödinger-Newton Equations* (Habiba Kalantarova, Koç University, 2007).
21. *Molecular Dynamics Simulation Study of Silicon Nanotubes* (Selçuk Yaşar, Koç University, 2008).
22. *Asymptotic Freedom of Non-Commutative Φ^3-Scalar Field Theory in 6-Dimensions* (Umut Büyükçam, Koç University, 2009).
23. *$f(R)$ Theories of Dark Gravity* (Neslihan Oflaz, Koç University, 2009).
24. *$(2+1)$-Dimensional Theories of Gravity* (Şeyda İpek, Koç University, 2010).
25. *Quantum Key Distribution Protocols* (Utkan Güngördü, Koç University, 2010).
26. *Anyonic Quantum Computation* (İsmail Enes Uysal, Koç University, 2011).
27. *Designing Qubits with Superconducting Circuits* (Fırat Solgun, Koç University, 2011).
28. *Non-Abelian Magnetic Monopoles and Dyons* (Seçil Gürkan, Koç University, 2012).
29. *Acoustic Black Holes* (Betül Demirkaya, Koç University, 2014).
30. *Hamilton-Jacobi Formulation of the Thermodynamics of Einstein-Born-Infeld-AdS Black Holes* (Kıvanç Ünlütürk, Koç University, 2019).
31. *Minisuperspace Models: Universe and Anti-Universe Pair Creation and Annihilation* (Elsad Cukali, Koç University, 2021).

Ph.D. Theses

1. *Kähler Spinors* (Metin Önder, Hacettepe University, 1985).
2. *Plane Waves in Supergravity* (Aytaç Beler, METU, 1988).
3. *Differential Geometric and Topological Description of Continuum Defects and Gauge Theories* (In Turkish) (Abdullah Verçin, Ankara University, 1988).

4. *An Exterior Calculus Package for REDUCE and Dimensional Reduction of Generalised Theories of Gravity* (Göktürk Üçoluk, METU, 1989).
5. *Constraints on the Number of Quark Generations Through Mass Matrices* (Recai Erdem, METU, 1991).
6. *Higher Dimensional Cosmological Models and Dynamical Reduction of Internal Dimensions (Kaluza-Klein Cosmologies)* (In Turkish) (Şuayyip Özkurt, Ankara University, 1991).
7. *The Relationship Between Scientific Productivity as Measured by Bibliometric Indicators and Individual, Social and Academic Characteristics of Turkish Physicists Working in Turkey During the Period 1966-1989* (Mehmet Sancar, METU, 1992).
8. *Euler-Poincaré Supergravity in Five Dimensions* (In Turkish) (Efkan Atasoy, Hacettepe University, 1996).
9. *Topological Effects in Classical and Quantum Physics and Some Applications* (In Turkish) (Adnan Teğmen, Ankara University, 1998).
10. *Relativistic Cosmology Models with Two Scalar Fields* (In Turkish) (Ayhan Yüzübenli, Hacettepe University, 2000).
11. *Non-Riemannian Theories of Gravitation and the Dirac Equation* (Muzaffer Adak, METU, 2001).
12. *Kerr Black Holes and Their Generalizations* (Hakan Cebeci, METU, 2003).
13. *Scherk-Schwarz Reduction of Effective String Theories in Even Dimensions* (Aybike Çatal-Özer, METU, 2003).
14. *Dualization of Supergravity Theories* (Nejat Tevfik Yılmaz, METU, 2004).
15. *Matrix Quantum Mechanics and Integrable Systems* (Yamaç Pehlivan, METU, 2004).
16. *Newton-Lévy-Leblond Equation* (In Turkish) (Murat Limoncu, Anadolu University, 2005).
17. *The Couplings of Electromagnetic and Dirac Spinor Fields to Gravity* (Özcan Sert, İstanbul Technical University, 2011).
18. *Z_2-Topological Insulators and the Quantum Spin Hall Effect on Triangular Optical Lattices* (Ahad K. Ardabili, Koç University, 2014).
19. *Cosmological Solutions of String Inspired Modified Theories of Gravity* (Neslihan Oflaz, Koç University, 2015).
20. *Masses and Mixings of Neutral Fermions in a Grand Unified E_6 Model with Pati-Salam Symmetry* (Seçil Gürkan, Koç University, 2017).
21. *Analog Black Holes and Energy Extraction by Super-Radiance from a Single-Vortex in Bose-Einstein Condensates* (Betül Demirkaya, Koç University, 2019).
22. *Bilayer Skyrmions in Chiral Magnets and Quantum Entanglement* (Walid al-Masri, Koç University, 2020).
23. *Generalizations of Differential Geometric Structures on Pre-Leibniz Algebroids* (Keremcan Doğan, Koç University, 2021).
24. *Three Dimensional Models of Gravity and Supergravity* (Cem Yetişmişoğlu, Koç University, *Expected* 2022).
25. Ongoing Ph.D. Work, (Kıvanç Ünlütürk, Koç University, *Expected* 2023).
26. Ongoing Ph.D. Work, (Ekin Sıla Yörük, Koç University, *Expected* 2023).

5. Alphabetical List of Contributors

- Muzaffer Adak (Department of Physics, Pamukkale University, Denizli, Turkey)
  madak@pamukkale.edu.tr
- Özgür Akarsu (Department of Physics, İstanbul Technical University, İstanbul, Turkey)
  akarsuo@itu.edu.tr
- Metin Arık (Department of Physics, Boğaziçi Uniersity, İstanbul, Turkey)
  metin.arik@boun.edu.tr
- Baha Balantekin (Department of Physics, University of Wisconsin-Madison, USA)
  baha@physics.wisc.edu

- Ithzak Bars (Department of Physics and Astronomy, University of Southern California, USA) bars@usc.edu
- Ahmet Baykal (Department of Physics, Niğde Ömer Halisdemir University, Niğde, Turkey) abaykal@ohu.edu.tr
- Ayşe Hümeyra Bilge (Faculty of Engineering and Basic Sciences, Kadir Has University, İstanbul, Turkey) ayse.bilge@khas.edu.tr
- Sultan Catto (Physics Department, The Graduate School, City University of New York, USA) scatto@gc.cuny.edu
- Tahir Çağın (Department of Materials Science and Engineering, Texas A&M University, USA) Tahir.Cagin@chemail.tamu.edu
- Aybike Çatal-Özer (Department of Mathematics, İstanbul Technical University, İstanbul, Turkey) ozerayb@itu.edu.tr
- Nihat Sadık Değer (Department of Mathematics, Boğaziçi University, İstanbul, Turkey) sadik.deger@boun.edu.tr
- Keremcan Doğan (Department of Mathematics, İstanbul Technical University, İstanbul, Turkey) kedogan@ku.edu.tr
- Behzat Ergun (Department of Physics, Technion, Haifa, Israel) bergun@utexas.edu
- Murat Günaydın (Department of Physics, Penn State University, Pennsylvania, USA) mgunaydin@psu.edu
- Metin Gürses (Department pf Mathematics, Bilkent University, Ankara, Turkey) gurses@fen.bilkent.edu.tr
- Umut Gürsoy (Institute for Theoretical Physics, Utrecht University, Utrecht, Netherlands) U.Gursoy@uu.nl
- Kaan Güven (Department of Physics, Koç University, İstanbul, Turkey) kguven@ku.edu.tr
- Mustafa Halilsoy (Department of Physics, Eastern Mediterranean University, Gazimağusa, Northern Cyprus) mustafa.halilsoy@emu.edu.tr
- Mahmut Hortaçsu (Department of Physics, Mimar Sinan University of Fine Arts, İstanbul, Turkey) hortacsu@itu.edu.tr
- Şeyda İpek (Department of Physics, Carleton University, Ottawa, Canada) ipekseyda@gmail.com
- Varga K. Kalantarov (Department of Mathematics, Koç University, İstanbul, Turkey) vkalantarov@ku.edu.tr
- Alper Kiraz (Department of Physics, Koç University, İstanbul, Turkey) akiraz@ku.edu.tr
- Şahin Koçak (Department of Mathematics, Anadolu University, Eskişehir, Turkey) skocak@anadolu.edu.tr
- Ali Mostafazadeh (Departments of Mathematics and Physics, Koç University, İstanbul, Turkey) amostafazadeh@ku.edu.tr
- Özgür Müstecaplıoğlu (Department of Physics, Koç University, İstanbul, Turkey) omustecap@ku.edu.tr
- Yuri N. Obukhov (Theoretical Physics Laboratory, Nuclear Safety Institute, Russian Academy of Sciences, Moscow, Russia) ynobukhov@gmail.com
- Yamaç Pehlivan (Department of Physics, Mimar Sinan University of Fine Arts, İstanbul, Turkey) yamac.pehlivan@gmail.com

- Todor Popov (Institute for Nuclear Researches and Nuclear Energy, Bulgarian Academy of Sciences, Sofia, Bulgaria) todor-popov@yahoo.fr
- Alphan Sennaroğlu (Department of Physics, Koç University, İstanbul, Turkey) asennar@ku.edu.tr
- Ergin Sezgin (Department of Physics, Texas A&M University, Texas, USA) sezgin@physics.tamu.edu
- Yorgo Şenikoğlu (Faculty of Engineering and Basic Sciences, Maltepe University, İstanbul, Turkey) yorgo.senikoglu@maltepe.edu.tr
- Bayram Tekin (Department of Physics, Middle East Technical University, Ankara, Turkey) btekin@metu.edu.tr
- Robin W. Tucker (Department of Physics, University of Lancaster, Lancaster, UK) robin.tucker@lancaster.ac.uk
- Teoman Turgut (Department of Physics, Boğaziçi University, İstanbul, Turkey) turgutte@boun.edu.tr
- Kıvanç Ünlütürk (Department of Physics, Koç University, İstanbul, Turkey) kunluturk@ku.edu.tr
- Cem Yetişmişoğlu (Department of Physics, Koç University, İstanbul, Turkey) cyetismisoglu@ku.edu.tr
- Ersin Yurtsever (Department of Chemistry, Koç University, İstanbul, Turkey) eyurtsever@ku.edu.tr

6. Organising Committee

- Ali Mostafazadeh (Departments of Mathematics and Physics, Koç University, İstanbul, Turkey)
- Özgür Müstecaplıoğlu (Department of Physics, Koç University, İstanbul, Turkey)
- Keremcan Doğan (Department of Mathematics, İstanbul Technical University, İstanbul, Turkey)
- Kıvanç Ünlütürk (Department of Physics, Koç University, İstanbul, Turkey)
- Cem Yetişmişoğlu (Department of Physics, Koç University, İstanbul, Turkey)

Detailed information about the conference can be found in the link below:

https://sites.google.com/ku.edu.tr/tekindereliconference/home?authuser=0

Journal of Physics: Conference Series **2191** (2022) 011002 doi:10.1088/1742-6596/2191/1/011002

Peer Review Statement

All papers published in this volume have been reviewed through processes administered by the Editors. Reviews were conducted by expert referees to the professional and scientific standards expected of a proceedings journal published by IOP Publishing.

- **Type of peer review:** Single Anonymous
- **Conference submission management system:** Morressier
- **Number of submissions received:** 30
- **Number of submissions sent for review:** 24
- **Number of submissions accepted:** 21
- **Acceptance Rate (Submissions Accepted / Submissions Received × 100):** 70
- **Average number of reviews per paper:** 1
- **Total number of reviewers involved:** 18
- **Contact person for queries:**
  Name:
  Email: tdereli@ku.edu.tr
  Affiliation: Koc University - Physics

Journal of Physics: Conference Series **2191** (2022) 012001 doi:10.1088/1742-6596/2191/1/012001

ΛCDM cosmology with a quiescent anisotropy in a higher dimensional steady state universe

Özgür Akarsu[1], **Tekin Dereli**[2] **and Nihan Katırcı**[3]

[1] Department of Physics, Istanbul Technical University, Maslak 34469, Istanbul, Turkey
[2] Department of Physics, Koç University, Sarıyer 34450, Istanbul, Turkey
[3] Department of Electrical and Electronics Engineering, Doğuş University, Ümraniye 34775, Istanbul, Turkey

E-mail: akarsuo@itu.edu.tr, tdereli@ku.edu.tr, nkatirci@dogus.edu.tr

Abstract. In this work, which follows a series of studies on the higher-dimensional steady state universe idea and prepared for Professor Tekin Dereli's Festschrift, we show the influence of the dynamical internal (unobservable) space on the evolution of the possible anisotropy of the external (observable) space. We obtain mathematically exactly the same Friedmann equation of the standard ΛCDM model for the external space, but with some remarkable physical differences. In particular, the higher-dimensional negative cosmological constant plays the role of the four-dimensional positive cosmological constant and the expansion anisotropy, viz., the shear scalar, of the external space mimics a negative cosmological constant; it would mimic a stiff fluid when allowed on top of the standard ΛCDM model. This latter feature gives us the opportunity to manipulate the CMB quadrupole temperature fluctuation, suggesting a possible answer to the fact that its observed value is lower than that predicted by the standard ΛCDM model.

1. Introduction

The idea that the spacetime actually has more than four dimensions, but appears four dimensional because the extra spatial dimensions are so small that would escape local detection goes back to years Einstein's general theory of gravity was introduced in 1915. Kaluza and Klein's attempt to unify gravitation and electromagnetism in the 1920s was based on the assumption that the universe we live in is five dimensional, but since the compact fifth dimension remains small, it effectively appears four dimensional [1, 2] (see Ref. [3] for a review). Unifying the fundamental interactions achieved in higher dimensions in more recent years provides a strong motivation for serious consideration of this possibility; anomaly-free superstring models of all fundamental interactions require a spacetime of ten dimensions for consistency, and the M-theory in which they are embedded lives in eleven dimensional spacetime, see, e.g., Ref. [4]. It is generally assumed that all but four of the spacetime dimensions are compactified on an unobservable internal manifold, leaving an observable $(1+3)$–dimensional spacetime. On the other hand, in the early 1980s, the dynamical (cosmological) reduction of internal dimensions to unobservable scales, that is, the idea that the external (observable) physical dimensions expand while the internal dimensions contract, began to be discussed in the cosmological context [5–7]. Since then, higher-dimensional cosmological models have been extensively studied in various approaches and contexts, and different possibilities have been explored; cosmological models where the external dimensions expand while the internal dimensions contract, remain static, expand at a much slower rate than the external dimensions, or exhibit some combinations of these possibilities, see, for instance, Refs. [8–38].

There is an interesting class in which the external and internal dimensions are dynamical but the total volume of the higher-dimensional space remains constant [6, 7, 17, 18, 24, 27, 30]. In particular, Dereli and Tucker [7], inspired by the inflationary model introduced by Guth [39], constructed a higher-dimensional general relativistic cosmological model assuming both external and internal spaces are flat, homogeneous and isotropic, but while the external space expands exponentially and the internal space contracts exponentially. It is found that the volume of the higher-dimensional space remains constant, and also that the energy density of the higher-dimensional fluid remains constant. These findings are interpreted as a leakage of mass from the contracting n–dimensional internal space to the expanding 3–dimensional external space, given that the 3–dimensional effective energy density measured in the external space also remains constant. According to this, the matter we observe is actually neither created nor exhausted, but is redistributed between the external and internal spaces. The effective $(1+3)$–dimensional universe (with an unobservable n–dimensional internal space) expanding exponentially adheres to the Perfect Cosmological Principle and simulates the Steady-State Theory of the Expanding Universe [40, 41] with a natural mechanism for maintaining a constant density of the matter without modifying the mathematical structure of the general theory of relativity, in contrast to the original the Steady-State Theory of the Expanding Universe, which suggests a continuous creation of matter via the introduction of the 'creation field' into the Einstein field equations.

Dereli along with his collaborators has studied various extensions of the model in Ref. [7], by borrowing the higher-dimensional constant volume feature as an ansatz, to explore possibility of obtaining dynamics beyond the exponential expansion, say, an effectively $(1+3)$–dimensional universe model that resembles the standard Lambda-cold dark matter (ΛCDM) model [42]. In Ref. [25], Akarsu and Dereli an extend it by allowing non-flat internal/external space, and explore a class of general relativistic cosmological solutions in the presence of a higher-dimensional ideal fluid that yields constant energy density along with a cosmological constant. Various dynamics for the external space were obtained; solutions with time varying deceleration parameter (for which, $q=-1$ corresponds to the exponential expansion/contraction), including oscillating and bouncing solutions when the flat/curved external and curved/flat internal spaces are considered. In Ref. [27], the same authors extend it by considering the dilaton gravity (inspired by the low-energy effective string theory, which already requires the presence of extra spatial dimensions [4]) and present $(1+3+n)$–dimensional cosmological solutions with an arbitrary dilaton coupling constant ω and exponential dilaton self-interaction potentials in the string frame. In particular, to determine the cases that are consistent with the observed universe and the primordial nucleosynthesis, they focus on the class in which the external space expands with a time varying deceleration parameter and discuss the effect of the number of the internal dimensions and of the value of the dilaton coupling constant. It is found that the external space starts with a decelerated expansion rate and evolves into an accelerated expansion phase at late times tuned by the values of ω and n, but ends with a Big Rip [43] in all cases. They then discuss the cosmological evolution in further detail for the case $\omega=1$, which corresponds to the anomaly-free supertring theory for $n=6$, and to the anomaly-free bosonic string theory for $n=22$ [4]; and find results quite consistent with the observed universe we know today. In Ref. [30], Akarsu, Dereli, and Oflaz consider a higher-dimensional modified gravity theory with an action that includes dimensionally continued Euler-Poincaré forms up to second order (known also as Gauss-Bonnet term) [44], which live in the presence of extra dimensions, assuming a matter field that yields different dynamical pressures in the external and internal spaces. It is shown that the second order Euler-Poincaré term in the constructions of higher-dimensional steady state cosmologies could be crucial in finding effectively $(1+3)$–dimensional universe models resembling the standard ΛCDM model.

In the current paper, which follows the above-mentioned series of studies on the higher-dimensional steady state universe idea and prepared for Professor Tekin Dereli's Festschrift, we briefly investigate the influence of the internal space on the evolution of the possible anisotropy of the external space. We obtain mathematically exactly the same Friedmann equation of the standard ΛCDM model for the external space, but with some interesting physical differences. In particular, the higher-dimensional negative cosmological constant plays the role of the four-dimensional positive cosmological constant and the expansion anisotropy, namely, the shear scalar, of the external space mimics a negative cosmological constant; it would mimic stiff fluid [45, 46] when allowed on top of the standard ΛCDM model [47–49]. This latter feature of the model gives us the opportunity to manipulate the cosmic microwave background (CMB) quadrupole temperature fluctuation, suggesting a possible answer to the fact that its observed value is lower than that predicted by the standard ΛCDM model [50–52].

2. The model

As the theory of gravitation, we consider the extension of the conventional general theory of relativity with a cosmological constant defined in 4-dimensional spacetime to $(4+n)$–dimensional spacetime while preserving its mathematical structure;

$$\tilde{R}_{\mu\nu} - \frac{1}{2}\tilde{g}_{\mu\nu}\tilde{R} + \tilde{\Lambda}\tilde{g}_{\mu\nu} = -\tilde{\kappa}\tilde{T}_{\mu\nu}, \tag{1}$$

where the indices μ, ν run through $0, 1, ..., n+4$ and 0 is reserved for the cosmic (proper) time t. $\tilde{R}_{\mu\nu}$, $\tilde{R}$, and $\tilde{g}_{\mu\nu}$ are, respectively, the Ricci tensor, Ricci scalar, and the metric tensor of the $(4+n)$–dimensional spacetime. Here $\tilde{\kappa} = 8\pi\tilde{G}$ with $\tilde{G}$ being the $(4+n)$–dimensional gravitational coupling constant and $\tilde{\Lambda}$ is a $(4+n)$–dimensional cosmological constant.

As the $(4+n)$–dimensional background geometry, we consider a spatially homogeneous but not necessarily isotropic $(1+3+n)$–dimensional spacetime metric that involves a 3–dimensional flat but not necessarily isotropic external (observable) space [described by the spatial section of the locally rotationally symmetric (LRS) Bianchi type I spacetime metric [47]], and a flat and isotropic compact internal (unobservable) space [described by the spatial section of the spatially flat Robertson-Walker (RW) metric]:

$$dS^2 = -dt^2 + a(t)^2(dx^2 + dy^2) + b(t)^2dz^2 + s(t)^2\left(d\theta_1^2 + ... + d\theta_n^2\right), \tag{2}$$

where $a(t)$ is the scale factor along the x- and y-axes, $b(t)$ is the scale factor along the z-axis of the 3–dimensional external space, and $s(t)$ is the scale factor of the n–dimensional internal space.

As the $(4+n)$–dimensional fluid, we consider a fluid described by an anisotropic energy-momentum tensor that allows the pressures associated with the external and internal spaces to be different;

$$\tilde{T}_{\mu}{}^{\nu} = \mathrm{diag}[-\tilde{\rho}, \tilde{p}_{\mathrm{ext}}, \tilde{p}_{\mathrm{ext}}, \tilde{p}_{\mathrm{ext}}, \tilde{p}_{\mathrm{int}}, ...], \tag{3}$$

where $\tilde{\rho} = \tilde{\rho}(t)$ is the $(4+n)$–dimensional energy density, $\tilde{p}_{\mathrm{ext}} = \tilde{p}_{\mathrm{ext}}(t)$ and $\tilde{p}_{\mathrm{int}} = \tilde{p}_{\mathrm{int}}(t)$ are the pressures that are associated with the external and internal dimensions, respectively. Since we do not know the nature of the higher-dimensional physical ingredients of universe, we are allowed to consider the possibility of describing it with an energy-momentum tensor that yields distinct and dynamical pressures in the external and internal spaces.

The higher-dimensional Einstein field equations (1) for the metric (2) in the presence of the energy-momentum tensor given in Eq. (3) lead to the following set of differential equations:

$$\frac{\dot{a}^2}{a^2}+\frac{2\dot{b}\dot{a}}{ba}+2n\frac{\dot{s}\,\dot{a}}{sa}+n\frac{\dot{s}\,\dot{b}}{sb}+\frac{1}{2}n(n-1)\frac{\dot{s}^2}{s^2}+\tilde{\Lambda} = \tilde{\kappa}\tilde{\rho}, \tag{4}$$

$$\frac{\ddot{a}}{a}+\frac{\ddot{b}}{b}+n\frac{\ddot{s}}{s}+\frac{\dot{b}\,\dot{a}}{ab}+n\frac{\dot{s}}{s}\left(\frac{\dot{a}}{a}+\frac{\dot{b}}{b}\right)+\frac{1}{2}n(n-1)\frac{\dot{s}^2}{s^2}+\tilde{\Lambda} = -\tilde{\kappa}\tilde{p}_{\rm ext}, \tag{5}$$

$$2\frac{\ddot{a}}{a}+n\frac{\ddot{s}}{s}+\frac{\dot{a}^2}{a^2}+2n\frac{\dot{s}\,\dot{a}}{sa}+\frac{1}{2}n(n-1)\frac{\dot{s}^2}{s^2}+\tilde{\Lambda} = -\tilde{\kappa}\tilde{p}_{\rm ext}, \tag{6}$$

$$(n-1)\frac{\ddot{s}}{s}+2(n-1)\frac{\dot{s}\,\dot{a}}{sa}+(n-1)\frac{\dot{s}\,\dot{b}}{sb}+\frac{\dot{s}^2}{s^2}+2\frac{\ddot{a}}{a}+\frac{\ddot{b}}{b}+\frac{\dot{a}^2}{a^2}+\frac{\dot{b}\,\dot{a}}{ab}+\tilde{\Lambda} = -\tilde{\kappa}\tilde{p}_{\rm int}. \tag{7}$$

The set of field equations (4)–(7) should be satisfied by six unknown functions a, b, s, $\tilde{\rho}$, $\tilde{p}_{\rm ext}$, and $\tilde{p}_{\rm int}$, and therefore is not fully determined. We must provide two constraint equations for a full determination of the system. In this work, we are interested in the higher-dimensional cosmologies that yield constant higher-dimensional volume, and accordingly we proceed with the following ansatz, as done in Refs. [25, 27, 30];

$$V_{3+n} = V_{\rm ext}V_{\rm int} = a^2bs^n = {\rm const}, \tag{8}$$

where V_{3+n} is the volume element of the $(3+n)$–dimensional space, while $V_{\rm ext} = a^2b$ and $V_{\rm int} = s^n$ are the volume elements of the 3–dimensional external and n–dimensional internal spaces, respectively.

In particular, we are interested in the evolution of the external space (the observable space, though is affected by the unobservable internal space), namely, the evolution of the average expansion rate $H_{\rm ext}$ and the shear scalar $\sigma^2_{\rm ext}$ (quantifying the expansion anisotropy) of the external space.

Therefore, we will first proceed with the definitions of these two parameters. We define the mean scale factor of the external space as $\upsilon_{\rm ext} \equiv V_{\rm ext}^{1/3}$, which in turn gives the average Hubble parameter of the external space as follows;

$$H_{\rm ext} \equiv \frac{\dot{\upsilon}_{\rm ext}}{\upsilon_{\rm ext}} = \frac{1}{3}(2H_x + H_z), \tag{9}$$

where $H_x = H_y = \frac{\dot{a}}{a}$ and $H_z = \frac{\dot{b}}{b}$ are the directional Hubble parameters along the x– (or y)–axis and the z–axis, respectively. Similarly, the Hubble parameter of internal space reads $H_{\rm int} = \frac{\dot{s}}{s}$ from the mean scale factor of the internal space defined as $\upsilon_{\rm int} \equiv V_{\rm int}^{1/n}$. Note that the higher-dimensional constant volume condition (8) implies $3H_{\rm ext} + nH_{\rm int} = 0$, which in turn guarantees that an expanding external space ($H_{\rm ext} > 0$) is accompanied by a contracting internal space ($H_{\rm int} < 0$). This is in line with that, in contrast to the external space, the internal space must have been never expanded to observable scales or have contracted to the unobservable scales.

The measure of the anisotropic expansion in the external space section can be quantified by the shear scalar (see Ref. [47]) defined specifically for the external space as follows;

$$\sigma^2_{\rm ext} \equiv \frac{1}{2}\sum_{i=1}^{3}(H_i - H_{\rm ext})^2 = \frac{1}{3}(H_x - H_z)^2. \tag{10}$$

Then, subtracting Eq. (5) from Eq. (6), and after some manipulations and integration, it turns out that

$$\sigma^2_{\rm ext} = \frac{\Gamma}{3\,(V_{\rm ext}V_{\rm int})^2}, \tag{11}$$

Journal of Physics: Conference Series **2191** (2022) 012001 doi:10.1088/1742-6596/2191/1/012001

where $\Gamma \geq 0$ is a non-negative definite integration constant. We notice that the shear scalar associated with the external space is controlled by the total volume element of the higher-dimensional space; i.e., its evolution is controlled by not only the external space but also the unobservable internal space. For instance, when we consider the usual 4–dimensional general relativity, in the simplest anisotropic (LRS Bianchi type I spacetime) extension of the standard ΛCDM model, the shear scalar is inversely proportional to the square of the volume element, $\sigma^2 \propto V^{-2}$, see, for instance, Refs. [47–49]. Indeed, in consistency with that, if we assume that the internal space is static, i.e., $V_{\rm int} = {\rm const}$, we obtain $\sigma^2_{\rm ext} \propto V^{-2}_{\rm ext}$ from Eq. (11). However, any dynamical internal space will lead $\sigma^2_{\rm ext}$ to deviate from this behavior.

Finally, using the constant higher-dimensional volume ansatz (8), along with the definitions of $H_{\rm ext}$ (9) and $\sigma^2_{\rm ext}$ (11), in the set of field equations (4)–(7), we reach the following set of equations that describes our model;

$$3H^2_{\rm ext} = -\frac{2n}{n+3}\left(\tilde{\kappa}\tilde{\rho} + \tilde{\Lambda} + \sigma^2_{\rm ext}\right), \tag{12}$$

$$-3H^2_{\rm ext} + \frac{2n}{n+3}\dot{H}_{\rm ext} = \frac{2n}{n+3}\left(\tilde{\kappa}\tilde{p}_{\rm ext} + \tilde{\Lambda} + \sigma^2_{\rm ext}\right), \tag{13}$$

$$-\tilde{\kappa}\tilde{p}_{\rm int} = \frac{14(n-1)}{n}H^2_{\rm ext} + \frac{3}{n}\dot{H}_{\rm ext} + \frac{1}{\sqrt{3}}H_{\rm ext}\sigma_{\rm ext} + \tilde{\Lambda} + \frac{5}{3}\sigma^2_{\rm ext}, \tag{14}$$

with the shear scalar of the external space is found to be constant,

$$\sigma^2_{\rm ext} = {\rm const}. \tag{15}$$

We have now a set of field equations (12)-(14) consisting of three equations with four unknown functions, $H_{\rm ext}$, $\tilde{\rho}$, $\tilde{p}_{\rm ext}$, and $\tilde{p}_{\rm int}$. Thus, we have freedom to introduce one more ansatz. To do so, we first recall the Friedmann equation of the standard ΛCDM model described by the usual spatially flat Friedmann-Lemaître-Robertson-Walker (FLRW) spacetime;

$$3H^2 = \Lambda + \kappa\rho_{\rm m0}(1+z)^3 + \kappa\rho_{\rm r0}(1+z)^4, \tag{16}$$

where H is the Hubble parameter, Λ is the positive cosmological constant, $\rho_{\rm m0}$ and $\rho_{\rm r0}$ are, respectively, the present-day energy densities of dust and radiation, and z is the redshift; here, all defined in the usual way in 4-dimensional spacetime. Here and onward, a subscript 0 attached to any quantity implies its value in the present-day ($z = 0$) universe.

Comparing Eq. (12) and Eq. (16), so as to achieve mathematically exactly the same form with the Friedmann equation of the standard ΛCDM model, assuming $\tilde{\kappa} > 0$, we suppose the higher-dimensional energy density is negative definite ($\tilde{\rho} < 0$) and is of the form

$$\tilde{\rho} = -c_1(1+z)^3 - c_2(1+z)^4, \tag{17}$$

where $c_1 > 0$ and $c_2 > 0$ are constants, and z is the redshift defined within the external space from $z = -1 + \frac{1}{v_{\rm ext}}$ using the mean scale factor of the external space. Using this in Eq. (12), we reach the following Friedmann equation describing our model in terms of redshift;

$$3H^2_{\rm ext} = \frac{2n\tilde{\kappa}}{n+3}c_1(1+z)^3 + \frac{2n\tilde{\kappa}}{n+3}c_2(1+z)^4 - \frac{2n}{n+3}\tilde{\Lambda} - \frac{2n}{n+3}\sigma^2_{\rm ext}. \tag{18}$$

This can be written in a more useful way, in terms of the density parameters and the Hubble constant of the external space $H_{\rm ext0}$, as follows;

$$\frac{H^2_{\rm ext}}{H^2_{\rm ext0}} = \Omega_{\rm m0}(1+z)^3 + \Omega_{\rm r0}(1+z)^4 - \Omega_{\tilde{\Lambda}0} - \Omega_{\sigma^2_{\rm ext}0}, \tag{19}$$

where we defined

$$\begin{aligned}\Omega_{\mathrm{m0}} &= \frac{2nc_1}{3(n+3)H_{\mathrm{ext0}}^2} \quad , \quad \Omega_{\mathrm{r0}} = \frac{2nc_2}{3(n+3)H_{\mathrm{ext0}}^2} \\ \Omega_{\tilde{\Lambda}0} &= \frac{2n\tilde{\Lambda}}{3(n+3)H_{\mathrm{ext0}}^2} \quad \text{and} \quad \Omega_{\sigma_{\mathrm{ext}}^2 0} = \frac{2n\sigma_{\mathrm{ext}}^2}{3(n+3)H_{\mathrm{ext0}}^2} \geq 0.\end{aligned} \tag{20}$$

First, we notice that, due to the negative definite coefficient in front of it, the shear scalar σ_{ext}^2 associated with the external space contributes to the expansion rate of the external space like a negative cosmological constant. Second, the higher-dimensional cosmological constant $\tilde{\Lambda}$ should yield negative values to play the role of the four-dimensional usual positive cosmological constant in the external space and be smaller than the negative of the shear scalar associated with the external space to derive late time acceleration in the external space;

$$\tilde{\Lambda} < -\sigma_{\mathrm{ext}}^2 \leq 0. \tag{21}$$

It is worth noting that, in contrast to a positive cosmological constant [53, 54], a negative cosmological constant is not only ubiquitous in the fundamental theoretical physics without any complication, but also a theoretical sweet spot; an anti-de Sitter (AdS) background, provided by a negative cosmological constant, is welcome due to the celebrated AdS/CFT (conformal field theory) correspondence [55] and is preferred by string theory (which requires the presence of extra dimensions) and string-theory-motivated supergravities [56]. It is also interesting to observe that the shear scalar of the external space resembles a negative cosmological constant.

Note that we reproduce mathematically exactly the same form of the Friedmann equation of the standard ΛCDM model for the external space, but with an important difference that the external space is now allowed to exhibit an anisotropic expansion that yields specifically a constant shear scalar, see Eqs. (15) and (19). This feature of our model, which allows late-time ellipsoidal expansion on top of the standard ΛCDM model, provides us with opportunity to manipulate the CMB quadrupole (multipole $\ell = 2$ corresponding to the angular scale $\theta = \pi/2$ of the power spectrum) temperature fluctuation, ΔT, with no consequences on the higher multipoles [57–61].

The evolution of the photon temperature along the ith–axis ($i = x, y, z$) is given by $T_i/T_0 = e^{-\int H_i \mathrm{d}t}$, where $T_0 = 2.7255 \pm 0.0006\,\mathrm{K}$ [62] is the present-day CMB monopole temperature [63]. Accordingly, using Eq. (10), the difference between the photon temperatures along the x–(or y–) axis and the z–axis since the recombination ($z_{\mathrm{rec}} = 1090$) to the present time ($z = 0$) due to the anisotropic expansion, $\Delta T_{\sigma_{\mathrm{ext}}^2} \equiv T_x - T_z$, reads

$$\Delta T_{\sigma_{\mathrm{ext}}^2} = T_0 \int_{t_{\mathrm{rec}}}^{t_0} (H_x - H_z)\,\mathrm{d}t = 3T_0 \sqrt{\frac{n+3}{2n}\Omega_{\sigma_{\mathrm{ext}}^2 0}} \int_{z=0}^{z_{\mathrm{rec}}} \frac{H_{\mathrm{ext0}}}{H_{\mathrm{ext}}} \frac{\mathrm{d}z}{1+z}, \tag{22}$$

for small anisotropies (so $e^{-\int H_x dt} \simeq 1 - \int H_x \mathrm{d}z$ etc). We use $\mathrm{d}t = -\frac{\mathrm{d}z}{H_{\mathrm{ext}}(1+z)}$ and assume that CMB was last scattered at the recombination redshift (era) z_{rec} (t_{rec}). We can adopt the observational best fit values from the recent Planck release [64]; $H_{\mathrm{ext0}} = 67.32\ \mathrm{km\,sec^{-1}Mpc^{-1}}$, $\Omega_{\mathrm{m0}} = 0.3158$, and $\Omega_{\mathrm{r0}} \approx 10^{-4}$. The model-independent upper bounds on the present-day expansion anisotropy read $\Omega_{\sigma_{\mathrm{ext}}^2 0} \lesssim 10^{-3}$, e.g., from type Ia Supernovae [65–71], and therefore its contribution to H_{ext} in Eq. (22) can be neglected. Thus, while retaining exactly the same expansion history for the comoving volume element of the external space with that of the standard ΛCDM model, we have

$$\Delta T_{\sigma_{\mathrm{ext}}^2} = 21\sqrt{\frac{n+3}{2n}\Omega_{\sigma_{\mathrm{ext}}^2 0}}\ K \sim 21\sqrt{\Omega_{\sigma_{\mathrm{ext}}^2 0}}\ K. \tag{23}$$

Journal of Physics: Conference Series **2191** (2022) 012001 doi:10.1088/1742-6596/2191/1/012001

Consequently, setting $\Omega_{\sigma^2_{\rm ext}0} \simeq 5.80 \times 10^{-7}$ we obtain $\Delta T_{\sigma^2_{\rm ext}} = 16\,\mu K$, which, provided that the orientation of the expansion anisotropy is set suitably, on top of the best fit standard ΛCDM model predicted value $\Delta T_{\rm std} \approx 32\,\mu K$ can be used to bring it to the observed value by the Planck satellite $\Delta T_{\rm PLK} \approx 15\,\mu K$ [64]. We would not have this opportunity in a simple LRS Bianchi type I spacetime extension of the standard ΛCDM model; in that case, $\sigma^2 \propto (1+z)^6$ and the upper bound on $\Omega_{\sigma^2 0}$ from the big bang nuclesynthesis (BBN) light element abundances, 10^{-23}, is one order of magnitude tighter than even the tightest upper bounds obtained from the CMB data [48, 49, 63, 72, 73].

Finally, as the 4-dimensional effective gravitational coupling strength measured by the observers who are aware of only the external space (three spatial dimensions) is inversely proportional to the volume of the internal space, $G_4 \propto \frac{1}{V_{\rm int}}$, the constant higher-dimensional volume assumption (8) leads to

$$\frac{\dot{G}_4}{G_4} \propto -\frac{\dot{V}_{\rm int}}{V_{\rm int}} = \frac{\dot{V}_{\rm ext}}{V_{\rm ext}} = 3H_{\rm ext}. \tag{24}$$

Using the Planck 2018 best-fit value $H_{\rm ext0} = 67.32\ {\rm km\,sec^{-1}Mpc^{-1}}$ we find

$$\left.\frac{\dot{G}_4}{G_4}\right|_{z=0} = 3H_{\rm ext0} = 2.05 \times 10^{-10}\ {\rm yr}^{-1} \tag{25}$$

for the present-day value of the rate of change of G_4. We note that this value is consistent with the observational bounds on the possibly varying Newton's constant, e.g., $|\frac{\dot{G}}{G}|_{z=0} \leq 10^{-10} - 10^{-12}\ {\rm yr}^{-1}$, see Ref. [74].

3. Conclusion

In this paper, which follows a series of studies on the higher-dimensional steady state universe idea, we have presented a brief exploration of the influence of the dynamics of the compact internal (unobservable) space on the evolution of the possible anisotropy of the external (observable) space. We have obtained mathematically exactly the same Friedmann equation of the standard ΛCDM model for the external space, but with some interesting physical differences. In particular, we have shown that the higher-dimensional negative cosmological constant plays the role of the usual four-dimensional positive cosmological constant and the shear scalar, which quantifies the expansion anisotropy, of the external space resembles a negative cosmological constant, as a consequence specific to the assumption of constant higher-dimensional volume. This latter feature of the model has given us the opportunity to manipulate the CMB quadrupole temperature fluctuation to bring its value predicted by the standard ΛCDM model to that observed by the Planck satellite. Our study also provides an example of how future precision observations that can reveal the properties of expansion anisotropy would be important in testing the higher-dimensional cosmological models with a dynamical internal space.

References

[1] Kaluza T 1921 *Sitzungsber Preuss Akad Wiss Berlin (Math Phys)* **966**
[2] Klein O 1926 *Z Phys* **37** 895
[3] Overduin J M and Wesson P S 1997 *Phys Rep* **283** 303
[4] Lidsey J E, Wands D and Copeland E J 2000 *Phys Rep* **337** 343
[5] Chodos A and Detweiler S 1980 *Phys Rev D* **21** 2167
[6] Freund P G O *Nucl Phys B* 1982 **209** 146
[7] Dereli T and Tucker R W 1983 *Phys Lett B* **125** 133
[8] Sahdev D 1984 *Phys Rev D* **30** 2495
[9] Tosa Y 1984 *Phys Rev D* **30** 2054 [Erratum 1985 *Phys Rev D* **31** 2697]
[10] Sato H 1984 *Prog Theor Phys* **72** 98

[11] Ishihara H 1984 *Prog Theor Phys* **72** 376
[12] Demaret J and Hanquin J L 1985 *Phys Rev D* **31** 258
[13] Okada Y 1986 *Nucl Phys B* **264** 197
[14] Barrow J D and Stein-Schabes J 1986 *Phys Lett B* **167** 173
[15] de Leon J P and Wesson P S 1993 *J Math Phys* **34** 4080
[16] Yearsley J and Barrow J D 1996 *Class Quant Grav* **13** 2693
[17] Bleyer U and Zhuk A 1996 *Astronomische Nachrichten* **317** 161
[18] Rainer M and Zhuk A 2000 *Gen Relativ Grav* **32** 79
[19] Mohammedi N 2002 *Phys Rev D* **65** 104018
[20] Darabi F 2003 Class Quant Grav **20** 3385
[21] Bringmann T, Eriksson M and Gustafsson M 2003 *Phys Rev D* **68** 063516
[22] Carroll S M, Johnson M C and Randall L 2009 *JHEP* **11** 094
[23] Reyes L M, Madriz-Aguilar J E and Urena-Lopez L A 2011 *Phys Rev D* **84** 027503
[24] Ho C M and Kephart T W 2012 *Int J Mod Phys A* **27** 1250151
[25] Akarsu Ö and Dereli T 2013 *Gen Relativ Grav* **45** 959
[26] Akarsu Ö and Dereli T 2013 *Gen Rel Grav* **45** 1211
[27] Akarsu Ö and Dereli T 2013 *JCAP* **02** 050
[28] Fiorini F, Gonzalez P A and Vasquez Y 2014 *Phys Rev D* **89** 024028
[29] Akarsu Ö, Dereli T, Katırcı N and Sheftel M B 2015 *Gen Rel Grav* **47** 61
[30] Akarsu Ö, Dereli T and Oflaz N 2015 *Class Quant Grav* **32** 215009
[31] Sloan D and Ferreira P 2017 *Phys Rev D* **96** 043527
[32] Akarsu Ö, Chopovsky A and Zhuk A 2018 *Phys Lett B* **778** 190
[33] Benisty D and Guendelman E I 2018 *Phys Rev D* **98** 043522
[34] Rasouli S M M and Vargas Moniz P 2019 *Class Quant Grav* **36** 075010
[35] Kim K K, Koh S and Yang H S 2018 *JHEP* **12** 085
[36] Russo J G and Townsend P K 2019 *Class Quant Grav* **36** 095008
[37] Middleton C, Brouse B A and Jackson S D 2019 *Eur Phys J C* **79** 982
[38] Akarsu Ö, Chopovsky A, Shulga V, Yalçınkaya E and Zhuk A 2020 *Phys Rev D* **101** 024004
[39] Guth A H 1981 *Phys Rev D* **23** 347
[40] Bondi H and Gold T 1948 *Mon Not Roy Astron Soc* **108** 252
[41] Hoyle F 1948 *Mon Not Roy Astron Soc* **108** 372
[42] Dodelson S and Schmidt F 2021 *Modern Cosmology* 2nd Edition Acad Press, Elsevier Sci.
[43] Caldwell R R, Kamionkowski M and Weinberg N N 2003 *Phys Rev Lett* **91** 071301
[44] Lovelock D 1971 *J Math Phys* **12** 498
[45] Zel'dovich Y B 1961 *Zh Eksp Teor Fiz* **41**, 1609 [1962 *J Exp Theor Phys* **14** 1143]
[46] Barrow J D 1978 *Nature* **272** 211
[47] Ellis G F R, Maartens R and MacCallum M A H 2012 *Relativistic Cosmology* Cambridge Uni. Press England
[48] Akarsu Ö, Kumar S, Sharma S and Tedesco L, 2019 *Phys Rev D* **100** 023532
[49] Akarsu Ö, Di Valentino E, Kumar S, Ozyigit M and Sharma S, arXiv:2112.07807
[50] Bennett C L *et al* 2011 *Astrophys J Suppl Ser* **192** 17
[51] Schwarz D J, Copi C J, Huterer D and Starkman G D 2016 *Class Quant Grav* **33** 184001
[52] Akrami Y *et al* 2020 *Astron Astrophys* **641** A7
[53] Weinberg S 1989 *Reviews of Modern Physics* **61** 1
[54] Sahni V and Starobinsky A 2000 *Int J Mod Phys D* **9** 373
[55] Maldacena J M 1998 *Adv Theor Math Phys* **2** 231
[56] Bousso R and Polchinski J 2000 *JHEP* **06** 006.
[57] Berera A, Buniy R V and Kephart T W 2004 JCAP **0410** 016
[58] Campanelli L, Cea P and Tedesco L 2006 *Phys Rev Lett* **97** 131302 [Erratum-ibid. **97** 209903]
[59] Campanelli L, Cea P and Tedesco L 2007 *Phys Rev D* **76** 063007
[60] Beltran Jimenez J and Maroto A L 2007 *Phys Rev D* **76** 023003
[61] Koivisto T and Mota D F 2008 *Astrophys J* **679** 1
[62] Fixsen D J 2009 *Astrophys J* **707** 916
[63] Barrow J D 1997 *Phys Rev D* **55** 7451
[64] Aghanim N *et al* 2020 *Astron Astrophys* **641** A6
[65] Campanelli L, Cea P, Fogli G L and Marrone A 2011 *Phys Rev D* **83** 103503
[66] Kalus B, Schwarz D J, Seikel M and Wiegand A 2013 *Astron Astrophys* **553** A56
[67] Lin H N, Wang S, Chang Z and Li X 2016 *Mon Not R Astron Soc* **456** 1881
[68] Wang Y Y and Wang F Y 2018 *Mon Not R Astron Soc* **474** 3516
[69] Soltis J, Farahi A, Huterer D and Liberato C M 2019 *Phys Rev Lett* **122** 091301

Journal of Physics: Conference Series **2191** (2022) 012001 doi:10.1088/1742-6596/2191/1/012001

[70] Zhao D, Zhou Y and Chang Z 2019 *Mon Not R Astron Soc* **486** 5679
[71] Hu J P, Wang Y Y and Wang F Y 2020 *Astron Astrophys* **643** A93
[72] Barrow J D 1976 *Mon Not R Astron Soc* **175** 359
[73] Saadeh D, Feeney S M, Pontzen A, Peiris H V and McEwen J D 2016 *Phys Rev Lett* **117** 131302
[74] Uzan J P 2011 *Living Rev Rel* **14** 2

Basics of Apparent horizons in black hole physics

E Altas [1], B Tekin [2]

[1] Department of Physics, Karamanoglu Mehmetbey University, 70100, Karaman, Turkey
[2] Department of Physics, Middle East Technical University, 06800, Ankara, Turkey

E-mail: [1] emelaltas@kmu.edu.tr

E-mail: [2] btekin@metu.edu.tr

Abstract. [1]Event Horizon, a null hypersurface defining the boundary of the black hole region of a spacetime, is not particularly useful for evolving black holes since it is non-local in time. Instead, one uses the more tangible concept of Apparent Horizon for dynamical black holes out there in the sky that do all sorts of things: evolve, merge and feed on the environment. Event Horizon, being a gauge-independent, global property of the total spacetime is easy to define and locate in the stationary case; on the other hand, Apparent Horizon depends on the embedding of the surface in spacetime and hence it is somewhat tricky to define. But for numerical simulations in General Relativity, locating the Apparent Horizon helps one to excise the black hole region and the singularity to have a stable computation. Moreover, for stationary solutions cross-sections of these horizons match. Here we give a detailed pedagogical exposition of the subject and work out the non-trivial case of a slowly moving and spinning black hole.

1. Introduction

Stationary (Kerr) and static (Schwarzschild) black hole solutions of General Relativity have rather dull lives: stationary ones do the same thing, static ones do nothing as observed by an observer outside the black hole. While these vacuum solutions obtained in an isolated universe serve as our starting point for a more physical and detailed understanding of actual astrophysical black holes, the latter are almost never isolated: the black holes out in the sky have accretion disks, companion stars, neutron stars or black holes. Black holes feed on their environment and grow; in fact they are the most dynamical parts of the vacuum. As the first LIGO/VIRGO gravitational wave detection showed [1], black holes can grow feeding on other black holes: cannibalistic behavior of these objects-highly curved vacua-could explain the existence of intermediate mass black holes.

As the astrophysical black holes evolve, concepts such as the Event Horizon defined easily for eternal black holes are not clearly adequate for us, the transient observers. Recall that the Event Horizon ($\mathcal{H}$) of a stationary black hole is a co-dimension one null hypersurface in the totality of the spacetime defined as the *boundary of the black hole region from which time-like or light-like geodesics cannot reach future null infinity* [2]. Stated in a different way: it is the boundary of the region which is not in the causal past of the future null-infinity. This says that the Event Horizon is a global property of the totality of events which is all of the spacetime. Therefore, one cannot locate the Event Horizon with local experiments in a finite interval of time. In this respect, it is apt to say that the Event Horizon Telescope detected the environment

[1] This paper is written for Prof. Tekin Dereli's 70th Birthday Festschrift.

of the black hole from which one can see at best the cross-section of the Event Horizon, not the Event Horizon itself.

For dynamical black holes one invents the more useful concept of the "Apparent Horizon" [3], a co-dimension two spatial surface (hence local in time), which, unfortunately, in general does not carry geometric invariant data as the Event Horizon but it contains sufficient information regarding the possible formation of an Event Horizon in the future that it pays to describe it in detail. In numerical relativity computations, detection of a black hole region is best done with Apparent Horizons. Within the context of General Relativity, existence of an Apparent Horizon implies the appearance of a future Event Horizon outside of it. Therefore, one can excise the region inside the Apparent Horizon (that also includes the singularity) for the stability of the computation since nothing will come out of that region in classical physics. For modified gravity theories, an Apparent Horizon need not be inside the Event Horizon (See the discussion and references in [4]).

Our task in this work is to give a detailed definition of the Apparent Horizon and some related concepts and apply it to slowly rotating and moving initial data which was recently given in [5]. The layout of this work is follow: in section II we introduce the necessary tools for the defining equation of an Apparent Horizon as a co-dimension two spatial hypersurface in n dimensions and use the ADM decomposition of the metric to arrive at an equation in local coordinates, in section III we consider a conformally flat initial data for $n = 1+3$ dimensions for which the momentum constraints can be solved exactly following the Bowen-York construction [6]; and we solve the Hamiltonian constraint for slowly moving and spinning initial data and compute the properties of the Apparent Horizon. In the Appendix we expound upon some technical points alluded to in the text.

2. Derivation of the apparent horizon equation

As stated above, the Event Horizon of a black hole, as a null hypersurface, cannot be determined locally: one has to know the total spacetime to define it. On the other hand, the Apparent Horizon can be determined locally in time. For this purpose, we need to define a congruence of null geodesics and its expansion. Our notations will be similar to those of the excellent lecture notes [2, 7]. We invite the reader to see [9] for detailed information about the apparent horizons.

As shown in Figure 1, we have an n dimensional spacetime manifold $\mathscr{M}$, with a co-dimension one spacelike hypersurface Σ, that is $\dim \Sigma = n-1$; and we introduce a co-dimension two subspace $\mathcal{S}$, $\dim \mathcal{S} = n-2$. Let n^μ be a timelike unit vector orthogonal to Σ:

$$n^\mu n_\mu = -1, \tag{1}$$

and s^μ be a spacelike unit vector orthogonal to $\mathcal{S}$

$$s^\mu s_\mu = 1. \tag{2}$$

We impose the condition that n and s-vectors are perpendicular to each other

$$n^\mu s_\mu = 0. \tag{3}$$

Instead of these two vectors, one can also work with the ingoing null vector k^μ and the outgoing null vector ℓ^μ, defined respectively as follows (see Figure 1)

$$k^\mu := \frac{1}{2}\left(n^\mu - s^\mu\right), \qquad \ell^\mu := n^\mu + s^\mu. \tag{4}$$

Let g denote the spacetime metric, then the induced metric on the hypersurface Σ is

$$\gamma_{\mu\nu} = g_{\mu\nu} + n_\mu n_\nu, \tag{5}$$

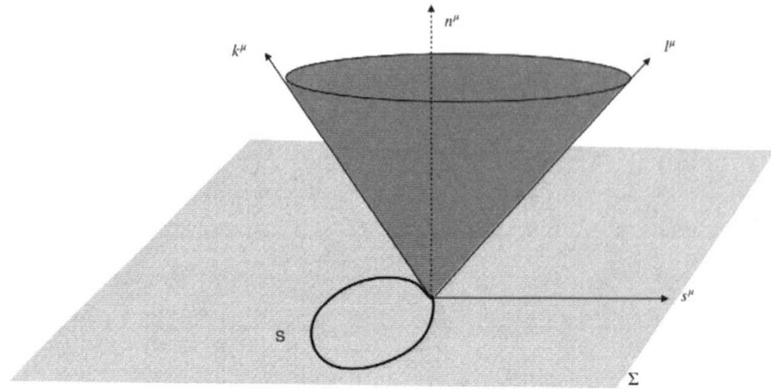

Figure 1. The unit vectors n^μ, s^μ together with the ingoing null vector k^μ and outgoing null vector ℓ^μ are shown.

while the induced metric on the subspace $\mathcal{S}$ reads

$$q_{\mu\nu} = \gamma_{\mu\nu} - s_\mu s_\nu = g_{\mu\nu} + n_\mu n_\nu - s_\mu s_\nu, \tag{6}$$

where μ, ν run over the spacetime directions. The important concept here is the *extrinsic curvature* of both of these surfaces. For the hypersurface Σ, we have

$$K_{\mu\nu} := -\gamma_{\mu\sigma}\gamma_{\nu\rho}\nabla^\sigma n^\rho, \tag{7}$$

where ∇_μ denotes the covariant derivative compatible with the spacetime metric, $\nabla_\mu g_{\nu\rho} = 0$. From a more geometric vantage point, our definition is as follows: given two vectors (X, Y) on the tangent space at the point p, that is $T_p\Sigma$, and n being the unit normal to Σ, then the extrinsic curvature of Σ is defined as $K(X,Y) := -\gamma(\nabla_X n, Y)$. So in local coordinates, one can take $X = \partial_\mu$, $Y = \partial_\nu$ to get $K_{\mu\nu} := K(\partial_\mu, \partial_\nu) = -\gamma(\nabla_{\partial_\mu} n, \partial_\nu)$ which matches (7). The minus sign is a convention. Equivalently, one has

$$K_{\mu\nu} = -\nabla_\mu n_\nu - n_\mu n^\sigma \nabla_\sigma n_\nu. \tag{8}$$

Similarly, we define the extrinsic curvature of the $(n-2)$-dimensional space $\mathcal{S}$ as

$$k_{\mu\nu} := -q_{\mu\sigma} q_{\nu\rho} \nabla^\sigma s^\rho, \tag{9}$$

and using the definition of the induced metric (6) one obtains

$$k_{\mu\nu} = -\nabla_\mu s_\nu - n_\nu n^\sigma \nabla_\mu s_\sigma - n_\mu n^\sigma \nabla_\sigma s_\nu - n_\mu n_\nu n_\sigma n_\rho \nabla^\sigma s^\rho + s_\mu s^\sigma \nabla_\sigma s_\nu + s_\mu n_\nu s_\sigma n_\rho \nabla^\sigma s^\rho. \tag{10}$$

One defines the *expansion* of the out-going null geodesic congruence as

$$\Theta_{(\ell)} := q^{\mu\nu} \nabla_\mu \ell_\nu, \tag{11}$$

which is the divergence of the null geodesic congruence along its propagation in the outgoing null direction. Using (4) and the extrinsic curvatures of the hypersurface and the surface, we can recast the expansion of the null geodesic congruence as

$$\Theta_{(\ell)} = K + k + (n^\mu n^\nu - s^\mu s^\nu)(K_{\mu\nu} + k_{\mu\nu}), \tag{12}$$

where $K = g^{\mu\nu} K_{\mu\nu}$ and $k = g^{\mu\nu} k_{\mu\nu}$. Since $n^\mu n^\nu K_{\mu\nu} = 0 = n^\mu n^\nu k_{\mu\nu} = s^\mu s^\nu k_{\mu\nu}$, $\Theta_{(\ell)}$ reduces to the following neat equation as

$$\Theta_{(\ell)} = -K - k + s^\mu s^\nu K_{\mu\nu}. \tag{13}$$

Equivalently one has

$$\Theta_{(\ell)} = -q^{\mu\nu}\Big(K_{\mu\nu} + k_{\mu\nu}\Big), \tag{14}$$

or

$$\Theta_{(\ell)} = -q^{ij}\Big(K_{ij} + k_{ij}\Big), \tag{15}$$

where the i, j indices run over coordinates on the hypersurface Σ.

The expansion $\Theta_{(\ell)}$ is employed to define the very important concept of a *trapped surface.* An outer trapped surface on Σ is a *closed* (that is compact without a boundary) co-dimension two surface such that for outgoing null geodesics orthogonal to the surface, one has $\Theta_{(\ell)} < 0$ *everywhere* on the surface. The subset of Σ that contains the trapped surfaces is called the *trapped region* $\mathcal{T}$, a co-dimension one surface. Finally, Apparent Horizon is the boundary of the trapped region (an obviously spatial surface) which we shall denote by $\mathcal{S} := \partial\mathcal{T}$. By definition Apparent Horizon is a marginally outer trapped surface (MOTS) and satisfies the Apparent Horizon equation:

$$\Theta_{(\ell)} = -K - k + s^\mu s^\nu K_{\mu\nu} = 0. \tag{16}$$

It is clear that for the case of time-symmetric initial data ($K_{\mu\nu} = 0$), the Apparent Horizon becomes a minimal surface since $k = 0$.

Now that we have defined the Apparent Horizon, given a metric in some coordinates, to proceed we need to lay out in detail how (16) is expressed in terms of the metric functions. For this purpose we choose the ADM decomposition of the metric [8].

Let $N = N(t, x^i)$ be the lapse function and $N^i = N^i(t, x^j)$ be the shift vector, then the line-element reads

$$ds^2 = (N_i N^i - N^2)dt^2 + 2N_i dt\, dx^i + \gamma_{ij} dx^i\, dx^j, \tag{17}$$

or in components one has

$$g_{00} = N_i N^i - N^2, \qquad g_{0i} = N_i, \qquad g_{ij} = \gamma_{ij}, \tag{18}$$

with the inverses given as

$$g^{00} = -N^{-2}, \qquad g^{0i} = N^i N^{-2}, \qquad g^{ij} = \gamma^{ij} - N^i N^j N^{-2}. \tag{19}$$

Using the definition (5), one has

$$g_{ij} = \gamma_{ij} - n_i n_j = \gamma_{ij}, \tag{20}$$

hence $n_i = 0$. Similarly the relation $g^{ij} = \gamma^{ij} - n^i n^j = \gamma^{ij} - N^i N^j N^{-2}$ yields $n^i = \pm N^i/N$. Since n^μ is a timelike vector, using $n_i = 0$, one has $n_0 = \pm N$ and choosing for $N > 0$, we choose the plus sign for the future-directed time-like vector to arrive at

$$n^\mu = \left(\frac{1}{N}, -\frac{N^i}{N}\right), \qquad n_\mu = (-N, \vec{0}). \tag{21}$$

We can work out the additional relations between the spacetime metric g and the metric of the hypersurface γ as follows

$$g_{00} = \gamma_{00} - n_0 n_0 = N_i N^i - N^2, \tag{22}$$

which yields $\gamma_{00} = N_i N^i$. And similarly

$$g_{0i} = \gamma_{0i} - n_0 n_i = N_i \tag{23}$$

yields $\gamma_{0i} = N_i$; and from the inverse metric relations, one obtains $\gamma^{0\mu} = 0$.

Similar computations for the co-dimension 2 spatial subspace $\mathcal{S}$, after using the condition $n_\mu s^\mu = 0$, yield $s^0 = 0$ and $q^{0\mu} = 0$.

Now we go back to (11) and express it for the apparent horizon as

$$q^{ij}\left(\nabla_i n_j + \nabla_i s_j\right) = 0. \tag{24}$$

From (8), one has $K_{ij} = -\nabla_i n_j = -N\Gamma^0_{ij}$ and we obtain[2]

$$q^{ij}\left(K_{ij} - \partial_i s_j + \Gamma^0_{ij}s_0 + \Gamma^k_{ij}s_k\right) = 0. \tag{25}$$

We denote the Christoffel connection of the induced metric γ as ${}^{\Sigma}\Gamma^k_{ij}$. Then substituting the corresponding components of the Christoffel connection one has

$$\Gamma^k_{ij} = {}^{\Sigma}\Gamma^k_{ij} + \frac{N^k}{N}K_{ij}, \tag{26}$$

and we arrive at

$$q^{ij}\left(K_{ij} - D_i s_j\right) = 0, \tag{27}$$

where D_i denotes the covariant derivative compatible with the spatial metric γ, $D_i\gamma_{jk} = 0$.

Before we start working out an example, let us note that there is another simple expression of the $\theta_{(\ell)}$ in (11) and hence equation (27). One can show that (see section A of the Appendix for the proof) Lie-dragging the metric on $\mathcal{S}$ along ℓ yields exactly the expansion: namely, one has

$$\theta_{(\ell)} = q^{\mu\nu}\nabla_\mu l_\nu = \frac{1}{2}q^{\mu\nu}\mathcal{L}_\ell q_{\mu\nu}, \tag{28}$$

where $\mathcal{L}_\ell$ denotes the Lie-derivative along the vector ℓ. In section B of the Appendix, $\theta_{(\ell)}$ is derived from the minimization of the area along the outing null direction which also yields a complementary physical picture.

3. Apparent horizon detection

3.1. The equation in explicit form

From now on we shall work in $n = 1 + 3$ dimensions. Assume now that the local coordinates on Σ are denoted as (r, θ, ϕ) and that the location of the Apparent Horizon depends both on θ and ϕ. The equation to be solved is

$$q^{ij}\left(\partial_i s_j - {}^{\Sigma}\Gamma^k_{ij}s_k - K_{ij}\right) = 0. \tag{29}$$

Assume that the surface $\mathcal{S}$ can be parameterized as a level set such that

$$\Phi(r,\theta,\phi) := r - h(\theta,\phi) = 0, \tag{30}$$

with h being a sufficiently differentiable function of its arguments. Since s^i is normal to the surface, one has $s_i \sim \partial_i\Phi$; and because it is a normal vector one can express $s_i := \lambda\partial_i\Phi$, which yields

$$s_i = \lambda\left(1, -\partial_\theta h, -\partial_\phi h\right). \tag{31}$$

To solve the equation defining the apparent horizon together with the constraint equations, let us take the metric on Σ to be conformally flat as in [6]

$$\gamma_{ij} = \psi^4 \begin{pmatrix} 1 & 0 & 0 \\ 0 & r^2 & 0 \\ 0 & 0 & r^2\sin^2\theta \end{pmatrix}, \tag{32}$$

[2] Note that, explicitly, we have $K_{ij} = \frac{1}{2N}\left(D_i N_j + D_j N_i - \partial_t\gamma_{ij}\right)$.

then one has

$$s^i = \lambda \left(\gamma^{rr}, -\gamma^{\theta\theta}\partial_\theta h, -\gamma^{\phi\phi}\partial_\phi h\right), \tag{33}$$

with the normalization factor given as

$$\lambda = \left(\gamma^{rr} + \gamma^{\theta\theta}(\partial_\theta h)^2 + \gamma^{\phi\phi}(\partial_\phi h)^2\right)^{-1/2}. \tag{34}$$

As should be clear at this stage, the Apparent Horizon equation will be a rather complicated non-linear partial differential equation with little hope to yield an exact analytical solution. Let us further assume (following [6]) $\gamma^{ij}K_{ij} = K = 0$, which is called the maximal slicing gauge. Then (29) reads more explicitly as

$$\gamma^{ij}\partial_i m_j - \gamma^{ij}\Gamma^k_{ij}m_k - \lambda^2 m^i m^j \partial_i m_j + \lambda^2 m^i m^j m_k \Gamma^k_{ij} + \lambda m^i m^j K_{ij} = 0, \tag{35}$$

where we defined $m_i := \partial_i \Phi$. After working out each piece in a somewhat tedious computation, one arrives at

$$\begin{aligned}
&-\gamma^{\theta\theta}\partial^2_\theta h - \gamma^{\phi\phi}\partial^2_\phi h - \frac{1}{2}\Big((\gamma^{rr})^2\partial_r\gamma_{rr} - \gamma^{\theta\theta}\gamma^{rr}\partial_r\gamma_{\theta\theta} - \gamma^{\phi\phi}\gamma^{rr}\partial_r\gamma_{\phi\phi} + \partial_\theta h\gamma^{\phi\phi}\gamma^{\theta\theta}\partial_\theta\gamma_{\phi\phi}\Big) \\
&+\lambda^2\Big((\gamma^{\theta\theta})^2(\partial_\theta h)^2\partial^2_\theta h + (\gamma^{\phi\phi})^2(\partial_\phi h)^2\partial^2_\phi h + 2\gamma^{\phi\phi}\gamma^{\theta\theta}\partial_\phi h\partial_\theta h\partial_\theta\partial_\phi h\Big) \\
&+\frac{\lambda^2}{2}\Big((\gamma^{rr})^3\partial_r\gamma_{rr} + (\gamma^{\theta\theta})^2\gamma^{rr}(\partial_\theta h)^2\partial_r\gamma_{\theta\theta} + (\gamma^{\phi\phi})^2\gamma^{rr}(\partial_\phi h)^2\partial_r\gamma_{\phi\phi} \\
&\qquad - (\partial_\phi h)^2\partial_\theta h(\gamma^{\phi\phi})^2\gamma^{\theta\theta}\partial_\theta\gamma_{\phi\phi}\Big) \\
&+\lambda\Big((\gamma^{rr})^2K_{rr} + (\gamma^{\theta\theta})^2(\partial_\theta h)^2K_{\theta\theta} + (\gamma^{\phi\phi})^2(\partial_\phi h)^2K_{\phi\phi} - 2\gamma^{rr}\gamma^{\theta\theta}\partial_\theta hK_{r\theta} \\
&\qquad - 2\gamma^{rr}\gamma^{\phi\phi}\partial_\phi hK_{r\phi} + 2\gamma^{\theta\theta}\gamma^{\phi\phi}\partial_\theta h\partial_\phi hK_{\theta\phi}\Big) = 0.
\end{aligned} \tag{36}$$

Given the metric γ_{ij} and the extrinsic curvature K_{ij}, one can find numerical solutions of this equation up to the desired accuracy. Our goal here is to find approximate analytical solutions to some physically reasonable initial data which must satisfy the Hamiltonian and the momentum constraints on the hypersurface Σ which we discuss next.

3.2. 1+3 form of Einstein equations

The Hamiltonian and the momentum constraints on the hypersurface Σ follow from Einstein's equations as

$$\begin{aligned}
&-{}^\Sigma R - K^2 + K_{ij}K^{ij} - 2\kappa T_{nn} = 0, \\
&2D_k K^k_i - 2D_i K - 2\kappa T_{ni} = 0.
\end{aligned} \tag{37}$$

We chosen $K = 0$ and consider the vacuum case with $T_{\mu\nu} = 0$. Of course this initial data evolves in time and the remaining parts of the Einstein equations written as a dynamical system are given as

$$\frac{\partial}{\partial t}\gamma_{ij} = -2NK_{ij} + D_i N_j + D_j N_i, \tag{38}$$

$$\frac{\partial}{\partial t}K_{ij} = -N\left(R_{ij} - {}^\Sigma R_{ij} - KK_{ij} + 2K_{ik}K^k_j\right) + \mathscr{L}_{\vec{N}}K_{ij} - D_i D_j N, \tag{39}$$

where $\mathscr{L}_{\vec{N}}$ is the Lie derivative along the shift vector N^i. Derivation of these well-known equations can be found in many textbooks, see our derivation in [10].

3.3. Conformally flat Bowen-York type data

For a conformally flat hypersurface Σ ($\gamma_{\rm ij} = \psi^4 f_{ij}$ with f being the flat metric in some coordinates), the constraint equations (37) (together with the "maximal slicing" condition $K = 0$) reduce to a non-linear elliptic equation and an easily solvable linear equation, respectively given as

$$\hat{D}_i\hat{D}^i\psi = -\frac{1}{8}\psi^{-7}\hat{K}^2_{ij}, \tag{40}$$

$$\hat{D}^i\hat{K}_{ij} = 0, \tag{41}$$

with $\hat{D}_i f_{jk} = 0$ and $K_{ij} = \psi^{-2}\hat{K}_{ij}$.

Bowen and York [6] gave the following 7-parameter $(p_i, c, \mathcal{J}_i)$ solution to (41) on $\mathbb{R}^3$ whose origin is removed:

$$\begin{aligned}\hat{K}_{ij} = &\frac{3}{2r^2}\Big(p_i n_j + p_j n_i + (n_i n_j - f_{ij})p\cdot n\Big) + \epsilon\frac{3c^2}{2r^4}\Big(p_i n_j + p_j n_i + (f_{ij} - 5n_i n_j)p\cdot n\Big)\\ &+ \frac{3}{r^3}\mathcal{J}^l n^k\Big(\varepsilon_{kil}n_j + \varepsilon_{kjl}n_i\Big),\end{aligned} \tag{42}$$

where $r > 0$ is the radial coordinate, n^i is the unit normal on a sphere of radius r (not related to the unit normal to Σ); $\epsilon = \pm 1$ and $p\cdot n = p^k n_k$. At this stage, one should note that the physical meaning of the parameters $(p_i, a, \mathcal{J}_i)$ is not clear; secondly, linearity of (41) means that each bracketed term solves the equation separately. For the sake of simplicity, we shall choose $a = 0$ in what follows.

Here we follow [5]. We shall need the following expression for the right-hand side of (40)

$$\hat{K}_{ij}\hat{K}^{ij} = \frac{9}{2r^4}\left(p^2 + 2(\vec{p}\cdot\vec{n})^2\right) + \frac{18}{r^5}\left(\vec{J}\times\vec{n}\right)\cdot\vec{p} + \frac{18}{r^6}\left(\vec{J}\times\vec{n}\right)\cdot\left(\vec{J}\times\vec{n}\right). \tag{43}$$

Inserting this expression to (40), one arrives at the complicated Hamiltonian constraint which can only be solved exactly after making several assumptions. We shall not go into that discussion which was given in [11] in some detail.

3.4. Conserved quantities

To understand the physical meaning of the parameters in the solution, we shall assume that the spacetime is asymptotically flat, hence the conformal factor behaves as

$$\psi(r) = 1 + \mathcal{O}(1/r), \qquad \text{as } r\to\infty. \tag{44}$$

Then one has the conserved *total momentum* associated to Σ easily written as a boundary integral on a sphere at spatial infinity:

$$P_i = \frac{1}{8\pi}\int_{S^2_\infty} dS\, n^j\, K_{ij} = \frac{1}{8\pi}\int_{S^2_\infty} dS\, n^j\, \hat{K}_{ij}. \tag{45}$$

Observe, from the second equality, that only the leading term in the conformal factor is relevant for this and the following computation. The total conserved *total angular momentum* can also be found easily as

$$J_i = \frac{1}{8\pi}\varepsilon_{ijk}\int_{S^2_\infty} dS\, n_l\, x^j K^{kl} = \frac{1}{8\pi}\varepsilon_{ijk}\int_{S^2_\infty} dS\, n_l\, x^j\hat{K}^{kl}. \tag{46}$$

Given (45) and (46), it is straightforward to compute the integrals for the extrinsic curvature (42) which at the end yield $P_i = p_i$ and $J_i = \mathcal{J}_i$. So for the computation of these two quantities, let us note once again that, the full form of the conformal factor is not needed; one only needs to know its behavior at infinity, that is the $\mathcal{O}(1)$ term.

From these two conserved quantities, one can see that physically the assumed extrinsic curvature (42) belongs to a a self-gravitating system (a curved vacuum) with non-zero momentum and angular momentum. To compute the total mass-energy, the ADM energy, of the system, the $\mathcal{O}(1)$ term of the conformal factor is not sufficient. For that computation we keep the next order term and assume

$$\psi(r) = 1 + \frac{E}{2r} + \mathcal{O}(1/r^2) \qquad \text{as } r \to \infty. \tag{47}$$

Then defining the deviation from the background as $h_{ij} := (\psi^4 - 1)\delta_{ij}$, the ADM energy simplifies as

$$E_{ADM} = \frac{1}{16\pi} \int_{S^2_\infty} dS\, n_i \left(\partial_j h^{ij} - \partial_i h^j_j\right) = -\frac{1}{2\pi} \int_{S^2_\infty} dS\, n^i\, \partial_i \psi, \tag{48}$$

whose explicit evaluation for (47) yields $E_{ADM} = E$, which of course at this stage is almost a tautology: we have to find the constant E by solving the Hamiltonian constraint.

3.5. Approximate solution of the Hamiltonian constraint for a boosted slowly rotating gravitating system

To solve the elliptic equation (40) using (43), let us take $\hat{k}$ to be the direction of the conserved angular momentum and choose $\vec{p}$ to be in the xz plane (this is just a choice of the orientation of the coordinates and no generality is lost)

$$\vec{J} = J\hat{k}, \qquad \vec{p} = p \sin\theta_0 \hat{i} + p\cos\theta_0 \hat{k}, \tag{49}$$

with θ_0 a fixed, conserved angle. Then the Hamiltonian constraint (40) becomes

$$\hat{D}_i \hat{D}^i \psi = \psi^{-7} \left(\frac{9Jp}{4r^5} c_1 \sin\theta \sin\phi - \frac{9J^2}{4r^6} \sin^2\theta - \frac{9p^2}{16r^4}(1 + 2(c_1 \sin\theta\cos\phi + c_2\cos\theta)^2) \right), \tag{50}$$

where $c_1 := \sin\theta_0, c_2 := \cos\theta_0$.

Needless to say, an exact solution of this equation is hopeless, therefore we shall search for the lowest order perturbative solution assuming an expansion in terms of the momentum and spin which corresponds to a curved 3-surface with a small linear and small angular momentum. In [12] the slowly spinning case with no linear momentum was solved in the leading order; and in [13] slowly moving without spin was solved and in [5], both motions were considered at the leading order. We now present this solution.

A cursory inspection on the right-hand side suggests that one should have a double series of the form

$$\psi(r,\theta,\phi) := \psi^{(0)} + J^2\psi^{(J)} + p^2\psi^{(p)} + Jp\psi^{(Jp)} + \mathcal{O}(p^4, J^4, p^2J^2), \tag{51}$$

where the functions on the right-hand side depend on (r, θ, ϕ). At the zeroth order, one has the usual Laplace equation

$$\hat{D}_i \hat{D}^i \psi^{(0)} = 0, \tag{52}$$

which needs boundary conditions to be uniquely solved. The following boundary conditions as employed by [13] are apt for the problem at hand: at spatial infinity one demands

$$\lim_{r\to\infty} \psi(r) = 1, \qquad \psi(r) > 0 \tag{53}$$

Journal of Physics: Conference Series **2191** (2022) 012002 doi:10.1088/1742-6596/2191/1/012002

and near the origin one has

$$\lim_{r\to 0}\psi(r)=\psi^{(0)}, \tag{54}$$

where $\psi^{(0)}$ might have a singularity at the origin. In fact the zeroth order solution satisfying these boundary conditions reads

$$\psi^{(0)}=1+\frac{a}{r}. \tag{55}$$

The equations at the next order are

$$\hat{D}_i\hat{D}^i\psi^{(J)} = -\frac{9}{4}\sin^2\theta\frac{r}{(r+a)^7}, \tag{56}$$

$$\hat{D}_i\hat{D}^i\psi^{(Jp)} = \frac{9}{4}c_1\sin\theta\sin\phi\frac{r^2}{(r+a)^7}, \tag{57}$$

$$\hat{D}_i\hat{D}^i\psi^{(p)} = -\frac{9}{16}\left(1+2(c_1\sin\theta\cos\phi+c_2\cos\theta)^2\right)\frac{r^3}{(r+a)^7}. \tag{58}$$

These are linear equations whose solutions can be found with the help of the following spherical harmonics :

$$Y_0^0(\theta,\phi)=\frac{1}{\sqrt{4\pi}},\qquad Y_1^0(\theta,\phi)=\sqrt{\frac{3}{4\pi}}\cos\theta,\qquad Y_2^0(\theta,\phi)=\sqrt{\frac{5}{16\pi}}(3\cos^2\theta-1),$$
$$Y_1^{-1}(\theta,\phi)=\sqrt{\frac{3}{4\pi}}\sin\theta\sin\phi,\quad Y_2^1(\theta,\phi)=\sqrt{\frac{15}{4\pi}}\sin\theta\cos\theta\cos\phi,\quad Y_1^1(\theta,\phi)=\sqrt{\frac{3}{4\pi}}\sin\theta\cos\phi.$$

Then a close inspection of (56) suggests that the proper ansatz for $\psi^{(J)}$ should be of the form

$$\psi^{(J)}(r,\theta,\phi)=\psi_0^{(J)}(r)Y_0^0(\theta,\phi)+\psi_1^{(J)}(r)Y_2^0(\theta,\phi),$$

from which the solution obeying the boundary conditions (53, 54) can be found to be

$$\psi^{(J)}(r,\theta,\phi)=\frac{(a^4+5a^3r+10a^2r^2+5ar^3+r^4)}{40a^3(a+r)^5}-\frac{r^2}{40a(a+r)^5}(3\cos^2\theta-1). \tag{59}$$

To solve (57) one should take

$$\psi^{(Jp)}(r,\theta,\phi)=\psi_0^{(Jp)}(r)Y_0^0(\theta,\phi)+\psi_1^{(Jp)}(r)Y_1^{-1}(\theta,\phi),$$

for which the solution obeying the boundary conditions is

$$\psi^{(Jp)}(r,\theta,\phi)=-\frac{c_1r\,(a^2+5ar+10r^2)}{80a(a+r)^5}\sin\theta\sin\phi. \tag{60}$$

The $\psi^{(p)}$ equation (58) is similar albeit slightly more complicated: the proper ansatz reads

$$\psi^{(p)}=\psi_0^{(p)}(r)Y_0^0(\theta,\phi)+\psi_1^{(p)}(r)Y_1^1(\theta,\phi)^2+\psi_2^{(p)}(r)Y_2^1(\theta,\phi)+\psi_3^{(p)}(r)Y_1^0(\theta,\phi)^2,$$

from which four equations follow whose solutions are as follows:

$$\psi_0^{(p)}(r) = -\frac{\sqrt{\pi}\,(84a^6+378a^5r+653a^4r^2+514a^3r^3+142a^2r^4-35ar^5-25r^6)}{80ar^2(a+r)^5}$$
$$-\frac{21\sqrt{\pi}a}{20r^3}\log\frac{a}{a+r}, \tag{61}$$

and

$$\psi_1^{(p)}(r) = \frac{\pi c_1^2\left(84a^5+378a^4r+658a^3r^2+539a^2r^3+192ar^4+15r^5\right)}{40r^2(a+r)^5} + \frac{21\pi a c_1^2}{10r^3}\log\frac{a}{r+a}. \tag{62}$$

$\psi_2^{(p)}(r)$ can be obtained from (62) with the replacement $c_1^2 \to \sqrt{\frac{3}{5\pi}}c_1c_2$ and $\psi_3^{(p)}(r)$ can be obtained from (62) with the replacement $c_1^2 \to c_2^2$. All these pieces can be combined to get $\psi^{(p)}$ at this stage, but a depiction of the final result is redundant since all the parts are given above and the final expression is cumbersome. We have now all the information at our disposal to compute the relevant quantities defined on Σ including the location of the Apparent Horizon.

First let us revisit the ADM energy computation which we started above: We need the dominant terms up to and including $\mathcal{O}(\frac{1}{r})$ in $\psi(r,\theta,\phi)$. A quick power series expansion yields

$$\psi(r) = 1 + \frac{a}{r} + \frac{J^2}{40a^3r} + \frac{5p^2}{32ar} + \mathcal{O}(\frac{1}{r^2}), \tag{63}$$

in which the Jp term appears at $\mathcal{O}(\frac{1}{r^2})$ and therefore makes no contribution to the energy. Then from (47), the ADM energy of the solution follows as

$$E_{\text{ADM}} = 2a + \frac{J^2}{20a^3} + \frac{5p^2}{16a}. \tag{64}$$

So one can immediately see that for vanishing spin and vanishing linear momentum (that is the case of the Schwarschild black hole written in the isotropic coordinates) the constant a is related to the mass of the Schwarschild black hole mass as $a = M/2$.

3.6. Apparent Horizion area and the irreducible mass

While studying the efficient processes of extracting energy from rotating black holes, Christodoulou [14] realized[3] that there is an irreducible mass M_{irr} which is related to the area A_{EH} of a *section* of the event horizon via

$$M_{\text{irr}} := \sqrt{\frac{A_{\text{EH}}}{16\pi}}. \tag{65}$$

For a moving, rotating black hole, the total energy was obtained in [14] as

$$E^2 = M_{\text{irr}}^2 + p^2 + \frac{J^2}{4M_{\text{irr}}^2}, \tag{66}$$

in which the physical meaning of each part is clear.

Since we have a dynamical, evolving system, we have at our disposal the area of the Apparent Horizon, not a section of the Event Horizon. But, following [13], a good approximation to M_{irr} can be given with the help of the area of the Apparent Horizon via

$$M_{\text{irr}} := \sqrt{\frac{A_{\text{AH}}}{16\pi}}. \tag{67}$$

[3] Note that after Hawking's area theorem [15] which came later than Christodoulou's observation, it became clear that there must be an irreducible mass at the classical level.

As we shall see, this definition yields the correct expression for the energy of our system obtained from an expansion of (66). But first we need to find the location of the Apparent Horizon, namely solve (36) up to the accuracy we have been working with. That area is given simply as

$$A = \int_0^{2\pi} d\phi \int_0^{\pi} d\theta \sqrt{\det q}, \tag{68}$$

which yields the following *exact* form:

$$A_{\rm AH} = \int_0^{2\pi} d\phi \int_0^{\pi} d\theta \, \sin\theta \, \psi^4 \, h^2 \left(1 + \frac{1}{h^2} \left(\partial_\theta h\right)^2 + \frac{1}{h^2 \sin^2\theta} \left(\partial_\phi h\right)^2\right)^{1/2}. \tag{69}$$

Hence to get the area, all we need is to find the location of the Apparent Horizon up to first order in the spin and momentum. This suggests the following ansatz:

$$h(\theta, \phi) = h^0 + p h^p + J h^J + \mathcal{O}(p^2, J^2, Jp), \tag{70}$$

where

$$\partial_r h = 0, \qquad \partial_r h^0 = 0 = \partial_\theta h^0 = \partial_\phi h^0. \tag{71}$$

Ignoring the terms such as $(\partial_\theta h)^2$, $(\partial_\phi h)^2$ and $\partial_\theta h \partial_\phi h$, (36) reduces to

$$\partial_\theta^2 h + \frac{1}{\sin^2\theta} \partial_\phi^2 h + \cot\theta \partial_\theta h - 2r - 4r^2 \frac{\partial_r \psi}{\psi} + \frac{6J}{\psi^4 r^2} \partial_\phi h - \frac{3p}{\psi^4} \Big(c_1 \sin\theta \cos\phi + c_2 \cos\theta\Big) = 0. \tag{72}$$

At the zeroth order, $\mathcal{O}(p^0, J^0)$, it yields

$$1 + 2r \frac{\partial_r \psi}{\psi} = 0, \tag{73}$$

with $\psi = 1 + \frac{a}{r}$; setting $r = h$, one finds

$$h^0 = a. \tag{74}$$

This solution identifies the parameter a as the location of the apparent horizon at the lowest, dominant, order. For example, for the Schwarzschild black hole $h = 2M$ (as noted above) would be the exact solution for which the Event Horizon and the Apparent horizon coincide in these conformally flat, isotropic coordinates.

At $\mathcal{O}(p)$ we have an inhomogeneous, linear Helmholtz equation on a sphere (S^2),

$$\left(\partial_\theta^2 + \frac{1}{\sin^2\theta} \partial_\phi^2 + \cot\theta \partial_\theta - 1\right) h^p = \frac{3}{16} \Big(c_1 \sin\theta \cos\phi + c_2 \cos\theta\Big), \tag{75}$$

while at $\mathcal{O}(J)$, we have a homogeneous one:

$$\left(\partial_\theta^2 + \frac{1}{\sin^2\theta} \partial_\phi^2 + \cot\theta \partial_\theta - 1\right) h^J = 0. \tag{76}$$

Therefore, we have to find *everywhere* finite solutions of the following equation

$$\left(\vec{\nabla}_{S^2}^2 + k\right) f\left(\theta, \phi\right) = g\left(\theta, \phi\right), \tag{77}$$

Journal of Physics: Conference Series **2191** (2022) 012002 doi:10.1088/1742-6596/2191/1/012002

where $\vec{\nabla}^2_{S^2}$ is the Laplacian on S^2 given as

$$\vec{\nabla}^2_{S^2} := \partial^2_\theta + \cot\theta\partial_\theta + \frac{1}{\sin^2\theta}\partial^2_\phi. \tag{78}$$

One can employ the Green's function technique to solve this problem. For the Helmholtz operator on the sphere, the Green function $G(\hat{x}, \hat{x}')$ is defined as

$$\left(\vec{\nabla}^2_{S^2} + \kappa(\kappa+1)\right) G(\hat{x}, \hat{x}') = \delta^{(2)}(\hat{x} - \hat{x}'), \tag{79}$$

which can be found as an infinite series expansion (for example, see [16])

$$G(\hat{x}, \hat{x}') = \frac{1}{4\sin\pi\kappa}\sum_{n=0}^{\infty}\frac{1}{(n!)^2}\frac{\Gamma(n-\kappa)}{\Gamma(-\kappa)}\frac{\Gamma(n+\kappa+1)}{\Gamma(\kappa+1)}\left(\frac{1+\hat{x}\cdot\hat{x}'}{2}\right)^n, \tag{80}$$

where $\hat{x} = \sin\theta\cos\phi\hat{i} + \sin\theta\sin\phi\hat{j} + \cos\theta\hat{k}$ and $\hat{x}'$ is a similar expression with some other θ and ϕ. Employing this Green's function with $\kappa = \frac{-1+i\sqrt{3}}{2}$, one finds the first non-trivial correction to the location of the Apparent Horizon as

$$h^p = -\frac{1}{16}\left(c_1\sin\theta\cos\phi + c_2\cos\theta\right), \tag{81}$$

and $h^J = 0$. Therefore the apparent horizon is perturbed from the zeroth order expansion to

$$r = h(\theta, \phi) = a - \frac{p}{16}\Big(\sin\theta_0\sin\theta\cos\phi + \cos\theta_0\cos\theta\Big), \tag{82}$$

where, recall that, θ_0 is the angle between the linear momentum and the spin vectors. So the magnitude of the spin vector is irrelevant at this order for the location of the Apparent Horizon, but the angle it makes with the momentum vector is relevant. In Figure 2, we plotted and example of how the shape of the horizon looks like. There is a dimple on the sphere whose size depends on the ratio p/a which we took to be large to see the dimple. In the limit $\theta_0 = 0$, h reduces to the form given in [13].

Let us now evaluate the area of the Apparent Horizon from (69) which at the end yields

$$A_{\text{AH}} = 64\pi a^2 + 4\pi p^2 + \frac{11\pi J^2}{5a^2}. \tag{83}$$

Thus the irreducible mass M_{irr} turns out to be

$$M_{\text{irr}} = 2a + \frac{p^2}{16a} + \frac{11J^2}{320a^3}, \tag{84}$$

so comparing with the energy, E_{ADM}, we have

$$E_{\text{ADM}} = M_{\text{irr}} + \frac{p^2}{2M_{\text{irr}}} + \frac{J^2}{8M^3_{\text{irr}}}, \tag{85}$$

which matches the result (66) of Christodoulou at this order.

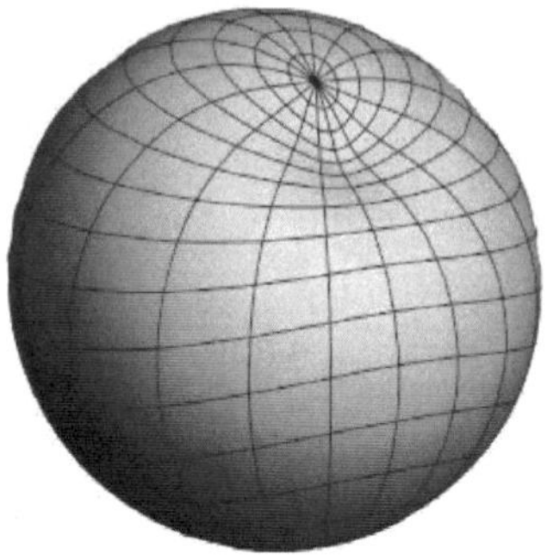

Figure 2. Shape of the Apparent Horizon when the angle between $\vec{p}$ and $\vec{J}$ is 45 degrees; and $p/a = 8\sqrt{2}$ which is outside the validity of the approximation we have worked with.

4. Conclusions

We have presented a step-by-step construction of the Apparent Horizon equation which is of extreme importance in black hole physics; and described in detail how it correctly yields the expected results, such as the irreducible mass, for a slowly moving and spinning black hole. For stationary black holes cross-sections of the event Horizon and the Apparent Horizon coincide. This exposition is of a pedagogical nature with details given in the Appendix including the derivation of the null Raychaudhuri equation which we have not used in the text, but added for more insight for the expansion of a null geodesic. We have skipped some interesting issues such as: numerically solving the case with no symmetry; multi black hole initial data; the proof that when the dominant energy condition is satisfied, the topology of the Apparent Horizon is that of S^2. For other nice expositions regarding horizons and related concepts see [7, 17, 18].

Acknowledgments

The work of E.A. is partially supported by the TUBITAK Grant No. 120F253. This work was written to celebrate the 70th birthday of Tekin Dereli who has been extremely influential in gravity research in Turkey.

Appendix A. An equivalent definition of the expansion $\Theta_{(l)}$

Here we give a proof of the second equality in (28): we have

$$\Theta_{(\ell)} = q^{\mu\nu}\nabla_\mu \ell_\nu = \frac{1}{2} q^{\mu\nu}\mathcal{L}_\ell q_{\mu\nu}, \tag{A.1}$$

where the first equality is identical to the definition of the $\Theta_{(\ell)}$. Starting from $q^{\mu\nu}\mathcal{L}_\ell q_{\mu\nu}$, one can easily arrive at the expansion $\Theta_{(\ell)}$. The construction is as follows:

$$q^{\mu\nu}\mathcal{L}_\ell q_{\mu\nu} = q^{\mu\nu}\left(\ell^\sigma \nabla_\sigma q_{\mu\nu} + q_{\sigma\nu}\nabla_\mu \ell^\sigma + q_{\sigma\mu}\nabla_\nu \ell^\sigma\right). \tag{A.2}$$

The first term on the right hand side automatically vanishes. To be able to see this explicitly, we express the metric $q_{\mu\nu}$ in terms of the spacetime metric $g_{\mu\nu}$

$$q^{\mu\nu}\ell^\sigma \nabla_\sigma q_{\mu\nu} = q^{\mu\nu}\ell^\sigma \nabla_\sigma \left(g_{\mu\nu} + k_\mu \ell_\nu + k_\nu \ell_\mu\right). \tag{A.3}$$

Since $\nabla_\sigma g_{\mu\nu} = 0$, the non vanishing terms are

$$q^{\mu\nu}\ell^\sigma\nabla_\sigma q_{\mu\nu} = q^{\mu\nu}\ell^\sigma\left(k_\mu\nabla_\sigma l_\nu + \ell_\nu\nabla_\sigma k_\mu + k_\nu\nabla_\sigma\ell_\mu + \ell_\mu\nabla_\sigma k_\nu\right), \tag{A.4}$$

where $q^{\mu\nu}k_\mu = 0 = q^{\mu\nu}\ell_\mu$, and so one gets

$$q^{\mu\nu}l^\sigma\nabla_\sigma q_{\mu\nu} = 0. \tag{A.5}$$

Now let us evaluate the second and third terms in (A.2) (which contribute equally). We can write

$$q^{\mu\nu}q_{\sigma\nu}\nabla_\mu\ell^\sigma = q^{\mu\nu}\left(g_{\sigma\nu} + k_\sigma\ell_\nu + k_\nu\ell_\sigma\right)\nabla_\mu\ell^\sigma. \tag{A.6}$$

Using $q^{\mu\nu}k_\mu = 0 = q^{\mu\nu}\ell_\mu$ again, the last expression reduces to the following

$$q^{\mu\nu}q_{\sigma\nu}\nabla_\mu\ell^\sigma = q^{\mu\nu}g_{\sigma\nu}\nabla_\mu\ell^\sigma = q^{\mu\nu}\nabla_\mu\ell_\nu. \tag{A.7}$$

Then (A.2) becomes

$$q^{\mu\nu}\mathcal{L}_\ell q_{\mu\nu} = 2q^{\mu\nu}\nabla_\mu\ell_\nu, \tag{A.8}$$

and one ends up with

$$\frac{1}{2}q^{\mu\nu}\mathcal{L}_\ell q_{\mu\nu} = q^{\mu\nu}\nabla_\mu\ell_\nu = \Theta_{(\ell)}, \tag{A.9}$$

which is the expression we wanted to prove.

Appendix B. Derivative of the area along the null vector field ℓ^μ

To gain a better physical insight to the expansion of the null geodesic congruence, let us show that when one takes the derivative of the area of the cross section along the null vector field ℓ^μ, the expansion $\Theta_{(\ell)}$ can be directly obtained as the integrand. On the surface $\mathcal{S}$, let us start with the area formula

$$A = \int dS\sqrt{q}, \tag{B.1}$$

of which the derivative along ℓ yields

$$\ell^\mu\partial_\mu A = \int dS\,\ell^\mu\partial_\mu\sqrt{q}, = \frac{1}{2}\int dS\,\sqrt{q}\,\ell^\mu q^{ab}\partial_\mu q_{ab}. \tag{B.2}$$

Equivalently one can express the result in terms of the Lie derivative along the vector field ℓ^μ using

$$q^{ab}\mathcal{L}_\ell q_{ab} = q^{ab}\ell^\mu\partial_\mu q_{ab} + 2q^{ab}q_{\mu b}\partial_a\ell^\mu. \tag{B.3}$$

Since the null vectors ℓ^μ and k^μ are the elements of the compliment of the subspace $\mathcal{S}$, one has $k^a = 0 = \ell^a$. By definition (6) we obtain

$$q^a_\mu = \delta^a_\mu + k_\mu\ell^a + k^a\ell_\mu = \delta^a_\mu, \tag{B.4}$$

and similarly

$$q^{a\mu} = g^{a\mu} + k^\mu\ell^a + k^a\ell^\mu = g^{a\mu}. \tag{B.5}$$

So that $q^{ab}q_{\mu b} = \delta^a_\mu$; and the last term in (B.3) becomes

$$q^{ab}q_{\mu b}\partial_a\ell^\mu = \delta^a_\mu\partial_a\ell^\mu = \partial_a\ell^a = 0. \tag{B.6}$$

Then $q^{ab}\mathcal{L}_\ell q_{ab}$ reduces to

$$q^{ab}\mathcal{L}_\ell q_{ab} = q^{ab}\ell^\mu\partial_\mu q_{ab}, \tag{B.7}$$

which can be related to $\ell^\mu \partial_\mu \sqrt{q}$ via

$$\ell^\mu \partial_\mu \sqrt{q} = \frac{1}{2} \sqrt{q}\, q^{ab} \mathcal{L}_\ell q_{ab}. \tag{B.8}$$

Now we can rewrite (B.2) in terms of Lie derivative

$$\ell^\mu \partial_\mu A = \int dS\, \sqrt{q} \frac{1}{2}\, q^{ab} \mathcal{L}_\ell q_{ab}. \tag{B.9}$$

In order to show the appearance of the expansion $\Theta_{(\ell)}$ explicitly, we should use the spacetime coordinates, recall that we have $\Theta_{(l)} = q^{\mu\nu} \mathcal{L}_\ell q_{\mu\nu}/2$, instead of the coordinates on the co-dimension two surface $\mathcal{S}$. It is straightforward to write

$$q^a_\mu q^b_v \mathcal{L}_\ell q_{ab} = \delta^a_\mu \delta^b_v \mathcal{L}_\ell q_{ab} = \mathcal{L}_\ell q_{\mu\nu}. \tag{B.10}$$

Multiplying this with $q^\mu_\sigma q^\nu_\rho$ one obtains

$$q^\mu_\sigma q^\nu_\rho q^a_\mu q^b_v \mathcal{L}_\ell q_{ab} = q^\mu_\sigma q^\nu_\rho \mathcal{L}_\ell q_{\mu\nu}, \tag{B.11}$$

which in terms of Kronecker delta functions reads

$$q^\mu_\sigma q^\nu_\rho \delta^a_\mu \delta^b_v \mathcal{L}_\ell q_{ab} = q^\mu_\sigma q^\nu_\rho \mathcal{L}_\ell q_{\mu\nu}, \tag{B.12}$$

and yields

$$q^a_\sigma q^b_\rho \mathcal{L}_\ell q_{ab} = q^\mu_\sigma q^\nu_\rho \mathcal{L}_\ell q_{\mu\nu}. \tag{B.13}$$

We multiply the last identity with $q^{\sigma\rho}$. Then we find the identity

$$q^{\sigma\rho} \delta^a_\sigma \delta^b_\rho \mathcal{L}_\ell q_{ab} = q^{\sigma\rho} q^\mu_\sigma q^\nu_\rho \mathcal{L}_\ell q_{\mu\nu}, \tag{B.14}$$

and so one arrives at

$$q^{ab} \mathcal{L}_\ell q_{ab} = q^{\sigma\rho} q^\mu_\sigma q^\nu_\rho \mathcal{L}_\ell q_{\mu\nu}, \tag{B.15}$$

where $q^{\sigma\rho} q^\mu_\sigma q^\nu_\rho = q^{\mu\nu}$. Finally we end up with

$$q^{ab} \mathcal{L}_\ell q_{ab} = q^{\mu\nu} \mathcal{L}_\ell q_{\mu\nu}. \tag{B.16}$$

This proves that the expansion $\Theta_{(\ell)}$ appears in the change of the area along the vector field ℓ^μ. The final expression is therefore

$$\ell^\mu \partial_\mu A = \int dS\, \sqrt{q} \frac{1}{2}\, q^{\mu\nu} \mathcal{L}_\ell q_{\mu\nu} = \int dS\, \sqrt{q} \frac{1}{2}\, q^{ab} \mathcal{L}_\ell q_{ab} = \int dS\, \sqrt{q}\, \Theta_{(l)}. \tag{B.17}$$

So setting $\Theta_{(\ell)} = 0$ to define the Apparent Horizon boils down to setting $\ell^\mu \partial_\mu A = 0$.

Appendix C. Null Raychaudhuri equation

The form of (28) already suggests that one can define a tensor whose trace is the expansion. Here we explore this tensor and obtain an expression for the change of the null expansion along the null direction ℓ as well as the null Raychaudhuri equation [2]. So let us introduce the *deformation tensor* $\Theta_{\mu\nu}$ as

$$\Theta_{\mu\nu} := \frac{1}{2} q^\sigma_\mu q^\rho_\nu \mathcal{L}_\ell q_{\sigma\rho}, \tag{C.1}$$

Journal of Physics: Conference Series **2191** (2022) 012002 doi:10.1088/1742-6596/2191/1/012002

such that

$$\Theta_{(\ell)} = g^{\mu\nu}\Theta_{\mu\nu} = q^{\mu\nu}\nabla_\mu \ell_\nu. \tag{C.2}$$

Carrying out the Lie-derivative in (C.1), one has

$$\Theta_{\mu\nu} = \nabla_\mu \ell_\nu - \omega_\mu \ell_\nu + \ell_\mu k^\sigma \nabla_\sigma \ell_\nu, \tag{C.3}$$

with

$$\omega_\mu := -k^\sigma \nabla_\mu \ell_\sigma - k^\sigma k^\rho \ell_\mu \nabla_\sigma \ell_\rho \tag{C.4}$$

which is called the rotation one form.

In what follows, we will make use of the Ricci identity

$$\nabla_\mu \nabla_\nu \ell^\mu - \nabla_\nu \nabla_\mu \ell^\mu = R_{\nu\lambda}\ell^\lambda. \tag{C.5}$$

From (C.2), one has

$$\Theta_{(\ell)} = \nabla_\mu \ell^\mu + k^\nu \ell^\mu \nabla_\mu \ell_\nu. \tag{C.6}$$

Here we assume that ℓ is a geodesic null vector but not necessarily affinely parameterized so that

$$\ell^\mu \nabla_\mu \ell_\nu = \kappa \ell_\nu, \tag{C.7}$$

where κ is a function on spacetime. Using $k^\nu \ell_\nu = -1$, one has

$$\Theta_{(l)} = \nabla_\mu \ell^\mu - \kappa. \tag{C.8}$$

So we have the following two equations:

$$\nabla_\mu \ell^\mu = \Theta_{(\ell)} + \kappa, \qquad \nabla_\mu \ell_\nu = \Theta_{\mu\nu} + \omega_\mu \ell_\nu - \ell_\mu k^\sigma \nabla_\sigma \ell_\nu. \tag{C.9}$$

Substituting these in (C.5), one has

$$\nabla_\mu \left(\Theta_\nu{}^\mu + \omega_\nu \ell^\mu - \ell_\nu k^\sigma \nabla_\sigma \ell^\mu\right) - \nabla_\nu(\Theta_{(\ell)} + \kappa) = R_{\nu\lambda}\ell^\lambda, \tag{C.10}$$

which more explicitly becomes

$$\nabla_\mu \Theta_\nu{}^\mu + \ell^\mu \nabla_\mu \omega_\nu + \omega_\nu \nabla_\mu \ell^\mu - k^\sigma \nabla_\sigma \ell^\mu \nabla_\mu \ell_\nu - \ell_\nu \nabla_\mu (k^\sigma \nabla_\sigma \ell^\mu) - \nabla_\nu(\Theta_{(\ell)} + \kappa) = R_{\nu\lambda}\ell^\lambda. \tag{C.11}$$

We use the expressions (C.9) one more time and reexpress the third and the fourth terms to obtain

$$\begin{aligned} \nabla_\mu \Theta_\nu{}^\mu + \ell^\mu \nabla_\mu \omega_\nu + \omega_\nu(\Theta_{(\ell)} + \kappa) - k^\sigma \nabla_\sigma \ell^\mu \left(\Theta_{\mu\nu} + \omega_\mu \ell_\nu - \ell_\mu k^\gamma \nabla_\gamma \ell_\nu\right) \\ -\ell_\nu \nabla_\mu (k^\sigma \nabla_\sigma \ell^\mu) - \nabla_\nu(\Theta_{(l)} + \kappa) = R_{\nu\lambda}\ell^\lambda. \end{aligned} \tag{C.12}$$

Since $\ell_\mu \nabla_\sigma \ell^\mu = 0$, the last term on the first line automatically vanishes. Contracting the final expression with ℓ^ν and using the fact that it is a null vector, one arrives at

$$\ell^\nu \nabla_\mu \Theta_\nu{}^\mu + \ell^\nu \ell^\mu \nabla_\mu \omega_\nu + \ell^\nu \omega_\nu(\Theta_{(\ell)} + \kappa) - \ell^\nu \Theta_{\mu\nu} k^\sigma \nabla_\sigma \ell^\mu - \ell^\nu \nabla_\nu(\Theta_{(\ell)} + \kappa) = R_{\nu\lambda}\ell^\lambda \ell^\nu. \tag{C.13}$$

It is easy to show that the contraction of the null vector ℓ^ν and the deformation tensor identically vanishes, $\ell^\nu \Theta_\nu{}^\mu = 0$. Then one has

$$\ell^\nu \nabla_\mu \Theta_\nu{}^\mu = -\Theta_{\mu\nu}\Theta^{\mu\nu}, \tag{C.14}$$

and also

$$\ell^\nu \ell^\mu \nabla_\mu \omega_\nu = \ell^\mu \nabla_\mu \kappa - \kappa^2. \tag{C.15}$$

Inserting these expressions in (C.13) we get

$$-\Theta_{\mu\nu}\Theta^{\mu\nu} + \kappa\Theta_{(l)} - \ell^\nu \nabla_\nu \Theta_{(l)} = R_{\nu\lambda}\ell^\lambda \ell^\nu. \tag{C.16}$$

The first term $\Theta_{\mu\nu}\Theta^{\mu\nu}$ can be written in terms of the trace free shear tensor $\sigma_{\mu\nu}$

$$\sigma_{\mu\nu} := \Theta_{\mu\nu} - \frac{1}{n-2} q_{\mu\nu}\Theta_{(\ell)} \tag{C.17}$$

as follows

$$\Theta_{\mu\nu}\Theta^{\mu\nu} = \sigma_{\mu\nu}\sigma^{\mu\nu} + \frac{1}{n-2}\Theta^2_{(\ell)}, \tag{C.18}$$

where $\sigma_{\mu\nu}\sigma^{\mu\nu} = \sigma_{ab}\sigma^{ab}$. Then (C.16) becomes

$$\ell^\mu \nabla_\mu \Theta_{(\ell)} = \nabla_\ell \Theta_{(\ell)} = \kappa\Theta_{(l)} - R_{\mu\nu}\ell^\mu \ell^\nu - \sigma_{ab}\sigma^{ab} - \frac{1}{n-2}\Theta^2_{(\ell)}. \tag{C.19}$$

The null vector field ℓ^μ is oriented in the future direction. Therefore the last equation, known as the null Raychaudhuri equation, is an evolution equation for the expansion $\Theta_{(\ell)}$.

References

[1] Abbott B P *et al.* 2016 *Phys. Rev. Lett.* **116** 061102
[2] Gourgoulhon E, *Geometry and physics of black holes*, https://luth.obspm.fr/ luthier/gourgoulhon/bh16
[3] Hawking S W and Ellis G F R 1973 (New York: Cambridge University Press)
[4] Baumgarte T and Shapiro S 2010 Numerical Relativity: Solving Einstein's Equations on the Computer (New York: Cambridge University Press)
[5] Altas E and Tekin B 2021 *Phys. Rev. D* **103** 084036
[6] Bowen J M and York J W 1980 *Phys. Rev. D* **21** 2047-56
[7] Gourgoulhon E 2012 (Springer)
[8] Arnowitt R, Deser S and Misner C 1959 *Phys. Rev.*; **116**, 1322; 1060 **117** 1595; 1962 ed L. Witten (Wiley, New York)
[9] Faraoni V 2015 *Cosmological and Black Hole Apparent Horizons*, Springer Lecture Notes in Physics
[10] Altas E and Tekin B 2020 *Phys. Rev. D* **101** 024035
[11] Altas E and Tekin B 2021 *The European Physical Journal C* **81** 328
[12] Gleiser R J, Nicasio C O, Price R H and Pullin J 1998 *Phys. Rev. D* **57** 3401
[13] Denniso K A, Baumgarte T W, and Pfeiffer H P 2006 *Phys. Rev. D* **74** 064016
[14] Christodoulou D 1970 *Phys. Rev. Lett.* **25** 1596-97
[15] Hawking S W 1971 *Phys. Rev. Lett.* **26** 1344-46
[16] Szmytkowski R 2006 *Journal of Mathematical Physics* **47** 063506
[17] Booth I 2005 *Can. J. Phys.* **83** 1073-99
[18] Curiel E 2019 *Nature Astron.* **3** no.1, 27-34

A Deep Learning Model for Automated Segmentation of Fluorescence Cell images

Musa Aydın[1*], Berna Kiraz[1,2], Furkan Eren[2], Yiğit Uysallı[2,3], Berna Morova[3], Selahattin Can Ozcan[4], Ceyda Acilan[4,5], and Alper Kiraz[2,3,4,6]

[1]Department of Computer Engineering, Fatih Sultan Mehmet Vakif University, Istanbul, Turkey
[2]Optofil Inc., Istanbul, Turkey
[3]Department of Physics, Koç University, Istanbul, Turkey
[4]Koç University Research Center for Translational Medicine (KUTTAM), Istanbul, 34450, Turkey
[5]School of Medicine, Koç University, Istanbul, 34450, Turkey
[6]Department of Electrical and Electronics Engineering, Koç University, Istanbul, Turkey

E-mail: maydin@fsm.edu.tr[1*], bkiraz@fsm.edu.tr[1], akiraz@ku.edu.tr[3]

Abstract. Deep learning techniques bring together key advantages in biomedical image segmentation. They speed up the process, increase the reproducibility, and reduce the workload in segmentation and classification. Deep learning techniques can be used for analysing cell concentration, cell viability, as well as the size and form of each cell. In this study, we develop a deep learning model for automated segmentation of fluorescence cell images, and apply it to fluorescence images recorded with a home-built epi-fluorescence microscope. A deep neural network model based on U-Net architecture was built using a publicly available dataset of cell nuclei images [1]. A model accuracy of 97.3% was reached at the end of model training. Fluorescence cell images acquired with our home-built microscope were then segmented using the developed model. 141 of 151 cells in 5 images were successfully segmented, revealing a segmentation success rate of 93.4%. This deep learning model can be extended to the analysis of different cell types and cell viability.

1. Introduction

Manual image segmentation performed by an expert is considered as the gold standard; however, it is a time-consuming process, which constantly requires expert intervention and can result in high variations due to different interpretations of experts. Besides, it is prone to human errors due to the background noise that may reduce contrast of the objects to be segmented. These shortcomings of manual segmentation can be overcome by deep learning techniques which have been widely used for biomedical image segmentation to speed up the process, increase the reproducibility, and reduce the workload in manual segmentation.

Convolutional neural network (CNN) is a deep neural network that extracts important features of an input with the application of different kernels. CNNs have been applied to many biomedical image segmentation problems such as cell segmentation, brain vessel segmentation, and brain tumor segmentation [2, 3, 4, 5, 6]. Among different CNN architectures, U-Net model is often selected as the architecture of choice for biomedical image segmentation due to its high performance and accuracy [7, 8].

In this study, we develop a deep learning model based on U-Net architecture for automated segmentation of fluorescence cell images. We trained our model using a publicly available dataset that contains images of cell nuclei as well as corresponding ground truths [1], and obtained a training accuracy of 97.3%. We then applied the trained model to automated segmentation of cell images obtained with a home-built epi-fluorescence microscope. Fluorescence images were acquired using U2OS cells stained with DAPI (6-diamidino-2-phenylindole) that enables imaging of nuclear DNA. 141 of 151 cells were successfully segmented in 5 fluorescence images, corresponding to a segmentation success rate of 93.4%.

2. Experimental Study

In this section we describe the dataset used for segmentation of cell images with the deep neural network together with the deep learning method and algorithm. For the images in the dataset, image pre-processing was applied before training in the deep neural network. The pre-processing steps were histogram matching and median filtering. Histogram matching is the process of transforming the histogram of one image to match with the histogram of another image [9]. In our implementation, the histogram of each image in the dataset was matched to the histogram function distribution of the image with the highest average light intensity value. This way, contrast differences between images were reduced. After that, a 3×3 median filter was applied to all images in order to eliminate the background noise. The dataset used in this study and all the developed codes can be accessed at github[1].

2.1. Dataset

In this study we used the publicly available dataset of cell nuclei images [1] which consisted of 670 images encoded with Portable Network Graphics format (*.png*) defined in RGB color space with corresponding binary ground truths that are manually created by experts. Figure 1 shows sample images and ground truths from the dataset.

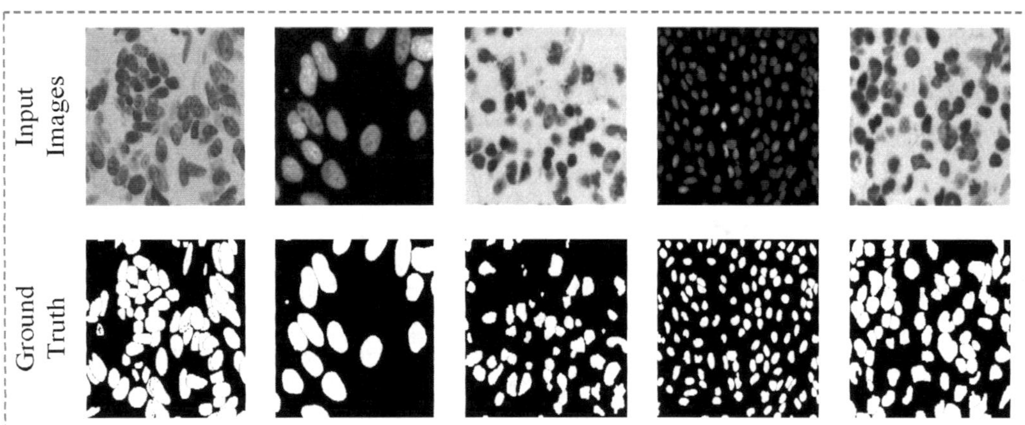

Figure 1. Exemplary input images and the corresponding ground truths from the dataset used for cell segmentation.

The sizes of the images in the cell dataset are different from each other. Thus, all images were resized to 256×256 in the image pre-processing step. 80% of the 670 images in the cell dataset

[1] https://github.com/msaaydin/U-net-Segmentation

 2191 (2022) 012003 doi:10.1088/1742-6596/2191/1/012003

was randomly selected for model training while the remaining 20% of was used for tests. Each image selected for training or tests also has ground truths available in the dataset. For ground truths, merging was done in the image pre-processing step. An image found in the original data set was manually segmented as follows. Initially, multiple ground truth images were created and saved corresponding to each cell in a reference image. As a result, more than one segmented ground truth images were created for a reference image. The ground truths corresponding to all images were then merged into a single image during the image pre-processing step, and the data set was updated. This way a new updates dataset was obtained. The new updated dataset can be accessed on github[1].

2.2. Method

In this study we employed the U-Net convolutional neural network that has a proven track record in numerous biomedical image segmentation studies in the literature [10]. The U-Net model employed, consisted of two main parts: Encoder and decoder. The main purpose of the encoder and decoder architecture used in the U-Net model is to enable concatenation of the information at different abstraction levels. Encoder reduces the dimensions of the features maps produced by convolutional layers using max pooling operations. This way low resolution, higher level information maps are constructed. These maps contain detailed features of the objects in the image which are sufficient for defining them while lacking many details. These details are then recovered by concatenating the information layers generated by the encoder with high resolution, lower level information maps generating with upsampling operations of the decoder.

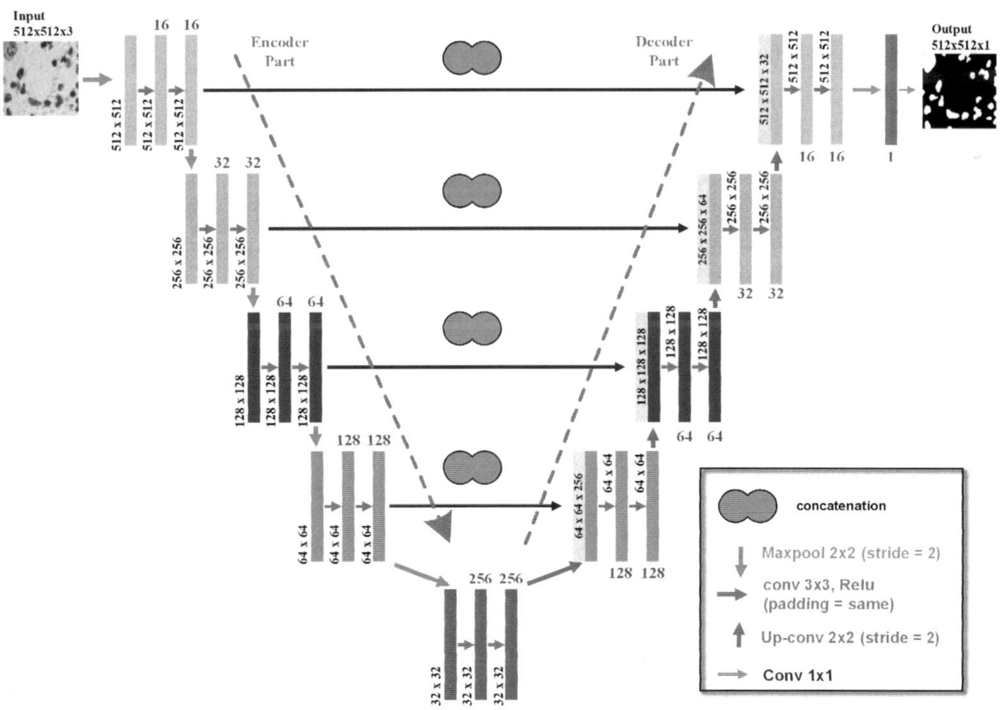

Figure 2. Diagram showing the encoder and decoder parts of the U-Net model developed for cell segmentation.

The first part, i.e. encoder, is known as the contracting path. Contracting path is a convolutional network containing steps where convolutions are performed, reducing the input dimensions. At each step of the contracting path, the data size is downsampled with max pooling operations (stride=2). Each step in the contracting path contains 2 convolutional layers with 3×3 filter size, employing rectified linear unit (ReLU) activation with zero padding. Feature size at the input layer is 16 and this number is doubled at each step of the contracting path. The second part of the U-Net model, i.e. decoder, is the expansive path where, at each step, the results obtained from 2×2 upsampling (stride=2) are concatenated with feature channels of the contructing path that are obtained from the same steps. Following concatenation, each step of the expansive path employs 2 convolutional layers with 3×3 filter size, employing ReLU activation with zero padding. After each upsampling step in the expansive path, the number of feature channels is halved. Hence, the contracting and expansive paths are symmetrical. In the final step of the U-Net model, a 1×1 convolution is implemented in order to extract the segmentation map from 16 feature maps. Due to the limited data size, each convolutional layer is followed by dropout operation with 0.2 ratio in order to prevent overfitting. All the operations (e.g. convolution, max pooling, upsampling, donwsapmling) used in the U-Net model constructed in this work are shown in Fig. 2.

Computational experiments were performed using the Google colab infrastructure and keras-tensorflow (version 2.4.1, python version = 3.7) library. In our U-Net network architecture, 16 feature maps were extracted in the first step, and propagated with 4 downsampling steps where the size of the feature map is doubled at each step (Fig. 2). U-Net model hyperparameters were selected as: Optimizer: Adam, epoch: 40, loss: binary crossentropy, metrics: [accuracy], learning rate: 0.001. 10% of the dataset was used for model validation together with a batch size of 4 at each iteration.

3. Results and Discussion

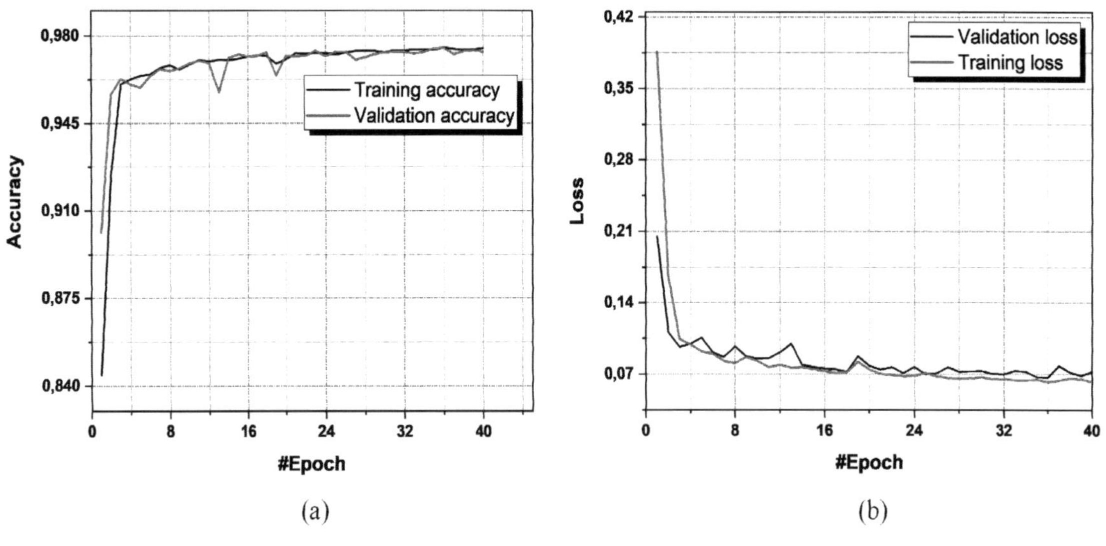

Figure 3. Changes of (a) accuracy and (b) loss with number of batches in cell segmentation during U-Net model training.

Figure 3 shows the accuracy and loss curves as a function of number of batches obtained during training and validation cycles. A very high accuracy of 97.3% was obtained with tests at the end of model training. The accuracy value was obtained as the average of accuracies obtained from 10 tests conducted.

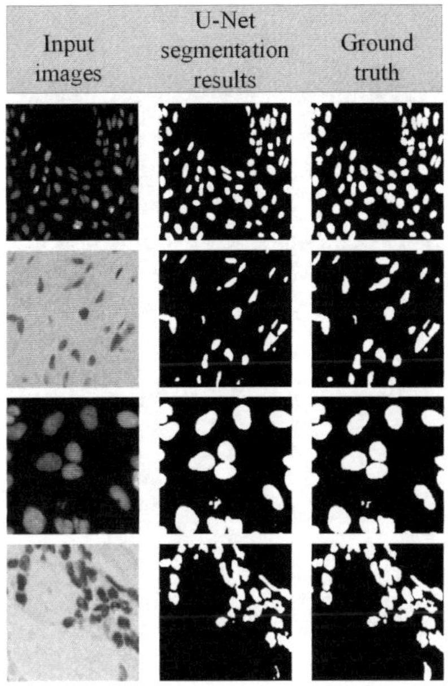

Figure 4. Exemplary input images and ground truths used for U-Net model training together with U-Net segmentation results.

Exemplary input images and U-Net segmentation results obtained with our model together with corresponding ground truths are shown in Fig. 4. A very good match is visible between the U-Net segmentation results and ground truths, as expected from the high accuracy values obtained in model training. We then applied our trained model to five fluorescence microscopy cell images recorded in our laboratory. These fluorescence images are recorded using a custom epi-fluorescence microscope which employs LED excitation at 4 colors (wavelengths of LEDs are 405, 488, 532, 633 nm), a 20X microscope objective (Nikon 20X, NA=0.4), and a CMOS camera detector (Basler daA3840-45um). Fluorescence images were collected from U2OS cells which

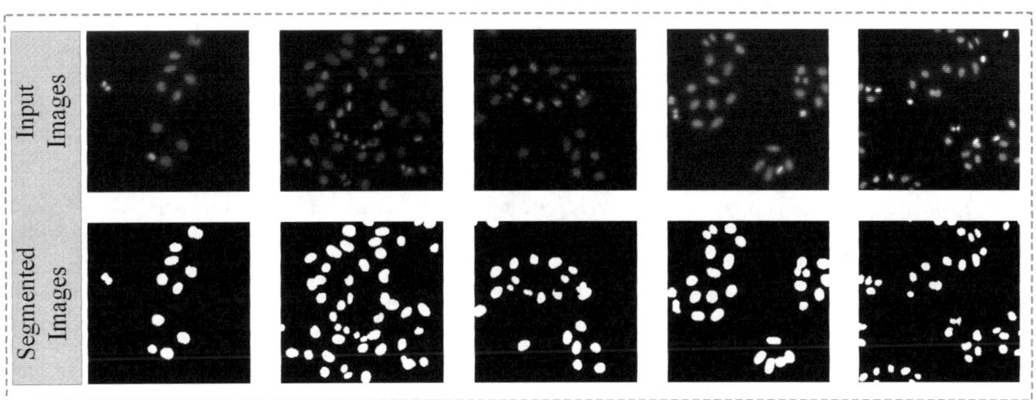

Figure 5. Input images obtained using our home-built epi-fluorescence microscope and the corresponding segmented images obtained as a result of segmentation with the U-Net model developed in this study.

were grown on coverslips and fixed with ice cold methanol at -20°C for 10 min, washed 3 times with PBS, and mounted on glass slides using Vectashield mounting medium containing DAPI (4,6-diamidino-2-phenylindole) (Vector Laboratories; H-1000-10). Segmentation results of these images are shown in Fig. 5. A very high accuracy in cell segmentation was also achieved with this dataset of 5 images. Out of a total of 151 cells in the images shown in Fig. 5, 141 of them were correctly segmented, corresponding to a segmentation success rate of 93.4%.

In conclusion, in this study we developed a U-Net model for cell segmentation and trained our model using a publicly available dataset of cell images. An accuracy of 97.3% was obtained at the end of model training. We then employed this trained model to cell segmentation using 5 fluorescence images recorded in our experimental setup, and successfully segmented 141 of 151 cells, corresponding to a segmentation success rate of 93.4%. This work can be extended to segmentation of images that contain different cell types and to classification for cell viability analysis. We are also planning to use such a U-Net based image segmentation approach to other biomedical image processing problems including brain vasculature analysis.

Acknowledgments

We acknowledge financial supports from TÜBİTAK (Project No: 7190434) and KOSGEB. A. Kiraz acknowledges partial support from the Turkish Academy of Sciences (TÜBA).

References

[1] "Cell nuclei data set," https://www.kaggle.com/c/data-science-bowl-2018/data, last access: 01-03-2021.

[2] R. Hollandi, A. Szkalisity, T. Toth, E. Tasnadi, C. Molnar, B. Mathe, I. Grexa, J. Molnar, A. Balind, M. Gorbe *et al.*, "nucleaizer: a parameter-free deep learning framework for nucleus segmentation using image style transfer," *Cell Systems*, vol. 10, no. 5, pp. 453–458, 2020.

[3] M. Habibzadeh, M. Jannesari, Z. Rezaei, H. Baharvand, and M. Totonchi, "Automatic white blood cell classification using pre-trained deep learning models: Resnet and inception," in *Tenth international conference on machine vision (ICMV 2017)*, vol. 10696. International Society for Optics and Photonics, 2018, p. 1069612.

[4] M. Delgado-Ortet, A. Molina, S. Alférez, J. Rodellar, and A. Merino, "A deep learning approach for segmentation of red blood cell images and malaria detection," *Entropy*, vol. 22, no. 6, p. 657, 2020.

[5] M. R. K. Mookiah, S. Hogg, T. J. MacGillivray, V. Prathiba, R. Pradeepa, V. Mohan, R. M. Anjana, A. S. Doney, C. N. Palmer, and E. Trucco, "A review of machine learning methods for retinal blood vessel segmentation and artery/vein classification," *Medical Image Analysis*, p. 101905, 2020.

[6] M. Havaei, A. Davy, D. Warde-Farley, A. Biard, A. Courville, Y. Bengio, C. Pal, P.-M. Jodoin, and H. Larochelle, "Brain tumor segmentation with deep neural networks," *Medical image analysis*, vol. 35, pp. 18–31, 2017.

[7] O. Ronneberger, P. Fischer, and T. Brox, "U-net: Convolutional networks for biomedical image segmentation," in *International Conference on Medical image computing and computer-assisted intervention*. Springer, 2015, pp. 234–241.

[8] M. Livne, J. Rieger, O. U. Aydin, A. A. Taha, E. M. Akay, T. Kossen, J. Sobesky, J. D. Kelleher, K. Hildebrand, D. Frey *et al.*, "A u-net deep learning framework for high performance vessel segmentation in patients with cerebrovascular disease," *Frontiers in neuroscience*, vol. 13, p. 97, 2019.

[9] S. P. Ehsani, H. S. Mousavi, and B. H. Khalaj, "Iterative histogram matching algorithm for chromosome image enhancement based on statistical moments," in *2012 9th IEEE International Symposium on Biomedical Imaging (ISBI)*. IEEE, 2012, pp. 214–217.

[10] Z. Liu, "Retinal vessel segmentation based on fully convolutional networks," *arXiv preprint arXiv:1911.09915*, 2019.

Quantum Entanglement and Neutrino Many-Body Systems

A.B. Balantekin
Physics Department, University of Wisconsin, Madison, WI 53706 USA
E-mail: baha@physics.wisc.edu

Abstract. Entanglement of constituents of a many-body system is a recurrent feature of quantum behaviour. Quantum information science provides tools, such as the entanglement entropy, to help assess the amount of entanglement in such systems. Many-neutrino systems are present in core-collapse supernovae, neutron star mergers, and the Early Universe. Recent work in applying the tools of quantum information science to the description of the entanglement in astrophysical many-neutrino systems is reviewed.

1. Characterization of Quantum Entanglement

Complex many-body systems are ubiquitous in physics. Calculating energy eigenvalues and eigenstates or time-evolution of such systems is notoriously difficult except for a few oversimplified cases. A significant limitation is the size of the Hilbert space in which the Hamiltonian describing such a system needs to be diagonalized. This Hilbert space can get very large very quickly with the increasing number of particles. One example is provided by the many-neutrino systems that can be formed in various astrophysical settings. For example, the proto-neutron star formed following the core-collapse of a type II supernova retains almost all the gravitational binding energy of the pre-supernova star, and primarily cools by emitting this excess energy in the form of neutrino-antineutrino pairs. Of the order of 10^{58} neutrinos are emitted from each such supernova. When so many neutrinos are present in a space with a linear dimension of a few hundred kilometers, Standard Model neutrino-neutrino interactions are no longer negligible. Neutrino propagation including this non-linear component is dubbed collective neutrino oscillations and it is well-reviewed in the literature [1, 2, 3, 4, 5]. Consequently the dynamics of such a system is determined not only by the one-body interaction of neutrinos with other particles in the background, but also the two-body interaction between neutrinos. For a system of N neutrinos and N antineutrinos, each of which can be in one of the three active neutrino flavor states (ν_e, ν_μ, or ν_τ states) and three active antineutrino states, the dimension of the Hilbert space is 6^N. Even in a simplified case where one only considers two flavors and ignores the presence of antineutrinos, the dimension of the Hilbert space is 2^N. Clearly even for relatively low values of N, such a calculation would be beyond the capabilities of present computers. Indeed this rapid increase of the dimension of the Hilbert space with increasing particle number illustrates one of the main motivations for the efforts to build quantum computers. Hence approximate treatments of the many-body problems are currently inevitable.

One common approximation is the mean-field approximation. To begin with the initial neutrino energy spectra is divided into energy bins. In the mean-field approximation one picks a particular energy bin and assumes that neutrinos in that bin interact with a suitably chosen "mean-field" representing all the other neutrinos in other energy bins. It should be pointed out that mean-field is not unique and, depending on the methods used, one can obtain different mean fields [6, 7, 8, 9, 10, 11]. Mean-field treatment of the collective neutrino oscillations in many different settings has been extensively studied in the literature.

Since collective neutrino oscillations occur in settings as diverse as the Early Universe, core-collapse supernovae, and neutron-star mergers it should be natural to ask how well the mean-field approximation describes the physics of neutrino propagation in those environments. In general a complete description of the many-body systems include both entangled and unentangled states. However, adopting the mean-field approximation means eliminating entangled states from the description of the many-body problem. In general if one has N particles distributed over k states, dimension of the Hilbert space is k^N. (In the above example with three neutrinos and three antineutrinos $k = 6$). In contrast, the Hamiltonian matrix in the mean-field basis has a dimension of kN. Earlier work exploring various aspects of collective neutrino oscillations differed on the question of entanglement [12, 13, 14, 15], illustrating the subtlety of this issue.

One way to quantify the degree of entanglement in a many-body system is to introduce the entanglement entropy. It is defined as

$$S = -\mathrm{Tr}(\rho \log \rho) \tag{1}$$

where ρ is the density matrix of the system. Density matrices are normalized so that $\mathrm{Tr}\,\rho = 1$. For a pure state, which satisfies the condition $\rho^2 = \rho$, the entanglement entropy vanishes. This would be the case if ρ represents the exact density matrix for the whole system. Consider two subsystems A and B of a larger system. If there is no entanglement between A and B the density matrix for the combined system is the direct product of the density matrices representing these two subsystems:

$$\rho = \rho_A \otimes \rho_B. \tag{2}$$

For a concrete example we can consider the neutrino flavor isospin. If we limit ourselves to two flavors ν_e and ν_x, we can take them as the upper and lower components in the neutrino flavor isospin SU(2) algebra (or a qubit). Since the density matrices need to be Hermitian, the density matrix for one single qubit can be written as the most general 2×2 Hermitian matrix of unit trace:

$$\rho = \frac{1}{2}(1 + \boldsymbol{\sigma} \cdot \mathbf{P}) \tag{3}$$

using the fundamental representation of SU(2), the Pauli matrices. In Eq. (3) 1 is the 2×2 unit matrix and the vector $\mathbf{P}$ is referred to as the polarization vector. Next we consider two neutrinos with energies E_A and E_B. If they are not entangled their density matrices satisfy the conditions $\rho_A^2 = \rho_A$ and $\rho_B^2 = \rho_B$. Consequently both polarization vectors are on the Bloch sphere $|\mathbf{P}^{(A)}|^2 = 1 = |\mathbf{P}^{(B)}|^2$ and we can write the density matrix for the combined system of these two unentangled neutrinos as

$$\rho = \rho_A \otimes \rho_B = \frac{1}{4}\left[(1 \otimes 1) + (\sigma_i^{(A)} \otimes 1)\mathbf{P}_i^{(A)} + (1 \otimes \sigma_i^{(B)})\mathbf{P}_i^{(B)} + (\sigma_i^{(A)} \otimes \sigma_j^{(B)})\mathbf{P}_i^{(A)}\mathbf{P}_j^{(B)}\right]. \tag{4}$$

Note that Eq. (4) for the two neutrino system is now written using the fundamental representation of SU(4). If instead these two neutrinos are entangled, one needs to write down the most general density matrix

$$\rho_{2\nu} = \frac{1}{4}\left[(1 \otimes 1) + (\sigma_i^{(1)} \otimes 1)x_i + (1 \otimes \sigma_i^{(2)})y_i + (\sigma_i^{(1)} \otimes \sigma_j^{(2)})\omega_{ij}\right] \tag{5}$$

Here x_i and y_j are three-vectors and ω_{ij} can be considered as elements of a 3×3 matrix. Of course, these parameters need to be calculated by solving the equation

$$i\frac{\partial}{\partial t}\rho_{2\nu} = [H_{2\nu}, \rho_{2\nu}] \tag{6}$$

where the Hamiltonian $H_{2\nu}$ includes interactions between the two neutrinos. For Eq. (5) to describe a physical density operator (i.e., to be a solution of the Eq. (6)) the condition $\det \omega \leq 0$ should be satisfied. Comparing Eqs. (4) and (5) we see that if the two qubits are not entangled $\omega_{ij} = x_i y_j$ and $\det \omega = 0$. If these two qubits are entangled with each other, but with no other particles then $\rho_{2\nu}$ in Eq. (5) describes a pure state: $\rho_{2\nu} = \rho_{2\nu}^2$. Imposing this requirement results in four conditions:

$$1 + |\mathbf{x}|^2 + |\mathbf{y}|^2 + \omega_{ij}\omega_{ij} = 4, \tag{7}$$

$$x_i = \omega_{ij} y_j, \tag{8}$$

$$y_i = x_j \omega_{ji}, \tag{9}$$

$$\omega_{ij} = x_i y_j - \frac{1}{2}\delta_{ij}\left[(\mathrm{Tr}\,\omega)^2 - \mathrm{Tr}\,\omega^2\right] + \omega_{ji}(\mathrm{Tr}\,\omega) - (\omega^2)_{ji}. \tag{10}$$

Eqs. (8) and (9) immediately yield

$$|\mathbf{x}|^2 = |\mathbf{y}|^2. \tag{11}$$

Eq. (10) connects the matrix ω and and its adjugate matrix Z. (These two matrices satisfy the condition $\boldsymbol{\omega}\mathbf{Z} = (\det \boldsymbol{\omega})\mathbf{I}$). It yields

$$\omega_{ij} = x_i y_j - Z_{ji}. \tag{12}$$

Inserting Eqs. (12) and (11) into Eq. (7) we finally obtain

$$|\mathbf{x}|^2 = |\mathbf{y}|^2 = 1 + \det \boldsymbol{\omega}. \tag{13}$$

To assess the amount of entanglement we first introduce the reduced density matrix for the first neutrino by tracing over the second neutrino degrees of freedom:

$$\rho_{\mathrm{red}}^{(1)} = \mathrm{Tr}_2\,\rho_{2\nu} = \frac{1}{2}\left(1 + \sigma_i^{(1)} x_i\right). \tag{14}$$

One can similarly introduce the reduced density matrix for the second neutrino by tracing over the first neutrino degrees of freedom:

$$\rho_{\mathrm{red}}^{(2)} = \mathrm{Tr}_1\,\rho_{2\nu} = \frac{1}{2}\left(1 + \sigma_i^{(2)} y_i\right). \tag{15}$$

Then the entanglement entropies for each neutrino can be written as

$$S^{(1)} = -\mathrm{Tr}_1\left(\rho_{\mathrm{red}}^{(1)} \log \rho_{\mathrm{red}}^{(1)}\right), \; S^{(2)} = -\mathrm{Tr}_2\left(\rho_{\mathrm{red}}^{(2)} \log \rho_{\mathrm{red}}^{(2)}\right) \tag{16}$$

It is straightforward to calculate these entropies by diagonalizing the density matrices given in Eqs. (14) and (15). One gets

$$S^{(1)} = -\frac{1 - |\mathbf{x}|}{2}\log\left(\frac{1 - |\mathbf{x}|}{2}\right) - \frac{1 + |\mathbf{x}|}{2}\log\left(\frac{1 + |\mathbf{x}|}{2}\right), \tag{17}$$

and

$$S^{(2)} = -\frac{1-|\mathbf{y}|}{2}\log\left(\frac{1-|\mathbf{y}|}{2}\right) - \frac{1+|\mathbf{y}|}{2}\log\left(\frac{1+|\mathbf{y}|}{2}\right). \tag{18}$$

Eq. (11) implies that these two entropies are equal. Note that for the unentangled case, $\omega_{ij} = x_i y_j$, we have $\det w = 0$ and both entropies vanish. Eq. (12) also shows that the adjugate matrix $\mathbf{Z}$ represents the correlations between two qubits [16]. Equality of the two entropies in Eqs. (17) and (18) is an example of a more generally applicable result. If a complex system can be divided into two pieces, these two parts would have equal entanglement entropies if certain conditions are satisfied [17, 18].

2. Many-neutrino systems in astrophysics

For neutrinos in astrophysical settings we adopt the single angle approximation for two flavors and no antineutrinos resulting in the Hamiltonian

$$H = \sum_p \omega_p \, \mathcal{B} \cdot \mathbf{J}(\mathbf{p}) + \mu \sum_{p,q,p\neq q} \mathbf{J}(\mathbf{p}) \cdot \mathbf{J}(\mathbf{q}) \tag{19}$$

where $\mathbf{J}(\mathbf{p})$ are the neutrino flavor isospin (SU(2)) operators, $\mathbf{p}$ are the neutrino momenta, δm^2 is the difference between squares of the masses associated with the two mass eigenstates, $\omega_p = \delta m^2/2p$ and an auxiliary vector $\mathcal{B}$ was introduced. Since we bin neutrino energies, in the discussion below we use discrete labels $A, B, \cdots$ to indicate these energy bins as in $\omega_A = \delta m^2/2p_A$ and write these labels as superscripts $(A), (B), \cdots$ to the neutrino isospin operators. In the discussion below for simplicity we assume that there is a single neutrino at each energy bin. This Hamiltonian in Eq. (19) can be written either in the mass or flavor basis; Hamiltonians in the two bases are related by a unitary transformation. The auxiliary vector in the mass basis is given by

$$\mathcal{B}_{\rm mass} = (0, 0, -1) \tag{20}$$

and in the flavor basis by

$$\mathcal{B}_{\rm flavor} = (\sin 2\theta, 0, -\cos 2\theta). \tag{21}$$

In the single-angle approximation the quantity μ represents an average over the entire neutrino ensemble, $\mu = (\sqrt{2}G_F/V)\langle 1 - \cos\vartheta_{\mathbf{pq}}\rangle$, where $\vartheta_{\mathbf{pq}}$ is the angle between neutrino momenta $\mathbf{p}$ and $\mathbf{q}$, G_F is the Fermi constant, and V is the quantization volume. Further details are given in the reviews listed in the previous section.

The density matrix can be obtained by solving an equation of the form Eq. (6) with the Hamiltonian in Eq. (19). Expectation values of the neutrino flavor isospin operators can then be found using

$$i\frac{\partial}{\partial t}\mathrm{Tr}\,(\mathbf{J}^{(A)}\rho) = \mathrm{Tr}\,([\mathbf{J}^{(A)}, H]\rho). \tag{22}$$

Defining

$$\mathbf{L}^{(A)} = \sum_{B,B\neq A} \mathbf{J}^{(B)}. \tag{23}$$

and

$$Q_n^{(A)} = \epsilon_{nmk} J_m^{(A)} L_k^{(A)} \tag{24}$$

one gets

$$\frac{\partial}{\partial t}\mathrm{Tr}(\rho\mathbf{J}^{(A)}) = \mathrm{Tr}\left(\rho(\omega_A \, \mathcal{B} \times \mathbf{J}^{(A)} - \mu\, \mathbf{Q}^{(A)})\right). \tag{25}$$

Journal of Physics: Conference Series **2191** (2022) 012004 doi:10.1088/1742-6596/2191/1/012004

For N neutrinos the density matrix can be written using the fundamental representation of $\mathrm{SU}(2^N)$:

$$\begin{aligned}
\rho &= \frac{1}{N}[(1\otimes 1\otimes 1\otimes\cdots\otimes 1)\\
&+ x_i^{(A)}(\sigma_i^{(A)}\otimes 1\otimes 1\otimes\cdots\otimes 1)+y_i^{(B)}(1\otimes\sigma_i^{(B)}\otimes 1\otimes\cdots\otimes 1)+\cdots\\
&+ \omega_{ij}^{(AB)}(\sigma_i^{(A)}\otimes\sigma_j^{(B)}\otimes 1\otimes\cdots\otimes 1)+\Omega_{ij}^{(BC)}(1\otimes\sigma_i^{(B)}\otimes\sigma_j^{(C)}\otimes\cdots\otimes 1)+\cdots\\
&+ z_{ijk}^{(ABC)}(\sigma_i^{(A)}\otimes\sigma_j^{(B)}\otimes\sigma_k^{(C)}\otimes 1\cdots\otimes 1)+\zeta_{ijk}^{(BCD)}(1\otimes\sigma_i^{(B)}\otimes\sigma_j^{(C)}\otimes\sigma_k^{(D)}\otimes\cdots\otimes 1)+\cdots\\
&+ \cdots]\,.
\end{aligned}\tag{26}$$

Note that individual neutrino qubits are represented as

$$\mathbf{J}_i^{(A)}=\frac{1}{2}(\sigma_i^{(A)}\otimes 1\otimes 1\otimes\cdots 1),\tag{27}$$

$$\mathbf{J}_i^{(B)}=\frac{1}{2}(1\otimes\sigma_i^{(B)}\otimes 1\otimes\cdots\otimes 1),\tag{28}$$

and so on. Hence the reduced density matrix, for example, for the neutrino with energy label A can be obtained by tracing over the neutrinos labeled $B, C, D, \cdots$:

$$\rho_{\rm red}^{(A)}=\mathrm{Tr}_{(B,C,D,\ldots)}\rho=\frac{1}{2}\left(1+\boldsymbol{\sigma}\cdot\mathbf{x}\right).\tag{29}$$

One can also note that

$$2\,\mathrm{Tr}\,(\rho\mathbf{J}^{(A)})=2\,\mathrm{Tr}_A\,(\rho_{\rm red}^{(A)}\mathbf{J}^{(A)})=\mathbf{x}^{(A)}.\tag{30}$$

If one adopts the mean-field picture by replacing the two-body term in that Hamiltonian with a one-body term representing one neutrino interacting with the mean-field, neutrinos with different energies become uncorrelated and the density matrix takes the form

$$\rho_{\rm mf}=\rho_A\otimes\rho_B\otimes\rho_C\otimes\cdots=\left[\frac{1}{2}(1+\boldsymbol{\sigma}\cdot\mathbf{P}^{(A)})\right]\otimes\left[\frac{1}{2}(1+\boldsymbol{\sigma}\cdot\mathbf{P}^{(B)})\right]\otimes\cdots\tag{31}$$

where $A, B, C, \cdots$ are the energy labels and each ρ on the right-hand side of the equation is of the form of Eq. (3).

Eq. (25) is valid for either mean-field or many-body treatments. A more convenient way to write down this equation is to combine terms multiplying neutrino flavor isospin operators

$$\frac{\partial}{\partial t}\mathrm{Tr}\,(\rho\mathbf{J}^{(A)})=\mathrm{Tr}\,[\rho((\omega_A\,\mathcal{B}+\mu\,\mathbf{L}^{(A)})\times\mathbf{J}^{(A)})].\tag{32}$$

Inserting Eq. (31) into Eq. (32) one obtains

$$\frac{\partial}{\partial t}\mathbf{P}^{(A)}=(\omega_A\mathcal{B}+\mu\mathbf{P})\times\mathbf{P}^{(A)}\tag{33}$$

where

$$\mathbf{P}=\sum_A\mathbf{P}^{(A)}.\tag{34}$$

It is easy to see that $\mathbf{P}$ satisfies the equation

$$\frac{\partial}{\partial t}\mathbf{P}=\mathcal{B}\times\left(\sum_A\omega_A\mathbf{P}^{(A)}\right)\tag{35}$$

and that $\mathcal{B} \cdot \mathbf{P}$ is a constant of motion.

One is interested in the adiabatic solutions of Eq. (33). Clearly for the $\mathbf{P}^{(A)}$ not to change at all every one of them needs to be parallel to the vector $\omega_A \mathcal{B} + \mu \mathbf{P}$. Hence a suitable ansatz for the treatment of adiabatic evolution is to assume that the largest component of each $\mathbf{P}^{(A)}$ lies in the co-moving plane defined by $\mathcal{B}$ and $\mathbf{P}$ [19, 20, 21, 22], with a smaller component along the direction of $\mathcal{B} \times \mathbf{P}$:

$$\mathbf{P}^{(A)} = \alpha_A \mathcal{B} + \beta_A \mathbf{P} + \gamma_A (\mathcal{B} \times \mathbf{P}), \tag{36}$$

where, for consistency, we require

$$\sum_A \alpha_A = 0, \ \sum_A \beta_A = 1, \ \sum_A \gamma_A = 0. \tag{37}$$

The coefficients α, β, and γ above are necessarily time-dependent. If initially all N neutrinos have the same flavor, then their initial values in the mass basis would be $\alpha_0 = 0, \beta_0 = 1/N, \gamma_0 = 0$. Inserting Eq. (36) into Eq. (35) we obtain

$$\frac{\partial}{\partial t}\mathbf{P} = \left(\sum_A \beta_A \omega_A\right)(\mathcal{B} \times \mathbf{P}) + \left(\sum_A \gamma_A \omega_A\right)[(\mathcal{B} \cdot \mathbf{P})\mathcal{B} - \mathbf{P}] \tag{38}$$

We adopt for the mass basis for the discussion below. For convenience we define

$$\Gamma = \left(\sum_A \gamma_A \omega_A\right). \tag{39}$$

We observe that unless Γ is positive the solutions for P_x and P_y exponentially grow. (This provides another constraint on γ_As). Introducing new variables $\Pi_{x,y}$ via

$$P_{x,y} = \Pi_{x,y} \exp\left(-\int \Gamma(t) dt\right) \tag{40}$$

we see that they satisfy the equations

$$\frac{\partial}{\partial t}\Pi_x = \left(\sum_A \beta_A \omega_A\right)\Pi_y, \tag{41}$$

$$\frac{\partial}{\partial t}\Pi_y = -\left(\sum_A \beta_A \omega_A\right)\Pi_x. \tag{42}$$

It follows that in this mean-field approximation Π_x and Π_y precess around $\mathcal{B}$ with a time-dependent frequency (through the time-dependence of β_As). Then P_x and P_y also precess similarly while decaying due to the exponential term in Eq. (40). Hence asymptotically P_x and P_y tend to be very small. (In the mass basis we used $\mathcal{B} \cdot \mathbf{P} = P_z$ which is a constant of motion fixed by the initial conditions). Then Eq. (36) implies that x and y components of each $\mathbf{P}^{(A)}$ are asymptotically very small. Since $|\mathbf{P}^{(A)}|^2 = 1$ for uncorrelated neutrinos, it then follows that

$$\left(\mathbf{P}_z^{(A)}\right)^2 \sim 1$$

asymptotically. Consequently allowed asymptotic values of $\mathbf{P}_z^{(A)}$ are $\sim \pm 1$. Since the constant of motion $\sum_A \mathbf{P}_z^{(A)}$ is fixed by the initial conditions, some of the final $\mathbf{P}_z^{(A)}$ values will be $+1$

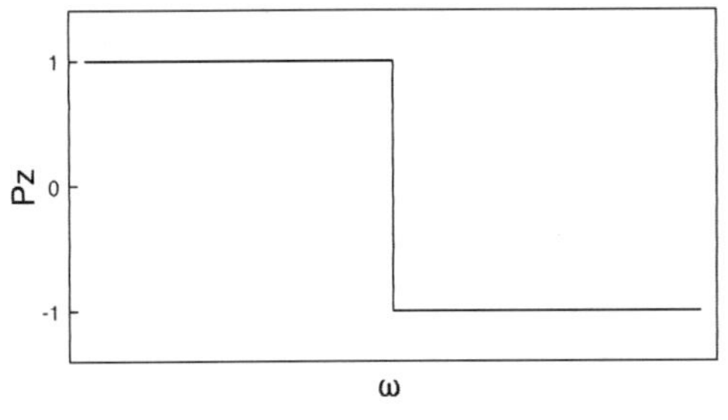

Figure 1. Schematic depiction of a single spectral split in the single-angle mean-field approximation: third component of the polarization vector for individual neutrinos as a function of their $\omega = \delta m^2/2E$.

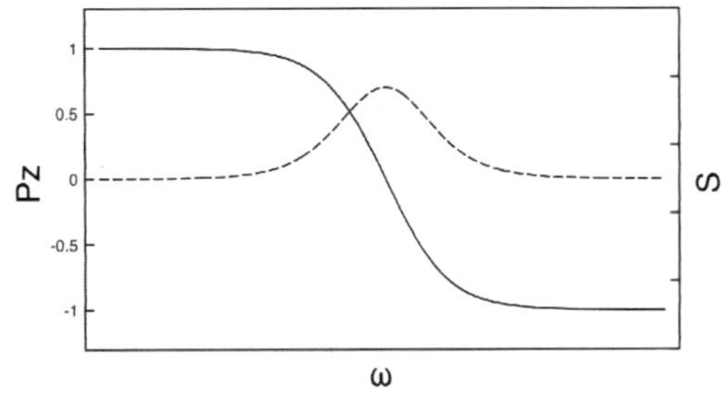

Figure 2. Schematic depiction of a single spectral split in the single-angle many body approach: third component of the polarization vector obtained from the reduced density matrices of individual neutrinos (solid line) as a function of their $\omega = \delta m^2/2E$. Also shown is their entanglement entropies (dashed line).

and some of them will be -1. This is the "spectral split" phenomenon as shown in Figure 1. It should be noted that, depending on the initial conditions, there may exist one or more spectral splits.

It is possible to calculate the many-body density matrix for a small number of neutrinos. For the number of neutrinos up to $N \sim 10$ it is possible to calculate the many-body density matrix using Bethe ansatz method to obtain the adiabatic eigenvalues [23, 24, 25, 26]. Runge-Kutta integration permits calculation for up to $N \sim 16$ neutrinos [27]. It is also possible to obtain exact solutions when the neutrinos are placed in a momentum lattice [28]. Preliminary work using quantum computers is encouraging [29, 30].

It turns out that presence of spectral splits is a good proxy for deviations from the mean-field picture. Once the density matrix for the entire ensemble is calculated obtaining reduced density matrices for individual neutrinos and resulting entanglement entropies is straightforward. One finds that, depending on the initial conditions, these entanglement entropies approach the maximal value of $\log 2$ for a good fraction of neutrinos [26]. In addition an interesting behavior is observed: in contrast to the single-angle mean-field case where the spectral split is essentially a step function (as schematically shown in Figure 1) at the split energy, the many-body spectral split happens smoothly over an energy range centered approximately at the mean-field split energy (as schematically shown in Figure 2) [27]. Away from the split energy many-body result is approximately same as the mean-field result. Consequently the entanglement entropy is essentially zero at energies away from the split energy, but reaches to sizable non-zero value at the split energy. Dependence of this general result on the initial conditions and conditions for the presence of multiple splits was elaborated in Ref. [27]. These intriguing results call for further study of entanglement and many-body effects in collective neutrino oscillations, especially for larger numbers of neutrinos. Recent work using matrix product states to explore spin models corresponding to collective neutrino oscillations found that entanglement entropies at most logarithmically scale with the system size, suggesting that tensor network methods could be efficient to this end [31].

Acknowledgments

I would like to express my gratitude to Tekin Dereli for many discussions on all aspects of physics covering almost five decades. I also would like to thank my collaborators S. Birol, M. Cervia, S.N. Coppersmith, C. Johnson, T. Kajino, A. Patwardhan, and Y. Pehlivan with whom much of the original work leading to this review was carried out. This work was supported in part by the U.S. Department of Energy, Office of Science, Office of High Energy Physics, under Award No. DE-SC0019465. It was also supported in part by the U.S. National Science Foundation Grants No. PHY-2020275 and PHY-2108339.

References

[1] Duan H and Kneller J P 2009 *J. Phys. G* **36** 113201
[2] Duan H, Fuller G M and Qian Y Z 2010 *Ann. Rev. Nucl. Part. Sci.* **60** 569
[3] Chakraborty S, Hansen R, Izaguirre I and Raffelt G 2016 *Nucl. Phys.* B **908** 366
[4] Balantekin, A B 2018 *J. Phys. G* **45** 113001
[5] Tamborra I and Shalgar S 2021 *Ann. Rev. Nucl. Part. Sci.* **71** 165
[6] Balantekin A B and Pehlivan Y 2007 *J. Phys. G* **34** 47
[7] Volpe C, Väänänen D and Espinoza C 2013 *Phys. Rev. D* **87** 113010
[8] Serreau J and Volpe C 2014 *Phys. Rev. D* **90** 125040
[9] Cirigliano V, Fuller G M, and Vlasenko A 2015 *Phys. Lett. B* **747** 27
[10] Vlasenko A, Fuller G M, and Cirigliano V 2014 Prospects for Neutrino-Antineutrino Transformation in Astrophysical Environments *Preprint* arXiv:1406.6724 [astro-ph.HE].
[11] Volpe C 2015 *Int. J. Mod. Phys. E* **24**, no. 09, 1541009
[12] Bell N F, Sawyer R F, Volkas R R, and Wong Y Y Y 2002 *Phys. Rev. A* **65** 042328
[13] Friedland A and Lunardini C 2003 *Phys. Rev. D* **68** 013007
[14] Bell N F, Rawlinson A A, and Sawyer R F 2003 *Phys. Lett. B* **573** 86
[15] Friedland A and Lunardini C 2003 *JHEP* **10** 043
[16] Fano U 1983 *Rev. Mod. Phys.* **55** 855
[17] Bombelli L, Koul R K, Lee J and Sorkin R D 1986 *Phys. Rev. D* **34** 373
[18] Srednicki M 1993 *Phys. Rev. Lett.* **71** 666
[19] Duan H, Fuller G M and Qian Y Z 2006 *Phys. Rev. D* **74** 123004
[20] Raffelt G G and Smirnov A Y 2007 *Phys. Rev. D* **76** 125008
[21] Raffelt G G and Smirnov A Y 2007 *Phys. Rev. D* **76** 081301 [erratum: *Phys. Rev. D* **77** (2008), 029903]
[22] Duan H, Fuller G M and Qian Y Z 2007 *Phys. Rev. D* **76** 085013
[23] Pehlivan Y, Balantekin A B, Kajino T and Yoshida T 2011 *Phys. Rev. D* **84** 065008
[24] Birol S, Pehlivan Y, Balantekin A B and Kajino T 2018 *Phys. Rev.* D **98** 083002
[25] Patwardhan A V, Cervia M J and Balantekin A B 2019 *Phys. Rev.* D **99** (2019) 123013
[26] Cervia M J, Patwardhan A V, Balantekin A B, Coppersmith S N and Johnson C W 2019 *Phys. Rev.* D **100** 083001
[27] Patwardhan A V, Cervia M J and Balantekin A B 2021 Spectral splits and entanglement entropy in collective neutrino oscillations *Preprint* arXiv:2109.08995 [hep-ph]
[28] Rrapaj E 2020 *Phys. Rev. C* **101** 065805
[29] Hall B, Roggero A, Baroni A and Carlson J 2021 *Phys. Rev. D* **104** 063009
[30] Yeter-Aydeniz K, Bangar S, Siopsis G and Pooser R C 2021 Collective Neutrino Oscillations on a Quantum Computer *Preprint* arXiv:2104.03273 [quant-ph].
[31] Roggero A 2021 Entanglement and Many-Body effects in Collective Neutrino Oscillations *Preprint* arXiv:2102.10188 [hep-ph].

Journal of Physics: Conference Series **2191** (2022) 012005 doi:10.1088/1742-6596/2191/1/012005

Kundt waves in modified gravitational models

A Baykal[†1], **T Dereli**[‡2]

[1]Niğde Ömer Halisdemir University, Faculty of Arts and Sciences, Department of Physics, Bor yolu üzeri, Merkez Yerleşke, 51240 Niğde, TURKEY

[2] Department of Physics, College of Sciences, Koç University, 34450 Sarıyer, İstanbul, TURKEY

E-mail: [†]abaykal@ohu.edu.tr [‡]tdereli@ku.edu.tr

Date: 12 January 2022

Abstract. Kundt's class of gravitational wave metrics belonging to Petrov type-N are constructed for generic $f(R)$ model with minimally coupled electromagnetic field, a Gauss-Bonnet extended gravitational model for vacuum in four spacetime dimensions, and RF^2-type nonminimally coupled Einstein-Maxwell models including a cosmological constant.

1. Introduction

The Kundt class of metrics is a well known class of solutions containing several distinct Petrov types of algebraically special metrics. Non-twisting Petrov type-N solutions with a cosmological constant and an expanding null geodesic congruence belong to the Robinson-Trautman class, whereas those with a non-expanding null geodesic congruence belong to Kundt's class. These classes of solutions have been classified and interpreted as the gravitational waves propagating in constant curvature spacetimes by Bičák, and Podolský [1] relatively recently despite the fact that Kundt's class of metrics was introduced by Kundt [2] long time ago.

More recently, the invariant classification of the Kundt's class of solutions for vacuum are given by McNutt et al.. [3] by using Cartan-Karlhede algorithm in four spacetime dimensions. Bičák and Pravda [4] also discussed curvature invariants of all type-N metrics and they showed that the curvature invariants belonging to the Kundt class of metrics are trivial (See, also Ref. [5]).

In a more general geometrical context, the plane-fronted gravitational wave metrics have also been investigated in metric-affine and in Riemann-Cartan geometries by a number of researchers [10–19]. The generalization of this class of metrics to higher spacetime dimensions was presented by Obukhov [14]. Garcia and Plebanski [20], and Salazar et al. [21] studied the nontwisting type-N solutions to Einstein-Maxwell theory including a cosmological constant, whereas the Kundt's class of metrics belonging to Petrov type-III with a cosmological constant have been studied by Griffiths et al.. [8]. Petrov type-N and type-II Kundt class of metrics in type-D and type-O backgrounds are studied by Podolsky and Ortaggio [9], who also showed that the Einstein-Maxwell theory does not admit a conformally-flat solution with a cosmological constant.

In higher dimensions, using a higher dimensional version of the Newman-Penrose formalism [25, 26] Ortaggio et al.. [29] constructed Kundt's class of solutions to Einstein vacuum field equations with a cosmological constant belonging to type-III and type-N solutions according to the classification scheme of the Weyl tensor in higher dimensions [26–30] that is based on alignment properties of Weyl tensor.

A higher dimensional classification scheme for the Kundt's class of metrics is also introduced by Podolský and Švarc [6]. Moreover, Podolskỳ and Žofka surveyed the constraints on the matter fields admitted by the Kundt class of metrics in higher dimensions [7]. Furthermore, by considering the higher order Euler-Poincare forms complementing the Einstein-Hilbert Lagrangian, Gleiser and Dotti constructed wave solutions [22] using this class of metrics. Macías and Lozano [23] constructed plane-fronted wave solutions to a four dimensional model that is obtained from a five dimensional Chern-Simons gravity in five spacetime dimensions by means of a Klauza-Klein reduction on a circle.

By making use of a Kerr-Schild type metric ansatz, Debney [31] constructed Kundt class of solutions to Einstein-Maxwell equations. More recently, Gürses, et al.. [32] have shown that Kerr-Schild type ansatz of the form $g = g_{backgr.} + V\lambda \otimes \lambda$ which involves a shear-free, twist-free, and nonexpanding null congruence $\lambda = \lambda_\alpha dx^\alpha$ on a given maximally symmetric background metric $g_{backgr.}$, are universal metrics [33–37]. The universal metrics solve the field equations of a given metric theory governed by a gravitational Lagrangian of the general form $f(g, R, \nabla R \dots)$ involving Riemann tensor (symbolically denoted by R here for simplicity), and also its covariant derivatives.

In three spacetime dimensions, Chow et al.. [38] studied the Petrov-Segre classification of the solutions to the topologically massive gravity [39] including a cosmological constant and they reduce all the solutions in the literature to three distinct cases: the biaxial timelike-squashed case and the spacelike-squashed AdS^3 case and the AdS pp-waves case, belonging to Petrov-Segre types D_t, D_s and N, respectively. In a subsequent paper [40], Chow et al. constructed new solutions belonging to the Kundt class of metrics having Petrov-Segre types D, N and III, and they also showed that such solutions have constant scalar curvature polynomial invariants.

The present work can be considered as extensions of the previous works by Mohseni [41] who previously discussed the plane fronted-wave solutions to some modified gravitational models, and the works of Dereli and Sert [42], and Gürses and Halil [43] that introduce pp-wave solutions to Prasanna's and the Horndeski's nonminimally coupled Einstein-Maxwell models, respectively. In the same vein, Macías and Lozano [23] previously studied Kundt class solutions to the models involving the nonminimal coupling of the Faraday tensor to gravitational theories obtained by kaluza-Klein reduction of the Chern-Simons curvature invariants.

The paper is organized as follows. We use a null tetrad formulation [44] and the exterior algebra of differential forms on pseudo-Riemannian manifolds. The metric ansatz and a null coframe that will be used in all subsequent sections are introduced in the following section. A technique, due to Osváth et al.. [45], to obtain particular solution to equation for the profile function is also discussed in this section. Sec. II also serves to set the geometrical notation and conventions as well. In Appendix B, the original of Newman-Penrose spin coefficients and curvature scalars are identified in terms of the geometrical quantities introduced.

The metric ansatz is used to construct wave solutions to generic $f(R)$ model [46] in Sec. III, Gauss-Bonnet (GB) extended vacuum Einstein-Maxwell model in four spacetime dimensions in Sec. IV, the nonminimally coupled Einstein Maxwell model [49] introduced by Prasanna in Sec. V and to the Horndeski's generalized Einstein-Maxwell model [50] in Sec. VI. We show that the equations for the wave profile for nonminimal couplings considered reduce to that of an effective Einstein-Maxwell equations for maximally symmetric background with a cosmological constant. For the Horndeski model, an explicit expression is derived for the particular solution.

A derivation of the Horndeski's unique model using a Kaluza-Klein ansatz to reduction of the $(n + 1)$-dimensional Euler-Poincarè invariant to n spacetime dimensions is presented. The field equations that follow from the reduced Lagrangian are derived by using an action principle in the Appendix A. The paper concludes with brief comments on the general properties of the solutions.

2. The metric ansatz

The class of metrics we study belongs to the general Kundt's class of metrics [2] which was introduced and studied by Osváth, et al.. [45]. In terms of some local null coordinates $\{x^\alpha\} = \{u, v, \zeta, \bar{\zeta}\}$, for $\alpha = 0, 1, 2, 3$, the metric ansatz [17] can be expressed in the form

$$g = -2\left(\frac{Q}{P}\right)^2 du(Sdu + dv) + \frac{2}{P^2} d\zeta d\bar{\zeta} \tag{1}$$

where the functions in the ansatz are defined as

$$P \equiv 1 + \frac{\lambda}{6}|\zeta|^2, \qquad Q \equiv 1 - \frac{\lambda}{6}|\zeta|^2, \qquad S \equiv -\frac{\lambda}{6}v^2 + \frac{Q}{2P}H(u, \zeta, \bar{\zeta}), \tag{2}$$

whereas the real profile function $H = H(u, \zeta, \bar{\zeta})$ and the constant parameter λ are to be determined by the gravitational field equations.

A simplified form of the ansatz (1) originally considered by Osvath et al.. [45] is adopted in the discussion below for a first orientations to Kundt wave type solutions in modified gravitational models (especially, for the nonminimally coupled RF^2 models) which do have fairly complicated field equations compared to the case with minimally coupled matter fields.

We make use of exterior algebra of differential forms in our calculations, also adopting a null coframe, however there are other choices of rigid frames as well. For example a semi-null coframe adopted in [14, 17] to study the same class of metrics.

We find it convenient to adopt the following Newman-Penrose (NP) null coframe

$$k = k_\alpha dx^\alpha = du, \quad l = l_\alpha dx^\alpha = \left(\frac{Q}{P}\right)^2 (Sdu + dv), \quad m = m_\alpha dx^\alpha = \frac{d\zeta}{P}, \tag{3}$$

for the metric ansatz given in eq. (1). In terms of the coframe 1-forms, the metric can be written in the form

$$g = \eta_{ab}\theta^a \otimes \theta^b \tag{4}$$

where the nonvanishing components of the flat metric are $-\eta_{01} = -\eta_{10} = \eta_{23} = \eta_{32} = +1$. A tensorial index belonging to the NP frame are raised/lowered by the flat metric η^{ab} and η_{ab} with $\{\theta^a\} = k, l, m, \bar{m}$ for $a = 0, 1, 2, 3$, respectively.

The coordinate components of tensorial quantities are labeled by the letters in the Greek alphabet, whereas the indices referring to an NP frame are labeled by the lowercase letters in the Latin alphabet. A numerical tensorial indice below refers exclusively to the null coframe defined in (3).

In terms of the NP formalism introduced, for example, in Refs. [51, 52], the frame fields associated to the coframe (3) can be written as

$$D = -(\partial_u - S\partial_v), \qquad \Delta = -\left(\frac{P}{Q}\right)^2 \partial_v, \qquad \delta = P\partial_{\bar{\zeta}}, \qquad \bar{\delta} = P\partial_\zeta. \tag{5}$$

The set of basis frame fields $\{e_a\}$ associated to (3) are explicitly given by $e_0 \equiv -\Delta = -l^\nu\partial_\mu$, $e_1 \equiv -D = -k^\mu\partial_\mu$, $e_2 \equiv \bar{\delta} = \bar{m}^\mu\partial_\mu$, and $e_3 \equiv \delta = m^\mu\partial_\mu$ with $k^\mu = g^{\mu\nu}k_\nu$, etc.. In terms of the frame fields (5), the exterior derivative acting on scalars can be written in the form

$$d = -k\Delta - lD + m\bar{\delta} + \bar{m}\delta \tag{6}$$

that is of practical value in calculations.

Because we make exclusive use of the differential forms, the symbols for the basis frame fields do not appear explicitly in the presentation below, and particularly the basis frame field $D = -e_1$ is not to be confused with the covariant exterior derivative, almost universally denoted also by D as well.

In terms of the NP null basis coframe, the invariant volume 4-form with a particular orientation is defined as $*1 = ik \wedge l \wedge m \wedge \bar{m}$ $(i^2 = -1)$, and consequently, the Hodge dual of a p-form can be calculated by introducing completely antisymmetric permutation symbol $\epsilon_{0123} = +i$ relative to the null coframe basis defined in eq. (4).

With the indices referring to the NP null coframe, the connection 1-forms ${\omega^0}_3$ can be obtained by solving Cartan's first structure equations

$$d\theta^a + {\omega^a}_b \wedge \theta^b = 0 \tag{7}$$

for the metric (1), and they are explicitly given by

$${\omega^0}_3 = -\frac{\lambda}{3Q}\zeta k, \qquad {\omega^1}_2 = \frac{Q^2}{P}S_\zeta k - \frac{\lambda}{3Q}\bar{\zeta} l, \tag{8}$$

$${\omega^0}_0 - {\omega^3}_3 = \frac{\lambda}{3}vk - \frac{\lambda}{3Q}\left(1 + \frac{Q}{2}\right)\bar{\zeta} m - \frac{\lambda P}{6Q}\zeta\bar{m}, \tag{9}$$

where a coordinate subscript to a function denotes a partial derivative with respect to the coordinate.

The corresponding curvature 2-forms, which are defined in terms of the Riemann tensor as ${\Omega^a}_b = \frac{1}{2}{R^a}_{bcd}\theta^c \wedge \theta^d$, can be expressed in terms of the connection 1-forms by the Cartan's second structure equation in the form

$${\Omega^0}_3 = d{\omega^0}_3 + \left({\omega^0}_0 - {\omega^3}_3\right) \wedge {\omega^0}_3, \tag{10}$$

$${\Omega^1}_2 = d{\omega^1}_2 - \left({\omega^0}_0 - {\omega^3}_3\right) \wedge {\omega^1}_2, \tag{11}$$

$${\Omega^0}_0 - {\Omega^3}_3 = d\left({\omega^0}_0 - {\omega^3}_3\right) + 2{\omega^0}_3 \wedge {\omega^1}_2. \tag{12}$$

For the connection 1-forms given explicitly in eqs. (8) and (9), the corresponding curvature 2-forms take the form

$${\Omega^0}_3 = \frac{\lambda}{3}k \wedge m, \qquad {\Omega^0}_0 - {\Omega^3}_3 = -\frac{\lambda}{3}(k \wedge l - m \wedge \bar{m}), \tag{13}$$

$${\Omega^1}_2 = -\frac{PQ}{2}\left(H_{\zeta\zeta} + \frac{\lambda\bar{\zeta}}{3P}H_\zeta\right) k \wedge m - \frac{PQ}{2}\left(H_{\zeta\bar{\zeta}} + \frac{\lambda}{3P^2}H\right) k \wedge \bar{m} + \frac{\lambda}{3} l \wedge \bar{m}. \tag{14}$$

From the curvature expressions above, one can readily verify that $R = 6\lambda$. Consequently, one can show that the Einstein forms corresponding to the metric ansatz (1) have the form

$$*G^0 = \lambda * k, \qquad *G^2 = \lambda * m, \qquad *G^3 = \lambda * \bar{m}, \tag{15}$$

$$*G^1 = -PQ\left(H_{\zeta\bar{\zeta}} + \frac{\lambda}{3P^2}H\right) * k + \lambda * l, \tag{16}$$

by making use of the useful contraction formula $*G^a = -\frac{1}{2}\Omega_{bc} \wedge *\left(\theta^a \wedge \theta^b \wedge \theta^c\right)$.

Consequently, for the metric ansatz (1), the Einstein field equations are satisfied with a cosmological constant by construction, and the only off-diagonal Ricci component relative to the null coframe (3) turns out to be

$$R_{00} = PQ\left(H_{\zeta\bar{\zeta}} + \frac{\lambda}{3P^2}H\right), \tag{17}$$

and that this particular Ricci component corresponds to the real Ricci spinor Φ_{22} in the NP notation (See the definitions provided in Appendix B).

Note that the numerical indices refer to the null coframe, and therefore the cosmological vacuum equations reduces to $l^\alpha l^\beta R_{\alpha\beta} = R_{00} = 2\Phi_{22} = 0$. This is a *linear* partial differential equation of the form

$$\left(\mathbf{\Delta} + \frac{2\lambda}{3}\right) H = 0, \tag{18}$$

for the profile function H in terms of the Laplacian operator $\mathbf{\Delta} \equiv 2P^2\partial_\zeta\partial_{\bar{\zeta}}$ on the wave-fronts defined by the transverse metric $2P^{-2}d\zeta d\bar{\zeta}$.

The general solution to the homogeneous equation (18) can be written in the form

$$H_h(u,\zeta,\bar{\zeta}) = h_\zeta + \bar{h}_{\bar{\zeta}} - \frac{\lambda}{3P}\left(\zeta\bar{h} + \bar{\zeta}h\right) \tag{19}$$

where $h = h(u,\zeta)$ is an analytic function of the complex coordinate variable ζ and u-dependence remains determined by the metric field equations. The only non-vanishing component of the Weyl tensor relative to a NP null coframe (3) is $\Psi_4 = l^\alpha \bar{m}^\beta l^\mu \bar{m}^\nu C_{\alpha\beta\mu\nu}$ representing a transverse component propagating in the $k^\alpha\partial_\alpha$ direction [52]. With the help of the definitions given in Appendix B and the curvature 2-form expression (14), one concludes that the only non-vanishing Weyl curvature component has the expression

$$\Psi_4 = -\frac{PQ}{2}\left(H_{\zeta\zeta} + \frac{\lambda\bar{\zeta}}{3P^2}H_\zeta\right). \tag{20}$$

The result in eq. (20) evidently implies that for a non-vanishing expression on the right-hand side, the metric is of Petrov type-N. The vector field $k^\alpha\partial_\alpha$ is along the four-fold principle null direction and for the conformally-flat solutions can be obtained by requiring $\Psi_4 = 0$ in addition to the metric field equation $\Phi_{22} = 0$.

For the vacuum, the expression for the Weyl component can be simplified by feeding the general solution (19) into the expression in eq. (20). Thus, by making use of the general solution to the vacuum field equations, one finds that

$$\Psi_4 = -\frac{PQ}{2}h_{\zeta\zeta\zeta}. \tag{21}$$

The Weyl curvature expression (21) implies that the metric ansatz (1) is conformally-flat in the case where $h(u,\zeta)$ is at most a quadratic polynomial in ζ.

For $h = \alpha(u)\zeta$ in the general homogeneous solution given in eq. (19), H_h takes the form

$$H_h(u,\zeta,\bar{\zeta}) = (\alpha(u) + \bar{\alpha}(u))\frac{Q}{P} \tag{22}$$

where α is an arbitrary complex function of the coordinate u. On the other hand, for the profile function of the form (22), the Weyl component Ψ_4 also vanishes identically so that the solution (22) is conformally-flat (See, for example, [17], for a discussion on this particular solution).

Sec. 5 and Sec. 6 are devoted to study of nonminimally coupled Einstein-Maxwell models, and therefore the definitions and ansatz involving electromagnetic field variables associated to the metric ansatz (1) for the GR case are now in order.

Because the only non-vanishing Ricci tensor is Φ_{22}, the metric ansatz in eq. (1) admits a null electromagnetic field with $\Phi_2 = F_{\alpha\beta}\bar{m}^\alpha l^\beta$ component only. The Faraday 2-form is closed $dF = 0$, and therefore it is locally exact. Such a Faraday 2-form can be derived from the gauge potential of the form $A = a(u, \zeta, \bar{\zeta})du$ defined in terms of a real function $a = a(u, \zeta, \bar{\zeta})$. It has the explicit form

$$F = -a_\zeta du \wedge d\zeta - a_{\bar{\zeta}} du \wedge d\bar{\zeta}. \tag{23}$$

F defined in eq. (23) satisfies the relations $k \wedge F = 0 = k^\alpha i_\alpha F$ in terms of the null coframe 1-form k, and also it defines a principle null direction for the electromagnetic field. The source-free Maxwell's equations reduce to

$$d * F = -2P^2 a_{\zeta\bar{\zeta}} * k = 0. \tag{24}$$

Consequently, the Maxwell's equations are satisfied identically provided that one has $a_{\zeta\bar{\zeta}} = 0$ outside a four-current source in the geometry defined by the metric ansatz.

The canonical energy momentum form that follows from the Lagrangian

$$L_F[F, g] = -\frac{1}{2} F \wedge *F \tag{25}$$

by a coframe variation can be written in the form

$$* T_a[F] = i_a F \wedge *F - \frac{1}{2} i_a (F \wedge *F) \tag{26}$$

where $i_a \equiv i_{e_a}$ stands for the contraction operator with respect to the basis frame field e_a.

Note that for the null electromagnetic field defined in eq. (23) with $F \wedge *F = 0 = F \wedge F$, the energy momentum form simplifies further, leaving only the first term on the right-hand side in eq. (26). Thus, for the Faraday 2-form defined in eq. (23), there is only one non-vanishing component of energy-momentum form, and it is explicitly given by

$$* T^1[F] = -P^2 a_\zeta a_{\bar{\zeta}} * k. \tag{27}$$

Consequently, Einstein-Maxwell equations for vacuum with a cosmological constant yields the inhomogeneous linear equation

$$H_{\zeta\bar{\zeta}} + \frac{\lambda}{3P^2} H = \frac{2\kappa P}{Q} a_\zeta a_{\bar{\zeta}}, \tag{28}$$

for the profile function H.

The particular solution to the inhomogeneous equation (28) is given by Osváth et al.. [45]. Explicitly, one can construct the particular solution of (28) by assuming that it has the form

$$H_p \equiv \mu_\zeta + \bar{\mu}_{\bar{\zeta}} - \frac{\lambda}{3P} \left(\zeta\bar{\mu} + \bar{\zeta}\mu \right), \tag{29}$$

which is formally akin to the general homogeneous solution expression given in (19), expressed in terms of a complex function $\mu = \mu(u, \zeta, \bar{\zeta})$. The function $H^{(1)}$ defined by

$$H^{(1)} \equiv \mu_\zeta - \frac{\lambda\bar{\zeta}}{3P} \mu \tag{30}$$

satisfies the equation

$$H^{(1)}_{\zeta\bar{\zeta}} + \frac{\lambda}{3P^2} H^{(1)} = \frac{\kappa P}{Q} a_\zeta a_{\bar{\zeta}}. \tag{31}$$

Consequently, eq. (31) can be written in terms of the function μ by feeding the expression in eq. (30) into eq. (31) as

$$\left[P^2 \left(\frac{\mu_{\bar{\zeta}}}{P^2} \right)_\zeta \right]_\zeta = \kappa \frac{P}{Q} a_\zeta a_{\bar{\zeta}}. \tag{32}$$

Finally, the resulting equation can directly be integrated to have

$$\mu(u,\zeta,\bar{\zeta}) = \kappa \int^{\bar{\zeta}} d\bar{\zeta}' P^2 \int^{\zeta} \frac{d\zeta'}{P^2} \int^{\zeta'} \frac{d\zeta'' P}{Q} a_{\zeta''} a_{\bar{\zeta}'}. \tag{33}$$

The result is obtained by carrying out integrations indicated in (33) by treating ζ and $\bar{\zeta}$ as independent complex variables. Subsequently one uses them to construct the general solution of the equation (28) in the form $H = H_h + H_p$ as a linear superposition of the homogeneous solution given in eq. (19). The particular solution H_p can be expressed in terms of the complex function μ with the help of eq. (31,) and the definition $H_p = H^{(1)} + \bar{H}^{(1)}$.

Although the particular solution (33) has an expression formally similar to the the general homogeneous solution (19), the form of the particular solution H_p does not lead to a simplified expression for the Weyl component as in (21) in terms of the function μ. In this particular case, one finds the general expression

$$\Psi_4 = -\frac{PQ}{2} h_{\zeta\zeta\zeta} - \frac{Q}{2P} \left(P^2 (H_p)_\zeta \right)_\zeta \tag{34}$$

for the Weyl spinor.

Note that the construction of the general form the particular solution, and consequently the expressions for Ψ_4 for the modified gravitational models discussed below proceed in the same way as in the vacuum case. Therefore, the expression for the Weyl component in eq. (34) remains valid for the modified gravitational models studied below with the corresponding particular solutions. On the other hand, because of the term corresponding to the particular solution H_p, the $\times$ and $+$ polarization modes persist even for a conformally-flat homogeneous part of the exact solution. In this regard, the form of a particular solution for a given modified gravitational model with electromagnetic coupling allows one to compare the properties of the gravitational wave solutions for such models.

It is worthwhile to investigate whether the metric ansatz (1) can be used to construct gravitational wave solution to a given modified gravitational models, in particular the models that alter the minimal coupling between the gravitational and electromagnetic fields. An example of such models arises in the context of gravitational models modified by terms originating from quantum field theoretical considerations on a curved background [65]. For such models involving nonminimal RF^2 couplings discussed below, it can be shown that H_p part of the profile function for the ansatz (1) gets modified. Consequently, the method of obtaining a particular solution introduced above can also be applied to such models.

Moreover, in addition to such nonminimally coupled models, we also discuss some modified gravitational models popular in the context of cosmology admitting maximally-symmetric vacuum solutions.

3. The Gravitational waves in $f(R)$ models

In order to facilitate a direct comparison of the plane-fronted wave solutions for a given $f(R)$ model with the well-known solutions, we find it convenient to work with the original form of the $f(R)$ field equations [46–48, 53], although it is possible to work with Brans-Dicke-type scalar-tensor equivalents with a particular potential term for the scalar field. The gravitational wave solutions in the linearised approximation to generic $f(R)$ model have previously been studied by Capozziello et al.. [54] in the flat background.

3.1. The metric field equations

It is convenient to start from a Lagrangian 4-form of the form

$$L = \frac{1}{2\kappa} \left(f(R) * 1 + 2\Lambda * 1 \right) - \frac{1}{2} F \wedge *F \tag{35}$$

where f is assumed to be a sufficiently smooth algebraic function of scalar curvature R, and κ and Λ are the gravitational coupling and cosmological constants, respectively.

The metric field equations that follow from the total Lagrangian (35) can be obtained by a suitable coframe variation $\delta L/\delta\theta_a \equiv *E^a = 0$ with $E^a \equiv E^a{}_b\theta^b$. With respect to an orthonormal or a null NP coframe, they have the same form with $*E^a$ explicitly given by

$$* E^a = -f' * R^a + \frac{1}{2} f * \theta^a + D * (df' \wedge \theta^a) + \Lambda * \theta^a + \kappa * T^a[F], \tag{36}$$

where a prime denotes a derivative with respect to scalar curvature, $f' \equiv \frac{df}{dR}$, and $*T^a[F] \equiv \delta L_m/\delta\theta_a$ stands for the canonical energy-momentum form for electromagnetic field with the explicitly expression given in eq. (26) [47]. The terms that are third and fourth order in the derivatives of the metric components can conveniently be expressed in the form

$$D * (df' \wedge \theta^a) = f'' \left[d * (dR \wedge \theta^a) + \omega^a{}_b \wedge *(dR \wedge \theta^b)\right] + f''' dR \wedge *(dR \wedge \theta^a) \tag{37}$$

where $df' = f'' dR$.

It is straightforward to see that, even in the case where $\Lambda = 0$, the vacuum field equations (36) admit a maximally symmetric solutions in addition to the flat Minkowski solution, provided that the function f satisfies the constraint $f(0) = 0$. Assuming that the scalar curvature in this case is equal to a non-vanishing constant $R = R_0$, the field equations then reduce to form

$$f'(R_0)R^a = \left(\tfrac{1}{2} f(R_0) + \Lambda\right)\theta^a \tag{38}$$

by making use of the fact that $df' = f''(R_0)dR_0 \equiv 0$ in this case, and consequently, the higher order terms all of which are displayed in (37) vanish identically. In this case, the Ricci 1-forms R^a and, consequently by calculating the contraction of the Ricci 1-forms (38), one obtains an algebraic constraint equation satisfied by the curvature scalar as

$$R_0 = \frac{2\left(f(R_0) + 2\Lambda\right)}{f'(R_0)}. \tag{39}$$

We shall assume that the algebraic function f has a form such that the constraint (39) yields at least a non-vanishing root, such that $f(R_0) \neq 0$. On the other hand, depending on the explicit form of the function $f(R)$, eq. (39) admits several distinct roots, or else, multiple root(s) as well. Also note that for the case with $f(R) = R^2$, the constraint equation (39) equation implies that $\Lambda = 0$. Moreover, for the case $f(R) = R^2$, the vacuum field equations with $\Lambda = 0$ are identically satisfied for the pp-wave metric ansatz.

For $\lambda \mapsto 0$ in the metric ansatz, the metric reduces to the pp-wave metric for which $R = 0$ identically. in this case, for the pp-wave ansatz to be a valid ansatz for generic $f(R)$ gravity, $f(0) = 0$ is to be satisfied, and the vacuum equations then yield the same equation for the profile function since the Einstein tensor is multiplied by the constant $f'(0)$. Such gravitational wave solutions in flat background have been discussed previously by Mohseni [41]. On the other hand, the vacuum field equations (36) for a generic $f(R)$ model allow gravitational wave solutions on a curved background.

3.2. The solutions for the electro-vacuum case

By comparing the diagonal components of the field equations $*E^a = 0$ with the Einstein forms given explicitly in eq. (16), it is straightforward to obtain the algebraic equation satisfied by the constant parameter λ in the metric ansatz (1). One finds

$$\Lambda + \frac{1}{2}\left(f(R_0) - R_0 f'(R_0)\right) - f'(R_0)\lambda = 0. \tag{40}$$

Journal of Physics: Conference Series **2191** (2022) 012005 doi:10.1088/1742-6596/2191/1/012005

Assuming that there exists a real root R_0 satisfying the constraint in eq. (40) for the scalar curvature, one can construct the solutions to the equations $E_{00} = 0$ for the profile function as in the GR case. The corresponding solution to generic $f(R)$ model equations with the Lagrangian 4-form given in eq. (35) can explicitly be written in the form given in eq. (1) with the profile function given by

$$H(u,\zeta,\bar{\zeta}) = h_\zeta + \bar{h}_{\bar{\zeta}} - \frac{\tilde{\lambda}}{3P^2}\left(\zeta\bar{h} + \bar{\zeta}h\right) \tag{41}$$

with the constant parameter $\tilde{\lambda}$ is given by

$$\tilde{\lambda} = \frac{\Lambda}{f'(R_0)} + \frac{f(R_0) - R_0 f'(R_0)}{2f'(R_0)}. \tag{42}$$

Note that, the constant parameter $\tilde{\lambda}$ appearing in the equation for the profile function is now determined by an algebraic equation determining the background scalar curvature R_0. Because generic $f(R)$ models admit maximally symmetric vacuum solutions as briefly discussed above, $\tilde{\lambda}$ survives even if one assumes that $\Lambda = 0$. For $\Lambda = \lambda = 0$, and also assuming $f(0) = 0$ (with the consistency condition $f'(0) \neq 0$), one recovers the general pp-wave solutions presented in [41] (for example, the Aichelburg-Sexl type solutions correspond to $h(u,\zeta) \sim \delta(u)\log\zeta$ which require an appropriate null particle source) by specializing the complex function $h(u,\zeta)$ in (41).

One can immediately find the equations for the profile function for $f(R)$ gravity coupled to electromagnetic field with the Lagrangian of the form given in (35). In this case, one readily finds

$$H_{\zeta\bar{\zeta}} + \frac{\tilde{\lambda}}{3P^2}H = \frac{2\kappa}{f'(R_0)}\frac{Q}{P}a_\zeta a_{\bar{\zeta}}. \tag{43}$$

The general solution to eq. in (43) can be written as a superposition of the homogeneous solution of the form in (19) and the particular solution of the form (29) with a corresponding μ defined in (33) by replacing the constants κ and λ with $\kappa/f'(R_0)$ and $\tilde{\lambda}$, respectively.

It is also possible to consider the models based on a slightly more general Lagrangian density of the form $f(R, T, R_{ab}T^{ab}) * 1$, involving nonminimal matter momentum tensor T_{ab} coupled to Ricci tensor R_{ab} as well as trace of the matter energy momentum $T \equiv T^a{}_a$ coupling to curvature scalar [55–57]. Any gravitational model based on these interaction lead to fairly involved field equations. In particular, note that for the electromagnetic energy-momentum tensor of the form given in (26), the nonminimal $R_{ab}T^{ab}[F]$ coupling can be expressed as a superposition of RF^2-type nonminimal couplings as

$$R_{ab}T^{ab}[F] * 1 = R^a \wedge *T^a[F] = -F_a \wedge R^a \wedge *F - \frac{1}{2}RF \wedge *F \tag{44}$$

involving Ricci tensor R_{ab} and scalar curvature R.

The nonminimal coupling of type $RF\wedge *F$ is not particularly interesting for the metric ansatz (1) and in Sections V and VI, the gravitational wave solutions based on the ansatz (1) for the some simpler models with other RF^2-type nonminimal couplings involving Riemann and Ricci tensors will be discussed.

4. Gravitational waves in a Gauss-Bonnet extended Einstein gravity

In four spacetime dimensions, the Euler-Poincaré invariant, more often referred to as Gauss-Bonnet scalar $\mathcal{G}$, can conveniently be defined in terms of the particular contraction of the curvature 2-forms of the form

$$\mathcal{G} * 1 = \frac{1}{4}\Omega_{ab} \wedge \Omega_{bc} * \theta^{abcd} = (R^{abcd}R_{abcd} - 4R_{ab}R^{ab} + R^2) * 1. \tag{45}$$

One can show that, relative to an orthonormal or a null coframe, $\mathcal{G} * 1$ is an exact form given by

$$\mathcal{G} * 1 = d\left[\frac{1}{4}\left(\omega_{ab} \wedge d\omega_{cd} + \tfrac{2}{3}\omega_{ab} \wedge \omega_{ce} \wedge \omega^{e}{}_{d}\right)\epsilon^{abcd}\right], \tag{46}$$

and therefore such a term in gravitational Lagrangian 4-form does not contribute to the metric field equations. In spacetime dimensions greater than four, the term of the form (45) yields metric field equations that are of second order in the partial derivatives of the metric components [58].

A way to incorporate the GB term in (45) into the gravitational field equations in four spacetime dimensions, is to couple it with a dynamic scalar field. On the other hand, one can also consider modified GB term of the algebraic form $f(\mathcal{G})$. We follow the latter prescription, and obtain the ensuing field equations in some generality.

4.1. The metric field equations

In four spacetime dimensions, the general gravitational Lagrangian 4-form involving a second order GB term can be written of the form [59, 60]

$$L = \frac{1}{2\kappa}F(R, \mathcal{G}) * 1 + L_m[g, \psi] \tag{47}$$

where F is now assumed to be sufficiently smooth algebraic function of the GB scalar $\mathcal{G}$ and the scalar curvature R and L_m is a Lagrangian 4-form for a matter field ψ minimally coupled to metric field g. The metric field equations that follow from the Lagrangian 4-form (47) by a coframe variational derivative can be written in the form [41, 53, 61]

$$\frac{1}{2}F_R\Omega_{bc} \wedge *\theta^{abc} + \frac{1}{2}(F - RF_R - \mathcal{G}F_{\mathcal{G}}) * \theta^a + D\left[*(dF_R \wedge \theta^a) - \Omega_{bc} * (dF_{\mathcal{G}} \wedge \theta^{abc})\right] + \kappa * T^a[\psi] = 0 \tag{48}$$

where we introduced the short hand notations for the partial derivatives:

$$F_R \equiv \frac{\partial F}{\partial R}, \qquad F_{\mathcal{G}} \equiv \frac{\partial F}{\partial \mathcal{G}}. \tag{49}$$

Note that for $F(R, \mathcal{G}) = f(R) + \mathcal{G}$, the metric field equations (48) reduce to those of $f(R)$ equations given in (36) above as the consistency of the equations for the models studied above requires.

Because no additional insight is to be gained by increasing the complexity of the algebraic equation satisfied by the parameter λ, we find it convenient to discuss the vacuum field equations following from a particular form of the Lagrangian density of the form $F(R, \mathcal{G}) = \frac{1}{2}R + f(\mathcal{G})$. In this case, the general field metric equations in eq. (48) reduce to a simplified form as

$$-*G^a + (f - \mathcal{G}f_{\mathcal{G}}) * \theta^a - \Omega_{bc} \wedge D * (df_{\mathcal{G}} \wedge \theta^{abc}) = 0, \tag{50}$$

where the fourth-order derivatives of the metric are contained in the third term on the left-hand side of eq. (50) with $df_{\mathcal{G}} = \frac{df}{d\mathcal{G}}d\mathcal{G}$ and

$$D * (df_{\mathcal{G}} \wedge \theta^{abc}) = -\epsilon^{abcd}D(i_d df_{\mathcal{G}}). \tag{51}$$

It is straightforward to deduce that the reduced vacuum equations given in eq. (50) admit a maximally symmetric solution, for which case $\mathcal{G}$ and consequently $f(\mathcal{G})$ become constants, similar to the case of generic $f(R)$ model discussed in the previous section. For the metric ansatz (1), the Gauss-Bonnet scalar $\mathcal{G}$ can be calculated to have the value $\mathcal{G} = \lambda^2/9$ with the help of the expression (45).

4.2. The gravitational wave solutions

For a constant curvature spacetime to be a vacuum solution in the GB modified GR model above, the constant parameter λ has to satisfy the algebraic relation

$$f\left(\frac{\lambda^2}{9}\right) - \frac{\lambda^2}{9} f'\left(\frac{\lambda^2}{9}\right) - \lambda = 0, \tag{52}$$

for consistency, and the function $f(\mathcal{G})$ is such that eq. (52) has at least a real root. The equation for the profile function then is same as in eq. (18), and consequently, the solution also has the same form as the homogeneous solution given in eq. (19), provided that λ_0 is a root of the constraint equation (52). Finally, compared to the GR solution, only the curvature parameter of the background spacetime is replaced by λ_0 for the metric ansatz (1) in accordance with the algebraic constraint in eq. (52).

5. Prasanna's nonminimally coupled model

One can consider the coupling the Faraday 2-form to the curvature tensor in a mathematically consistent manner [62], as for example, RF^2-type couplings listed in [42] by Dereli and Sert (See, also, Ref. [63]). Without a cosmological term, Dereli and Sert studied *pp*-wave solutions to the nonminimally coupled Einstein-Maxwell model that follow from the Lagrangian (53).

Although such nonminimal coupling terms are not compatible with the strong principle of equivalence [62,64], Drummond and Hathrell showed in a remarkable paper [65] that the effective action for the photon involves nonminimal RF^2 interaction terms in a curved background arising from the one-loop polarization calculations, and such nonminimal curvature interactions allow superluminal propagation for low energy photons.

RF^2-type nonminimal coupling terms also naturally arise from the Kaluza-Klein reduction of quadratic-curvature terms, and in particular, the quadratic Gauss-Bonnet term studied previously by Dereli and Üçoluk [66].

The electro-vacuum pp-waves with a nonminimal coupling of type $RF \wedge *F$, or more generally the nonminimal interaction of type $f(R)F \wedge *F$, have previously been considered by Mohseni, concluding that they are identical to the corresponding solutions in GR [41].

It is also interesting to note that the differential equation for the profile function for pp-wave metric in Horndeski's generalized Einstein-Maxwell model [50] with the nonminimal coupling terms of the form $F \wedge F_{ab} * \tilde{\Omega}^{ab}$ where $\tilde{\Omega}^{ab}$ stands for the double-dual curvature 2-form involving the coupling terms $RF \wedge *F$, $R_a \wedge F^a \wedge *F$ and $F \wedge F_{ab} * \tilde{\Omega}_{ab}$ presented in [43] is identical to the one in a nonminimally coupled theory with the nonminimal coupling term $F \wedge F_{ab} * \Omega^{ab}$ [42].

We first consider the simplest possible nonmiminal RF^2 coupling defined by $F \wedge F_{ab} * \Omega^{ab}$ in order to gain some insight for the gravitational wave solutions in a curved background for such models.

5.1. The Lagrangian and the field equations

The nonminimally coupled Einstein-Maxwell model with a cosmological constant λ we want to study is governed by the Lagrangian 4-form

$$L = \frac{1}{2\kappa}(R + 2\lambda) * 1 - \frac{1}{2} F \wedge *F + \frac{\gamma}{2} F \wedge F_{ab} * \Omega^{ab} \tag{53}$$

where γ is a constant multiplying the nonminimal coupling term, namely the term

$$F \wedge F_{ab} * \Omega^{ab} = \tfrac{1}{2} F_{ab} R^{ab}{}_{cd} F^{cd} * 1, \tag{54}$$

first considered by Prasanna [49]. The field equations for the Faraday 2-form F that follow from the nonminimally coupled Einstein-Maxwell Lagrangian (53) are

$$dF = 0, \qquad d * (F - \gamma F_{ab}\Omega^{ab}) = 0. \tag{55}$$

One can readily verify that the ansatz (23) for the Faraday 2-form and for the metric ansatz (1), the modified Maxwell's equations (55) are satisfied.

The metric field equations $*E^a = 0$ which follow from a coframe variation of the Lagrangian density (53), $*E_a \equiv \delta L/\delta\theta^a$ can be written explicitly in the convenient form as [42]

$$* E^a = - * G^a + \lambda * \theta^a + \kappa * T^a[F] + \kappa\gamma(\tfrac{1}{2}F^a{}_c F_b \wedge *\Omega^{cb} - *T^a[F, F_{bc}\Omega^{bc}] + D\lambda^a) = 0 \quad (56)$$

where the Lagrange multiplier 2-form λ^a is explicitly given by

$$\lambda^a = 2i_b D(F^{ba} * F) + \tfrac{1}{2}\theta^a \wedge i_b i_c D(F^{bc} * F). \quad (57)$$

Consequently, the coframe field equations containing $D\lambda^a$ term involve, in general, the third order derivatives of the gauge potential for the model.

The energy-momentum 3-forms $*T_a[F, F_{cd}\Omega^{cd}] = T_{ab}[F, F_{cd}\Omega^{cd}] * \theta^b$ that arises from the nonminimal interaction term is explicitly given by

$$* T_a[F, F_{cd}\Omega^{cd}] \equiv \tfrac{1}{4}(F_a \wedge F_{bc} * \Omega^{bc} + F_{bc}(i_a\Omega^{bc}) \wedge *F - F \wedge F_{bc} i_a * \Omega^{bc} - F_{bc}\Omega^{bc} \wedge i_a * F). \quad (58)$$

The expression in eq. (58) for the 1-form $*T_a[F, F_{cd}\Omega^{cd}]$ has an explicit form similar to that of the canonical energy-momentum form $*T^a[F]$ defined in eq. (26). By definition, $*T_a[F, F_{cd}\Omega^{cd}]$ has a vanishing trace, and it is symmetrical under the interchange $F \leftrightarrow F_{ab}\Omega^{ab}$: $*T_a[F, F_{cd}\Omega^{cd}] = *T_a[F_{cd}\Omega^{cd}, F]$. It is also symmetric with respect to its indices: $\theta^a \wedge *T^b[F, F_{cd}\Omega^{cd}] = \theta^b \wedge *T^a[F, F_{cd}\Omega^{cd}]$. In the expression for the energy-momentum forms, the short-hand notations $F_a \equiv i_a F$ and $F_{ab} \equiv i_b i_a F$ are used for convenience.

5.2. The Gravitational wave solutions

Before presenting the differential equation satisfied by the profile function obtained by inserting the metric ansatz (1) into the field equations (56), we first note the following relation

$$F_{ab}\Omega^{ab} = \frac{2\lambda}{3}F \quad (59)$$

which is of some practical importance in simplifying the metric field equations as well. Consequently, the energy-momentum 3-forms defined in (58) reduce to

$$* T_a[F, F_{bc}\Omega^{bc}] = *T_a\left[F, \frac{2\lambda}{3}F\right] = \frac{2\lambda}{3} * T_a[F] \quad (60)$$

for the metric ansatz (1) and for the Faraday 2-form of the form (23). Moreover, by making use of the gravitational Bianchi identity satisfied by the curvature 2-form, $\Omega^a{}_b \wedge \theta^b = 0$, and the relation (59), one can show that

$$F_{ac}F_b \wedge *\Omega^{cb} = -\frac{1}{2}F_a \wedge F_{bc} * \Omega^{bc} = -\frac{\lambda}{2}F_a \wedge *F. \quad (61)$$

By making use of the expression in equation (57), it is straightforward to verify that the only non-vanishing component of the Lagrange multiplier 2-form is λ^1, and it is given by

$$\lambda^1 = -(i_a DF^{1a}) * F + DF^{1a} \wedge *(F \wedge \theta_a) \quad (62)$$

More conveniently, the non-vanishing component of the Lagrange multiplier 2-form λ^1 reduces to the form

$$\lambda^1 = i\mathcal{A}k \wedge m - i\bar{\mathcal{A}}k \wedge \bar{m}, \quad (63)$$

where the complex function $\mathcal{A}$ introduced in the equation for convenience has the explicitly expression

$$\mathcal{A} \equiv P^3 a_{\zeta\zeta} a_{\bar{\zeta}} + \frac{\lambda P^2(Q+1)}{3Q} \bar{\zeta} a_{\bar{\zeta}} a_{\zeta} + \frac{\lambda P^2}{3Q} \zeta a_{\zeta} a_{\zeta}. \tag{64}$$

Thus, the components of the exterior derivative expression for $D\lambda^1$ turn out to contain at most second derivatives of the gauge potential for the ansatz (1). In addition to a shift in the constant multiplying the electromagnetic energy momentum form, the effect of the nonminimal interaction comes from the Lagrange multiplier term (63). Consequently, the contribution of the Lagrange multiplier term $D\lambda^a$ to the metric equations originates from $D\lambda^1$ term which can be calculated by making use of the simplifications $\lambda^2 = 0 = \lambda^3$. Thus, the expression for the covariant exterior derivative reduces to

$$D\lambda^1 = d\lambda^1 - \omega^0{}_0 \wedge \lambda^1. \tag{65}$$

Consequently, with the help of the explicit expressions for the connection 1-forms, this formula takes the explicit form

$$D\lambda^1 = 2Re\left(P\mathcal{A}_{\bar{\zeta}} + \frac{\lambda\zeta}{2Q}\mathcal{A}\right) * k. \tag{66}$$

Eventually, in parallel to the GR case, $l^\alpha l^\beta E_{\alpha\beta} = E_{00} = 0$ gives the inhomogeneous equation of the form

$$H_{\zeta\bar{\zeta}} + \frac{\lambda}{3P^2}H = \kappa\left(2 - \frac{\gamma\lambda}{3}\right)\frac{P}{Q}a_{\zeta}a_{\bar{\zeta}} - \frac{2\kappa\gamma}{PQ}Re\left(P\mathcal{A}_{\bar{\zeta}} + \frac{\lambda\zeta}{2Q}\mathcal{A}\right) \tag{67}$$

for the profile function H. Here $Re(\cdot)$ denotes a real part of an expression inside the bracket.

For both of the nonminimal coupling Einstein-Maxwell models we considered, the ansatz (1) leads to an inhomogeneous equation of type similar to (67). The particular solution of an equation of this type can be constructed as a contour integral [45] in the same way as the particular solution H_p is constructed as in the Einstein-Maxwell case in terms of the complex function μ in eq. (33) and with the general expression in eq. (34).

6. Horndeski's nonminimally coupled model

Gürses and Halil [43] previously studied pp-wave solutions to the Horndeski model. Horndeski [67] himself studied Petrov type III and Petrov type-N solutions investigating the extension of the Mariot-Robinson and Kundt-Trumper theorems to the generalized Einstein-Maxwell model [50].

We construct type-N gravitational waves by taking a cosmological constant into account for this model.

6.1. The Horndeski's Lagrangian and the field equations

In terms of differential forms, the original Lagrangian introduced by Horndeski [50] can be written out as a Lagrangian 4-form of the form

$$L = \tfrac{1}{2\kappa}(R + 2\lambda) * 1 - \tfrac{1}{2}F \wedge *F + \tfrac{\gamma}{2}F \wedge F_{ab} * \tilde{\Omega}^{ab}, \tag{68}$$

where $\tilde{\Omega}^{ab}$ is the double dual curvature 2-form defined in terms of double dual Riemann tensor ${}^*R^{*a}{}_{bcd}$ as

$$\tilde{\Omega}^a{}_b = \tfrac{1}{2}\,{}^*R^{*a}{}_{bcd}\theta^c \wedge \theta^d, \tag{69}$$

and that the double-dual Riemann tensor has components expressed in terms of the permutation symbols as

$${}^*R^{*ab}{}_{cd} \equiv \tfrac{1}{4}\epsilon^{ab}{}_{mn}R^{mn}{}_{pr}\epsilon^{pr}{}_{cd}. \tag{70}$$

Note that the particular nonminimal coupling in (68) is a special linear combination of the nonminimal couplings

$$F \wedge F_{ab} * \tilde{\Omega}^{ab} = -F \wedge *(RF + 2F_a \wedge R^a + F_{ab}\Omega^{ab}) \tag{71}$$

where $R_a \equiv R_{ab}\theta^b = i_b\Omega^b{}_a$ is the Ricci 1-form and R is the scalar curvature. The individual terms on the right-hand side in (71) can be expressed in the component form as

$$RF \wedge *F = \tfrac{1}{2}F^{ab}F_{ab}R * 1, \tag{72}$$
$$F_a \wedge R^a \wedge *F = F_{ac}F^c_{\ b}R^{ab} * 1, \tag{73}$$
$$\Omega_{ab}F^{ab} \wedge *F = \tfrac{1}{2}F_{ab}F^{cd}R^{ab}{}_{cd} * 1, \tag{74}$$

in terms of the scalar curvature, Ricci and Riemann tensors, respectively. Note also that the nonminimal coupling term in the Horndeski Lagrangian (68) can also be written in the following equivalent forms:

$$F \wedge F_{ab} * \tilde{\Omega}^{ab} = -F \wedge \tilde{F}_{ab}\Omega^{ab} = -F \wedge \Omega_{ab} * (F \wedge \theta^a \wedge \theta^b). \tag{75}$$

The field equation that follow from the Lagrangian (68) by a coframe variation by using the first order formalism is relegated to Appendix A.

As in the previous models discussed above, the metric field equations that follow from the Horndeski Lagrangian (68) can be written as $*E^a = \delta L/\delta\theta_a = 0$ with

$$* E^a = - * G^a + \lambda * \theta^a + \kappa * T^a[F] - \tfrac{\gamma\kappa}{2}F^a \wedge F_{bc} * \tilde{\Omega}^{bc} + \gamma\kappa D(F_b \wedge D\tilde{F}^{ab}) \tag{76}$$

and $\tilde{F}^{ab} \equiv i^b i^a * F$ stand for the components of the dual Faraday 2-form.

The source-free field equation for the Faraday 2-form $F = dA$ takes the form

$$dF = 0, \qquad d * (F - kF_{ab}\tilde{\Omega}^{ab}) = 0. \tag{77}$$

In contrast to the Prasanna's model studied in the previous section, for the Horndeski's model with the set of dynamic variables as $\{g, A\}$, the system of field equations (76) and (77) are all of second order. In this regard, it is the unique RF^2-type coupling among all the other couplings mentioned.

Note that the modified electromagnetic field equations in eq. (77) contain only the second order derivatives of metric components as a consequence of the second Bianchi identity, namely the identity $D * \tilde{\Omega}^{ab} = -\frac{1}{2}\epsilon^{ab}{}_{cd}D\Omega^{cd} \equiv 0$.

On the other hand, one can verify, from the modified electromagnetic equations (55) and (77) with a source term of the form $*J$ with a four current 1-form $J = J_\alpha dx^\alpha$, that the local charge conservation is satisfied for both of the models as a consequence of the operator identity $d^2 \equiv 0$. Therefore, the local conservation of charge is consistent with both the Horndeski's second order model and the Prasanna's third order RF^2 models.

6.2. The gravitational wave solutions
For the Faraday 2-form of the form (23), and the metric ansatz (1), the nonminimal coupling term reduces to

$$F_{ab}\tilde{\Omega}^{ab} = -\tfrac{\lambda}{3}F. \tag{78}$$

Consequently, the modified Maxwell's equations (77) are satisfied identically, in parallel to the case in the Prasanna model. Such a simplifying relation between the curvature and the Faraday 2-forms in both of the nonmimimal coupling cases is peculiar to the maximally symmetric background, and hence to the metric ansatz (1).

As another consequence of the relation in given eq. (78), one has

$$F_a \wedge F_{bc} * \tilde{\Omega}^{cb} = -\tfrac{\lambda}{3} * T_a[F] \tag{79}$$

as in the non-mimimal case studied above.

The key relation in eq. (78) also helps reduce the metric field equations (76) to a more manageable form. For the metric ansatz (1), the only non-vanishing component of the Lagrange multiplier 2-form, as before, is given by

$$\lambda^1 = Pa_{\bar{\zeta}} \left(P^2 a_\zeta\right)_\zeta ik \wedge m - Pa_\zeta \left(P^2 a_{\bar{\zeta}}\right)_{\bar{\zeta}} ik \wedge \bar{m} \tag{80}$$

for this case as well. Consequently, by making use of the expression in (65), the equation for the profile function takes the form

$$H_{\zeta\bar{\zeta}} + \frac{\lambda}{3P^2} H = \kappa \left(2 + \frac{\lambda\gamma}{3}\right) \frac{Q}{P} a_\zeta a_{\bar{\zeta}} + \frac{\gamma\kappa}{P} \left\{ \left[\frac{P^2 a_{\bar{\zeta}}}{Q} \left(P^2 a_\zeta\right)_\zeta \right]_{\bar{\zeta}} + \left[\frac{P^2 a_\zeta}{Q} \left(P^2 a_{\bar{\zeta}}\right)_{\bar{\zeta}} \right]_\zeta \right\}. \tag{81}$$

In the limit $\lambda \mapsto 0$, the background becomes flat and the metric (1) takes the form of a pp-wave metric. Consequently, one recovers the result obtained by Gürses and Halil [43], and Dereli and Sert [42]. In this limit, eq.(81) becomes

$$H_{\zeta\bar{\zeta}} = \kappa a_\zeta a_{\bar{\zeta}} + \kappa\gamma a_{\zeta\zeta} a_{\bar{\zeta}\bar{\zeta}}. \tag{82}$$

One can readily verify that the general solution to this equation can be expressed in terms of functions $h(u,\zeta)$ and $f(u,\zeta)$ which are analytic in the variable ζ in the form

$$H(u,\zeta,\bar{\zeta}) = h_\zeta + h_{\bar{\zeta}} + \kappa(f\bar{f} + \gamma f_\zeta \bar{f}_{\bar{\zeta}}) \tag{83}$$

with $a(u,\zeta,\bar{\zeta}) \equiv f(u,\zeta) + \bar{f}(u,\bar{\zeta})$.

The particular solution to (81) cannot be expressed in a simple form as in eq. (83). Moreover, the solution to eq. (81) does not allow one to interpret the resulting metric as describing a partially massive spin-2 graviton and partially massless spin-2 photon with the assumption that $f(u,\zeta) = \alpha(u)\zeta + \beta(u)\zeta^2$ in the flat background [42]. However, in parallel to the GR case, the particular solution to eq. (81) can be constructed by using an appropriate function μ which is defined in eq. (29). In this case, for a given admissible gauge potential, $\mu(u,\zeta,\bar{\zeta})$ can be expressed in the form

$$\mu = \kappa \int^{\bar{\zeta}} d\bar{\zeta}' P^2 \int^{\zeta} \frac{d\zeta'}{P^2} \int^{\zeta'} d\zeta'' \left\{ \left(1 + \frac{\lambda\gamma}{6}\right) \frac{Q}{P} a_{\zeta''} a_{\bar{\zeta}'} + Re \left(\frac{\gamma\kappa}{P^2 Q} \left[\frac{P^2 a_{\bar{\zeta}'}}{Q} \left(P^2 a_{\zeta''}\right)_{\zeta''} \right]_{\bar{\zeta}'} \right) \right\}. \tag{84}$$

The general expression for the complex function $\mu(\zeta,\bar{\zeta})$ in eq. (84) is valid for an electromagnetic potential which satisfies eq. (77).

7. Concluding remarks

For a given profile function H as a solution, one can calculate the corresponding Weyl spinor Ψ_4 with the explicit expression given in eq. (20). Thus, the way the plane-fronted gravitational wave solutions presented above for each model allows a direct comparison of the models discussed in a unified manner.

In a generic metric theory of gravity, for which the metric is assumed to be determined by some field equations, the possible polarization modes gravitational waves in the far zone are attributed to some particular components of the Riemann tensor, namely R_{0i0j} components of the Riemann curvature tensor. Thus, for i, j standing for the spatial components, there are at most six distinct polarization modes that are independent of the field equations that determine the metric. In a previous study of these polarization modes, and their invariance under null frame rotations in [68, 69], these polarization modes are identified with the real and imaginary parts of the NP scalars $\Psi_4(u), \Psi_2(u), \Psi_3(u)$ and $\Phi_{22}(u)$, as a function of the retarded real null coordinate u. In the recent works [70–73], the properties of gravitational waves are studied for some particular metric theories. These polarization modes are also relevant to all the models discussed in the current work.

The technical approach adopted here can also be used to introduce a more refined linearised approximation scheme relative to a null coframe that allows some additional gravitational wave features to be added to the metric perturbations in a convenient manner.

The geodesic deviation equation for test particles following neighbouring geodesic curves can be decomposed into, for example, the effects arising from the curved background, and the effects arising from a type-N gravitational waves [1]. In all the modified gravitational models discussed above, both of these effects are present. In particular, $\times$ and $+$ polarization modes registered by the test particles correspond to the real and the imaginary parts of the Weyl spinor Ψ_4, respectively. Consequently, the part of Ψ_4 which is determined by the particular solution H_p has an effect on the transverse polarizations modes, and the corresponding amplitude.

For both of the nonminimally coupled electromagnetic cases discussed above, the field equations lead to an effective T_{00} component for the nonminimally coupled electromagnetic field. Consequently, the $\Phi_{22} = R_{00}$ curvature components corresponding to transverse breathing modes [68, 69] are modified in contrast to the GR case.

Finally, note that the gravitational wave solutions constructed in this work follow the pattern introduced in [32], and therefore they imply that the universal metrics of Kerr-Schild-Kundt type of metrics can be extended to the cases that include the general RF^2-type nonminimal couplings as well.

Appendix A: A derivation of the Horndeski's generalized Lagrangian and the field equations

It follows from the Lovelock's theorem [58] that the field equations for the Einstein-Hilbert Lagrangian supplemented by a GB term (45)in five dimensions contain only the second order derivatives of the five dimensional pseudo-Riemannian metric. By introducing a Kaluza-Klein ansatz into the five dimensional metric, the dimensional reduction of the GB term (45) on a circle is bound to yield RF^2- and F^4-type gauge field couplings in four spacetime dimensions [66] that are second order in all the field variables. In this manner, the Horndeski Lagrangian given in eq. (68) was rederived by Buchdahl [74] a few years after Horndeski introduced it [50] following a completely different approach.

Following Buchdahl, it is sufficient to introduce a simple $(n+1)$-dimensional Kaluza-Klein metric ansatz of the form

$$g^{(n+1)} = \eta_{AB}\mathbf{\Theta}^A \otimes \mathbf{\Theta}^B = g^{(n)} + \mathbf{\Theta}^n \otimes \mathbf{\Theta}^n \tag{85}$$

adapted to the local product form $U^n \times S^1$ of the total manifold. Here, $\mathbf{\Theta}^n = dy + A(x)$ in terms of the local coordinates $\{x^\mu, y\}$ on some chart $U^n \subset M^n$, and y denotes the compactified coordinate on S^1. The indices have the range $A, B, C, \ldots = 0, 1, 2, \ldots (n-1), n$, and one can introduce a suitable local orthonormal coframe θ^a on some coordinate neighbourhood on $M^{(n)}$ of the form

$$g^{(n)} = \eta_{ab}\theta^a \otimes \theta^b$$

with $\mathbf{\Theta}^a(x, y) = \theta^a(x)$. The lowercase Latin indices refer to the pseudo-Riemannian manifold M^n and run over the range $a, b, c, \ldots = 0, 1, 2, \ldots, (n-1)$.

The Levi-Civita connection 1-forms on $M^{(n+1)}$ satisfying the metricity condition, $\mathbf{\Gamma}_{AB} + \mathbf{\Gamma}_{BA} = 0$, and the structure equations with vanishing torsion

$$d\mathbf{\Theta}^A + \mathbf{\Gamma}^A{}_B \wedge \mathbf{\Theta}^B = 0$$

can be expressed in terms of the Levi-Civita connection 1-forms ω_{ab} on base manifold $M^{(n)}$ and the field strength $F = dA$ as

$$\mathbf{\Gamma}^a{}_b = \omega^a{}_b - \tfrac{1}{2}F^a{}_b\mathbf{\Theta}^n, \qquad \mathbf{\Gamma}^a{}_n = -\tfrac{1}{2}F^a. \tag{86}$$

Consequently, the curvature 2-forms $\mathbf{\Omega}^{AB} = d\mathbf{\Gamma}^{AB} + \mathbf{\Gamma}^A{}_C \wedge \mathbf{\Gamma}^{CB}$ can be expressed in the form

$$\mathbf{\Omega}^{ab} = \pi^{ab} + \tau^{ab} \wedge \mathbf{\Theta}^n, \qquad \mathbf{\Omega}^{an} = -\rho^a - \sigma^a \wedge \mathbf{\Theta}^n, \tag{87}$$

in terms of convenient tensor-valued forms

$$\begin{aligned} &\pi^{ab} = \Omega^{ab} - \tfrac{1}{2}F^{ab}F - \tfrac{1}{4}F^a \wedge F^b, \qquad \tau^{ab} = -\tfrac{1}{2}DF^{ab}, \\ &\rho^a = \tfrac{1}{2}DF^a, \qquad \sigma^a = -\tfrac{1}{4}F^a{}_b F^b \end{aligned}$$

that were introduced in [66] previously. In $n+1 \geq 5$ spacetime dimensions, the metric field equations that follow from the dimensionally-continued Euler-Poincare (EP) form explicitly read

$$L^{(n+1)} = \tfrac{1}{2}\mathbf{\Omega}_{AB} \wedge \star\mathbf{\Theta}^{AB} + \tfrac{1}{4}\mathbf{\Omega}_{AB} \wedge \mathbf{\Omega}_{CD} \wedge \star\mathbf{\Theta}^{ABCD}. \tag{88}$$

The metric field equations that follow from the Lagrangian in (88) containing the quadratic curvature term, can be obtained straightforwardly by a coframe variational derivative, $\delta L^{(n+1)}/\delta\mathbf{\Theta}_A = 0$. They explicitly read

$$\tfrac{1}{2}\mathbf{\Omega}_{BC} \wedge \star\mathbf{\Theta}^{ABC} + \tfrac{1}{4}\mathbf{\Omega}_{BC} \wedge \mathbf{\Omega}_{DE} \wedge \star\mathbf{\Theta}^{ABCDE} = 0. \tag{89}$$

In the equations above $\star$ denotes the Hodge dual operator defined on $M^{(n+1)}$ which determines the volume element on the base manifold by $\star 1 = *1 \wedge \mathbf{\Theta}^n$. Moreover, by feeding the Kaluza-Klein metric ansatz (85) into the field equations (89), one ends up with the equations involving only the second order derivatives of the set of variables $\{g^{(n)}, A\}$ of the ansatz (85) that parametrises the $g^{(n+1)}$ metric components, as a consequence of the fact that the metric filed equations in eq. (89) are manifestly second order in the derivatives.

Equivalently, one can also reduce the action (88), and subsequently derive the field equations therefrom. More precisely, by inserting the ansatz (85) into the Lagrangian density (88), one finds that the reduced Lagrangian n-form $L^{(n)}_{red.}$ defined by

$$L^{(n+1)} = L^{(n)}_{red.} \wedge dy$$

has the explicit expression

$$\begin{aligned} L^{(n)}_{red.} &= \tfrac{1}{2}\pi_{ab} \wedge *\theta^{ab} - \sigma_a \wedge *\theta^a + \tfrac{1}{4}\pi_{ab} \wedge \pi_{cd} \wedge *\theta^{abcd} - (\pi_{ab} \wedge \sigma_c + \tau_{ab} \wedge \rho_c) \wedge *\theta^{abc} \\ &= \tfrac{1}{2}\Omega_{ab} \wedge *\theta^{ab} - \tfrac{1}{2}F \wedge *F + \tfrac{1}{4}\Omega_{ab} \wedge \Omega_{cd} \wedge *\theta^{abcd} - \tfrac{1}{4}\Omega_{ab} \wedge F \wedge *(\theta^{ab} \wedge F) \\ &\quad + \tfrac{3}{32}F \wedge F \wedge *(F \wedge F) \end{aligned}$$

up to a disregarded exact n-form. Each term in $L^{(n)}_{red.}$ separately leads to second order field equations in the set of metric field variables of $M^{(n+1)}$, namely, $g^{(n)}_{\mu\nu}$, and the gauge potential A.

In four spacetime dimensions, the GB term is an exact form, and therefore it can be disregarded from the total Lagrangian. Consequently, up to redefinitions of the coupling constants, the reduced Lagrangian $L^{(n)}_{red.}$ corresponds to the Lagrangian density of the Horndeski's model after discarding the F^4 self-interaction terms for $n = 4$. In four spacetime dimensions, one can verify that the resulting RF^2 coupling can be written in terms of the double-dual curvature 2-form $\tilde{\Omega}^{ab}$, or else the linear combination of nonlinear couplings to the curvature components as

$$\Omega_{ab} \wedge F \wedge *(\theta^{ab} \wedge F) = -F \wedge F_{ab} * \tilde{\Omega}^{ab} = F \wedge F_{ab} * \Omega^{ab} - 2F \wedge *(R_a \wedge F^a) + RF \wedge *F. \quad (90)$$

The coupling to the double-dual curvature on the left-hand side can be considered as a generalization of the Horndeski-type RF^2 nonminimal interaction term to higher dimensions. The Kaluza-Klein reduction mechanism can be used to derive more general nonminimal couplings [75] that leads to second order equations for the field variables.

Returning back to the Horndeski's Lagrangian, the second order field equations that follow from the Lagrangian given in eqn. (68) can explicitly be found by a constrained variational derivative as follows.

In deriving the field equation by making use of the first order formalism (see, for example, Ref. [42] in the context of nonminimal coupling considered above), the connection and the coframe 1-forms are assumed to be independent gravitational variables, and the Riemannian case is recovered by introducing vanishing torsion ($\Theta^a = d\theta^a + \omega^a{}_b \wedge \theta^b = 0$), and vanishing non-metricity ($D\eta_{ab} = -\eta_{ac}\omega^c{}_b - \eta_{bc}\omega^c{}_a = 0$) constraints for the independent connection 1-forms.

For the metricity constraint, we assume that $\omega_{ab} = -\omega_{ba}$, and the zero-torsion constraint $\Theta^a = 0$ is imposed on the connection by introducing a Lagrange multiplier terms of the form $\lambda_a \wedge (d\theta^a + \omega^a{}_b \wedge \theta^b)$ to the original Lagrangian 4-form. Eventually, the Lagrange multiplier 2-form λ^a is to be eliminated in favour of the remaining variables to obtain the metric field equations.

In the first order formalism, only the first order exterior derivatives of the independent variables appear. For example, the Lagrangian 4-form is assumed to depend the gravitational variables θ^a and $\omega^a{}_b$ and their derivatives $d\theta^a$ and $d\omega_{ab}$, which are to be replaced by more convenient counterparts Θ^a and $\Omega^a{}_b$, respectively. Note that all the field equations studied above for the gravitational models cobsidered above can conveniently be discussed using this approach. In particular, some other RF^2 couplings the field equations were derived using the first order formalism [47, 76–79] previously. The derivation of the field equations for the Horndeski's model, which involves subtle technical points, is presented in a streamlined manner as follows.

To this end, one starts with the original Lagrangian 4-form extended by the constraint terms of the form

$$L'[\theta^a, \omega^a{}_b, F, \lambda^a, \mu] = \tfrac{1}{2\kappa}(R + 2\lambda) * 1 - \tfrac{1}{2}F \wedge *F - \tfrac{\gamma}{2}F \wedge F_{ab} * \tilde{\Omega}^{ab} + \lambda_a \wedge \Theta^a + \mu \wedge dF. \quad (91)$$

The variational derivatives commute with exterior derivative and also with an integral, thus for simplicity of notation, we suppress the integral symbol before the Lagrangian 4-form, and that the variational derivative is applied to an action integral $I = \int_U L'$ over a compact region U. The Lagrange multiplier 2-form μ is introduced to render the Faraday 2-form closed: $dF = 0$.

After a straightforward exterior algebra calculation, one can show that the total variational derivative of the Lagrangian 4-form with respect to the independent variables takes the form

$$\begin{aligned} \delta L' =& \delta\theta_a \wedge (-\tfrac{1}{\kappa} * G^a + \lambda * \theta^a + *T^a[F] + \gamma D\lambda^a - \tfrac{\gamma}{2}F^a \wedge F_{bc} * \tilde{\Omega}^{bc}) \\ &+ \delta\omega_{ab} \wedge \frac{1}{2}\left[D\left(\tfrac{1}{\kappa} * (\theta^a \wedge \theta^b) - \gamma\tilde{F}^{ab}F\right) - \theta^a \wedge \lambda^b + \theta^b \wedge \lambda^a\right] \\ &+ \delta F \wedge (-d\mu - *F + \gamma F_{ab} * \tilde{\Omega}^{ab}) + \delta\mu \wedge dF + \delta\lambda_a \wedge \Theta^a, \end{aligned} \quad (92)$$

up to a discarded exact form. Therefore, subject to the constraints $\delta L'/\delta\lambda_a = \Theta^a = 0$ and $\delta L'/\delta\mu = dF = 0$, the equations for the connection 1-forms, $\delta L'/\delta\omega_{ab} = 0$, can uniquely be solved for the Lagrange multiplier 2-form as

$$\lambda^a = F_b \wedge D\tilde{F}^{ab} + \theta^a \wedge d * (F \wedge F). \tag{93}$$

In the calculations above, note that it is essential to have a closed Faraday 2-form because the terms of the form $i_a D\tilde{F}^a = i_a \nabla^a \tilde{F} = -d^\dagger * \tilde{F} = *dF = 0$ show up in the derivation of the expression for the Lagrange multiplier 2-form (93). Consequently, the metric field equations (76) follow from (93) and (92).

Finally, it is worth to emphasize that among the F^2R-type nonminimal couplings given in eqs. (72)-(74), only for the particular case of the Horndeski's Lagrangian (68), Lagrange multiplier terms, namely $D\lambda^a$ terms, contain at most second order derivatives of the field variables $\{g, A\}$ as a consequence of vanishing torsion $\Theta^a = D\theta^a = 0$, and the identities $d^2 \equiv 0$ and

$$D^2\tilde{F}^{ab} = \Omega^a{}_c \tilde{F}^{cb} + \Omega^b{}_c \tilde{F}^{ac} \tag{94}$$

for the exterior derivative and the covariant exterior derivatives, respectively. In this regard, the expression for the Lagrange multiplier 2-form in equation (93) is to be compared to the one, for example, given in equation (57) for the Prasanna's model.

Appendix B: Newman-Penrose scalars

In the text, the NP quantities are expressed as contractions of the associated tensors with the null frame/coframe components with labels referring to a coordinate basis [52]. More conveniently, and perhaps in a manner that is of considerable practical use, the NP quantities can be identified as components of differential forms belonging to a null coframe [51].

Note however that, in contrast to the presentation of the NP formalism by making use of a null frame fields, for example in Refs. [51] and [52], the discussion above makes exclusive use of the null coframe fields $k, l, m, \bar{m}$ and refers an associated null frame fields to facilitate the comparison. The use of the coframe fields lead to a slightly differences in the expressions for the associated the basis frame fields and vice versa.

A technical advantage of our approach is that it allows us to obtain the metric field equations with the tensorial indices identified in terms of the NP scalars at once with the help of the definitions provided in this Appendix.

In general, the field equations derived from modified gravitational Lagrangian do not simply determine the anti-self dual part of the curvature 2-form interms of matter fields but they lead to higher order equations involving all the curvature components. The approach here is well-adapted for the application of the NP formalism to modified gravitational models as well. The expressions referring to a coordinate coframe, or an associated frame fields are not necessarily required along the calculations, except for an explicit expression for the exterior derivative in eq. (6). The NP scalars are expressed in terms of contractions only for convenience of the reader accustomed to the calculations using a coordinate coframe.

The spin coefficients defined originally by Newman and Penrose can be expressed as the components of the complex connection 1-forms, $\omega_{ab} = \eta_{ac}\omega^c{}_b = -\omega_{ba}$, introduced in the text by the 1-form equations

$$\omega^0{}_3 = -\tau k - \kappa l + \rho m + \sigma \bar{m}, \tag{95}$$

$$\tfrac{1}{2}\left(\omega^0{}_0 - \omega^3{}_3\right) = \gamma k + \epsilon l - \alpha m - \beta \bar{m}, \tag{96}$$

$$\omega^1{}_2 = \nu k + \pi l - \lambda m - \mu \bar{m}, \tag{97}$$

Journal of Physics: Conference Series **2191** (2022) 012005 doi:10.1088/1742-6596/2191/1/012005

The curvature scalars Φ_{ik} with $i, k = 0, 1, 2$ and Ψ_k with $k = 0, 1, 2, 3, 4$ can be defined by the expressions for the curvature 2-forms as

$$\begin{aligned}
\Omega^0{}_3 &= -\Phi_{00} l \wedge m - \Phi_{01}(k \wedge l + m \wedge \bar{m}) + \Phi_{02} k \wedge \bar{m} + \tfrac{1}{12} R k \wedge m \\
&\quad - \Psi_0 l \wedge \bar{m} - \Psi_1(k \wedge l - m \wedge \bar{m}) + \Psi_2 k \wedge m, && (98)\\
\tfrac{1}{2}\left(\Omega^0{}_0 - \Omega^3{}_3\right) &= \Phi_{10} l \wedge m + \Phi_{11}(k \wedge l + m \wedge \bar{m}) - \Phi_{12} k \wedge \bar{m} - \tfrac{1}{24} R(k \wedge l - m \wedge \bar{m}) \\
&\quad + \Psi_1 l \wedge \bar{m} + \Psi_2(k \wedge l - m \wedge \bar{m}) - \Psi_3 k \wedge m, && (99)\\
\Omega^1{}_2 &= \Phi_{20} l \wedge m + \Phi_{21}(k \wedge l + m \wedge \bar{m}) - \Phi_{22} k \wedge \bar{m} + \tfrac{1}{12} R l \wedge \bar{m} \\
&\quad + \Psi_2 l \wedge \bar{m} + \Psi_3(k \wedge l - m \wedge \bar{m}) - \Psi_4 k \wedge m. && (100)
\end{aligned}$$

where self-dual components are identified in terms of Ψ_{ik} and R, whereas anti-self-dual components are identified in terms of Φ_{ik}.

By using the curvature 2-form definitions (98)-(100), one can verify that the Weyl curvature spinors can be expressed in the form as contraction of associated frame fields with the Weyl curvature tensor in the form

$$\Psi_0 = k^\mu m^\nu k^\alpha m^\beta C_{\mu\nu\alpha\beta}, \tag{101}$$
$$\Psi_1 = k^\mu l^\nu k^\alpha m^\beta C_{\mu\nu\alpha\beta}, \tag{102}$$
$$\Psi_2 = -k^\mu m^\nu l^\alpha \bar{m}^\beta C_{\mu\nu\alpha\beta}, \tag{103}$$
$$\Psi_3 = l^\mu k^\nu l^\alpha \bar{m}^\beta C_{\mu\nu\alpha\beta}, \tag{104}$$
$$\Psi_4 = l^\mu \bar{m}^\nu l^\alpha \bar{m}^\beta C_{\mu\nu\alpha\beta}, \tag{105}$$

that are in accordance with the definitions given in Refs. [51, 52].

With the help of the Cartan's second structure equations in (10)-(12), the above curvature definitions in (98)-(100), and the convenient expression for exterior derivative (6), the complex scalar field equations originally derived by Newman and Penrose can be reproduced after some straightforward exterior algebra computations.

Only the Ricci spinors Φ_{ik} are determined by the Einstein field equations. For example, for the Einstein-Maxwell theory, they are determined in terms of Maxwell spinors using the present formalism as follows. The self-dual Faraday 2-form $\mathcal{F} = \frac{1}{2}(F - i * F)$ can be expanded into the basis self-dual 2-forms in terms of the Maxwell spinors as

$$\mathcal{F} = -\Phi_0 l \wedge \bar{m} - \Phi_2(k \wedge l - m \wedge \bar{m}) + \Phi_2 k \wedge m. \tag{106}$$

Using the expansion in eq (106), the Maxwell scalars Φ_k can be expressed in terms of the contractions as

$$\Phi_0 = F_{\mu\nu} k^\mu m^\nu, \qquad \Phi_2 = F_{\mu\nu} \bar{m}^\mu l^\nu, \qquad \Phi_1 = \tfrac{1}{2} F_{\mu\nu}(k^\mu l^\nu + \bar{m}^\mu m^\nu). \tag{107}$$

In terms of the self-dual 2-form $\mathcal{F} =$, the source-free Maxwell's equations take the form $d\mathcal{F} = 0$, and the vanishing of the individual 3-form components yield the NP form of the Maxwell's equations as a set of scalar field equations for Φ_k with $k = 0, 1, 2$.

The electromagnetic field ansatz (23) discussed above has only Φ_2 component corresponding to a null component that satisfies $\mathcal{F} \wedge \mathcal{F} = 2(F \wedge F - iF \wedge *F) = 0$. Consequently, the canonical energy momentum 1-form defined in (26) can also be written in the form

$$* T_a[F] = -2i i_a \mathcal{F} \wedge \bar{\mathcal{F}}, \tag{108}$$

and by making use of these curvature expressions (98)-(100) in the Einstein forms, some straightforward exterior algebra manipulations reproduce the Einstein-Maxwell field equations, namely the equations $*G^a = \kappa * T^a[F]$, in the scalar form as $\Phi_{ik} = \kappa \Phi_i \bar{\Phi}_k$. (Here $\kappa = 8\pi G/c^4$ denotes the gravitational coupling constant). In the same manner, the above definitions for the NP scalars in terms of tensor-valued forms allow one to construct other algebraically special solutions to the field equations discussed above using the first order formalism in terms of the original NP quantities.

Acknowledgments
It is a pleasure to contribute to the Festschrift in honour of Professor Tekin Dereli.

[1] Bičák J and Podolský J 1999 J. Math. Phys. **40** p 4495; Bičák J and Podolský J 1999 J. Math. Phys. **40** p 4506
[2] Kundt W 1961 Zeits. Phys. **163**, p 77; Ehlers J, Kundt W 1962 in "Exact solutions of the gravitational field equations, in: Gravitation: An introduction to current research", Ed. L. Witten (New York: John Wiley & Sons) p 49
[3] McNutt D, Milson R and Coley A 2013 Class. Quantum Grav. **30**, 055010
[4] Bičák J,and V. Pravda V 1998 Class. Quantum Grav. **15**, p 1539
[5] Coley A, Milson R, Pelavas N, Pravda V, Pravdovà A and Zalaletdinov R 2003 Phys. Rev D **67**, 104020
[6] Podolský J and Švarc R 2013 Class. Quantum Grav. **30**, 125007
[7] Podolský J and Žofka M 2009 Class. Quantum Grav. **29**, 105008
[8] Griffiths J B, P. Docherty P and Podolský J 2004 Class. Quantum Grav. **21**, p 207
[9] Podolský J and Ortaggio M 2003 Class. Quantum Grav. **20** p 1685
[10] Adamowicz W 1980 Gen. Relat. Gravit. **12**, p 677
[11] Babourova O V, Frolov B N and Klimova E A 1999 Class. Quantum Grav. **16** p 1149
[12] García A, Macías A, Puetzfeld D and Socorro J, 2000 Phys. Rev. D **62**, 044021.
[13] Puetzfeld D 2001 *A plane-fronted wave solution in metric-affine gravity*, in "Exact Solutions and Scalar Field in Gravity, Recent Developments", Eds. A. Macìas, J. Cervantes-Cota, and C. Lämmerzahl (Dordrecht: Kluwer) p 141
[14] Obukhov Y-N 2004 Phys. Rev. D **69** 024013
[15] Pasic V and Vassiliev D 2005 Class. Quantum Grav. **22** 3961-3975
[16] Pasic V and Barakovic E 2014 Gen. Relativ. Gravit. **46** p 1787
[17] Blagojevič M and Cvetkovič B 2017 Phys. Rev. D **95** p 104018.
[18] Obukhov Y-N 2017 Phys. Rev. D **95** 084028
[19] Blagojevič M, Cvetkovič B and Obukhov Y N, 2017 Phys. Rev. D **96**, 064031.
[20] García Díaz A and Plebański J F 1981 J. Math. Phys. **22** p 2655
[21] Salazar H I, García Díaz A and Plebanski J F 1983 J. Math. Phys. **24** p 2191
[22] Gleiser R J and Dotti G 2005 Phys. Rev. D **71** 124029
[23] Macías A and Lozano E 2003 Phys. Rev. D **67** 085009.
[24] Ortaggio M, Pravda V and Pravdovà A 2007 Class. Quantum Grav. **24** p 1657
[25] Pravdovà A and Pravda V 2008 Class. Quantum Grav. **25** 235008.
[26] Coley A, Milson R, Pravda V, and Pravdovà A 2004 Class. Quantum Grav. **21** p L35
[27] Coley A 2008 Class. Quantum Grav. **25** 033001.
[28] Coley A, Hervik S, Papadopoulos G and Pelavas N 2009 Class. Quantum Grav. **26** 105016
[29] Ortaggio M, Pravda V and Pravdovà A 2010 Phys. Rev. D **82**, 064043
[30] Ortaggio M, Pravda V and Pravdovà A 2013 Class. Quantum Grav. **30** 013001
[31] Debney G 1974 J. Math. Phys. **15** p 992
[32] Gürses M, Şişman T Ç, and Tekin B 2017 Class. Quantum Grav. **34** 075003
[33] Gibbons G W 1975 Commun. Math. Phys. **45** p 191
[34] Deser S 1975 J. Phys. A: Math. Gen. **8** p 1972
[35] Amati A and Klimčik C 1989 Phys. Lett. B **219** p 443
[36] Coley A A, Gibbons G W, Hervik S and Pope C N 2008 Class. Quantum Grav. **25** 145017
[37] Hervik S, V. Pravda V and Pravdovà A 2014 Class. Quantum Grav. **31** 215005
[38] Chow D D K, Pope C N, and Sezgin E 2010 Class. Quantum Grav. **27** 105001
[39] Deser S, Jackiw R and Templeton S 1982 Phys. Rev. Lett. **48**, p 975
[40] Chow D D K, Pope C N and Sezgin E 2010 Class. Quantum Grav. **27** 105002
[41] Mohseni M 2012 Phys. Rev. D **85** 064038
[42] Dereli T and Sert Ö, 2011 Phys. Rev. D **83** 065005
[43] Gürses M and Halil M 1978 Phys. Lett. A**68** p 182
[44] Debney G C, Kerr R P, and Schild A 1969 J. Math. Phys. **10** p 1842

[45] Osváth I, Robinson I, and Rózga K 1985 J. Math. Phys. **26** p 1755
[46] Buchdahl H A 1970 Mon. Not. Roy. Astron. Soc. **150** p 1
[47] Baykal A, 2013 Eur. Phys. J. Plus **128**: 125
[48] Baykal A, 2016 Turk. J. Phys. **40** pp 77-112; Baykal A, *pp-waves in modified gravity, Preprint*: arXiv:1510.00522v5 [gr-qc]
[49] Prasanna A R 1971 Phys. Lett. **37 A** p 331
[50] Horndeski G W 1976 J. Math. Phys. **17** p 1980
[51] Stephani H, Kramer D, MacCallum M, Hoenselaers C, and Herlt E 2003 *Exact solutions of Einstein's field equations, Cambridge Monographs on Mathematical Physics*, 2nd Ed. (Cambridge: Cambridge Univ. Press) Chapters 24 and 31
[52] Griffiths J B and Podolský J 2009 *Exact space-times in Einstein's General Relativity, Cambridge Monographs on Mathematical Physics*, (Cambridge: Cambridge Univ. Press) Chapter 18
[53] De Felice A, and Tsujikawa S 2010 Living Rev. Relativity **13** p 3
[54] Capozziello S, Corda C, and De Laurentis M F 2008 Phys. Lett. B**669** p 255
[55] Harko T, Lobo F S N, Nojiri S, and Odintsov S D 2011 Phys. Rev. D **84** 024020
[56] Haghani Z, Harko T, Lobo F S N, Sepangi H R, and Shahidi S 2013 Phys. Rev D **88** 044023
[57] Odintsov S D, and Sàez-Gòmez D 2013 Phys. Lett. B **725** p 437
[58] Lovelock D 1969 Arch. Ration. Mech. Anal. **33**, p 54; Lovelock D 1971 J. Math. Phys. **12**, p 498
[59] Cognola G, Elizalde E, Nojiri S, Odintsov S D and Zerbini S 2006 Phys. Rev. D **73** 084007.
[60] Elizalde E, Myrzakulov R, Obukhov V V and Sàez-Gòmez D 2010 Class. Quantum Grav. **27** 095007
[61] Özer H, Baykal A and Delice Ö 2016 Eur. Phys. J. Plus **131**:290
[62] Goenner H F M 1984 Found. Phys. **14** p 866
[63] Balakin A B and Lemos J P S 2005 Class. Quantum Grav. **22** p 1867
[64] Teyssandier P 2004 Annales Fond. Broglie **29** p 173
[65] Drummond I T and Hathrell S J 1980 Phys. Rev. D **22** p 343 Shore G M 1996 Nuc. Phys. B **460** p 379
[66] Dereli T and Üçoluk G 1990 Class. Quantum Grav. **7** p 1109; Dereli T and Üçoluk G 1990 Class. Quantum Grav. **7** p 533
[67] Horndeski G W 1979 J. Math. Phys. **20** p 726
[68] Eardley D M, Lee D L, Ligthman A P, Wagoner R V and Will C M 1973 Phys. Rev Lett. **30** p 884
[69] Eardley D M, Lee D L and Ligthman A P 1973 Phys. Rev D **8** p 3308
[70] Kausar H R, Philippoz L, and Jetzer P 2016 Phys. Rev. D **93** 124071
[71] Sagi E 2010 Phys. Rev. D **81** 064031
[72] Alves M E S, Miranda O D, and de Araujo J C N 2010 Class. Quant. Grav. **27** 145010
[73] Stephon A, Finn L S and Yunes N 2008 Phys. Rev D **78** 066005
[74] Buchdahl H A 1979 J. Math. Phys. A: Math. Gen. **12** p 1037
[75] Müller-Hoissen F 1990 Nucl. Phys. B**337** p 709
[76] Dereli T and Tucker R W 1987 Class. Quantum Grav. **4** p 791
[77] Kopczysńki W 1990 Ann. Phys. (N. Y.) **203** p 308
[78] Hehl F W, McCrea J D, Mielke E W and Ne'eman Y 1995 Phys. Rept. **258** p 1
[79] Davydov E A 2017 Comparing metric and Palatini approaches to vector Horndeski theory, *Preprint*: arXiv:1708.09796v1 [hep-th].

Journal of Physics: Conference Series **2191** (2022) 012006 doi:10.1088/1742-6596/2191/1/012006

A trip around octonions

Ayşe Hümeyra Bilge[1], **Tekin Dereli**[2] **and Şahin Koçak**[3]

[1] Kadir Has University, Istanbul, Turkey
[2] Koç University, Istanbul, Turkey
[3] Anadolu University (emeritus), Eskişehir, Turkey

E-mail: ayse.bilge@khas.edu.tr, tdereli@ku.edu.tr, skocak@anadolu.edu.tr

Abstract. In these expository notes, after a contemplation on the dawn of octonions, we give proofs for the Frobenius theorem and the Hurwitz theorem, we review the basics of Clifford algebras and spin groups, and exemplify the startling role played by the octonions in 7− and 8−dimensional phenomena such as the special 3− and 4−forms, the Bonan form, $Spin(7)$ and $Spin(8)$ groups and the mysterious triality.

1. The First Steps

The natural numbers were long around. Then came the rationals by the need to share. The Greeks took the lengths of segments as the reals. Everything positive, it goes without saying. And they made the terrifying discovery that there were "incommensurable" segments! Such as an edge and a diagonal of a regular pentagon. It might be this traumatic event that they stuck in the number mud and became masters of geometry.

After about a millennium the negative numbers (and zero!) made their appearance in the enigmatic India and made their way very slowly to Europe. In the Italy of the 16th Century cubic equations were solved in breathtaking contests among mathematicians and square-roots of negative numbers imposed themselves in hermetic formulas such as $\sqrt[3]{2+\sqrt{-121}}+\sqrt[3]{2-\sqrt{-121}}$ for an equation $x^3 = 15x + 4$ with an obvious root. This nightmare of Cardano was only partly relieved some decades later by Bombelli, who noticed that $(2+\sqrt{-1})^3 = 2+\sqrt{-121}$ and $(2-\sqrt{-1})^3 = 2-\sqrt{-121}$, giving the obvious root 4. Even for the minds of Leibniz or even Euler these new "imaginary" ghosts were incomprehensible. Atiyah notes bitterly that it took centuries to understand the square root of -1.

Then came the fateful 19th Century. Everything was being rethought and rebuilt from scratch. We cannot tell here the dramatic story of this heroic epoch. But let us only remark that early on in the century the glorious Euclidean geometry was shattered, the futility to solve (in radicals) the quintic (and higher degree) polynomials were grasped, foundations of a trustworthy calculus were laid, and as a no more postponable issue, a solid basis for the numbers (the naturals, rationals, reals, negatives and the imaginaries) was devised, the reals being more troublesome than the imaginaries! (At the end of the century the Euclidean geometry was however rehabilitated and assigned to a more modest corner of mathematics.)

A "complex number" (as they were unluckily dubbed) was nothing more than a pair of real numbers, say (a, b). Addition and multiplication were defined by $(a, b) + (c, d) = (a + c, b + d)$ and $(a, b) \cdot (c, d) = (ac - bd, ad + bc)$. The number $(1, 0)$ was the unit of multiplication and the spooky square-root of -1 was nothing else than $(0, 1)$ since $(0, 1) \cdot (0, 1) = (-1, 0) = -(1, 0)$.

Since the objects $(a,0)$ behave as the reals under addition and multiplication we can identify them with the reals (and write simply a again for them) and since by the above definitions $(a,b) = (a,0) + (b,0) \cdot (0,1)$ holds, we can write $(a,b) = a + b \cdot i$, if we denote $(0,1)$ by i, thus getting the more familiar form.

If you want to be more pragmatic, you can say that you introduce a new object i with $i^2 = -1$ and call the expressions $a + bi$ your new numbers which you add and multiply as $(a+bi)+(c+di) = (a+c)+(b+d)i$ and $(a+bi)(c+di) = (ac-bd)+(ad+bc)i$. Nobody can query you about what your new object is, because at the end of the 19th Century (and now), not the objects, but only the relations are important, i.e. how you add and multiply them (or whatever the relations are).

Surely the complex numbers were interpreted as points of the plane and it was an irresistible desire to extend the complex number system to the 3-dimensional space. No less geniuses than Gauss and Hamilton were unable to multiply the triplets! Gauss himself writes [11]:

"The writer has reserved for himself ... the question why the relations between things that make up a manifold of more than two dimensions cannot provide quantities admissible in universal arithmetic."

(The case of Hamilton is well documented and we heartily recommend [2].)

What is a number system? Today we might give different answers, but in those days, arithmetic operations and their properties were the main guide. You ought to be able to add, subtract, multiply and divide (with non-zero numbers) and the accustomed properties should hold. Gauss had made a great leap in declaring a non-zero natural number to be zero in modular arithmetic, but the commutativity and associativity of the addition and multiplication were subliminal and not yet on the discussion table. (But seemingly not for Gauss as we will note shortly). In a survey of early attempts on multiplying the triplets and the impossibility proofs thereof, Kenneth May [11] writes, that *"the student can easily discover for himself if he experiments without getting bogged down (as did Hamilton!) in trying particular definitions of multiplication"*. We find this assessment somewhat merciless since it is always easier to be smart afterwards; one should take the *Zeitgeist* into account. Today it is a one-line exercise that on an odd-dimensional $\mathbb{R}^n (n > 1)$ there can be no real division algebra by considering an element as a linear transformation via multiplication and noticing that there is a real eigenvalue producing a zero-divisor. Nevertheless, one is compelled to admit that the following argument recalled by May might possibly have been accessible even in those days:

If we introduce a second object j to extend the complex numbers to the space and consider the triplets $a + bi + cj$ (with a, b, c real), we should first of all declare what ij is!

Let $ij = \alpha + \beta i + \gamma j$. Then, $i(ij) = i(\alpha + \beta i + \gamma j) = \alpha i - \beta + \gamma ij = \alpha i - \beta + \gamma(\alpha + \beta i + \gamma j) = (\alpha\gamma - \beta) + (\alpha + \beta\gamma)i + \gamma^2 j$. On the other hand, $i(ij) = (ii)j = -j$ (by the subliminally assumed associativity). By comparing the two expressions, we get $c^2 = -1$ for a real c, which is impossible! (Thus the complex numbers cannot be extended to a three-dimensional real associative algebra at all; irrespective of the division property.)

It seems somehow, that if we want to multiply the triplets, ij requires a separate room for itself! Let us allocate it and call it k. Then we must admit quartets in form $a + bi + cj + dk$. To multiply triplets, we need quartets! But how to multiply them, and to which price? Their multiplication rules were carved by Hamilton on a stone of a bridge in Dublin in 1843: $i^2 = j^2 = k^2 = ijk = -1$.

Now ij is k as planned: $(ijk)k = -k$, so that $ij = k$ since $k^2 = -1$. (Hamilton's sole writing "ijk" shows that "associativity" is innate!)

But what is ji? If we multiply $ijk = -1$ with ji, we get $ji(ijk) = -ji$ and hence $ji = -k$! Similarly, $jk = -kj = i$ and $ki = -ik = j$. The price is the loss of commutativity! It must have cost Hamilton some pain of overcoming. You can now multiply any two quartets using the (never disputed) distributivity and real-linearity. In the resulting system, called *Quaternions* by Hamilton, you can do also division by non-zero quaternions. But be careful: How do you "divide", for example k by i? Due to the loss of commutativity, the solutions of the equations $ix = k$ and $xi = k$ are different! (The discovery of the quaternions is generally credited solely to Hamilton, but as became known half a century after his death, Gauss had dared to dispense with the commutativity of multiplication and defined the quaternions as early as 1819! [1])

During the extensions from the reals to the complex numbers, and from them to quaternions, there is an unexpected accompanying phenomenon behind the scene: We first wanted to take the square-roots of negative numbers, then we wanted innocently to extend the arithmetic operations to the 3-space and landed at the 4-dimensional space, but we are rewarded with more structure: The simple absolute value $|a|$ of a real number (with the obvious property $|ab| = |a||b|$) has rather natural counterparts for the complex numbers and quaternions. We can surely assign an "absolute value" (or magnitude, modulus, norm, or whatever you call) for a complex number $a+bi$, namely the usual Euclidean norm $\|(a,b)\| = (a^2+b^2)^{1/2}$ of $(a,b) \in \mathbb{R}^2$; and for a quaternion $a + bi + cj + dk$ the norm of $(a,b,c,d) \in \mathbb{R}^4$, but who can hope that the multiplicative property $\|ab\| = \|a\| \, \|b\|$ holds for them? Somewhat miraculously, they come to be true as reflected in the following identities:

We defined the multiplication of two complex numbers by $(a+bi)(c+di) = (ac-bd)+(ad+bc)i$ and it holds $(a^2 + b^2)(c^2 + d^2) = (ac - bd)^2 + (ad + bc)^2$; we defined the multiplication of two quaternions by

$$\begin{aligned}(a_0 + a_1 i + a_2 j + a_3 k)(b_0 + b_1 i + b_2 j + b_3 k) = {} & (a_0 b_0 - a_1 b_1 - a_2 b_2 - a_3 b_3) \\ & + (a_0 b_1 + a_1 b_0 + a_2 b_3 - a_3 b_2) i \\ & + (a_0 b_2 - a_1 b_3 + a_2 b_0 + a_3 b_1) j \\ & + (a_0 b_3 + a_1 b_2 - a_2 b_1 + a_3 b_0) k\end{aligned}$$

and it holds

$$\begin{aligned}(a_0^2 + a_1^2 + a_2^2 + a_3^2)(b_0^2 + b_1^2 + b_2^2 + b_3^2) = {} & (a_0 b_0 - a_1 b_1 - a_2 b_2 - a_3 b_3)^2 \\ & + (a_0 b_1 + a_1 b_0 + a_2 b_3 - a_3 b_2)^2 \\ & + (a_0 b_2 - a_1 b_3 + a_2 b_0 + a_3 b_1)^2 \\ & + (a_0 b_3 + a_1 b_2 - a_2 b_1 + a_3 b_0)^2 \ !\end{aligned}$$

These identities were surely known before complex numbers and quaternions; the first since ancient times, the second since Euler. We will see a last (and more formidable) instance of this miracle shortly.

We would like to note as a historical side-remark that a Turkish general and solitary mathematician Vidinli Hüseyin Tevfik Paşa was also possessed by the quest to multiply the triplets and in a book titled "Linear Algebra" (published in 1882 in Istanbul, in English [14]) he defined $ij = -ji = 1$ (and $i^2 = j^2 = -1$), thus creating a non-commutative and non-associative algebra extending the complex numbers to $\mathbb{R}^3$, where the non-zero elements have a two-sided inverse. A norm multiplicativity holds e.g. for elements of the special type $a + bi + bj$, reflected in the partial 3-squares identity

$$(a^2 + b^2 + b^2)(c^2 + d^2 + d^2) = (ac - 2bd)^2 + (ad + bc)^2 + (ad + bc)^2,$$

(which is in fact a 2-squares identity for the considered elements which constitute a subalgebra isomorphic to the complex numbers since $a + bi + bj = a + \sqrt{2}b(\frac{1}{\sqrt{2}}(i + j))$ with $(\frac{i+j}{\sqrt{2}})^2 = -1$). Norm multiplicativity holds more generally for elements of type $a + bri + bsj$ for fixed $r, s \in \mathbb{R}$

(which also constitute a subalgebra isomorphic to $\mathbb{C}$ via $(\frac{ri+sj}{\sqrt{r^2+s^2}})^2 = -1$), resulting in the identity

$$(a^2 + b^2r^2 + b^2s^2)(c^2 + d^2r^2 + d^2s^2) = (ac - (r^2 + s^2)bd)^2 + ((ad + bc)r)^2 + ((ad + bc)s)^2,$$

a narrow miss of the impossible 3-squares identity, might one think! He gives numerous geometric applications of this curious algebra. We hope to report on our forgotten fellow and his book elsewhere.

2. Where To Go From Here?

Having meanwhile three real associative algebras with division property on $\mathbb{R}$, $\mathbb{R}^2$ and $\mathbb{R}^4$, what might happen in further search? One is tempted to imitate the pattern of quaternions (i.e. two new objects i and j with $i^2 = j^2 = -1$ and a new room k for ij with $ji = -ij$) with more objects, say, for example, with three new objects i, j, m with $i^2 = j^2 = m^2 = -1$, $ij = -ji$, $im = -mi$, $jm = -mj$ and with new rooms for ij, im, jm and ijm. We have then a total of 8 rooms for $1, i, j, m, ij, im, jm, ijm$ (so that an arbitrary element of the candidate algebra will be a linear sum of these elements with real coefficients) and further multiplications are executed by associative (and distributive) simplification. For example, $(ij)(im) = i(ji)m = -i(ij)m = -(ii)jm = jm$, or, $(ijm)(ijm) = -jmjm = -mm = 1$.

We get on the whole an associative algebra on $\mathbb{R}^8$. Would this algebra allow divisions (with non-zero elements)? Unfortunately not! For example, $(1 + ijm)(1 - ijm) = 0$, so that there are zero-divisors! By the same reason, a generalization of this construction with n new objects yielding a real associative algebra on $\mathbb{R}^{2^n}$ will not be a division algebra either. However, this idea is not that bad and it will lead us to the Clifford algebras with their own merits.

Any further search is in vain, in fact. If we insist on associativity, there are no other real division algebras beyond the reals ($\mathbb{R}$), the complex numbers ($\mathbb{C}$) and the quaternions ($\mathbb{H}$).

Theorem (Frobenius 1878 [6]). *A finite-dimensional associative real division algebra is isomorphic either to* $\mathbb{R}$, *or to* $\mathbb{C}$ *or to* $\mathbb{H}$.

Before giving a proof of this theorem we want to define some notions we have been using more precisely:

Some Definitions: An algebra A over a field F is an F-vector space A with a bilinear product $A \times A \to A$, $(a, b) \mapsto ab$. (We then have $(a_1 + a_2)b = a_1b + a_2b$, $a(b_1 + b_2) = ab_1 + ab_2$, $(\lambda a)b = a(\lambda b) = \lambda(ab)$ for $\lambda \in F$)

The algebra is called associative if $a(bc) = (ab)c$ holds. A is called an algebra with unit if there exists $1 \in A$ with $1a = a1 = a$ (When we talk of an algebra with unit we assume that $A \neq \{0\}$ so that the unit is necessarily different from zero; we can anyway discard that trivial case from start). F is naturally an algebra over itself and the map $F \to A, \lambda \mapsto \lambda 1$ is an algebra isomorphism and thus F can be thought lying embedded in A.

An algebra $A \neq \{0\}$ is called a division algebra if for any a, b, $a \neq 0$, the equations $ax = b$ and $ya = b$ have unique solutions; in other words, the left and right multiplication maps $L_a : A \to A$, $x \mapsto ax$ and $R_a : A \to A$, $x \mapsto xa$ are linear-isomorphisms.

A division algebra has no zero-divisors: If $ab = 0$, then either a or b must be zero: Say $a \neq 0$, then the unique-solution property of $ay = 0$ would give $b = 0$ since $a0 = 0$. If the algebra A is finite-dimensional as a vector space over F, then the converse is also true: If there are no zero-divisors, then $ax = b$ and $ya = b$ have unique solutions (for proof consider the maps L_a and R_a: Injectivity implies surjectivity).

We will consider in these notes only real (i.e. $F = \mathbb{R}$), finite-dimensional division algebras with unit.

The algebra is full of little gems: An associative division algebra has automatically a unit! To see this, consider $ax = a$ for an $a \neq 0$. This has a unique solution. Call it e. Thus, $ae = a$ (and

$e \neq 0$). We have $(ae)e = ae$. By associativity, $a(ee) = ae$. Hence $a(ee - e) = 0$. As there are no zero-divisors, $ee - e = 0$. Now, for any x, $x(ee - e) = 0$, $x(ee) - xe = 0$, $(xe)e - xe = 0$, $(xe - x)e = 0$, $xe = x$, thus e is a right-unit. Similarly, $(ee - e)x = 0$ and and $ex = x$, and e is a unit. We used above a weaker version of associativity, called alternativity: $a(bc) = (ab)c$ holding when two of a, b, c are equal. Thus we have seen:

Remark: An alternative division algebra has necessarily a unit.

Now, we give a proof of the Frobenius theorem (which is a streamlined version of the scattered proof in the beautiful book *Numbers* [5]; for another elementary proof see [13]).

Proof of the Frobenius Theorem. Given a finite-dimensional, associative, real division algebra A, any element $a \in A$ can be decomposed as $a = r + x$, where r is real (since A has necessarily a unit, $\mathbb{R}$ is embedded in A) and x^2 is a non-positive real. To see this we can consider the powers of a (up to the dimension of A) and obtain a polynomial which splits into first and second degree factors over the reals. Since there are no zero-divisors, one of the factors must vanish. If a is not real, it must satisfy a quadratic polynomial, say, $a^2 - 2ra + s = 0$ (with $r^2 - s < 0$). Now we get, $(a - r)^2 = a^2 - 2ra + r^2 = r^2 - s < 0$.

The surprising thing is that elements of A, which, when squared, become non-positive reals, constitute a linear subspace of A! Calling this subset A', it is obvious that a multiple of an element of A' belongs to A', so let $u, v \in A'$ be two linearly independent elements. Then, u, v and 1 are also linearly independent (if $\lambda u + \mu v + \nu 1 = 0$, then $\lambda^2 u^2 = \mu^2 v^2 + 2\mu\nu v + \nu^2$ yielding $\lambda = \mu = \nu = 0$ since u^2 and v^2 are reals) and consequently, $u + v$ and $u - v$ are not reals. Let the quadratic polynomials they satisfy be $(u+v)^2 - 2p(u+v) + q = 0$ and $(u-v)^2 - 2p'(u-v) + q' = 0$. Opening the squares and adding up the equations gives, by linear-independence of u, v and 1, $p = p' = 0$, hence $u + v \in A'$, showing that A' is indeed a linear subspace of A. We thus get a fine decomposition of the algebra as $A = \mathbb{R} \oplus A'$ with $x^2 \leq 0$ for $x \in A'$.

We note that for $u, v \in A'$, $uv + vu$ is real: $uv + vu = (u + v)^2 - u^2 - v^2$.

Now, if $A' = \{0\}$, then $A = \mathbb{R}$; if $\dim(A') = 1$, choose $u \in A'$ with $u^2 = -1$ and we get $A = \mathbb{C}$. If $\dim(A') \geq 2$, then we can construct a so-called Hamiltonian triple, that is a triple of elements with the same multiplicative relationships as the i, j and k of the quaternions: First choose $u \in A'$ with $u^2 = -1$ and a $v' \in A'$ such that u and v' are linearly independent. Then $uv' + v'u = r$ with $r \in \mathbb{R}$. Now take $v = v' + \frac{r}{2}u$. We get, $uv + vu = uv' + \frac{r}{2}u^2 + v'u + \frac{r}{2}u^2 = 0$. Now it can be easily seen that u, v and $W = uv$ constitute a Hamiltonian triple. For example, $w^2 = (uv)(uv) = ((uv)u)v = (u(vu))v = -(u(uv))v = -(u^2 v)v = v^2 = -1$; or, $vw = v(uv) = -v(vu) = -v^2 u = u$, etc. This shows that there is an embedded $\mathbb{H}$ in A. Now let x be any element of A'. We have $xu + ux = r$, $xv + vx = s$ and $xw + wx = t$ for some $r, s, t \in \mathbb{R}$. From the first equation we get $(xu)v + (ux)v = rv$, from the second, $u(xv) + u(vx) = su$, and subtracting them (vital use of associativity! For Hamiltonian triples we could dispense with associativity and get by alternativity) we get $(xu)v - u(vx) = x(uv) - (uv)x = xw - wx = rv - su$. Adding up this equation with $xw + wx = t$ we find $2xw = rv - su + t$ and by multiplying from the right with w we obtain $x = \frac{1}{2}(ru + sv + tw)$! That means, there is nothing more in A' than the span of u, v and w, in other words, $A = \mathbb{H}$. □

3. Where Did The Octonions Come From?

A short (and rather naive) narrative like this can hardly follow the inscrutable ways of the actual discovery process. As we try to understand the division algebras and to see what could be done after Hamilton, we come to admit that the octonions were discovered long before Frobenius listed the associative division algebras. A few months after learning the discovery of the quaternions from a letter of Hamilton, J.T. Graves, a friend of Hamilton, had promptly extended the quaternions to the 8-dimensional space, and called them octaves, as the unhappily delayed attestment of Hamilton of the year 1847 shows [7]:

> My present object is to mention that J. T. Graves, to whom I had previously communicated my theory of *quaternions*, was early led, by his extension of Euler's theorem, to conceive an analogous theory of *octaves*, involving *seven* distinct imaginaries, or square roots of negative unity, namely, *four* new roots, which he denoted by the letters l, m, n, o, to be combined with my *three* letters, i, j, k, into one common imaginary or symbolic system. Thus, as I had already (in October and November, 1843) communicated to him and to the Academy the fundamental equations of quaternions, namely,
>
> $$\left.\begin{array}{l} i^2 = j^2 = k^2 = -1, \\ ij = k,\ jk = i,\ ki = j, \\ ji = -k,\ kj = -i,\ ik = -j, \end{array}\right\} \quad \text{(a)}$$
>
> which may be concisely summed up in the formula
>
> $$i^2 = j^2 = k^2 = ijk = -1; \quad \text{(b)}$$
>
> so he proposed to found a theory of octaves on the following equations,
>
> $$\left.\begin{array}{l} i^2 = j^2 = k^2 = l^2 = m^2 = n^2 = o^2 = -1, \\ i = jk = lm = on = -kj = -ml = -no, \\ j = ki = ln = mo = -ik = -nl = -om, \\ k = ij = lo = nm = -ji = -ol = -mn, \\ l = mi = nj = ok = -im = -jn = -ko, \\ m = il = oj = kn = -li = -jo = -nk, \\ n = jl = io = mk = -lj = -oi = -km, \\ o = ni = jm = kl = -in = -mj = -lk; \end{array}\right\} \quad \text{(c)}$$
>
> which he communicated to me, in a letter dated January 4, 1844, and which may be concisely expressed by the single but continued equation,
>
> $$\left.\begin{array}{l} i^2 = j^2 = k^2 = l^2 = m^2 = n^2 = o^2 = -1 \\ = ijk = ilm = ion = jln = jmo = klo = knm. \end{array}\right\} \quad \text{(d)}$$

Meanwhile, also apparently intrigued by the quaternions, Cayley has already published a cryptic note in 1845:

Cambridge, January 16, 1845. A. CAYLEY.

P.S. *On Quaternions.*

It is possible to form an analogous theory with seven imaginary roots of (-1) (? with $\nu=2^n-1$ roots when ν is a prime number). Thus if these be $\iota_1, \iota_2, \iota_3, \iota_4, \iota_5, \iota_6, \iota_7$, which group together according to the types

123, 145, 624, 653, 725, 734, 176,

i. e. the type 123 denotes the system of equations

$$\iota_1\iota_2=\iota_3,\quad \iota_2\iota_3=\iota_1,\quad \iota_3\iota_1=\iota_2,$$
$$\iota_2\iota_1=-\iota_3,\quad \iota_3\iota_2=-\iota_1,\quad \iota_1\iota_3=-\iota_2,$$

&c. We have the following expression for the product of two factors:

$$\begin{aligned}
&(X_0+X_1\iota_1+\dots X_7\iota_7)(X'_0+X'_1\iota_1+\dots X'_7\iota_7)\\
&=X_0X'_0-X_1X'_1-X_2X'_2\dots-X_7X'_7.\\
&+[\overline{23}+\overline{45}+\overline{76}+(01)]\iota_1\\
&+[\overline{31}+\overline{46}+\overline{57}+(02)]\iota_2\\
&+[\overline{12}+\overline{65}+\overline{47}+(03)]\iota_3\\
&+[\overline{51}+\overline{62}+\overline{47}+(04)]\iota_4\\
&+[\overline{14}+\overline{36}+\overline{72}+(05)]\iota_5\\
&+[\overline{24}+\overline{53}+\overline{17}+(06)]\iota_6\\
&+[\overline{25}+\overline{34}+\overline{61}+(07)]\iota_7
\end{aligned}$$

where $(01)=X_0X'_1+X_1X'_0$

:

$\overline{12}=X_1X'_2-X_2X'_1$

&c.

Both Graves and Cayley, took seven imaginaries, gave a multiplication table and derived as a bonus an eight-square identity! (or was it maybe the other way around?) Is it that easy? Why seven imaginaries? Why those multiplication tables? How does the mysterious 8-squares identity pop up? All in all, it seems that it was a clever cast. The ensuing intense search in higher (especially 2^n) dimensions for new number systems is an indication that there was no conceptual frame yet. Nevertheless, you can take the multiplication table of Graves or Cayley (but beware: there is a misprint in Cayley's table!), express a new number as, say,

$$a=(a_0+a_1i+a_2j+a_3k+a_4l+a_5m+a_6n+a_7o)\in\mathbb{R}^8,$$

take another number

$$b=(b_0+b_1i+b_2j+b_3k+b_4l+b_5m+b_6n+b_7o),$$

multiply them according to the table and get

$$ab=c=(c_0+c_1i+c_2j+c_3k+c_4l+c_5m+c_6n+c_7o),$$

and then you will discover the incredible identity

$$\begin{aligned}(a_0^2+a_1^2+a_2^2+a_3^2+a_4^2+a_5^2+a_6^2+a_7^2)(b_0^2+b_1^2+b_2^2+b_3^2+b_4^2+b_5^2+b_6^2+b_7^2)\\
=(c_0^2+c_1^2+c_2^2+c_3^2+c_4^2+c_5^2+c_6^2+c_7^2)!\end{aligned}$$

That is, $\|ab\|=\|a\|\,\|b\|$, where $\|a\|$ is the standard norm of a in $\mathbb{R}^8$. Moreover, you can do the divisions $ax=b$ and $ya=b$ within the system. You will at some point notice that the element $\overline{a}=(a_0-a_1i-a_2j-a_3k-a_4l-a_5m-a_6n-a_7o)$ is the analogue of the complex conjugate, as useful as that ($a\overline{a}=\overline{a}a=\|a\|^2$), making the division such easy:

$$ax=b,\quad \overline{a}(ax)=\overline{a}b,\quad (\overline{a}a)x=\overline{a}b,\quad \|a\|^2x=\overline{a}b,\quad x=\frac{1}{\|a\|^2}(\overline{a}b).$$

You were however lucky in this computation (using the associativity). You will soon sadly realize that the multiplication of the new number system is generally not associative. But if division is more important for you, you will accept unwillingly the loss of associativity.

We can elucidate at this point some of the mystery of n-square identities. If we have an identity of the form

$$(a_1^2 + a_2^2 + \cdots + a_n^2)(b_1^2 + b_2^2 + \cdots + b_n^2) = (c_1^2 + c_2^2 + \cdots + c_n^2),$$

where the tuple $(c_1, c_2, \ldots, c_n)$ is bilinear in $(a_1, a_2, \ldots, a_n)$ and $(b_1, b_2, \ldots, b_n)$, then we can define an algebra on $\mathbb{R}^n$ by

$$(a_1e_1 + a_2e_2 + \cdots + a_ne_n)(b_1e_1 + b_2e_2 + \cdots + b_ne_n) = (c_1e_1 + c_2e_2 + \cdots + c_ne_n),$$

where the e_i are the standard basis vectors of $\mathbb{R}^n$. The Euclidean norm is then multiplicative for this algebra. If, conversely, we can find an algebra on $\mathbb{R}^n$, for which the Euclidean norm is multiplicative, then we find an n-square identity via this multiplicativity.

This observation brings a bunch of problems which we cannot address here. If the algebra thus defined has a unit, life becomes easy. In that case, Hurwitz showed that the only possible cases are the already known $\mathbb{R}, \mathbb{C}, \mathbb{H}$ and $\mathbb{O}$. We will dwell on his theorem in more detail below. Otherwise, i.e. if there does not exist a unit, there are an infinitude of algebras with multiplicative Euclidean norm (but again in these somehow privileged four dimensions 1, 2, 4 and 8.) For example, the multiplication

$$(a_1e_1 + a_2e_2)(b_1e_1 + b_2e_2) = (a_1b_1 + a_2b_2)e_1 + (a_2b_1 - a_1b_2)e_2$$

defines an algebra on $\mathbb{R}^2$, without a unit, but still with multiplicative norm:

$$(a_1^2 + a_2^2)(b_1^2 + b_2^2) = (a_1b_1 + a_2b_2)^2 + (a_2b_1 - a_1b_2)^2.$$

We would like to remark that an algebra on $\mathbb{R}^n$ with a multiplicative norm is necessarily a division algebra since it can not have zero-divisors. If we forget about the norms and ask for division algebras on $\mathbb{R}^n$, then the possible dimensions are still only the 1, 2, 4 and 8! This was shown by Milnor and Kervaire (independently) in 1958 [12] by rather sophisticated topological means. The classification of division algebras in these dimensions is however subject of ongoing research and not yet completed.

Before going to the discussion of normed algebras, we want to recall another look on octonions. Dickson noticed in 1919 [4] that the octonions can be understood as pairs of quaternions, as the complex numbers were conceived as pairs of real numbers by Hamilton (likewise, the quaternions can be defined as pairs of complex numbers). Dickson defined the product of quaternion pairs as follows:

$$(a, b)(c, d) = (ac - \overline{d}b, da + b\overline{c}),$$

where a, b, c and d are quaternions, $\overline{d}, \overline{c}$ are the conjugates of d and c. (This product rule is a refinement of Hamilton's complex multiplication rule and can be applied verbatim to define the quaternions as complex number pairs.) If we denote the elements $(1, 0), (i, 0), (j, 0)$ and $(k, 0)$ by $e_0 = 1, e_1, e_2$ and e_3; and the elements $(0, 1), (0, i), (0, j)$ and $(0, k)$ by e_4, e_5, e_6 and e_7, we obtain the multiplication table on the standard basis of $\mathbb{R}^8$.

$e_0 = 1$	e_1	e_2	e_3	e_4	e_5	e_6	e_7
e_1	-1	e_3	$-e_2$	e_5	$-e_4$	$-e_7$	e_6
e_2	$-e_3$	-1	e_1	e_6	e_7	$-e_4$	$-e_5$
e_3	e_2	$-e_1$	-1	e_7	$-e_6$	e_5	$-e_4$
e_4	$-e_5$	$-e_6$	$-e_7$	-1	e_1	e_2	e_3
e_5	e_4	$-e_7$	e_6	$-e_1$	-1	$-e_3$	e_2
e_6	e_7	e_4	$-e_5$	$-e_2$	e_3	-1	$-e_1$
e_7	$-e_6$	e_5	e_4	$-e_3$	$-e_2$	e_1	-1

Journal of Physics: Conference Series **2191** (2022) 012006 doi:10.1088/1742-6596/2191/1/012006

Now, at the point we arrived, after sacrificing first commutativity and then associativity, we have four real division algebras $\mathbb{R}, \mathbb{C}, \mathbb{H}$ and $\mathbb{O}$, with a unit and with a multiplicative norm, which comes from the standard inner products on the respective Euclidean spaces. It turns out that this is the end of the road in the following sense:

Theorem (Hurwitz 1898 [10]). *A finite-dimensional, real algebra with unit and with a multiplicative norm coming from an (bilinear, symmetric and positive definite) inner product is isomorphic either to* $\mathbb{R}$*, or to* $\mathbb{C}$ *or to* $\mathbb{H}$ *or to* $\mathbb{O}$*.*

A normed algebra is free of zero-divisors and, by finite-dimensionality, a division algebra. Normed algebras have a simple and useful theory. The inner product simplifies some of the foregoing considerations and enables new nice relationships. We will now consider them and give a proof of the Hurwitz theorem following the beautiful references [9] and [8].

4. An Excellent Set-up: Normed Algebras Over the Reals

Definition. A normed algebra (over $\mathbb{R}$) is a finite dimensional algebra A ($\neq \{0\}$) over $\mathbb{R}$ with identity and with an inner product $\langle x, y\rangle$ such that the associated norm ($\|x\| = \langle x, x\rangle^{1/2}$) satisfies $\|xy\| = \|x\|\|y\|$.

The embedded $\mathbb{R} \subset A$, which consists of the real multiples of the identity of A, is called the real part of A and denoted by $Re(A)$ (sometimes by ReA). The norm of the identity $\mathbf{1}$ of A is 1 since $\|\mathbf{1}\|\|\mathbf{1}\| = \|\mathbf{1}\|$ (note that $\mathbf{1} \neq 0$ and thus $\|\mathbf{1}\| > 0$). We identify $Re(A)$ with $\mathbb{R}$ and denote the identity of A henceforth also with 1. The orthogonal complement of $Re(A)$ is called the imaginary part of A, denoted by $Im(A)$ (sometimes by ImA) and thus A is decomposed as $Re(A) \oplus Im(A) = \mathbb{R} \oplus Im(A)$. Any element $x \in A$ can be written uniquely as $x = x_1 + x'$ with $x_1 \in \mathbb{R}$ and $x' \in Im(A)$. We will write $x_1 = Re(x)$ and $x' = Im(x)$.

We define a conjugation by $\overline{x} = x_1 - x'$. Thus,

$$x_1 = Re(x) = \frac{1}{2}(x + \overline{x}), \qquad x' = Im(x) = \frac{1}{2}(x - \overline{x}).$$

The left and right multiplication maps L_a and R_a on A with $a \in A$ satisfy the following equations:

Lemma. *We have* $\langle L_a x, L_a y\rangle = \langle ax, ay\rangle = \langle x, y\rangle\|a\|^2$ *and* $\langle R_a x, R_a y\rangle = \langle xa, ya\rangle = \langle x, y\rangle\|a\|^2$.

Proof. Let us show the latter (the first being the same). We use repeatedly the multiplicative property of the norm.

$$\begin{aligned}
\|(x+y)a\|^2 &= \|x+y\|^2\|a\|^2 \\
\langle (x+y)a, (x+y)a\rangle &= \langle x+y, x+y\rangle\|a\|^2 \\
\langle xa, xa\rangle + 2\langle xa, ya\rangle + \langle ya, ya\rangle &= (\langle x, x\rangle + 2\langle x, y\rangle + \langle y, y\rangle)\|a\|^2 \\
\|xa\|^2 + 2\langle xa, ya\rangle + \|ya\|^2 &= (\|x\|^2 + 2\langle x, y\rangle + \|y\|^2)\|a\|^2 \\
\|x\|^2\|a\|^2 + 2\langle xa, ya\rangle + \|y\|^2\|a\|^2 &= (\|x\|^2 + 2\langle x, y\rangle + \|y\|^2)\|a\|^2 \\
\langle xa, ya\rangle &= \langle x, y\rangle\|a\|^2.
\end{aligned}$$

□

Lemma. *We have* $\langle ax, y\rangle = \langle x, \overline{a}y\rangle$ *and* $\langle xa, y\rangle = \langle x, y\overline{a}\rangle$*, in other words, the adjoint of* L_a *is* $L_{\overline{a}}$ *and the adjoint of* R_a *is* $R_{\overline{a}}$.

Proof. Consider the first. Inserting $a = a_1 + a'$, this reduces to $\langle a'x, y\rangle = \langle x, -a'y\rangle$. Now, working out the equality $\langle (1+a')x, (1+a')y\rangle = \langle x, y\rangle\|1 + a'\|^2$ (of the previous lemma), yields the desired result. □

The following lemma is very handy (and of omnipotent use!):

Lemma. *For $x, y \in A$, the followings hold:*

(i) $\overline{\overline{x}} = x$,

(ii) $\langle \overline{x}, \overline{y} \rangle = \langle x, y \rangle$,

(iii) $\langle x, y \rangle = Re(x\overline{y}) = Re(\overline{x}y)$,

(iv) $\overline{xy} = \overline{y}\,\overline{x}$,

(v) $Re(x\overline{y}) = \frac{1}{2}(x\overline{y} + y\overline{x})$,

(vi) $x\overline{x} = \overline{x}x = \|x\|^2 = x_1^2 - x'^2$.

Proof. *(i)* is obvious. For *(ii)*: insert $x = x_1 + x'$ and $y = y_1 + y'$. For *(iii)*: $\langle x, y \rangle = \langle x, 1y \rangle = \langle x\overline{y}, 1 \rangle = \langle Re(x\overline{y}) + Im(x\overline{y}), 1 \rangle = Re(x\overline{y})$; likewise, $\langle x, y \rangle = Re(\overline{x}y)$. For the surprising and important *(iv)*:

$$\langle \overline{xy}, z \rangle = \langle \overline{\overline{xy}}, \overline{z} \rangle = \langle xy, \overline{z} \rangle = \langle x, \overline{z}\,\overline{y} \rangle = \langle zx, \overline{y} \rangle = \langle z, \overline{y}\,\overline{x} \rangle = \langle \overline{y}\,\overline{x}, z \rangle.$$

For *(v)*: $Re(x\overline{y}) = \frac{1}{2}(x\overline{y} + \overline{x\overline{y}}) = \frac{1}{2}(x\overline{y} + y\overline{x})$. As a consequence we have $x\overline{y} = -y\overline{x}$ and $\overline{x}y = -\overline{y}x$ for $\langle x, y \rangle = 0$.
For *(vi)*: $\|x\|^2 = \langle x, x \rangle = Re(x\overline{x}) = \frac{1}{2}(x\overline{x} + x\overline{x}) = x\overline{x} = (x_1 + x')(x_1 - x') = x_1^2 - x'^2$. □

We will need also the following:

Lemma. *For $x, y, z, w \in A$ the followings hold:*

(i) $\langle xw, yz \rangle + \langle xz, yw \rangle = 2\langle x, y \rangle \langle w, z \rangle$,

(ii) $x(\overline{y}w) + y(\overline{x}w) = 2\langle x, y \rangle w$,

(iii) $(w\overline{y})x + (w\overline{x})y = 2\langle x, y \rangle w$.

Proof. *(i)*: Let us take the equality $\langle xw, yw \rangle = \langle x, y \rangle \|w\|^2$, and put $w + z$ for w; then,

$$\begin{aligned}
\langle x(w+z), y(w+z) \rangle &= \langle x, y \rangle \langle w+z, w+z \rangle \\
\langle xw + xz, yw + yz \rangle &= \langle x, y \rangle (\|w\|^2 + \|z\|^2 + 2\langle w, z \rangle) \\
\langle xw, yw \rangle + \langle xw, yz \rangle + \langle xz, yw \rangle + \langle xz, yz \rangle &= \langle x, y \rangle (\|w\|^2 + \|z\|^2 + 2\langle w, z \rangle) \\
\langle x, y \rangle \|w\|^2 + \langle xw, yz \rangle + \langle xz, yw \rangle + \langle x, y \rangle \|z\|^2 &= \langle x, y \rangle (\|w\|^2 + \|z\|^2 + 2\langle w, z \rangle) \\
\langle xw, yz \rangle + \langle xz, yw \rangle &= 2\langle x, y \rangle \langle w, z \rangle.
\end{aligned}$$

(ii): Using the equality in *(i)*,

$$\begin{aligned}
\langle xw, yz \rangle + \langle xz, yw \rangle &= 2\langle x, y \rangle \langle w, z \rangle \\
\langle \overline{y}(xw), z \rangle + \langle z, \overline{x}(yw) \rangle &= \langle 2\langle x, y \rangle w, z \rangle \\
\overline{y}(xw) + \overline{x}(yw) &= 2\langle x, y \rangle w.
\end{aligned}$$

Putting $\overline{x}$ for x and $\overline{y}$ for y we find *(ii)* and *(iii)* is similar. □

As a consequence we get $x(\overline{y}w) = -y(\overline{x}w)$ and $(w\overline{y})x = -(w\overline{x})y$ for $\langle x, y \rangle = 0$.
A normed algebra need not be associative as the example of octonions shows. But a weaker form associativity, the alternativity necessarily holds:

Lemma. *A normed real algebra A is alternative, i.e. the associativity $(xy)z = x(yz)$ holds if any two of x, y and z coincide.*

Proof. We first remark that if any one of x, y or z is real, then $(xy)z = x(yz)$ obviously holds. Now let us assume $x = y$. Inserting $x = x_1 + x'$ we see that it is enough to show $(x'x')z = x'(x'z)$, or what the same is, $(\overline{x'}x')z = \overline{x'}(x'z)$. It suffices to show that $\langle(\overline{x'}x')z - \overline{x'}(x'z), w\rangle = \langle\|x'\|^2 z - \overline{x'}(x'z), w\rangle = 0$ for any $w \in A$:

$$\begin{aligned}\langle\|x'\|^2 z - \overline{x'}(x'z), w\rangle &= \langle\|x'\|^2 z, w\rangle - \langle\overline{x'}(x'z), w\rangle \\ &= \|x'\|^2\langle z, w\rangle - \langle x'z, x'w\rangle \\ &= \|x'\|^2\langle z, w\rangle - \|x'\|^2\langle z, w\rangle = 0.\end{aligned}$$

The other cases ($x = z$ or $y = z$) are similar. Note that associativity holds even if any two of x, y and z coincide only up to conjugation. □

We already know that a normed algebra is a division algebra. But with the meanwhile developed apparatus we can do the divisions easily:

Lemma. *Let A be a normed real algebra (with unit). Then the equations $ax = b$ and $ya = b$ (with $a \neq 0$) have the following unique solutions:*

$$x = \frac{1}{\|a\|^2}\overline{a}b \quad \textit{and} \quad y = \frac{1}{\|a\|^2}b\overline{a}.$$

Furthermore, any non-zero $a \in A$ has a unique right and left inverse $a^{-1} = \frac{1}{\|a\|^2}\overline{a}$.

Proof. If $ax = b$, then $\overline{a}(ax) = \overline{a}b$ and by the above lemma $(\overline{a}a)x = \overline{a}b$, hence $\|a\|^2 x = \overline{a}b$ and $x = \frac{1}{\|a\|^2}\overline{a}b$ (and this is indeed a solution, thus the unique one). Similarly for $ya = b$. To find the right and left inverses of $a \neq 0$ take $b = 1$. □

The following lemma shows that the (Cayley-)Dickson product we met in the foregoing section is immanent in the structure of a normed algebra:

Lemma. *Suppose that B is a subalgebra (with $1 \in B$) of the normed real algebra A and $\varepsilon \in B^{\perp}$ with $\|\varepsilon\| = 1$. Then $B\varepsilon$ is orthogonal to B and*

$$(a + b\varepsilon)(c + d\varepsilon) = (ac - \overline{d}b) + (da + b\overline{c})\varepsilon$$

for all $a, b, c, d \in B$.

Proof. Since $1 \in B$, $x \in B$ if and only if $\overline{x} \in B$. $\langle a, b\varepsilon\rangle = \langle\overline{b}a, \varepsilon\rangle = 0$ since $\overline{b}a \in B$ for $a, b \in B$. This shows that $B \perp B\varepsilon$.

Since $\langle\varepsilon, 1\rangle = 0$, $\varepsilon \in Im(A)$, so that $\|\varepsilon\|^2 = \varepsilon\overline{\varepsilon} = -\varepsilon^2$ and thus $\varepsilon^2 = -1$. We examine the right-hand side of the equation

$$(a + b\varepsilon)(c + d\varepsilon) = ac + (b\varepsilon)(d\varepsilon) + a(d\varepsilon) + (b\varepsilon)c,$$

and show that $(b\varepsilon)(d\varepsilon) = -\overline{d}b$, $a(d\varepsilon) = (da)\varepsilon$ and $(b\varepsilon)c = (b\overline{c})\varepsilon$. We use in the following computations the properties $x(\overline{y}w) + y(\overline{x}w) = 0$ and $(w\overline{y})x + (w\overline{x})y = 0$ for $\langle x, y\rangle = 0$ proven above:

$$\begin{aligned}(b\varepsilon)(d\varepsilon) &= -\overline{d}((\overline{b\varepsilon})\varepsilon) = -\overline{d}((\overline{\varepsilon}\,\overline{b})\varepsilon) = \overline{d}((\varepsilon\overline{b})\varepsilon) = -\overline{d}((\varepsilon\overline{\varepsilon})b) = -\overline{d}b, \\ a(d\varepsilon) &= a(-\overline{\varepsilon}\,\overline{d}) = a(\varepsilon\overline{d}) = -\overline{\varepsilon}(\overline{a}\,\overline{d}) = \varepsilon(\overline{a}\,\overline{d}) = -(\overline{(\overline{a}\,\overline{d})})\overline{\varepsilon} = (da)\varepsilon, \\ (b\varepsilon)c &= -(b\overline{c})\overline{\varepsilon} = (b\overline{c})\varepsilon.\end{aligned}$$

□

We are now ready to prove the Hurwitz Theorem:

Proof of the Hurwitz Theorem. Let A be a normed algebra. Let $B_1 = Re(A) = \mathbb{R}$. If $B_1 = A$, we are done. If not, choose $\varepsilon_1 \in B_1^{\perp}$ with $\|\varepsilon_1\| = 1$ and let $B_2 = B_1 + B_1\varepsilon_1$, which is isomorphic to $\mathbb{C}$. If $B_2 = A$, we are done. If not, choose $\varepsilon_2 \in B_2^{\perp}$ with $\|\varepsilon_2\| = 1$ and $B_3 = B_2 + B_2\varepsilon_2$, which is isomorphic to $\mathbb{H}$. If $B_3 = A$, we are done. If not, choose $\varepsilon_3 \in B_3^{\perp}$ with $\|\varepsilon_3\| = 1$ and $B_4 = B_3 + B_3\varepsilon_3$, which is isomorphic to $\mathbb{O}$. We claim $B_4 = A$. If not, choose $\varepsilon_4 \in B_4^{\perp}$ with $\|\varepsilon_4\| = 1$ and let $B_5 = B_4 + B_4\varepsilon_4$. $B_5 \cong \mathbb{O} \oplus \mathbb{O}$, which is not alternative (a small exercise; you will hit probably by the first search upon a non-alternative pair!) and thus it could not be normed. □

5. A Short Look at the Clifford Algebras

You will recall that we were close to discover the Clifford algebras at the stage of looking around after the discovery of the quaternions as happened also historically. It seems that Clifford's work remained fragmentary after his untimely death in 1879 and these algebras were rediscovered independently by Lipschitz shortly thereafter.
The octonions are not a Clifford algebra since the latter are associative but the octonions are not. Nevertheless, there are intimate relationships between them and we give a short look at them.

Let V be a (real) vector space with a non-degenerate quadratic form $q : V \to \mathbb{R}$. (i.e. $q(v) = b(v, v)$ for a non-degenerate symmetric, bilinear map $b : V \times V \to \mathbb{R}$.) Clifford algebras can be characterized by a universal property:

The Clifford algebra on the vector space V (with the quadratic form q), denoted by $Cl(V, q)$, is an associative algebra with unit, together with a linear injection $i : V \to Cl(V, q)$ satisfying the property $i(v)^2 = -q(v) \cdot 1$, such that for any associative algebra A with unit and any linear map $f : V \to A$ satisfying $f(v)^2 = -q(v) \cdot 1$, there exists a *unique* algebra homomorphism $\tilde{f} : Cl(V, q) \to A$ such that the following diagram commutes:

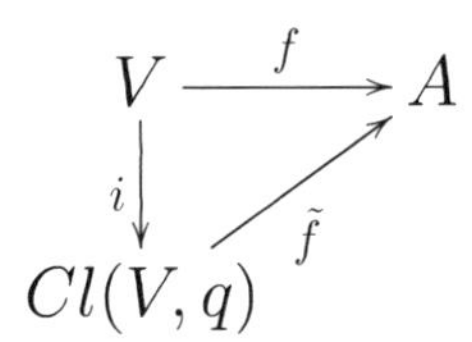

$$(\tilde{f}i = f, \quad \tilde{f} \ extends \ f)$$

$Cl(V, q)$ is uniquely determined by this property, i.e. if B is another associative algebra with unit, together with an injection $j : V \to B$ with $j(v)^2 = -q(v).1$ and the above universal property is satisfied also for B, then there exists an algebra isomorphism $h : Cl(V, q) \to B$ making the following diagram commutative:

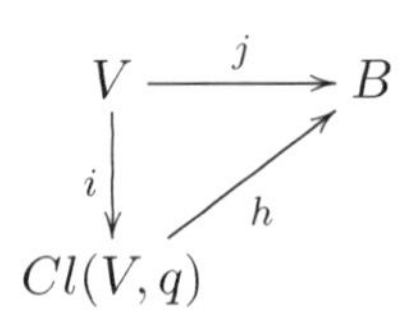

(The algebras are not only isomorphic, but there exists an isomorphism respecting the embeddings.)

Journal of Physics: Conference Series **2191** (2022) 012006 doi:10.1088/1742-6596/2191/1/012006

To see this, use i and j as test maps. By universality of both $Cl(V, q)$ and B we get the algebra homomorphisms $\tilde{i}$ and $\tilde{j}$:

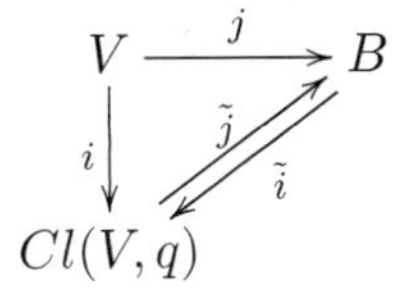

Note that $\tilde{i}$ and $\tilde{j}$ are inverse to each other: Use i as test map ($f = i$):

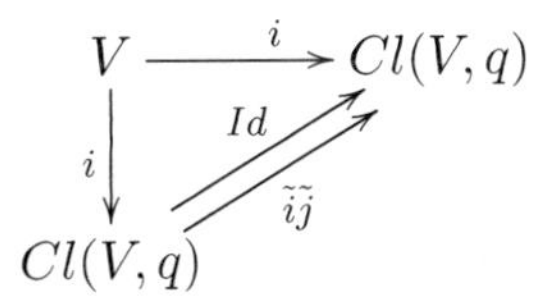

Both Id and $\tilde{i}\tilde{j}$ extend i:

$$(\tilde{i}\tilde{j})i = \tilde{i}(\tilde{j}i) = \tilde{i}j = i.$$

By the uniqueness of the extension we get $\tilde{i}\tilde{j} = Id_{Cl(V,q)}$. Similarly we get $\tilde{j}\tilde{i} = Id$.

Universal properties show that if a certain object exists, then it is unique up to isomorphism. But does the object exist? There are various ways to show this. We will give an intuitive, explicit construction.

Let $V = \mathbb{R}^{r,s}$ and $q(x) = x_1^2 + x_2^2 + \cdots + x_r^2 - x_{r+1}^2 - x_{r+2}^2 - \cdots - x_{r+s}^2$, with $r + s = n$. (The general case is equivalent to this. Choosing an appropriate base of $\mathbb{R}^{r,s}$, a non-degenerate quadratic form can be expressed as such.)

We will denote $Cl(\mathbb{R}^n, q)$ by $Cl_{r,s}$, $Cl_{n,0}$ by Cl_n and $Cl_{0,n}$ by $Cl_n^{'}$. To construct an algebra, it is enough to give a vector space, choose a basis and define the products of the basis elements in terms of these basis elements. One can then extend this product distributively. The emerging algebra is associative if and only if the product is associative on basis elements. (The same is true for commutativity and for the existence of a unit.)

In this spirit, to define $Cl_{r,s}$ we first consider the real vector space generated by the (abstract) basis elements

$$\begin{aligned}
&1, e_1, e_2, \ldots, e_n, \\
&e_1e_2, e_1e_3, \ldots, e_1e_n, e_2e_3, \ldots, e_2e_n, \ldots, e_{n-1}e_n, \\
&e_1e_2e_3, e_1e_2e_4, \ldots, e_{n-2}e_{n-1}e_n, \\
&\vdots \\
&e_1e_2e_3 \cdots e_n.
\end{aligned}$$

In short, we take the vector space

$$\langle e_{i_1}e_{i_2} \cdots e_{i_k} \mid i_1 < i_2 < \cdots < i_k \rangle_{\mathbb{R}},$$

where the element for the empty index set is taken as 1. We now define the product:

$$\begin{aligned}
&1 \text{ acts as unit}, \\
&e_i^2 = -1 \text{ for } 1 \le i \le r,\ e_i^2 = 1 \text{ for } r+1 \le i \le r+s = n, \\
&e_i \cdot e_j = e_ie_j \text{ for } i < j \text{ and } e_i \cdot e_j = -e_je_i \text{ for } i > j.
\end{aligned}$$

For any two basis elements $e_{i_1}e_{i_2} \cdots e_{i_k}$ and $e_{j_1}e_{j_2} \cdots e_{j_l}$, we write

$$e_{i_1}e_{i_2} \cdots e_{i_k}e_{j_1}e_{j_2} \cdots e_{j_l}$$

and cancel and rearrange according to the above rules, which results in an associative multiplication.

The injection $\mathbb{R}^n \to Cl_{r,s}$ is given by $e_i \mapsto e_i$. Image of $x = \sum x_i e_i$ satisfies $x^2 = -q(x).1$. We must verify the universal property. Let $f : \mathbb{R}^n \to A$ be a test map with $f(x)^2 = -q(x).1$. We must extend f to $\tilde{f} : Cl_{r,s} \to A$. First note that

$$\begin{aligned} f(x+y)^2 &= -q(x+y) \\ (f(x)+f(y))^2 &= -b(x+y, x+y) \\ f(x)^2 + f(y)^2 + f(x)f(y) + f(y)f(x) &= -q(x) - q(y) - 2b(x,y) \\ f(x)f(y) + f(y)f(x) &= -2b(x,y). \end{aligned}$$

Thus for $e_i, e_j \in \mathbb{R}^n$ $(i \neq j)$ it holds $f(e_i)f(e_j) = -f(e_j)f(e_i)$. Now we define $\tilde{f}$ on the linear generators and extend it linearly:

$$\tilde{f}(1) = 1, \quad \tilde{f}(e_i) = f(e_i), \quad \tilde{f}(e_{i_1} e_{i_2} \cdots e_{i_k}) = f(e_{i_1}) f(e_{i_2}) \cdots f(e_{i_k}).$$

It is easy to see that this map is multiplicatively homomorphic:

$$\begin{aligned} \tilde{f}(e_i e_j) =& f(e_i)f(e_j) = \tilde{f}(e_i)\tilde{f}(e_j) \text{ for } i<j, \\ \tilde{f}(e_j e_i) =& \tilde{f}(-e_i e_j) = -\tilde{f}(e_i e_j) \\ =& -f(e_i)f(e_j) = f(e_j)f(e_i) = \tilde{f}(e_j)\tilde{f}(e_i) \text{ for } i<j, \\ \tilde{f}(e_{i_1} e_{i_2} \cdots e_{i_k} \cdot e_{j_1} e_{j_2} \cdots e_{j_l}) =& \tilde{f}(\pm e_{h_1} e_{h_2} \cdots e_{h_m}) \\ =& \pm f(e_{h_1}) f(e_{h_2}) \cdots f(e_{h_m}) \\ =& f(e_{i_1}) \cdots f(e_{i_k}) \cdot f(e_{j_1}) \cdots f(e_{i_l}) \\ =& \tilde{f}(e_{i_1} e_{i_2} \cdots e_{i_k}) \cdot \tilde{f}(e_{j_1} e_{j_2} \cdots e_{j_l}). \end{aligned}$$

We have now the Clifford algebra $Cl_{r,s}$ at our disposal.

It is fun to see that we have $Cl_1 = \langle 1, e_1 \rangle_{\mathbb{R}}$ with $e_1^2 = -1$ so that $Cl_1 \cong \mathbb{C}$; and $Cl_2 = \langle 1, e_1, e_2, e_1 e_2 \rangle_{\mathbb{R}}$ with $e_1^2 = e_2^2 = -1, e_1 e_2 = -e_2 e_1$, so that $Cl_2 \cong \mathbb{H}$:

$$\begin{aligned} (e_1 e_2)^2 &= e_1 e_2 e_1 e_2 = -e_1 e_1 e_2 e_2 = -1, \\ e_1(e_1 e_2) &= -(e_1 e_2) e_1, \quad e_2(e_1 e_2) = -(e_1 e_2) e_2. \end{aligned}$$

so that the map

$$e_1 \mapsto i, \quad e_2 \mapsto j, \quad e_1 e_2 \mapsto k$$

gives an algebra isomorphism.

What about Cl_3, or Cl_1', Cl_2' or $Cl_{1,1}$? These are good exercises and we will possibly need some of them below.

For example, $Cl_1{}' = \langle 1, e_1 \rangle_{\mathbb{R}}$ with $e_1^2 = 1$ is isomorphic to $\mathbb{R} \oplus \mathbb{R}$ via $e_1 \mapsto (1, -1)$ and $Cl_2{}' = \langle 1, e_1, e_2, e_1 e_2 \rangle_{\mathbb{R}}$ with $e_1^2 = e_2^2 = 1$ is isomorphic to $\mathbb{R}(2)$ (the algebra of 2×2 matrices) via

$$e_1 \mapsto \begin{pmatrix} 0 & 1 \\ 1 & 0 \end{pmatrix}, \quad e_2 \mapsto \begin{pmatrix} 1 & 0 \\ 0 & -1 \end{pmatrix}.$$

Remark. The universality property enables us to construct an algebra homomorphism $Cl_{r,s} \to A$ by way of giving the images of $e_1, e_2, \ldots, e_n \in \mathbb{R}^n \subset Cl_{r,s}$ satisfying

$$\begin{aligned} f(e_i)^2 &= -1 \quad && \textit{for } \ 1 \leq i \leq r, \\ f(e_i)^2 &= 1 && \textit{for } \ r+1 \leq i \leq r+s = n, \\ f(e_i)f(e_j) + f(e_j)f(e_i) &= 0 && \textit{for } \ i \neq j, \end{aligned}$$

because, in such a case we can first extend f linearly to $\mathbb{R}^n$ and since

$$f(x)^2 = f\left(\sum x_i e_i\right)^2 = \left(\sum x_i f(e_i)\right)^2 = -q(x) ,$$

we can then extend f to $Cl_{r,s}$ by the universal property. This is a very useful device in Clifford algebra theory.

Mapping e_i to $-e_i$ gives rise to an involution $\alpha : Cl_{r,s} \longrightarrow Cl_{r,s}$ yielding a decomposition into *even* and *odd* parts, $Cl_{r,s} = Cl^0_{r,s} \oplus Cl^1_{r,s}$ with $\alpha(u) = u$ for $u \in Cl^0_{r,s}$ and $\alpha(u) = -u$ for $u \in Cl^1_{r,s}$. Note that $Cl^0_{r,s}$ is a subalgebra of $Cl_{r,s}$.

Proposition. *There exists an algebra isomorphism* $Cl_{r,s} \cong Cl^0_{r+1,s}$. *In particular,* $Cl_n \cong Cl^0_{n+1}$.

(For a proof of the case we need, $Cl_n \cong Cl^0_{n+1}$, consider $e_i \mapsto e_0 e_i$ for $\mathbb{R}^n = \langle e_1, e_2, \ldots, e_n \rangle$ and $\mathbb{R}^{n+1} = \langle e_0, e_1, e_2, \ldots, e_n \rangle$.)

The following reduction properties will enable us to compute all Clifford algebras.

Theorem (Reduction Theorem). *There exist isomorphisms*

(i) $Cl_{n,0} \otimes Cl_{0,2} \cong Cl_{0,n+2}$,

(ii) $Cl_{0,n} \otimes Cl_{2,0} \cong Cl_{n+2,0}$,

(iii) $Cl_{r,s} \otimes Cl_{1,1} \cong Cl_{r+1,s+1}$.

As an example, we give a proof for the first case in detail:

We will define an isomorphism $Cl_{0,n+2} \to Cl_{n,0} \otimes Cl_{0,2}$ which will come from a map $\mathbb{R}^{n+2} \to Cl_{n,0} \otimes Cl_{0,2}$. Choose the basis $e_1, e_2, \ldots, e_{n+2}$ for $\mathbb{R}^{n+2}$ with $q(e_i) = -1$ and denote the standard generators of $Cl_{n,0}$ by $e'_1, e'_2, \ldots, e'_n$, those of $Cl_{0,2}$ by e''_1, e''_2. Let f be given as

$$\begin{aligned} f : \ \mathbb{R}^{n+2} &\longrightarrow Cl_{n,0} \otimes Cl_{0,2} \\ e_i &\mapsto e'_i \otimes e''_1 e''_2 \ \textit{for}\ 1 \le i \le n \\ e_{n+1} &\mapsto 1 \otimes e''_1 \\ e_{n+2} &\mapsto 1 \otimes e''_2 . \end{aligned}$$

Let us check the necessary relations:

$$\begin{aligned} f(e_i)^2 &= (e'_i \otimes e''_1 e''_2) \cdot (e'_i \otimes e''_1 e''_2) = e'^2_i \otimes e''_1 e''_2 e''_1 e''_2 = (-1) \otimes (-1) = 1 \otimes 1, \\ f(e^2_{n+1}) &= (1 \otimes e''_1)(1 \otimes e''_1) = 1 \otimes e''^2_1 = 1 \otimes 1, \\ f(e^2_{n+2}) &= (1 \otimes e''_2)(1 \otimes e''_2) = 1 \otimes e''^2_2 = 1 \otimes 1. \end{aligned}$$

$$\begin{aligned} f(e_i)f(e_j) + f(e_j)f(e_i) &= (e'_i \otimes e''_1 e''_2)(e'_j \otimes e''_1 e''_2) + (e'_j \otimes e''_1 e''_2)(e'_i \otimes e''_1 e''_2) \\ &= e'_i e'_j \otimes (-1) + e'_j e'_i \otimes (-1) \\ &= 0 \ \ \textit{for}\ \ 1 \le i, j \le n, \ \ i \ne j. \end{aligned}$$

$$\begin{aligned} f(e_i)f(e_{n+1}) + f(e_{n+1})f(e_i) &= (e'_i \otimes e''_1 e''_2)(1 \otimes e''_1) + (1 \otimes e''_1)(e'_i \otimes e''_1 e''_2) \\ &= e'_i \otimes e''_1 e''_2 e''_1 + e'_i \otimes e''_1 e''_1 e''_2 \\ &= e'_i \otimes (-e''_2) + e'_i \otimes e''_2 = 0. \end{aligned}$$

$$\begin{aligned} f(e_i)f(e_{n+2}) + f(e_{n+2})f(e_i) &= (e'_i \otimes e''_1 e''_2)(1 \otimes e''_2) + (1 \otimes e''_2)(e'_i \otimes e''_1 e''_2) \\ &= e'_i \otimes e''_1 e''_2 e''_2 + e'_i \otimes e''_2 e''_1 e''_2 \\ &= e'_i \otimes e''_1 + e'_i \otimes (-e''_1) = 0. \end{aligned}$$

Journal of Physics: Conference Series **2191** (2022) 012006 doi:10.1088/1742-6596/2191/1/012006

$$\begin{aligned} f(e_{n+1})f(e_{n+2}) + f(e_{n+2})f(e_{n+1}) &= (1\otimes e_1'')(1\otimes e_2'') + (1\otimes e_2'')(1\otimes e_1'') \\ &= 1\otimes e_1''e_2'' + 1\otimes e_2''e_1'' = 0. \end{aligned}$$

Thus we get an algebra homomorphism $Cl_{0,n+2} \longrightarrow Cl_{n,0}\otimes Cl_{0,2}$.

The terms $1\otimes e_1''$, $1\otimes e_2''$ and $(e_i'\otimes e_1''e_2'')(1\otimes e_1'')(1\otimes e_2'') = e_i'\otimes(-1)$ appear in the image and they generate $Cl_{n,0}\otimes Cl_{0,2}$. The map is then surjective and by dimensional reasons an isomorphism.

With the help of the reduction theorem we can now prove the startling periodicity theorem:

Theorem (Periodicity Theorem). *We have*

(i) $Cl_{n+8} \cong Cl_n \otimes Cl_8 \cong Cl_n \otimes \mathbb{R}(16)$.

(ii) $Cl'_{n+8} \cong Cl'_n \otimes Cl'_8 \cong Cl'_n \otimes \mathbb{R}(16)$.

Proof. Let us show the first case (the second being similar). Using repeatedly the reduction theorem we get

$$\begin{aligned} Cl_{n+8} &\cong Cl'_{n+6}\otimes Cl_2 \\ &\cong Cl_{n+4}\otimes Cl'_2 \otimes Cl_2 \\ &\cong Cl'_{n+2}\otimes Cl_2\otimes Cl'_2\otimes Cl_2 \\ &\cong Cl_n \otimes \overbrace{Cl'_2\otimes Cl_2\otimes Cl'_2\otimes Cl_2}^{Cl_8} \\ &\cong Cl_n\otimes \mathbb{R}(2)\otimes \mathbb{H}\otimes \mathbb{R}(2)\otimes\mathbb{H} \\ &\cong Cl_n\otimes(\mathbb{H}\otimes\mathbb{H})\otimes\mathbb{R}(2)\otimes\mathbb{R}(2) \\ &\cong Cl_n\otimes\mathbb{R}(4)\otimes\mathbb{R}(4) \\ &\cong Cl_n\otimes\mathbb{R}(16). \end{aligned}$$

We owe to the reader a justification of the final steps: $\mathbb{R}(n)\otimes\mathbb{R}(m)\cong\mathbb{R}(nm)$ (where $\mathbb{R}(n)$ is the algebra of real $n\times n$ matrices) and $\mathbb{H}\otimes_{\mathbb{R}}\mathbb{H}\cong\mathbb{R}(4)$. For the first take

$$(a_{ij})\otimes B \mapsto \begin{pmatrix} a_{11}B & a_{12}B & \cdots & a_{1n}B \\ \vdots & \ddots & & \vdots \\ \vdots & & \ddots & \vdots \\ a_{n1}B & a_{n2}B & \cdots & a_{nn}B \end{pmatrix}$$

and for the second consider the $\mathbb{R}$-bilinear map

$$\begin{aligned} \psi:\ \mathbb{H}\times\mathbb{H} &\longrightarrow Hom_{\mathbb{R}}(\mathbb{H},\mathbb{H})\cong\mathbb{R}(4) \\ (q_1,q_2) &\longmapsto \psi(q_1,q_2)(x) = q_1 x\bar{q}_2, \end{aligned}$$

which gives

$$\begin{aligned} \tilde{\psi}:\ \mathbb{H}\otimes_{\mathbb{R}}\mathbb{H} &\longrightarrow Hom_{\mathbb{R}}(\mathbb{H},\mathbb{H}) \\ (q_1,q_2) &\longmapsto \tilde{\psi}(q_1\otimes q_2)(x) = q_1 x\bar{q}_2, \end{aligned}$$

which can be checked to be an algebra isomorphism. □

The Cl_7, which we will talk about more below, can be computed to be $Cl_7 \cong {Cl_1}'\otimes Cl_2\otimes {Cl_2}'\otimes Cl_2 = (\mathbb{R}\oplus\mathbb{R})\otimes\mathbb{H}\otimes\mathbb{R}(2)\otimes\mathbb{H} = (\mathbb{R}\oplus\mathbb{R})\otimes\mathbb{R}(8) = \mathbb{R}(8)\oplus\mathbb{R}(8)$.

Journal of Physics: Conference Series **2191** (2022) 012006 doi:10.1088/1742-6596/2191/1/012006

6. Spin Groups

We now want to give the definition of spin groups and a minimum of basic facts about them. We consider only the positive-definite case. Spin groups live in Clifford algebras and can be defined as

$$Spin(n) = \{v_1 v_2 \cdots v_{2k} \mid v_i \in \mathbb{R}^n, \|v_i\| = 1\} \subset Cl_n.$$

There is however another more structured and illuminating definition which we want to consider and reduce to the former.

Let us define

$$\Gamma = \{u \in Cl_n^* \mid \alpha(u) x u^{-1} \in \mathbb{R}^n \quad \text{for all} \quad x \in \mathbb{R}^n\}$$

and $\Gamma^0 = \Gamma \cap Cl_n^0$, where Cl_n^* denotes the group of invertible elements of Cl_n. α is the grading automorphism defined in the previous section.

Lemma. *Γ is a subgroup of Cl_n^*.*

Proof. For $u_1, u_2 \in \Gamma$, $u_1 \cdot u_2 \in \Gamma$:

$$\alpha(u_1 u_2) x (u_1 u_2)^{-1} = \alpha(u_1)\alpha(u_2) x u_2^{-1} u_1^{-1} = \alpha(u_1)(\alpha(u_2) x u_2^{-1}) u_1^{-1}$$

For $u \in \Gamma$, $u^{-1} \in \Gamma$: Define

$$\begin{aligned} \widetilde{Ad}(u): \quad \mathbb{R}^n &\longrightarrow \mathbb{R}^n \\ x &\longmapsto \alpha(u) x u^{-1}. \end{aligned}$$

This map is one-to-one: $\alpha(u) x u^{-1} = 0$ implies $x = 0$. So, it is onto. Given $y \in \mathbb{R}^n$, there exists $x \in \mathbb{R}^n$ with $\alpha(u) x u^{-1} = y$, i.e. $x = \alpha(u^{-1}) y u = \alpha(u^{-1}) y (u^{-1})^{-1}$. It means that $u^{-1} \in \Gamma$. □

Now we define the *twisted adjoint representation* (which is a group homomorphism):

$$\begin{aligned} \widetilde{Ad}: \quad \Gamma &\longrightarrow GL(\mathbb{R}^n) \\ u &\longmapsto \widetilde{Ad}(u). \end{aligned}$$

Lemma. $\widetilde{Ad}(u) \in O(n)$.

Proof.

$$\begin{aligned} \langle \widetilde{Ad}(u)(x), \widetilde{Ad}(u)(x) \rangle &= \langle \alpha(u) x u^{-1}, \alpha(u) x u^{-1} \rangle \\ &= -(\alpha(u) x u^{-1})^2 \quad \textit{since} \quad \alpha(u) x u^{-1} \in \mathbb{R}^n \\ &= -(\alpha(u) x u^{-1})(\alpha(u) x u^{-1}) \\ &= \alpha(u) x u^{-1} \alpha(\alpha(u) x u^{-1}) \quad \textit{since} \quad \alpha(v) = -v \textit{ for } v \in \mathbb{R}^n \\ &= \alpha(u) x u^{-1} u \alpha(x) \alpha(u^{-1}) \\ &= \alpha(u) x (-x) \alpha(u^{-1}) \\ &= \alpha(u) \langle x, x \rangle \alpha(u^{-1}) \\ &= \langle x, x \rangle. \end{aligned}$$

□

So we have a map $\widetilde{Ad}: \Gamma \longrightarrow O(n)$. We note two facts:

- $v \in \mathbb{R}^n, v \neq 0$, belongs to Γ:

$$\begin{aligned}\alpha(v)xv^{-1} &= (-v)xv^{-1} = -(vx)v^{-1} = -(-xv - 2\langle v, x\rangle)v^{-1} \\ &= x + 2\langle v, x\rangle v^{-1} = x - 2\frac{\langle v, x\rangle}{\langle v, v\rangle}v \in \mathbb{R}^n.\end{aligned}$$

- $\widetilde{Ad}(v)$ is a reflection across $v^\perp$ in $\mathbb{R}^n$:

$$\widetilde{Ad}(v)(v) = -v \text{ and } \widetilde{Ad}(v)(x) = x \text{ for } \langle v, x\rangle = 0.$$

  We denote this reflection along $v^\perp$ by $Ref(v)$.

We now want to compute the kernel of $\widetilde{Ad}$.

Lemma. *The kernel of* $\widetilde{Ad} : \Gamma \longrightarrow O(n)$ *consists of non-zero scalars.*

Proof. Let $\widetilde{Ad}(u) = Id$, i.e. $\alpha(u)xu^{-1} = x$ for all $x \in V$. Decompose u into even and odd parts: $u = u_0 + u_1$, so that we have $\alpha(u) = u_0 - u_1$. Then, inserting this into $\alpha(u)x = xu$ we get $u_0x = xu_0$ (odd degrees) and $-u_1x = xu_1$ (even degrees). Especially, $u_0e_i = e_iu_0$ and $-u_1e_i = e_iu_1$ for the standard basis $e_1, \ldots, e_n$ of $\mathbb{R}^n$.

Fixing one e_i temporarily, u_0 can be expressed as $u_0 = a_i + e_ib_i$, where the terms a_i and b_i do not contain e_i. Inserting this into $u_0e_i = e_iu_0$ and inspecting the degrees of the terms, one obtains $b_i = 0$ and thus $u_0 = a_i$. This means that u_0 does not contain e_i! As this is true for all i, u_0 must be a scalar.

Similarly, let $u_1 = c_i + e_id_i$, where e_i is fixed and c_i and d_i do not contain e_i. Inserting this into $-u_1e_i = e_iu_1$, one obtains $d_i = 0$. This means that u_1 does not contain any e_i and thus must be zero as an odd-degree element. Consequently, u is a scalar. As $u \in \Gamma$, it must be a non-zero scalar. □

Lemma. $\widetilde{Ad} : \Gamma \longrightarrow O(n)$ *is onto.*

Proof. Any element $f \in O(n)$ is, by Cartan-Dieudonné theorem, a product of reflections, say $f = Ref(v_1)Ref(v_2)\cdots Ref(v_k)$. As $Ref(v) = \widetilde{Ad}(v)$, we get

$$f = \widetilde{Ad}(v_1)\widetilde{Ad}(v_2)\cdots\widetilde{Ad}(v_k) = \widetilde{Ad}(v_1v_2\cdots v_k).$$ □

We are now in a position to clarify the structure of Γ. Let $u \in \Gamma$. We have $\widetilde{Ad}(u) = \widetilde{Ad}(v_1v_2\cdots v_k)$ for suitable $v_i \in \mathbb{R}^n$. Then

$$\begin{aligned}\widetilde{Ad}(u^{-1}v_1v_2\cdots v_k) &= Id \\ u^{-1}v_1v_2\cdots v_k &= \text{non-zero scalar!} \\ u &= \lambda v_1v_2\cdots v_k, \quad v_i \in \mathbb{R}^n,\ v_i \neq 0,\ \lambda \in \mathbb{R},\ \lambda \neq 0 \\ u &= \underbrace{\lambda\|v_1\|\|v_2\|\cdots\|v_k\|}_{\mu}\frac{v_1}{\|v_1\|}\frac{v_2}{\|v_2\|}\cdots\frac{v_k}{\|v_k\|}, \quad v_i \neq 0, \lambda \neq 0 \\ u &= \mu w_1w_2\cdots w_k, \quad w_i \in \mathbb{R}^n,\ \|w_i\| = 1,\ \mu \neq 0.\end{aligned}$$

So we can write

$$\Gamma = \{\lambda v_1v_2\cdots v_k \mid v_i \in \mathbb{R}^n, \|v_i\| = 1, \lambda \in \mathbb{R}^*\}$$

and

$$\Gamma^0 = \Gamma \cap Cl_n^0 = \{\lambda v_1v_2\cdots v_{2k} \mid v_i \in \mathbb{R}^n, \|v_i\| = 1, \lambda \in \mathbb{R}^*\}.$$

Journal of Physics: Conference Series **2191** (2022) 012006 doi:10.1088/1742-6596/2191/1/012006

As a consequence we see that Γ^0 is mapped by $\widetilde{Ad}$ onto $SO(n)$ with kernel consisting of the non-zero scalars. We want to get rid of the factor λ in an invariant way.

We define the *norm* function

$$\begin{aligned} N: \ Cl_n &\longrightarrow Cl_n \\ u &\longmapsto u.\alpha(u^t) \end{aligned}$$

where t is the transpose-antiinvolution of Cl_n, i.e. $(e_{i_1}e_{i_2}\cdots e_{i_k})^t = e_{i_k}e_{i_{k-1}}\cdots e_{i_1}$.

Lemma. *$N(u) \in \mathbb{R}^*$ for $u \in \Gamma$.*

Proof.

$$\begin{aligned} u &= \lambda v_1 v_2 \cdots v_k \quad with\ v_i \in \mathbb{R}^n, \|v_i\| = 1, \lambda \neq 0 \\ u^t &= \lambda v_k v_{k-1} \cdots v_1 \\ \alpha(u^t) &= \lambda(-v_k)(-v_{k-1})\cdots(-v_1) \\ u \cdot \alpha(u^t) &= \lambda v_1 v_2 \cdots v_k . \lambda(-v_k)(-v_{k-1})\cdots(-v_1) = \lambda^2. \end{aligned}$$

□

Now consider the set

$$\{u \in Cl_n^* \mid \alpha(u)xu^{-1} \in \mathbb{R}^n \ for\ x \in \mathbb{R}^n,\ N(u) = 1\} \cap Cl_n^0.$$

This set coincides with the $Spin(n) = \{v_1 v_2 \cdots v_{2k} \mid v_i \in \mathbb{R}^n, \|v_i\| = 1\} \subset Cl_n$, given as an ad hoc definition in the beginning of this section. Because, in the first place, u must have the form $u = \lambda v_1 v_2 \cdots v_k$. The condition $N(u) = 1$ necessitates $\lambda = \pm 1$ (and $\lambda = -1$ can be put in v_1). As $u \in Cl_n^0$, k must be even. We have thus arrived at a better understanding of $Spin(n)$. Since $\alpha(u) = u$ and $N(u) = uu^t$ for $u \in Cl^0$, we can express $Spin(n)$ also as the set

$$\{u \in Cl_n^0 \mid uxu^{-1} \in \mathbb{R}^n \ \ for\ \ x \in \mathbb{R}^n,\ uu^t = 1\}$$

(the condition $uu^t = 1$ implies that u is invertible). The restriction of $\widetilde{Ad}$ thus gives a surjective homomorphism $Spin(n) \longrightarrow SO(n)$ with kernel $\{\pm 1\}$.

We remark that, as a topological group, $Spin(n)$ is a 2:1 covering space of $SO(n)$ and for $n \geq 3$ it is simply connected and thus the universal covering space of $SO(n)$.

7. A Closer Look at Seven and Eight Dimensions

In this section we want to give an octonionic model for Cl_7 and Cl_8.

To apply the universal property of Clifford algebras, we consider the 8-dimensional real vector space $V = \mathbb{O} \cong \mathbb{R}^8$, the algebra $A = End_{\mathbb{R}}(\mathbb{O} \oplus \mathbb{O})$, and the map

$$f: \mathbb{O} \longrightarrow End_{\mathbb{R}}(\mathbb{O} \oplus \mathbb{O}),\ f(u) = \begin{pmatrix} 0 & R_u \\ -R_{\overline{u}} & 0 \end{pmatrix},$$

R_u being the right octonionic multiplication $R_u : \mathbb{O} \to \mathbb{O},\ R_u(x) = xu$. We have

$$f(u)^2 = \begin{pmatrix} -R_u R_{\overline{u}} & 0 \\ 0 & -R_{\overline{u}} R_u \end{pmatrix}$$

Journal of Physics: Conference Series **2191** (2022) 012006 doi:10.1088/1742-6596/2191/1/012006

and since $R_u R_{\overline{u}}(x) = (x\overline{u})u = x(\overline{u}u) = \|u\|^2 x$ for $x \in \mathbb{O}$ (and $R_{\overline{u}} R_u(x) = \|u\|^2 x$) we get

$$f(u)^2 = \begin{pmatrix} -\|u\|^2 I_8 & 0 \\ 0 & -\|u\|^2 I_8 \end{pmatrix} = -\|u\|^2 I_{16}$$

(where $I_8 = Id_{\mathbb{O}}$ and $I_{16} = Id_{\mathbb{O}\oplus\mathbb{O}}$). Thus we obtain by the universal property an algebra homomorphism

$$\tilde{f} : Cl_8 = Cl(\mathbb{O}) \longrightarrow End_{\mathbb{R}}(\mathbb{O} \oplus \mathbb{O}).$$

We already know that Cl_8 is isomorphic to the matrix algebra $\mathbb{R}(16)$ and as such, a simple algebra (i.e. it has no non-trivial two-sided ideals. This is a good exercise.) The kernel of $\tilde{f}$, being a two-sided ideal, must be zero, as it cannot be the whole of Cl_8 because $\tilde{f}$ is non-zero. Hence $\tilde{f}$ is one-to-one and by dimensional reasons, it is an isomorphism.

Let $e_0, e_1, \ldots, e_7$ be the standard basis of $\mathbb{O}$. We should make careful distinction between octonionic multiplication and Clifford multiplication, e.g. e_0 is the unit of octonionic multiplication, but the unit of Clifford multiplication is the external $1 \in \mathbb{R}$. Note that $\mathbb{O} = \langle e_0, e_1, \ldots, e_7 \rangle_{\mathbb{R}}$ and

$$Cl_8 = \langle 1, e_0, e_1, \ldots, e_7, e_0e_1, \ldots, e_6e_7, \ldots, e_0e_1 \ldots e_7 \rangle_{\mathbb{R}}.$$

We have

$$f(e_0) = \tilde{f}(e_0) = \begin{pmatrix} 0 & I_8 \\ -I_8 & 0 \end{pmatrix} \text{ and } \tilde{f}(1) = \begin{pmatrix} I_8 & 0 \\ 0 & I_8 \end{pmatrix} = I_{16}.$$

We know that Cl_7 is isomorphic to the even part Cl_8^0 of Cl_8. What is it under the isomorphism $\tilde{f} : Cl_8 \longrightarrow End_{\mathbb{R}}(\mathbb{O} \oplus \mathbb{O})$?

We first remark that, in any even-dimensional Clifford algebra Cl_n, the even part Cl_n^0 consists exactly of elements commuting with the *volume element* $e_1e_2 \cdots e_n$, in our case, with $e_0e_1 \cdots e_7$, since

$$e_i \cdot (e_0e_1 \cdots e_7) = -(e_0e_1 \cdots e_7) \cdot e_i \text{ and } (e_ie_j) \cdot (e_0e_1 \cdots e_7) = (e_0e_1 \cdots e_7) \cdot (e_ie_j).$$

Now, what is the image of $e_0e_1 \cdots e_7$ under $\tilde{f}$ in $End_{\mathbb{R}}(\mathbb{O} \oplus \mathbb{O})$?

$$\begin{aligned} \tilde{f}(e_0e_1 \cdots e_7) &= \tilde{f}(e_0)\tilde{f}(e_1) \cdots \tilde{f}(e_7) \\ &= \begin{pmatrix} 0 & I_8 \\ -I_8 & 0 \end{pmatrix} \begin{pmatrix} 0 & R_{e_1} \\ -R_{\overline{e}_1} & 0 \end{pmatrix} \cdots \begin{pmatrix} 0 & R_{e_7} \\ -R_{\overline{e}_7} & 0 \end{pmatrix} \\ &= \begin{pmatrix} R_{e_1}R_{e_2} \cdots R_{e_7} & 0 \\ 0 & -R_{e_1}R_{e_2} \cdots R_{e_7} \end{pmatrix} \end{aligned}$$

A straightforward check shows that $R_{e_1}R_{e_2} \cdots R_{e_7} = I_8$! We thus get

$$\tilde{f}(e_0e_1 \cdots e_7) = \begin{pmatrix} I_8 & 0 \\ 0 & -I_8 \end{pmatrix}.$$

Now we can identify Cl_8^0 inside $End_{\mathbb{R}}(\mathbb{O} \oplus \mathbb{O})$: A matrix commuting with this one must be of type $\begin{pmatrix} A & 0 \\ 0 & B \end{pmatrix}$. We can express this fact also as $Cl_8^0 \cong End_{\mathbb{R}}(\mathbb{O}) \oplus End_{\mathbb{R}}(\mathbb{O})$.

Let us recall the identification of Cl_n with Cl_{n+1}^0 and apply it to the present case:

$$\begin{aligned} Cl_7 \longrightarrow Cl_8^0 &\longrightarrow End_{\mathbb{R}}(\mathbb{O}) \oplus End_{\mathbb{R}}(\mathbb{O}) \subset End_{\mathbb{R}}(\mathbb{O} \oplus \mathbb{O}), \\ e_i \longmapsto e_0e_i &\longmapsto \begin{pmatrix} 0 & I \\ -I & 0 \end{pmatrix} \begin{pmatrix} 0 & R_{e_i} \\ R_{e_i} & 0 \end{pmatrix} = \begin{pmatrix} R_{e_i} & 0 \\ 0 & -R_{e_i} \end{pmatrix}, \\ v = \sum \lambda_i e_i \longmapsto \sum \lambda_i e_0 e_i &\longmapsto \begin{pmatrix} R_v & 0 \\ 0 & -R_v \end{pmatrix} \end{aligned}$$

The resulting isomorphism $Cl_7 \cong End_{\mathbb{R}}(\mathbb{O}) \oplus End_{\mathbb{R}}(\mathbb{O})$ can thus be viewed as the extension of the map

$$\begin{aligned} Im(\mathbb{O}) \cong \mathbb{R}^7 &\longrightarrow End_{\mathbb{R}}(\mathbb{O}) \oplus End_{\mathbb{R}}(\mathbb{O}) \\ v &\longmapsto (R_v, -R_v). \end{aligned}$$

(We could define this map from the start, but we wanted to relate the octonionic models of Cl_7 and Cl_8 and wanted to understand how Cl_7 sits in $End_{\mathbb{R}}(\mathbb{O} \oplus \mathbb{O})$.)

We now want to look at $Spin(7) \subset Cl_7^0$ and see how it sits in $End_{\mathbb{R}}(\mathbb{O}) \oplus End_{\mathbb{R}}(\mathbb{O})$. Recall that $Spin(7) = \{v_1 v_2 \cdots v_{2k} \mid v_i \in Im(\mathbb{O}), \|v_i\| = 1\}$ so that we get

$$\begin{aligned} v_1 \cdot v_2 \cdots v_{2k} \longmapsto (R_{v_1}, -R_{v_1})(R_{v_2}, -R_{v_2}) \cdots (R_{v_{2k}}, -R_{v_{2k}}) \\ = (R_{v_1} R_{v_2} \cdots R_{v_{2k}}, R_{v_1} R_{v_2} \cdots R_{v_{2k}}). \end{aligned}$$

Since R_{v_i} are orthogonal (by $\|v_i\| = 1$), we get a diagonal embedding

$$Spin(7) \subset O(8) \oplus O(8) \subset End_{\mathbb{R}}(\mathbb{O}) \oplus End_{\mathbb{R}}(\mathbb{O}).$$

Interestingly, $Spin(7)$ is generated by (R_v, R_v) (for $v \in Im(\mathbb{O}) \cong \mathbb{R}^7$ and $\|v\| = 1$) also, because R_v itself is a product of an even number of right multiplications, $R_v = R_{e_1} R_{e_2} \cdots R_{e_7} R_v$, since, as we remarked above, $R_{e_1} R_{e_2} \cdots R_{e_7} = Id_{\mathbb{O}}$.

Projecting onto the first factor we obtain a very useful result:

Proposition. *$Spin(7)$ is isomorphic to the subgroup of $O(8)$ generated by octonionic right multiplications R_v for $v \in Im(\mathbb{O})$, $\|v\| = 1$.*

8. Special Forms in Seven and Eight Dimensions

A calibrated n-manifold is a Riemannian manifold equipped with a closed,non-degenerate p-form for some $0 \leq p \leq n$.Theory of calibrations in general is due to Harvey and Lawson. However, much earlier in 1966 Edmond Bonan introduced G_2 and $Spin(7)$ holonomy manifolds that are seven and eight dimensional, respectively. He constructed all the parallel forms on these exceptional holonomy manifolds and showed that they are Ricci-flat. The "associative" 3- and "co-associative" 4-forms on $\mathbb{R}^7$ ($= Im(\mathbb{O})$) and the Bonan form on $\mathbb{R}^8$ ($= \mathbb{O}$), which play an eminent role in exceptional G_2 geometry (G_2 being the automorphism group of the octonions) can most succinctly be defined in terms of octonions, indicating that the octonions are their natural home:

$$\begin{aligned} \varphi \in \Lambda^3(Im\mathbb{O})^*, \ \varphi(x, y, z) &= \langle x, yz \rangle, \\ \psi \in \Lambda^4(Im\mathbb{O})^*, \ \psi(x, y, z, w) &= \langle x, y \times z \times w \rangle, \\ \Phi \in \Lambda^4\mathbb{O}^*, \ \Phi(x, y, z, w) &= \langle x, y \times z \times w \rangle \end{aligned}$$

(so that $\psi = \Phi \mid_{Im\mathbb{O}}$), where $y \times z \times w = \frac{1}{2}[y(\overline{z}w) - w(\overline{z}y)]$ is the triple cross product for $y, z, w \in \mathbb{O}$.

This triple cross product might seem strange and indeed it is not equal to $(y \times z) \times w$, where $y \times z$ is defined by $Im(\overline{z}y) = \frac{1}{2}(\overline{z}y - \overline{y}z)$.

For $y, z \in Im(\mathbb{O})$, we have $y \times z = \frac{1}{2}((-z)y - (-y)z) = \frac{1}{2}(yz - zy)$. We know for $y, z \in \mathbb{O}$, $\langle y, z \rangle = Re(y\overline{z}) = \frac{1}{2}(y\overline{z} + z\overline{y})$, so that for $y, z \in Im(\mathbb{O})$ it holds $\langle y, z \rangle = -\frac{1}{2}(yz + zy)$, whence we get $y \times z = yz + \langle y, z \rangle$.

For $y, z \in \mathbb{O}$ and $\langle y, z \rangle = 0$ we have $y \times z = \frac{1}{2}(\overline{z}y - \overline{y}z) = \overline{z}y$ so that it holds $\|y \times z\| = \|y\|\|z\|$. Now consider three pair-wise orthogonal elements $y, z, w \in \mathbb{O}$. Then, $y(\overline{z}w) = -z(\overline{y}w)$ since y

and z orthogonal; $z(\overline{y}w) = -z(\overline{w}y)$ since y and w are orthogonal; $z(\overline{w}y) = -w(\overline{z}y)$ since z and w orthogonal, so that we have $y(\overline{z}w) = -w(\overline{z}y)$, whence we get $y \times z \times w = y(\overline{z}w)$ and hence $\|y \times z \times w\| = \|y\|\|z\|\|w\|$. This might be seen as a rationale for the definition of the triple cross product. (This property can not be expected from the iterated double cross product: e.g. e_1, e_2 and $e_3 \in \mathbb{O}$ are pair-wise orthogonal, but $e_1 \times e_2 = Im(\overline{e_2}e_1) = Im(-e_2e_1) = Im(e_3) = e_3$ and $(e_1 \times e_2) \times e_3 = e_3 \times e_3 = Im(\overline{e}_3e_3) = Im(-e_3^2) = Im(1) = 0$. On the other hand, $e_1 \times e_2 \times e_3 = e_1(\overline{e}_2e_3) = -e_1(e_2e_3) = -e_1e_1 = 1$.)

The double and triple cross products can easily be seen to be alternating and likewise, the multi-linear functions φ, ψ and Φ can be seen to be alternating by showing that they vanish if two of the variables coincide. For many amazing details about these forms we refer to the seminal work of Harvey-Lawson [9] and to the book Harvey [8]; we will give below only a few indications of proofs.

To show, for example, that φ is alternating, consider the cases $x = y$, $x = z$ and $y = z$:

For $x = y$ we have $\langle x, xz\rangle = \|x\|^2\langle 1, z\rangle = 0$ since $z \in Im(\mathbb{O})$. Similarly for $x = z$. For $y = z$ we have $\langle x, y^2\rangle = \langle x, -\overline{y}y\rangle = -\langle x, \|y\|^2\rangle = -\|y\|^2\langle x, 1\rangle = 0$.

We recommend as an exercise to see whether $\psi = *\varphi$ with respect to the standard inner product on $Im(\mathbb{O}) = \mathbb{R}^7$.

$G_2 = Aut(\mathbb{O})$ is a compact, connected, simple Lie group of dimension 14. We can guess the dimension easily, but let us first note that G_2 can be viewed as a subgroup of the orthogonal group $O(Im(\mathbb{O})) = O(7)$: Let $g \in G_2$. $1 \in \mathbb{O}$ can be seen to go to 1 under g, and since $x^2 \in \mathbb{R}(= Re(\mathbb{O}))$ iff $x \in \mathbb{R}$ or $x \in Im(\mathbb{O})$, $g(x)$ must belong to $Im(\mathbb{O})$ for $x \in Im(\mathbb{O})$, so that g respects the conjugation, $\overline{g(x)} = g(\overline{x})$; and consequently, $\|g(x)\|^2 = g(x)\overline{g(x)} = g(x)g(\overline{x}) = g(x\overline{x}) = g(\|x\|^2) = \|x\|^2 g(1) = \|x\|^2$, showing $g \in O(7)$.

Now, to construct an automorphism of $\mathbb{O} = \langle 1, e_1, e_2, \ldots, e_7\rangle_{\mathbb{R}}$, where 1 goes necessarily to 1, we can map e_1 to any element of the unit sphere in $Im(\mathbb{O}) = \langle e_1, e_2, \ldots, e_7\rangle_{\mathbb{R}}$ (giving a 6-dimensional choice) and then map e_2 to any element in this sphere orthogonal to e_1 (an additional five dimensional choice). Now $e_3 = e_1e_2$ has to go to $g(e_1)g(e_2)$. We have a last 3-dimensional choice for e_4 (orthogonal to e_1, e_2 and e_3). The images of e_5, e_6 and e_7 are then determined by the multiplication table and we get 6+5+3=14.

The forms φ and ψ are invariant under G_2:

Let $g \in G_2$. Then, $g\,|_{Im(\mathbb{O})}\colon Im(\mathbb{O}) \longrightarrow Im(\mathbb{O})$ and we have

$$g^*(\varphi)(x, y, z) = \varphi(g(x), g(y), g(z)) = \langle g(x), g(y)g(z)\rangle = \langle g(x), g(yz)\rangle = \langle x, yz\rangle,$$

so that $g^*(\varphi) = \varphi$ and

$$\begin{aligned} g^*(\psi)(x, y, z, w) &= \psi(g(x), g(y), g(z), g(w)) = \langle g(x), g(y) \times g(z) \times g(w)\rangle \\ &= \langle g(x), \frac{1}{2}[g(y)(\overline{g(z)}g(w)) - g(w)(\overline{g(z)}g(y))]\rangle \\ &= \langle g(x), \frac{1}{2}[g(y(\overline{z}w)) - g(w(\overline{z}y))]\rangle \\ &= \langle g(x), g[\frac{1}{2}(y(\overline{z}w) - w(\overline{z}y))]\rangle \\ &= \langle x, \frac{1}{2}(y(\overline{z}w) - w(\overline{z}y))\rangle \\ &= \langle x, y \times z \times w\rangle, \end{aligned}$$

so that $g^*(\psi) = \psi$. Similarly, one sees $g^*(\Phi) = \Phi$ since g respects the octonionic multiplication and conjugation and it is obviously orthogonal also on $\mathbb{O} = Re(\mathbb{O}) \oplus Im(\mathbb{O})$ ($\mathbb{R}^8 = \mathbb{R} \oplus \mathbb{R}^7$), fixing the first component.

Surprisingly, the Bonan form Φ is invariant even under $Spin(7)$! We want to give only the proof-idea for this property. There is a four-fold cross product for octonions:

$$x \times y \times z \times w = \frac{1}{4}[\overline{x}(y \times z \times w) + \overline{y}(z \times x \times w) + \overline{z}(x \times y \times w) + \overline{w}(y \times x \times z)],$$

where $x, y, z, w \in \mathbb{O}$. This cross product is alternating and for pair-wise orthogonal vectors it can be expressed as $x \times y \times z \times w = \overline{x}(y(\overline{z}w))$. Moreover, it can be shown that $\langle x, y \times z \times w\rangle = Re(x \times y \times z \times w)$. Furthermore, it can be shown with the help of the so-called Moufang identities that, for $v \in Im(\mathbb{O})$, $\|v\| = 1$ and $x, y, z, w \in \mathbb{O}$ the following equality holds: $(xv) \times (yv) \times (zv) \times (wv) = v(x \times y \times z \times w)\overline{v}$.

Now we can see the invariance of the Bonan form (sometimes called the Cayley calibration) Φ under $Spin(7)$ as follows: Recall that $Spin(7) \subset O(\mathbb{O}) = O(8)$ is generated by octonionic right multiplications R_v for $v \in Im(\mathbb{O})$, $\|v\| = 1$. So, it will be enough to see $(R_v^*\Phi)(x, y, z, w) = \Phi(x, y, z, w)$ for $x, y, z, w \in \mathbb{O}$:

$$\begin{aligned}(R_v^*\Phi)(x,y,z,w) &= \Phi(xv, yv, zv, wv)\\ &= \langle xv, yv \times zv \times wv\rangle\\ &= Re(xv \times yv \times zv \times wv)\\ &= Re(v(x \times y \times z \times w)\overline{v})\\ &= Re(x \times y \times z \times w)\\ &= \langle x, y \times z \times w\rangle\\ &= \Phi(x,y,z,w),\end{aligned}$$

whereby we used the simple fact

$$Re(vt\overline{v}) = \frac{1}{2}[vt\overline{v} + \overline{vt\overline{v}}] = \frac{1}{2}[vt\overline{v} + v\overline{t}\overline{v}] = \frac{1}{2}v(t + \overline{t})\overline{v} = \frac{t + \overline{t}}{2}v\overline{v} = Re(t).$$

for any $t \in \mathbb{O}$.

For the interesting roles the forms φ, ψ and Φ play in the context of calibrations we refer again to [9].

9. A Glimpse into Triality

As a last topic we want to give a look at the interesting phenomenon of triality. Recall that in Section 7 we constructed an algebra isomorphism

$$\tilde{f} : Cl_8 = Cl(\mathbb{O}) \longrightarrow End_{\mathbb{R}}(\mathbb{O} \oplus \mathbb{O}) \text{ with } f(v) = \begin{pmatrix} 0 & R_v \\ -R_{\overline{v}} & 0 \end{pmatrix} \text{ for } v \in \mathbb{O}.$$

We have also seen that Cl_8^0 is identified with $End_{\mathbb{R}}(\mathbb{O}) \oplus End_{\mathbb{R}}(\mathbb{O}) \subset End_{\mathbb{R}}(\mathbb{O} \oplus \mathbb{O})$ under this isomorphism. Let us now determine $Spin(8) \subset Cl_8^0$. As we have seen,

$$Spin(8) = \{u \in Cl_8^0 \mid uxu^{-1} \in \mathbb{O} \;\; for \; x \in \mathbb{O} \;\; and \;\; uu^t = 1\}.$$

One can check that the transpose-antiinvolution in Cl_8^0 is translated by $\tilde{f}$ to taking the adjoint in the endomorphism-algebra: $\tilde{f}(u^t) = (\tilde{f}(u))^t$ for $u \in Cl_8^0$. This means that under the identification by $\tilde{f}$ we can write:

$$Spin(8) = \left\{ \begin{pmatrix} A & 0 \\ 0 & B \end{pmatrix} \in End_{\mathbb{R}}(\mathbb{O} \oplus \mathbb{O}) \mid \begin{pmatrix} A & 0 \\ 0 & B \end{pmatrix}\begin{pmatrix} A & 0 \\ 0 & B \end{pmatrix}^t = I_{16} \text{ and for } x \in \mathbb{O} \right.$$

$$\left. \text{there exists } v \in \mathbb{O} \text{ such that } \begin{pmatrix} A & 0 \\ 0 & B \end{pmatrix}\begin{pmatrix} 0 & R_x \\ -R_{\overline{x}} & 0 \end{pmatrix}\begin{pmatrix} A & 0 \\ 0 & B \end{pmatrix}^{-1} = \begin{pmatrix} 0 & R_v \\ -R_{\overline{v}} & 0 \end{pmatrix} \right\}.$$

Note that

$$\begin{pmatrix} A & 0 \\ 0 & B \end{pmatrix} \begin{pmatrix} A & 0 \\ 0 & B \end{pmatrix}^t = I_{16}$$

gives $AA^t = BB^t = I_8$ and we can write

$$Spin(8) = \{(A,B) \in O(8) \times O(8) \mid \begin{pmatrix} A & 0 \\ 0 & B \end{pmatrix} \begin{pmatrix} 0 & R_x \\ -R_{\overline{x}} & 0 \end{pmatrix} \begin{pmatrix} A^t & 0 \\ 0 & B^t \end{pmatrix} = \begin{pmatrix} 0 & R_v \\ -R_{\overline{v}} & 0 \end{pmatrix}\},$$

or,

$$Spin(8) = \{(A,B) \in O(8) \times O(8) \mid \text{for } x \in \mathbb{O} \text{ there exists } v \in \mathbb{O} \text{ such that } AR_xB^t = R_v \text{ and } BR_{\overline{x}}A^t = R_{\overline{v}}\}.$$

The second condition is the transpose of the first and by the connectedness of $Spin(8)$ we get

$$Spin(8) = \{(A,B) \in O(8) \times O(8) \mid \text{for } x \in \mathbb{O} \text{ there exists } v \in \mathbb{O} \text{ such that } AR_xB^t = R_v\}.$$

We note that the assignment $x \in \mathbb{O} \longmapsto v \in \mathbb{O}$ is nothing else than $\widetilde{Ad}(u)$, where $u \in Cl_8^0$ is now in the guise of $(A,B) \in End_{\mathbb{R}}(\mathbb{O}) \oplus End_{\mathbb{R}}(\mathbb{O}) \cong Cl_8^0$.

We can express v in terms of x as follows:

$$AR_xB^t(e_0) = R_v(e_0) = e_0 \cdot v = v$$

and thus $v = A(B^t(e_0) \cdot x)$, where the dots denote the octonionic multiplication. (A and B are called the positive and negative spinor representations and C the associated vector representation.)

If we define $C = A\, L_{B^t(e_0)}$, then $C \in SO(8)$ and $v = C(x)$.

Applying the condition $AR_xB^t = R_v$ to an arbitrary element $z \in \mathbb{O}$, we get $(AR_xB^t)(z) = z \cdot v$, or, $A(B^t(z) \cdot x) = z \cdot C(x)$. Denoting $B^t(z)$ by $y \in \mathbb{O}$ we obtain the following beautiful relationship between A, B and C:

For any $x, y \in \mathbb{O}$ it holds $A(y \cdot x) = B(y) \cdot C(x)$, where the dots denote the octonionic multiplication.

The same computation shows, that given A, B and $C \in SO(8)$ satisfying the relationship in this lemma, then $(A,B) \in Spin(8)$, with C the associated element to (A,B). We thus get the following result:

Proposition. *Let $A, B \in SO(8)$. Then $(A,B) \in Spin(8)$ if and only if there exists $C \in SO(8)$ such that $A(y \cdot x) = B(y) \cdot C(x)$ for all $x, y \in \mathbb{O}$, where the dots denote the octonionic multiplication.*

The actions of $Spin(8)$ on the first and second components of $\mathbb{O} \oplus \mathbb{O}$ via projections $(A,B) \mapsto A : \mathbb{O} \to \mathbb{O}$ and $(A,B) \mapsto B : \mathbb{O} \to \mathbb{O}$ are called the positive and negative spinor representations and the associated action on $\mathbb{O}$ via C is called the associated vector representation.

We now proceed to construct the triality automorphism of $Spin(8)$.

Let $E'(x) = \overline{E(\overline{x})}$ for $E \in End(\mathbb{O})$ and $x \in \mathbb{O}$. Now, let us be given $(A,B) \in Spin(8)$ with the associated C. Then,

$$A'(y \cdot x) = \overline{A(\overline{y \cdot x})} = \overline{A(\overline{x} \cdot \overline{y})} = \overline{B(\overline{x}) \cdot C(\overline{y})} = \overline{C(\overline{y})} \cdot \overline{B(\overline{x})} = C'(y) \cdot B'(x).$$

This means that (A', C') belongs to $Spin(8)$ and is associated with B'.

Lemma. *The map* $\alpha : Spin(8) \longrightarrow Spin(8), \quad (A, B) \longmapsto (A', C')$ *is an automorphism and* $\alpha^2 = Id$.

The proof is straightforward.

It might be practical to carry the associated vector representation as the third component of a triple and write $\alpha : (A, B, C) \mapsto (A', C', B')$.

Now start again with $(A, B) \in Spin(8)$ with the associated C satisfying $A(y \cdot x) = B(y) \cdot C(x)$ for $x, y \in \mathbb{O}$ and insert $y \cdot x$ for y and $\overline{x}$ for x:

$$
\begin{aligned}
A((y \cdot x) \cdot \overline{x}) &= B(y \cdot x) \cdot C(\overline{x}), \\
A(y \cdot \|x\|^2) &= B(y \cdot x) \cdot C(\overline{x}), \\
\|x\|^2 A(y) &= B(y \cdot x) \cdot C(\overline{x}), \\
B(y \cdot x) &= \|x\|^2 A(y) \cdot \overline{C(\overline{x})} / \|C(\overline{x})\|^2 = A(y) \cdot C'(x).
\end{aligned}
$$

This means that (B, A) also belongs to $Spin(8)$ with the associated representation C'. We get:

Lemma. *The map* $\beta : Spin(8) \longrightarrow Spin(8), \quad (A, B) \longmapsto (B, A)$ *is an automorphism and* $\beta^2 = Id$.

In this case we could write in the triple notation, $\beta : (A, B, C) \mapsto (B, A, C')$.

Combining these two automorphisms we find the triality automorphism:

Theorem. $\tau = \alpha\beta : Spin(8) \longrightarrow Spin(8)$ *is an automorphism with* $\tau^3 = Id$.

Indeed, $\tau(A, B, C) = (\alpha\beta)(A, B, C) = \alpha(B, A, C') = (B', C, A')$. Then we get $\tau(B', C, A') = (C', A', B)$ and finally, $\tau(C', A', B) = (A, B, C)$, so that we find $\tau^3 = Id$!

Acknowledgments

This work was part of the project "Self-Duality in Higher Dimensional Manifolds" (No: 106T558) supported by *The Scientific And Technological Research Council of Turkey (TUBITAK).*

References

[1] Altmann, S.L., 1989. *Hamilton, Rodrigues, and the quaternion scandal.* Mathematics Magazine, **62(5)**, pp.291-308.

[2] Baez, J., 2002. *The octonions.* Bulletin of the American Mathematical Society, **39(2)**, pp.145-205.

[3] Bonan,E., 1966. *Sur les varietés riemanniennes a groupe d'holonomie* G_2 *ou* $Spin(7)$. CR Acad.Scie.Paris. **262** pp.127-129

[4] Dickson, L.E., 1919. *On quaternions and their generalization and the history of the eight square theorem.* Annals of Mathematics, pp.155-171.

[5] Ebbinghaus H.D., Hermes H., Hirzebruch F., Koecher M., Mainzer K., Neukirch J., Prestel A., Remmert R., 1991. *Numbers*, Springer Verlag.

[6] Frobenius, H., 1878. *Über lineare Substitutionen und bilineare Formen.* Journal für die reine und angewandte Mathematik (Crelles Journal), **1878(84)**, pp.1-63.

[7] Hamilton W. R. 1847. *The Transactions of the Royal Irish Academy*, pp.338–341,

[8] Harvey, F.R., 1990. *Spinors and calibrations.* Elsevier.

[9] Harvey, R. and Lawson, H.B., 1982. *Calibrated geometries.* Acta Mathematica, **148**, pp.47-157.

[10] Hurwitz, A., 1898. *Über die Composition der quadratischen Formen von belibig vielen Variablen.* Nachrichten von der Gesellschaft der Wissenschaften zu Göttingen, Mathematisch-Physikalische Klasse, pp.309-316.

[11] May, K.O., 1966. *The impossibility of a division algebra of vectors in three dimensional space.* The American Mathematical Monthly, **73(3)**, pp.289-291.

[12] Milnor, J., 1958. *Some consequences of a theorem of Bott.* Annals of Mathematics, pp.444-449.

[13] Palais, R.S., 1968. *The classification of real division algebras.* The American Mathematical Monthly, **75(4)**, pp.366-368.

[14] Hüseyin Tevfik Paşa, 1882. *Linear Algebra*, Constantinople.

$O(d, d)$ covariant formulation of Type II supergravity and Scherk-Schwarz reduction

Aybike Çatal-Özer

Department of Mathematics, Istanbul Technical University, Istanbul, TURKEY

E-mail: ozerayb@itu.edu.tr

Abstract. T-duality is a stringy symmetry which relates string backgrounds with different space-time geometries. In the low energy limit, it manifests itself as a continuous $O(d, d)$ symmetry acting on supergravity fields, after dimensional reduction on a d dimensional torus. Double Field Theory (DFT) is a T-duality covariant extension of string theory which aims to realize $O(d, d)$ as a manifest symmetry for the low energy effective space-time actions of string theory without dimensional reduction. The mathematical framework needed to construct DFT goes beyond Riemannian geometry and is related to Hitchin's generalized geometry program. On the other hand, Scherk-Schwarz reduction of DFT of Type II strings with a duality twist in $O(d, d)$ yields Gauged Double Field Theory (GDFT), that can be regarded as an $O(d, d)$ covariant extension of gauged supergravity. The purpose of this contribution is to give a short review on Scherk-Schwarz reductions of DFT and its intriguing connections to integrable deformations of string sigma models.

1. Introduction

T-duality is a stringy symmetry which relates string backgrounds with different space-time geometries [1]. It is natural to expect that T-duality (along with other stringy dualities) should also be a symmetry of the corresponding supergravity theory, the effective theory which describes string theory in the low energy limit. Indeed, compactified supergravity theories possess non-compact global symmetries [2] as had been understood as early as in 70s [3, 4, 5, 6], and T-duality is realized as a discrete subgroup of these symmetries. Historically, these non-compact global symmetries are called hidden symmetries, as they arise after dimensional reduction on a torus and only after certain dualizations [7, 8]. Compactifying eleven dimensional supergravity on a $d+1$ dimensional torus and dualizing certain fields one ends up with a $10-d$ dimensional supergravity theory, with a global symmetry group which includes $O(d, d, R)$[1]. After quantization, the discrete subgroup $O(d, d, Z)$ is promoted to the T-duality symmetry group of the toroidally compactified type II string theories.

Double Field Theory (DFT) of Type II strings is an extension of massless Type II string theories, in which the T-duality symmetry is already manifest in higher dimensions without the requirement of dimensional reduction. In the low energy limit, this provides manifestly $O(d, d)$ invariant actions for the corresponding supergravity theories. This is achieved by extending the space-time by introducing dual, winding type coordinates [9, 10, 11, 12, 13]. In [14], a manifestly $O(d, d)$ invariant action was constructed on such a doubled space, based on two dynamical

[1] In the manuscript, $O(d, d, R)$ is simply referred to as $O(d, d)$.

fields called the generalized metric and the generalized dilaton field. The action comes with an $O(d,d)$ covariant constraint, called the strong constraint, imposition of which effectively halves the number of coordinates the fields and the gauge parameters can depend on. A trivial solution of this constraint is obtained when none of the fields or gauge parameters in the theory depend on the winding type coordinates. In this case, the fields are said to be in supergravity frame. As the name suggests, when this solution of the constraint is imposed, the DFT action constructed in [12] reduces to the standard NS-NS action for the massless fields of string theory. Later in [15], Zwiebach and Hohm constructed a manifestly $Spin^+(d,d)$ invariant action, which reduces in the supergravity frame to the Ramond-Ramond sector of the democratic formulation of Type II supergravity theory [16]. In this action, the dynamical fields are two spinor fields: χ and $\mathbb{S}$. The spinor χ encodes all the p-form gauge potentials and their Hodge duals that live in the RR sector of Type II theory. The field $\mathbb{S}$ is the spinor field which projects onto the generalized metric under the double covering homomorphism between $O(d,d)$ and $Pin(d,d)$. These actions will be given in Section 2.

As mentioned above, both the NS-NS and the RR sectors of the DFT action are invariant under the global symmetry group $Pin(d,d)$.[2] It is well known that theories with a global symmetry group can be dimensionally reduced by using a Scherk-Schwarz (SS) type ansatz [18, 19]. This provides a generalization of the Kaluza-Klein (KK) dimensional reduction scheme, which allows the higher dimensional fields to have a dependence on the coordinates of the internal space. The coordinate dependence is dictated by how the fields transform under the global symmetry group. This yields in lower dimensions gaugings, mass terms and a scalar potential, all determined by the duality twist matrix, which is an element of the global symmetry group. SS reduction is known to yield a consistent dimensional reduction in the sense that solutions of the lower dimensional field equations are also solutions of the field equations derived from the higher dimensional parent theory. This consistency is guaranteed by the fact that the reduction ansatz is determined by the global symmetry.

SS reduction of DFT was studied in [20, 21, 22, 23, 24] for the NS-NS sector and in [17] for the RR sector. The resulting theory dubbed Gauged Double Field Theory (GDFT) reduces in the supergravity frame to gauged supergravity. Interestingly, it is possible to obtain also the *non-geometric gaugings* this way [25]. This means that these gaugings cannot be obtained as a result of standard geometric compactification of supergravity or cannot be T-dualized to one of those. The origin of such gaugings had been previously related with the existence of the so-called *non-geometrix fluxes* in higher dimensions [26, 27, 28]. It is remarkable that GDFT provides a framework in which non-geometric fluxes can be treated on the same footing with the geometric fluxes [21].

Another remarkable feature of GDFT is that it sets a convenient stage for understanding certain solution generating duality transformations and integrable deformations of string sigma models. It has been understood recently that non-Abelian T-duality, Yang-Baxter deformations and Poisson Lie T-duality can be understood in terms of local $O(d,d)$ transformations [29, 30, 31, 32, 33, 34, 35, 36, 37, 38]. This makes DFT and GDFT a very convenient framework for describing these dualities and deformations.

The purpose of this article is to give a short review of SS reductions of DFT and its intriguing connections to integrable deformations of string sigma models. The structure of the review is as follows. In the next section, we start by introducing the basic ingredients in the construction of DFT. The $O(d,d)$ invariant actions of both the NS-NS and the RR sectors will be given in that section. In Section 3, we will describe the SS reduction of DFT, again for both sectors. We will introduce the fluxes in this section, which describe GDFT as a deformation of DFT. Section 4 explains how non-Abelian T-duality and Yang-Baxter deformations can be described within the

[2] In the RR sector, this is broken to $Spin(d,d)$ due to the chirality condition on χ and is further reduced to $Spin^+(d,d)$ due to the existence of a self-duality constraint. See [15, 17] for more details.

Journal of Physics: Conference Series **2191** (2022) 012007 doi:10.1088/1742-6596/2191/1/012007

framework of GDFT. We end in Section 5 with a discussion of results.

2. A brief review of double field theory

In closed string field theory, all fields depend on two types of coordinates: the usual space-time coordinates x and the dual winding type coordinates $\tilde{x}$. The dual coordinates are conjugate to the winding degrees of freedom of strings, in the same way space coordinates and momenta are conjugate variables in classical field theory. The aim of DFT is to realize this in the sector of massless fields in order to construct a manifestly T-duality invariant action describing this sector. In DFT, the space-time and winding type coordinates combine to form an $O(d,d)$ vector transforming as:

$$X'^M = h^M{}_N X^N, \qquad X^M = \begin{pmatrix} \tilde{x}_i \\ x^i \end{pmatrix}. \tag{1}$$

Here $\tilde{x}_i$ are the dual coordinates, and $h^M{}_N$ is a general $O(d,d)$ matrix. Remember that an $O(d,d)$ matrix h is of the form

$$h = \begin{pmatrix} a & b \\ c & d \end{pmatrix}, \tag{2}$$

where a, b, c and d are $d \times d$ matrices satisfying

$$a^t c + c^t a = 0, \quad b^t d + d^t b = 0, \quad a^t d + c^t b = I. \tag{3}$$

In what follows we will always decompose the indices M labelling the $O(d,d)$ representation as ${}^M = ({}_i, {}^i)$, where ^{i} and ${}_i$ label representations of the $GL(d)$ subgroup of $O(d,d)$. We will raise and lower indices by the $O(d,d)$ invariant metric η (presented in equation (4) below), so that $X_M = \eta_{MN} X^N$. The doubling of the coordinates is formal in the sense that the consistency of the DFT action that we will present shortly relies on imposition of a constraint, which effectively halves the number of coordinates that the fields and the gauge parameters can depend on. The constraint is $O(d,d)$ invariant and is given below:

$$\partial^M \partial_M A = \eta^{MN} \partial_M \partial_N A = 0\,, \qquad \partial^M A\, \partial_M B = 0\,, \qquad \eta^{MN} = \begin{pmatrix} 0 & 1 \\ 1 & 0 \end{pmatrix}. \tag{4}$$

Here, A and B represent any fields or parameters of the theory. The first of the above constraints is called the *weak constraint* and follows from the level matching constraint in closed string theory. The second constraint is stronger and is called the *strong constraint.* A trivial solution of the strong constraint is obtained when all the fields and gauge parameters in the theory depend only on the standard coordinates x and have no dependence on the dual coordinates $\tilde{x}$. In this case, the theory is said to be in the supergravity frame, since the DFT action reduces to the standard supergravity action in this frame. It can be shown that for any other solution of the strong constraint, there is a choice of coordinates $(x', \tilde{x}')$ related to $(x, \tilde{x})$ by an $O(d,d)$ transformation in such a way that the fields and gauge parameters have no dependence on the coordinates $\tilde{x}'$ [13]. This means that the theory is restricted to lie in a maximally isotropic subspace (with respect to the metric preserved by $O(d,d)$) of the total $2d$ dimensional doubled space [12].[3]

[3] Another important solution of the strong constraint is obtained by allowing the dilaton field to have a linear dependence on one of the winding type coordinates, say $\tilde{x}_i$. In order to satisfy (4), none of the fields (including the dilaton field) may have a dependence on the corresponding space coordinate x^i, or equivalently the background supports $U(1)$ isometry with the corresponding Killing vector field $\frac{\partial}{\partial x^i}$. In this case, the theory is said to be in the generalized supergravity frame, since the DFT equations are known to reduce to generalized supergravity equations of [39, 40] for this solution [41, 42].

The DFT action can be presented in terms of a generalized vielbein, as was first done in [9] (see [43] for later developments). Below we present the generalized metric formulation, which was first constructed by Hohm, Hull and Zwiebach for the NS-NS sector [14], and then by Hohm, Kwak and Zwiebach for the RR sector [15]:

$$\mathcal{S} = \int dx d\tilde{x} \left(\mathcal{L}_{\text{NS-NS}} + \mathcal{L}_{\text{RR}} \right), \tag{5}$$

where

$$\mathcal{L}_{\text{NS-NS}} = e^{-2d} \, \mathcal{R}(\mathcal{H}, d), \tag{6}$$

and

$$\mathcal{L}_{\text{RR}} = \frac{1}{4} (\partial\!\!\!/ \chi)^\dagger \, \mathbb{S} \, \partial\!\!\!/ \chi = \frac{1}{4} \langle \, \partial\!\!\!/ \chi, \; C^{-1} \mathbb{S} \, \partial\!\!\!/ \chi \rangle. \tag{7}$$

This action has to be complemented by the following self-duality constraint

$$\partial\!\!\!/ \chi = -\mathcal{K} \, \partial\!\!\!/ \chi, \quad \mathcal{K} \equiv C^{-1} \mathbb{S}. \tag{8}$$

Let us explain the ingredients in the equations (6) and (7). The field $\mathcal{H}$ is the generalized metric and it is constructed from the semi-Riemannian metric g and the B-field b, which is an antisymmetric 2-form field:

$$\mathcal{H}_{MN} = \begin{pmatrix} \mathcal{H}^{ij} & \mathcal{H}^{i}{}_{j} \\ \mathcal{H}_{i}{}^{j} & \mathcal{H}_{ij} \end{pmatrix} = \begin{pmatrix} g^{ij} & -g^{ik} b_{kj} \\ b_{ik} g^{kj} & g_{ij} - b_{ik} g^{kl} b_{lj} \end{pmatrix}. \tag{9}$$

$\mathcal{H}$ is a symmetric $O(d,d)$ matrix and as such it satisfies $\mathcal{H}_{MP} \eta^{PQ} \mathcal{H}_{QR} = \eta^{MR}$. The term $\mathcal{R}(\mathcal{H}, d)$ is the generalized Ricci scalar and its explicit form is:

$$\begin{aligned} \mathcal{R}(\mathcal{H}, d) \;=\; & 4\mathcal{H}^{MN} \partial_M \partial_N d - \partial_M \partial_N \mathcal{H}^{MN} - 4\mathcal{H}^{MN} \partial_M d \partial_N d + 4\partial_M \mathcal{H}^{MN} \partial_N d \\ & + \frac{1}{8} \mathcal{H}^{MN} \partial_M \mathcal{H}^{KL} \partial_N \mathcal{H}_{KL} - \frac{1}{2} \mathcal{H}^{MN} \partial_M \mathcal{H}^{KL} \partial_K \mathcal{H}_{NL} \\ & + \frac{1}{2} \partial_M \varepsilon^a{}_P \partial^M \varepsilon^b{}_Q S_{ab} \eta^{PQ}. \end{aligned} \tag{10}$$

Here, $\varepsilon^a{}_P$ is the generalized vielbein with $\mathcal{H}_{MN} = \varepsilon^a{}_M S_{ab} \varepsilon^b{}_N$, where $S_{ab} = diag(-1, 1, \cdots, 1; -1, 1, \cdots, 1)$ is the planar metric. The term in the last line is not in the original generalized metric formulation of DFT and vanishes when the strong constraint is imposed [22]. The field d in (6) is the generalized dilaton field and it is defined as

$$e^{-2d} = \sqrt{g} e^{-2\phi}. \tag{11}$$

The field χ in the action (7) is a $Clif(d,d)$ spinor field that packages the p-form gauge fields in the RR sector of Type II supergravity. For Type IIA one has odd degree forms, whereas for Type IIB χ packages even degree forms. The spinor field is differentiated with the Dirac operator

$$\partial\!\!\!/ \equiv \Gamma^M \partial_M = \Gamma^i \partial_i + \Gamma_i \tilde{\partial}^i, \tag{12}$$

where $\tilde{\partial}^i = \frac{\partial}{\partial \tilde{x}_i}$. Here the Gamma matrices $\Gamma^M = (\Gamma_i, \Gamma^i)$ are the matrix representations of the generators of the Clifford algebra $Clif(d,d) \equiv Clif(R^{2d}, \eta)$:

$$\{\Gamma^M, \Gamma^N\} = 2\eta^{MN}, \tag{13}$$

where the $O(d,d)$ invariant metric η is as in (4). The labelling is such that the first d elements span a maximally isotropic subsace with respect to the metric η, and the the remaining elements span the orthogonal complement. Then one has

$$\{\Gamma_i, \Gamma^j\} \;=\; 2\delta_i{}^j\,, \qquad \{\Gamma_i, \Gamma_j\} \;=\; 0\,, \qquad \{\Gamma^i, \Gamma^j\} \;=\; 0\,. \tag{14}$$

It is also useful to define $\psi^M \equiv \frac{1}{\sqrt{2}}\Gamma^M$. The spinorial action of the Clifford algebra elements Γ^M on spinor fields χ is best understood when one identifies the spinor fields with (not necessarily homogenous) differential forms (or with polyforms, as is commonly called in physics literature) of the exterior algebra $\bigwedge^{\bullet} R^d$ [44, 45, 46, 15, 17]. A generic spinor field φ

$$\varphi(x,\tilde{x}) = \sum_p \frac{1}{p!} C_{i_1\ldots i_p}(x,\tilde{x})\psi^{i_1}\ldots\psi^{i_p} \tag{15}$$

is identified with the differential form

$$\varphi(x,\tilde{x}) = \sum_p \frac{1}{p!} C_{i_1\ldots i_p}(x,\tilde{x})dx^{i_1}\wedge\ldots\wedge dx^{i_p}. \tag{16}$$

Then, the action of ψ^i and ψ_i on χ is by wedge product and by contraction, respectively. More precisely, one has

$$\psi^i.\chi = \psi^i \wedge \chi, \quad \psi_i.\chi = i_{\psi_i}\chi, \tag{17}$$

where one defines $i_{\psi_i}\psi^j = \delta_i{}^j$. More details can be found in [15, 17].

The Ramond-Ramond sector couples to the NS-NS sector via the spinor field $\mathbb{S}$. Here $\mathbb{S}$ is the spinor field which projects to the generalized metric $\mathcal{H}$ under the double covering homomorphism ρ between $O(d,d)$ and $Pin(d,d)$, that is, $\rho(\mathbb{S}) = \mathcal{H}$. In Lorentzian signature, the generalized metric $\mathcal{H}$ is in the coset $SO^-(d,d)$[4], and there are subtleties in lifting this to an element $\mathbb{S}$ of $Spin^-(d,d)$ (for a detailed discussion, see [15]). So, in [15] the following viewpoint was adopted: it is the spin field $\mathbb{S} \in Spin^-(d,d)$, rather than the generalized metric, which has to be regarded as the fundamental gravitational field. The generalized metric $\mathcal{H}$ is then constructed by projecting onto the corresponding unique element in $SO^-(d,d)$. Finally, the matrix C is called the charge conjugation matrix and is defined as

$$C = \Lambda_1^{\pm}\ldots\Lambda_d^{\pm} \tag{18}$$

with

$$\Lambda_i^{\pm} = (\psi^i \mp \psi_i), \tag{19}$$

where the upper plus sign is for even dimensions and the lower minus sign is for odd dimensions. It can be shown that the following useful relation to Hodge duality * (defined with respect to the semi-Riemannian metric g) holds:

$$*\lambda(A) = -C^{-1}S_g^{-1}A. \tag{20}$$

Here,

$$\lambda(A_n) \equiv (-1)^{Int[n/2]}A_n = (-1)^{n(n-1)/2}A_n \tag{21}$$

[4] When the space-time metric g is positive definite, so is the generalized metric $\mathcal{H}$, and hence its components form a matrix that lies in $SO^+(d,d)$. In this case, the corresponding spin group element $\mathbb{S}$ is in $Spin^+(d,d)$. However, when the semi-Riemannian metric g, has Lorentzian signature, then $\mathcal{H}$ is in $SO^-(d,d)$ and correspondingly $\mathbb{S}$ lives in $Spin^-(d,d)$. Here, $SO^+(d,d)$ is the component of $SO(d,d)$ connected to the identity. It is also a subgroup, whereas its complement, $SO^-(d,d)$ is a coset of $SO^+(d,d)$.

for an n-form A_n, and action of λ on a polyform A is defined by linear extension. Also, $S_g^{-1} = S_{g^{-1}}$ is the $Spin(d,d)$ element that projects onto the $SO(d,d)$ element

$$h_{g^{-1}} \equiv \begin{pmatrix} g^{-1} & 0 \\ 0 & g \end{pmatrix} \tag{22}$$

under the double covering homomorphism ρ that is, $\rho(S_{g^{-1}}) = h_{g^{-1}}$.

The inner product $\langle\ ,\ \rangle$ that determines the action in (7) is the Mukai pairing, which is a $Spin(d,d)$ invariant bilinear form on the space of spinors [47, 44]. It is defined as $\langle\ ,\ \rangle : S \otimes S \to \wedge^n T^*M$:

$$\langle \chi_1, \chi_2 \rangle = (\tau(\chi_1) \wedge \chi_2)_{\text{top}} = \sum_j (-1)^j (\chi_1^{2j} \wedge \chi_2^{n-2j} + \chi_1^{2j+1} \wedge \chi_2^{n-2j-1}), \quad \chi_1, \chi_2 \in \wedge^\bullet T^*M. \tag{23}$$

Here, $()_{\text{top}}$ denotes the top degree component of the form, and the superscript k denotes the k-form component of the form.

2.1. *Gauge symmetries of the action*

It can be shown that the DFT actions (6) and (7) are invariant under the following gauge transformations [13]

$$\begin{aligned} \delta_\xi \mathcal{H}_{MN} &= \widehat{\mathcal{L}}_\xi \mathcal{H}_{MN} \\ &\equiv \xi^P \partial_P \mathcal{H}_{MN} + \left(\partial_M \xi^P - \partial^P \xi_M\right) \mathcal{H}_{PN} + \left(\partial_N \xi^P - \partial^P \xi_N\right) \mathcal{H}_{MP}\,, \\ \delta d &= \xi^M \partial_M d - \frac{1}{2} \partial_M \xi^M\,, \\ \delta_\xi \chi &= \widehat{\mathcal{L}}_\xi \chi \equiv \xi^M \partial_M \chi + \frac{1}{\sqrt{2}} \partial\!\!\!/ \xi^M \Gamma_M \chi \\ &= \xi^M \partial_M \chi + \frac{1}{2} \partial_N \xi_M \Gamma^N \Gamma^M \chi\,, \\ \delta_\xi \mathcal{K} &= \xi^M \partial_M \mathcal{K} + \frac{1}{2} [\Gamma^{PQ}, \mathcal{K}] \partial_P \xi_Q\,, \end{aligned} \tag{24}$$

$$\delta_\lambda \chi = \partial\!\!\!/ \lambda = \frac{1}{\sqrt{2}} \Gamma^M \partial_M \lambda, \tag{25}$$

where λ is a space-time dependent spinor. Here, the generator $\xi^M = (\tilde{\xi}_i, \xi^i)$ is an $O(d,d)$ vector and Γ^{PQ} is defined as $\Gamma^{PQ} \equiv \frac{1}{2}[\Gamma^P, \Gamma^Q]$. In the frame $\tilde{\partial}^i = 0$, the gauge parameter $\xi^M = (\tilde{\xi}_i, \xi^i)$ combines the diffeomorphism parameter $\xi^i(x)$ and the Kalb-Ramond gauge parameter $\tilde{\xi}_i(x)$. The parameter λ generates the double field theory version of the abelian gauge symmetry of p-form gauge fields.

It can be shown that the gauge transformations close to form a gauge algebra:

$$\begin{aligned} \left[\delta_{\xi_1}, \delta_{\xi_2}\right] &= -\delta_{[\xi_1, \xi_2]_C}, \\ \left[\delta_\lambda, \delta_\xi\right] &= \delta_{\widehat{\mathcal{L}}_\xi}, \end{aligned} \tag{26}$$

provided that the strong constraint (4) is satisfied. Here, $[,]_C$ is the C-bracket, which is defined as

$$\left[\Sigma_1,\ \Sigma_2\right]_C^M = 2\Sigma_{[1}^N \partial_N \Sigma_{2]}^M - \Sigma_{[1}^P \partial^M \Sigma_{2]P}. \tag{27}$$

If we formally write $\Sigma = \xi + \tilde{\xi}$ and expand out the bracket in (27), we obtain

$$\begin{aligned}[\xi_1 + \tilde{\xi}_1, \xi_2 + \tilde{\xi}_2]_C &= [\xi_1, \xi_2] + \mathcal{L}_{\tilde{\xi}_1}\xi_2 - \mathcal{L}_{\tilde{\xi}_2}\xi_1 - \frac{1}{2}\tilde{d}(\tilde{i}_{\tilde{\xi}_1}\xi_2 - \tilde{i}_{\tilde{\xi}_2}\xi_1) \\ &+ [\tilde{\xi}_1, \tilde{\xi}_2] + \mathcal{L}_{\xi_1}\tilde{\xi}_2 - \mathcal{L}_{\xi_2}\tilde{\xi}_1 - \frac{1}{2}d(i_{\xi_1}\tilde{\xi}_2 - i_{\xi_2}\tilde{\xi}_1).\end{aligned} \tag{28}$$

Here, $\tilde{d}$ is defined with respect to the dual coordinates so that acting on a function f, one has $(\tilde{d}f)^i = \tilde{\partial}^i f$ and $(\tilde{d}f)_i = 0$. $\tilde{i}$ is defined so that $i_{\tilde{\xi}_1}\xi_2 = \tilde{i}_{\xi_2}\tilde{\xi}_1 = (\tilde{\xi}_1)_i \xi_2^i$ and $\tilde{\mathcal{L}} = \tilde{d}i + \tilde{i}d$.

We digress here in order to discuss Lie bialgebroids and Courant algebroids, where the bracket (28) above plays a fundamental role. A Lie algebroid A is a vector bundle over a base manifold M which comes with a Lie bracket $[,]_A$ on its sections that satisfies the Leibniz rule and the Jacobi identity and an *anchor*, which is a Lie algebra homomorphism $\rho : A \to TM$ so that

$$\rho([X,Y]_A) = [\rho(X), \rho(Y)],$$

where X, Y are sections of A and $[,]$ is the standard Lie bracket of vector fields. Now, suppose that the dual vector bundle A^* also carries a Lie algebroid structure with a Lie bracket $[,]_{A^*}$ on its sections and an anchor ρ^*. If the two algebroid structures on A and A^* are compatible (the technical definition of compatibility is given in [48]), then the pair (A, A^*) forms a Lie bialgebroid. In this case, it is possible to construct a Courant algebroid structure on the direct sum bundle $A \oplus A^*$ as was first shown in [48] [5]. A Courant algebroid is a vector bundle E over a manifold M equipped with a non-degenerate symmetric bilinear form (,) on the bundle, a skew-symmetric bracket $[,]_{Cour}$ on sections of E and a bundle map (anchor) $\rho : E \to TM$ with certain properties (a modifed Jacobi identity, a modified Leibniz rule for the bracket, the bracket and the bilinear form must be compatible in a certain sense and ρ is a Lie algebra homomorphism [49]). When (A, A^*) forms a Lie bialgebroid, one takes as the inner product the natural inner product $(X + \eta, Y + \xi) = \frac{1}{2}(\xi(X) + \eta(Y))$, with $X, Y \in A$ and $\xi, \eta \in A^*$ and as the anchor $\rho = \rho_A + \rho_{A^*}$, where ρ_A and ρ_{A^*} are the anchors on the Lie algebroids A and A^*, respectively. The skew symmetric bracket on $E = A \oplus A^*$ can be written in terms of the brackets on A and A^* [48]. Just like the Lie bracket $[,]_A$ induces a differential d_A on sections of $\bigwedge^\bullet A^*$, the bracket $[,]_{A^*}$ induces a differential d_{A^*} on sections of $\bigwedge^\bullet A$. Then $\mathcal{D} \equiv d_A + d_{A^*}$ is used to differentiate the sections of the exterior bundle of E. The compatibility between the Lie algebroid structures on A and A^* makes sure that the inner product, the bracket, the anchor and the differential constructed on E define a Courant algebroid. The bracket that appears in (28) is exactly of the same form as the bracket on a Courant algebroid $A \oplus A^*$ if we call the differentials on A and A^* d and $\tilde{d}$, respectively. An interesting case is when $A = TM$ with the standard Lie bracket and $\rho = Id$ (which obviously constitute a Lie algebroid structure) and $A^* = T^*M$ is the cotangent bundle with zero Lie bracket (also a Lie algebroid structure trivially compatible with the one on TM). In this case, the bracket that defines a Courant algebroid structure on the direct sum bundle $TM \oplus T^*M$ is the famous Courant bracket

$$[X + \xi, Y + \eta] = [X, Y] + L_X\eta - L_Y\xi + \frac{1}{2}d\left(i_X\eta - i_Y\xi\right), \tag{29}$$

which was first studied in [50], where Courant used it to define a new geometrical structure called Dirac structure, which unifies Poisson geometry and presymplectic geometry. The same

[5] A special case of this construction is when the base manifold M is point, in which case the Lie algebroid becomes a Lie algebra g. In this particular case, one can construct a Lie algebra structure on the direct sum of the underlying vector space of g and its dual g^* provided that they form a Lie bialgebra and the doubled structure which replaces the Courant algebroid is a Drinfeld double. The Drinfeld double construction plays an essential role in the discussion of YB deformations, as we will discuss in subsection 4.2.

Journal of Physics: Conference Series **2191** (2022) 012007 doi:10.1088/1742-6596/2191/1/012007

bracket plays a prominent role in generalized geometry program of Nigel Hitchin in defining a generalized complex structure on the complexification of the bundle $TM \oplus T^*M$, which is a geometric structure that unifies complex structure and symplectic structure [45, ?, 44].

The Courant bracket (29) is exactly what the bracket (27) reduces to when the trivial solution $\tilde{\partial}^i = 0$ of the strong constraint is imposed. Automorphisms of the Courant bracket are the diffeomorphisms and 2-form gauge transformations, which are the symmetries of string theory [44]. Correspondingly, the gauge transformations (25) and (24) reduce in the supergravity frame to diffeomorphisms of the metric g and the B-field b and the gauge transformations of the B-field, with ξ and $\tilde{\xi}$ being the parameters for these gauge transformations. On the other hand, when all the fields and gauge parameters are in the supergravity frame, that is when $\tilde{\partial}^i = 0$, the action given in (6) reduces to the standard NS-NS action for the massless fields of string theory and the action (7) reduces to the RR sector of the democratic formulation of Type II supergravity theories. The duality constraint (8) reduces to the conventional duality relation for p-form fields in the democratic formulation

$$F^{(p)} = (-1)^{(d-p)(d-p-1)/2} * F^{(d-p)}. \tag{30}$$

The field equations in the democratic formulation for the higher degree fields which are absent in the standard formulation become, after applying (30), the Bianchi identities for the lower degree fields in the standard formulation.

3. Scherk-Schwarz reduction of DFT

Twisted toroidal compactification or Scherk-Schwarz reduction is a useful way of introducing masses into supergravity and string compactifications [18, 19]. Standard toroidal compactification on a torus T^d involves expanding all the fields in the theory in terms of the harmonics of the torus and then truncating away all the massive fields (which are very heavy in the limit of vanishing volume for the torus). This is equivalent to imposing a Kaluza-Klein (KK) dimensional reduction anzats, that is, taking all the fields independent of the internal toroidal coordinates. If the higher dimensional theory possesses a global symmetry group, then it is possible to keep some of the massive modes in the truncation without violating the consistency of the compactification. This is equivalent to imposing a duality twisted reduction anzats which allows the fields to depend on the internal coordinates in a way dictated by the global symmetry group. In their seminal papers [18, 19], Scherk and Schwarz showed that this also yields a consistent compactification, meaning that the solutions of the lower dimensional field equations can be uplifted to solutions of the higher dimensional parent theory. If a higher dimensional field, call it ψ, transforms in a certain representation of G, the SS dimensional anzats imposed on ψ is

$$\psi(x, y) = U(y)\,[\psi(x)], \quad U(y) \in G, \tag{31}$$

where $U(y)$ acts on G in that particular representation. Here, x and y correspond to the external and internal coordinates, respectively. Requirement of consistency imposes certain restrictions on $U(y)$, as we will discuss shortly in the context of DFT. In compacitifications of supergravity, the global symmetry group exploited in the SS dimensional reduction usually comes from a previous toroidal reduction. For example, consider a theory compactified on a two-torus T^2. The lower dimensional theory has an $SL(2, R)$ symmetry as part of its global symmetry group, as $SL(2, R)$ is the large diffeomorphism group of T^2. In a further compactification on a circle S^1 one can introduce a duality-twisted ansatz for these fields as in (31), where $U(y)$ is in $SL(2, R)$ [51, 52, 53, 54, 55]. The total 3 dimensional internal space is usually called a twisted torus in the literature. In contrast, DFT possesses a global $O(d, d)$ symmetry, which is already manifest in higher dimensions, without requiring toroidal dimensional reduction. This makes it possible to introduce a duality twisted anzats for all the fields in the higher dimensional theory, yielding mass terms and gaugings that cannot be obtained by standard compactification schemes. For example, when DFT is reduced to 4 dimensions and the fields are restricted to the supergravity frame, one obtains the whole electric bosonic sector of four dimensional gauged N = 4 supergravity [56] as was shown in [21, 20, 22]. Prior to their work, only a certain class of gaugings present in the theory had been known to have a higher dimensional origin. Below we will give a quick description of the duality twisted reduction of DFT, not only in the NS-NS sector, but also in the RR sector, whose SS reduction was first studied in [17].

The DFT action presented in (7) is invariant under the following transformations:

$$\mathbb{S}(X) \longrightarrow \mathbb{S}'(X') = (S^{-1})^{\dagger}\,\mathbb{S}(X)\,S^{-1}\,, \quad \chi(X) \longrightarrow \chi(X') = S\chi(X). \tag{32}$$

Here $S \in Spin^{+}(d,d)$ and $X' = hX$, where $h = \rho(S) \in SO(d,d)$. The generalized dilaton field is invariant. The duality group is broken to $Spin(d,d)$ because the spinor field χ in (7) must have a fixed chirality, depending on whether it is related to Type IIA or Type IIB string theory, and the full $Pin(d,d)$ does not preserve the fixed chirality. Moreover, a general $Spin(d,d)$ transformation does not preserve the self-duality constraint (8) and hence the duality group is further reduced to the subgroup $Spin^{+}(d,d)$. The transformation of $\mathbb{S}$ implies the following transformation rule for the generalized metric $\mathcal{H} = \rho(\mathbb{S})$:

$$\mathcal{H}(X) \longrightarrow \mathcal{H}'(X') = (h^{-1})^{T}\,\mathcal{H}(X)\,h^{-1}\,. \tag{33}$$

These transformation rules dictate the anzats for the duality twisted reduction of the DFT action:

$$\mathbb{S}(X,Y) = (S^{-1})^{\dagger}(Y)\mathbb{S}(X)S^{-1}(Y), \tag{34}$$

$$\chi(X,Y) = S(Y)\chi(X). \tag{35}$$

Here, X denotes collectively the coordinates of the reduced theory, whereas Y denotes the internal coordinates, and the whole dependence on the internal coordinates is encoded in the duality twist matrix $S(Y)$. Note that the internal and external coordinates can be further decomposed as $Y = (\tilde{y}, y)$ and $X = (\tilde{x}, x)$, where the coordinates with a tilde are the dual coordinates associated with winding excitations. The anzats (34) implies the following anzats in the NS-NS sector:

$$\mathcal{H}_{MN}(X,Y) = U^{A}_{\ M}(Y)\mathcal{H}_{AB}(X)U^{B}_{\ N}(Y). \tag{36}$$

Although the generalized dilaton field is invariant under $O(d,d)$ transformations, it is possible to introduce the following anzats for it, due to its invariance under constant shifts:

$$d(X,Y) = d(X) + \rho(Y). \tag{37}$$

Imposing this non-trivial ansatz on the generalized dilaton field gives rise to an overall conformal scaling in the NS-NS sector and in order to induce this factor also in the RR sector, one needs to modify the anzats on the field χ as follows:

$$\chi(X,Y) = e^{-\rho(Y)}S(Y)\chi(X). \tag{38}$$

On the other hand, we have for the gauge parameters

$$\xi^{M}(X,Y) = (U^{-1})^{M}_{\ A}\hat{\xi}^{A}(X) \tag{39}$$

and

$$\lambda(X,Y) = e^{-\rho(Y)}S(Y)\hat{\lambda}(X). \tag{40}$$

When we plug in the ansatze (34, 36, 37, 38, 39) and (40) in the DFT action and the gauge transformation rules, we end up with a deformation of DFT, determined by the twist matrices $U(Y), S(Y)$ and the function $\rho(Y)$. For consistency, we need the following conditions to be satisfied: first of all, the Y dependence should drop both in the reduced action and the gauge algebra so that we have a truly dimensionally reduced theory. Also, we need the reduced action to be invariant under the deformed gauge transformation rules and also the gauge algebra must close. One finds that all these conditions are satisfied provided that the following conditions on the twist matrices hold:

Firstly, the external coordinates X must remain untwisted, that is, we should have

$$(U^{-1})^M_{\ A}\partial_M g(X) = \partial_A g(X). \tag{41}$$

The second condition is

$$\partial^P (U^{-1})^M_{\ A}\partial_P g(X) = 0. \tag{42}$$

If a given coordinate and its dual are either both external or both internal, then this condition is trivially satisfied. A similar condition must be imposed on ρ, too:

$$\partial^P \rho\partial_P g(X) = 0. \tag{43}$$

In addition, the weak and the strong constraint has to be imposed on the external space so that

$$\partial_A\partial^A V(X) = 0, \quad \partial_A V(X)\partial^A W(X) = 0, \tag{44}$$

for any fields or gauge parameters V, W that have a dependence on the coordinates of the external space only. Under these conditions, it can be checked that all the Y dependence of the resulting theory is encoded in the entities called *flux*, which are defined in terms of twisted matrices as in below:

$$f_{ABC} = 3\Omega_{[ABC]}, \quad \eta_A = \partial_M (U^{-1})^M_{\ A} - 2(U^{-1})^M_{\ A}\partial_M\rho, \tag{45}$$

where ρ is as in (37) and

$$\Omega_{ABC} = -(U^{-1})^M_{\ A}\partial_M (U^{-1})^N_{\ B} U^D_{\ N}\eta_{CD}. \tag{46}$$

Note that Ω_{ABC} are antisymmetric in the last two indices: $\Omega_{ABC} = -\Omega_{ACB}$. We also make the following definition

$$f_A = -\partial_M (U^{-1})^M_{\ A} = \Omega^C_{\ AC}. \tag{47}$$

Note that, due to complete antisymmetry of f_{ABC} in its indices, the only independent blocks of f_{ABC} out of the 8 possible combinations are $f^I_{\ JK}, f_{IJK}, f^{IJ}_{\ \ K}$ and f^{IJK}, $I = 1, \cdots, d$. It is customary to call these the geometric flux, the H-flux, the Q-flux and the R-flux, respectively [57].

It can be shown that the conditions (41) and (42) imply the following conditions on fluxes:

$$f^A_{\ BC}\partial_A g(X) = 0, \quad f^A\partial_A g(X) = 0. \tag{48}$$

Note that the second condition in (48) and (43) imply together that η_A should also satisfy

$$\eta^A\partial_A g(X) = 0. \tag{49}$$

In order for the Y dependence to drop from the reduced theory, the fluxes must be constant. Also, the following Jacobi identity and the orthogonality condition should be satisfied for the closure of the gauge algebra:

$$f_{E[AB} f_{C]D}^{\ \ \ \ E} = 0, \tag{50}$$

and

$$\eta^A f_{ABC} = 0. \tag{51}$$

These conditions are sufficient to ensure that the SS reduction of the DFT action of the NS-NS sector of Type II string theory is consistent. It was shown in [17] that the consistency of the SS reduction of the RR sector requires one more condition, which is given below:

$$f_{ABC} f^{ABC} = 0. \tag{52}$$

The theory obtained under these conditions is called Gauged Double Field Theory (GDFT). The reduction is consistent and the action and the gauge transformation rules of the reduced theory are determined completely by the constant fluxes. This leads to an important principle that will play a crucial role in section 4: *reductions with inequivalent twist matrices yield the same dimensionally reduced theory, if the fluxes associated with these twist matrices are equivalent.* As mentioned above, in the supergravity frame GDFT reduces to gauged supergravity with fluxes corresponding to gauge parameters. Indeed, it can be shown that the constraints to be obeyed by the fluxes determining the GDFT (of the NS-NS sector) are in one-to-one correspondence with the constraints of half-maximal gauged supergravities. The extra condition that comes from the requirement of the consistency of the reduction of the RR sector implies that the gauged theory in hand corresponds to a truncation of maximal supergavity [58].

We would like to note that it is not necessary to impose the strong constraint in the internal space, that is, one does not need to impose

$$\partial^P U^A{}_M \partial_P U^B{}_N. \tag{53}$$

Therefore, the duality twisted anzats (34)-(37) allow for a relaxation of the strong constraint on the total space. This idea was exploited in [59] to find massive deformations of Type IIA theory. It was shown there that the massive Type IIA theory can be obtained from a duality twisted reduction of DFT with a twist matrix that violates both the weak and the strong constraint explicitly, where the mass deformation is induced through the reduction of the RR sector. This is interesting because massive Type IIA theory cannot be obtained from a conventional dimensional reduction of eleven dimensional supergravity.

The GDFT action resulting from the duality twisted reduction of DFT is as below:

$$S_{GDFT} = v \int dx\, d\tilde{x} \left(e^{-2d}(\mathcal{R} + \mathcal{R}_f) + \frac{1}{4}\langle \nabla\!\!\!/\chi, C^{-1}\mathbb{S}\, \nabla\!\!\!/\chi\rangle \right), \tag{54}$$

where v is defined as

$$v = \int d^d Y e^{-2\rho(Y)}, \tag{55}$$

and

$$\mathcal{R} \to \mathcal{R}_{\text{def}} = \mathcal{R} + \mathcal{R}_f, \tag{56}$$

with

$$\begin{aligned} \mathcal{R}_f \;=\; & -\frac{1}{2} f^A{}_{BC} \mathcal{H}^{BD} \mathcal{H}^{CE} \partial_D \mathcal{H}_{AE} - \frac{1}{12} f^A{}_{BC} f^D{}_{EF} \mathcal{H}_{AD} \mathcal{H}^{BE} \mathcal{H}^{CF} \\ & -\frac{1}{4} f^A{}_{BC} f^B{}_{AD} \mathcal{H}^{CD} - 2\eta_A \partial_B \mathcal{H}^{AB} + 4\eta_A \mathcal{H}^{AB} \partial_B d - \eta_A \eta_B \mathcal{H}^{AB}. \end{aligned} \tag{57}$$

The Dirac operator $\nabla\!\!\!/$ in (54) is defined as

$$\begin{aligned} \nabla\!\!\!/ \;&\equiv\; \partial\!\!\!/ + \frac{1}{12\sqrt{2}} f_{ABC} \Gamma^A\, \Gamma^B\, \Gamma^C + \frac{1}{2\sqrt{2}} \eta_B \Gamma^B \\ &=\; \partial\!\!\!/ + \frac{1}{6} f_{ABC} \psi^A\, \psi^B\, \psi^C + \frac{1}{2} \eta_B \psi^B. \end{aligned} \tag{58}$$

The constraint (52) implies that it is a nilpotent operator: $\nabla\!\!\!/^2 = 0$. The fact that it is determined by the fluxes f_{ABC} and f_A follows from the fact that the spinorial twist matrix $S(Y)$ projects under the double covering homomorphism ρ between $Pin(d,d)$ and $O(d,d)$ onto the twist matrix $U(Y)$, and hence these elements are generated by the same Lie algebra element in different representations [17].

3.1. More on fluxes

In the next section, where we will discuss string dualities and integrable deformations of string sigma models in the context of GDFT, we will see that invariance of the fluxes under the relevant $O(d,d)$ transformation will be of crucial importance. Due to its important role in that section, we would like to give a more detailed picture of their algebraic structure in this subsection.

A generic element T of $O(d,d)$ has the following form

$$T = \begin{pmatrix} e^A & e^A B \\ \beta e^A & \beta e^A B + (e^{-A})^T \end{pmatrix}. \tag{59}$$

Any such matrix can be generated as a product of the following $O(d,d)$ matrices:

$$h_B = \begin{pmatrix} 1 & -B \\ 0 & 1 \end{pmatrix} \quad B^T = -B\,, \tag{60}$$

$$h_\beta = \begin{pmatrix} 1 & 0 \\ \beta & 1 \end{pmatrix}, \quad \beta^T = -\beta \tag{61}$$

$$h_A = \begin{pmatrix} e^A & 0 \\ 0 & (e^{-A})^T \end{pmatrix}. \tag{62}$$

If the corresponding $Spin(d,d)$ elements are denoted by S_B, S_β and S_A, then their spinorial action on a spinor field α (regarded as a polyform as discussed in previous sections) is

$$S_B: \ \alpha \longmapsto e^{-B}\wedge\alpha = (1 - B + \frac{1}{2}B\wedge B - \ldots)\wedge\alpha, \tag{63}$$

$$S_\beta: \ \alpha \longmapsto e^{\beta}\alpha = (1 + i_\beta + \frac{1}{2}i_\beta^2 + \cdots)\alpha, \tag{64}$$

$$S_A: \ \alpha \longmapsto \frac{1}{\sqrt{detR}}(e^A)^*\alpha. \tag{65}$$

Here $B = \frac{1}{2}B_{kl}\psi^k\wedge\psi^l$, and $\beta = \frac{1}{2}\beta^{kl}\psi_k\wedge\psi_l$. Also, $i_\beta\alpha = \frac{1}{2}\beta^{ij}i_{\psi_i}(i_{\psi_j}\alpha)$ and $R^*\alpha = R_j^{\ i}\psi^j\wedge i_{\psi_i}\alpha$, with $R = e^A$, which is the usual action of $GL^+(d)$ on forms, where $GL^+(d)$ is the space of (orientation preserving) linear transformations of strictly positive determinant.

The flux associated with $U(Y)$ in (59) can be computed from (45). We do not present the result here, but rather refer the reader to [57] (see their equation (5.16) for the general expressions and equation (5.18) for the form of the fluxes in the supergravity frame, that is, when $\tilde{\partial}^i = 0$.) It can be checked that in the supergravity frame, the fluxes become the structure functions for the Roytenberg bracket $[,]_{Royt}$ [49, 60, 61, 62], which is obtained from the Courant bracket by twisting it with a 2-form field B and a bivector field β. Equivalently, one can take (e_i, e^i) as a non-holonomic basis for the direct sum bundle $TM\oplus T^*M$ satisfying

$$[e_i, e_j] = f_{ij}^{\ \ k}e_k \tag{66}$$

$$de^k = \frac{1}{2}f_{ij}^{\ \ k}e^i\wedge e^j, \tag{67}$$

and twist them as

$$e_i \to \hat{e}_i = e_i + B_{ij} e^j, \qquad e^i \to \hat{e}^i = e^i + \beta^{ij} e_j.$$

Calculating the Courant bracket of the resulting basis elements gives

$$\begin{aligned} [e_i, e_j]_{Royt} &= [\hat{e}_i, \hat{e}_j]_{Cour} = f_{ij}^{\ \ k} \hat{e}_k + H_{ijk} \hat{e}^k, \\ \left[e_i, e^j\right]_{Royt} &= [\hat{e}_i, \hat{e}^j]_{Cour} = f_{ik}^{\ \ j} \hat{e}^k + Q_i^{\ jk} \hat{e}_k, \\ \left[e^i, e^j\right]_{Royt} &= \left[\hat{e}^i, \hat{e}^j\right]_{Cour} = Q_k^{\ ij} \hat{e}^k + R^{ijk} \hat{e}_k. \end{aligned} \tag{68}$$

Explicit expressions or $f_{ij}^{\ \ k}, H_{ijk}, Q_i^{\ jk}$ and R^{ijk} are given in equation (5.18) of [57]. When $H = R = 0$ and f, Q are constant, (68) becomes the Lie algebra on a Drinfeld double with f and Q being the structure constants of the Lie algebra on g and dual of g (see footnote 5). Imposing the generalized Jacobi identity to be obeyed by the Courant bracket, one ends up with a set of constraints to be obeyed by the fluxes. When these conditions are satisfied, the bracket is a Courant bracket and $(\hat{e}_i, \hat{e}^i)$ forms a basis for the sections of a Courant algebroid. These non-constant fluxes play a fundamental role in the so-called flux formulation of DFT [57, 43]. In this formulation, the twist matrix $U(y)$ plays the role of a generalized vielbein used to define flat derivatives: $\mathcal{D}_A = U_A^{\ M} \partial_M$ and the action and the field equations are written in terms of the fluxes (see 2.18, 2.58 and 2.59 of [57]). For the consistency of the theory (e.g. for gauge invariance of the action, closure of the gauge algebra, covariance of generalized tensors), the fluxes must obey certain Bianchi identities (see 1.5 of [57] for the identities coming from the NS sector). When the strong constraint is imposed, these Bianchi identities are exactly the same as the constraints fluxes should obey so that they form the structure functions of the Roytenberg algebra (68) [61]. Therefore, we see that the underlying mathematical structure best suited to describe the non-constant fluxes is a Courant algebroid. Indeed, it was shown in [63] that the axioms for a Courant algebroid can be written in terms of three equations, which take the following form in local coordinates [64, 65]:

$$\eta^{IJ} \rho_I^i \rho_J^j = 0, \tag{69}$$

$$\rho_I^i \partial_i \rho_J^j - \rho_J^i \partial_i \rho_I^j - \eta^{KL} \rho_K^j F_{LIJ} = 0, \tag{70}$$

$$4 \rho_{[L}^i \partial_i F_{IJK]} + 3 \eta^{MN} F_{M[IJ} F_{KL]N} = 0. \tag{71}$$

Here e^I is a local basis for the sections of the Courant algebroid, F_{IJK} are the structure functions of the Courant bracket:

$$\left[e^I, e^J\right] = \eta^{IK} \eta^{JL} F_{KLM} e^M, \tag{72}$$

and ρ is the anchor with components $\rho = (\rho^i_{\ j}, \rho^{ij})$. If we identify the components of ρ with the twist matrix $U(y)$ in (45), (70) becomes the definition for the DFT fluxes (in the supergavity frame) and (71) is just a compact way of writing the Bianchi identities for fluxes [57, 64]. For constant fluxes, the Bianchi identities reduce to the constraints to be obeyed by the fluxes for the consistency of the SS reduction, which we discussed in the previous section. As an interesting remark, we note that generalizing the set of axioms (69-71) in such a way that they yield the expressions and Bianchi identities for the DFT fluxes (without imposing the constraint), an algebroid structure for DFT was proposed in [64].

4. Integrable deformations of string sigma models and GDFT

Recently, it has been understood that certain interesting integrable deformations of string sigma models can be described as coordinate dependent $O(d,d)$ transformations [29, 30, 31, 32, 33, 34, 35, 36]. Moreover, it was shown in general terms that $O(d,d)$ transformations preserve classical integrability [37]. This makes DFT and GDFT a very convenient setting for discussing such deformations. In this final section of this review paper, we will discuss non-Abelian T-duality and Yang-Baxter deformations, and describe how regarding them as transformations in DFT makes the analysis of the deformed theories easier, by following closely the papers [34] and [36].

4.1. Non-Abelian T-duality

Non-Abelian T-duality (NATD) is a generalization of T-duality for strings on backgrounds with non-Abelian isometries [66, 67, 68, 69, 70, 71, 72]. NAT dual background of a given string background can be found by using the standard tools of the Buscher method. For a generic non-linear sigma model with isometry group G, one starts with gauging the symmetry group (or a subgroup of it) and introduces Lagrange multiplier terms which constrain the gauge field to be pure gauge. Integrating out the Lagrange multipliers, one obtains the original model. Integrating out the gauge field gives the NATD model, for which the Lagrange multiplier terms play the role of coordinates on the dual manifold. Recently, it was shown that NATD can be described as a coordinate dependent $O(d,d)$ transformation [29, 30, 31, 34, 35, 73]. The transformation under NATD of the metric, the B-field and the dilaton field are determined by the coordinate dependent $O(d,d)$ matrix (which we also call T_{NATD}) obtained by embedding the following $O(n,n)$ matrix in $O(d,d)$ in the usual way (see 4.2.28-4.2.29 of [74]):

$$T_{\text{NATD}} = \begin{pmatrix} 0 & 1_n \\ 1_n & \theta_{IJ} \end{pmatrix}, \quad \theta_{IJ} = \nu_K C_{IJ}{}^K. \tag{73}$$

Here C_{IJ}^K are the structure constants of the Lie algebra of the isometry group G with respect to a fixed basis T_I. We assume that G is n dimensional and acts without isotropy so that $I, J = 1, \cdots, n$. The corresponding $Spin(d,d)$ element is

$$S_{\text{NATD}} = C_n S_\theta = S_\beta C_n. \tag{74}$$

The factors S_θ and S_β in S_{NATD} are as in (63), (64) with $\theta_{ij} = \nu_k C_{ij}{}^k$ and $\beta_{ij} = \nu_k C_{ij}{}^k$, respectively. C_n is the charge conjugation matrix

$$C_n = \Lambda_1 \cdots \Lambda_n, \tag{75}$$

where

$$\Lambda_i = (\psi^i - \psi_i). \tag{76}$$

Now suppose that a supergravity background with non-Abelian isometry G is given and we would like to find the NAT dual of it by dualizing with respect to G by transforming it with T_{NATD} and S_{NATD}. For this, DFT is a very convenient framework since all the DFT fields transform naturally under $O(d,d)$ and its spinorial counterpart. So, instead of working with the standard supergravity fields, we work with the corresponding DFT fields: the generalized metric, the generalized dilaton field and the spinorial field χ that encodes the RR gauge potentials and their Hodge duals. To be more precise, the spinorial field that transforms is F (that encodes the field strengths, not the gauge potentials), which is related to χ as $F = e^{-B} \not{\partial} \chi$. Due to the isometry respected by the background, these fields can be written in the following form:

$$\begin{aligned} \mathcal{H}^{MN}(x,\theta) &= L^M{}_A(\theta) \mathcal{H}^{AB}(x) L^N{}_B(\theta), & (77) \\ \mathcal{K}(x,\theta) &= S_L(\theta) \mathcal{K}(x) S_L^{-1}(\theta), & (78) \\ F(x,\theta) &= e^{-B'} S_L(\theta) e^B F(x) = S_L(\theta) F(x), & (79) \end{aligned}$$

where

$$L = \begin{pmatrix} l^T & 0 \\ 0 & l^{-1} \end{pmatrix}. \tag{80}$$

Here, $\theta = \{\theta^i, i = 1, \cdots, n\}$ are coordinates for G and components of l are the coefficients of the left invariant 1-forms $\sigma^I = l^I{}_i d\theta^i$ on G defined from the Maurer-Cartan form: $g^{-1}dg = \sigma^I T_I$. The second equality in equation (79) is due to the special diagonal form of L.

Note that the isometry of the background has made it possible to write all the fields in a separated form as in a SS reduction, where the matrix $L(\theta)$ plays the role of a duality twist. For this reason, we will refer to the fields $\mathcal{H}(x), \mathcal{K}(x), F(x)$ on the right hand side that depend only on the x coordinates *untwisted fields*, and in this context the background fields will be referred to as *twisted fields*. These twisted fields satisfy the field equations of DFT in the supergravity frame since they form a supergravity background. What about the untwisted fields? As discussed in detail in [34], it can be shown that these fields satisfy a set of equations which is a deformation of the DFT equations and the deformation is determined solely by the fluxes associated with the twist matrix.[6] When the twist matrix is as in (80) one can easily compute that the associated flux is the geometric flux with $f_{ij}^{\ k} = C_{ij}^{\ k}$. The untwisted fields can now be twisted by a a different twist matrix $U(Y)$, which satisfies the constraints (41), (42) to obtain DFT fields $\mathcal{H}(x, Y), \mathcal{K}(x, Y)$ and $F(x, Y)$. These new twisted fields (different from the DFT fields associated with the seed background since $U(Y) \neq L(\theta)$) satisfy the field equations of DFT (and hence of supergravity if we identify the new coordinates Y with space-time coordinates) if and only if $U(Y)$ produces the same constant flux as with L, that is geometric flux with $f_{ij}^{\ k} = C_{ij}^{\ k}$ (remember our discussion in section 3, on page 12). Now comes the key point: our matrix T_{NATD} which generates the NAT dual backgrounds is exactly of this form, that is, the flux associated with it is just the geometric flux with the same components! (In fact with a generic matrix of the form (73) one would also have H-flux, but it vanishes in our case due to the identity $C_{L[I}^{\ \ H} C_{JK]}^{\ \ \ L} = 0$ which follows from the Jacobi identity for the Lie algebra of G.) As a result, one immediately concludes that the fields of the NAT dual background must also satisfy the supergravity equations, that is, NATD is a solution generating transformation. The dual supergravity fields can be read off from the transformed DFT fields given below:

$$\mathcal{H}'(x, \nu) = T_{\mathrm{NATD}}(\nu)\mathcal{H}(x)(T_{\mathrm{NATD}})^t(\nu), \tag{81}$$

$$\mathcal{K}'(x, \nu) = S_{\mathrm{NATD}}(\nu)\mathcal{K}(x)(S_{\mathrm{NATD}})^{-1}(\nu), \tag{82}$$

$$F'(x, \nu) = e^{-\sigma(\nu)} e^{-B'(x,\nu)} S_{\mathrm{NATD}}(\nu) e^{B(x)} F(x), \tag{83}$$

$$d'(x, \nu) = d(x) + \sigma(\nu). \tag{84}$$

$B'(x, \nu)$ that appears in (83) is read off from $\mathcal{H}'(x, \nu)$ in (81). The field $\sigma(\nu)$ in (84) and (83) is non-vanishing only when the isometry group is non-unimodular.

As we mentioned in section 3, the invariance of fluxes under the NATD transformation plays a crucial role in showing that NATD is a solution generating transformation. Interestingly, this idea is also sufficient to prove that NATD also preserves supersymmetry of the background, at least for backgrounds of a certain type. Indeed, very recently in [75] transformation of supersymmetry equations under NATD was discussed also using the framework GDFT for backgrounds which are of the form $Mink_{3,1} \times M$, where M is six dimensional. Demanding that the four dimensional solution preserves at least $\mathcal{N} = 1$ supersymmetry implies that the structure group of the generalized tangent bundle $TM \oplus T^*M$ of the six dimensional internal manifold M is reduced from $SO(6,6)$ to $SU(3) \times SU(3)$. This topological condition on the internal manifold implies the existence of two globally defined compatible pure spinors Φ_1 and Φ_2 of non-vanishing norm. These $Clif(6,6)$ spinors can be constructed from the internal spinors arising from the 10 dimensonal Killing spinors generating the supersymmetry transformations in

[6] These are almost the same as the field equations of GDFT, except for the equations coming from variations of the field χ. There are also some subtleties when the twist matrix is not in $Spin^+(d, d)$. See [34] for a detailed discussion.

10 dimensions. For preservation of supersymmetry, these pure spinors (polyforms) must satisfy the following set of first order differential equations, as was first shown in [76]:

$$d(e^{2A-\phi}e^B \wedge \Phi_1) = 0, \tag{85}$$

$$d(e^{2A-\phi}e^B \wedge \Phi_2) = e^{2A-\phi}dA \wedge e^B \wedge \bar{\Phi}_2 + \frac{i}{8}e^{3A}e^B \wedge \lambda(*_6 F). \tag{86}$$

For detailed information and derivation of these equations, see Appendix A of [77] and Appendix B of [78]. Under NATD, the pure spinors transform as

$$\Phi \to \Phi' = \sqrt{G}\; S_{\rm NATD}.\Phi = \sqrt{G}\; e^{-B'} S_{\rm NATD} e^B \Phi, \tag{87}$$

where

$$G = det((g+B)+\theta)^{-1}. \tag{88}$$

In [75] these equations were embedded in DFT and it was shown that (87) maps solutions to solutions by using the framework of GDFT, the key idea again being the preservation of fluxes.

4.2. Yang-Baxter deformations

Yang-Baxter (YB) sigma model was introduced by Klimcik in [79] as a particular type of deformation for Principal Chiral Models (PCM) with simple compact Lie group G as the target manifold. The deformation is based on solutions of modified classical Yang-Baxter equation (mCYBE) associated with the Lie algebra of G. The integrability of the YB sigma model was shown again by Klimcik later in [80]. The applicability of YB deformations was extended to symmetric coset spaces in [81], which in turn was applied to $AdS_5 \times S^5$ in [82]. The NS-NS sector of the corresponding deformed supergravity background was found in [83]. When applied to AdS backgrounds, such deformations are usually called η-deformations. It is also possible to consider similar deformations based on the classical Yang-Baxter equation (CYBE), as opposed to mCYBE and the resulting models are called homogeneous Yang-Baxter models. Such deformations of the $AdS_5 \times S^5$ string were first studied in [84]. These deformations are particularly interesting, as they provide a natural generalization of the Lunin-Maldacena (LM) deformations [85], obtained by T-duality-shift-T-duality (TsT) transformations [86, 87, 88]. Recently it has been understood that YB deformations, like NATD, can also be described as a coordinate dependent $O(d,d)$ transformation. In fact, this is not surprising, since it is by now well understood that YB deformation is related to NATD [89, 90, 91, 30]. The first work in this direction was [32], where it was proposed that the homogeneous YB deformation could be obtained by the action of a β-transformation. This proposal is based on observation and is supported by working out the example of the $AdS_5 \times S^5$ background. Later in [33], a proof of the proposal of [32] was given for deformations of the $AdS_5 \times S^5$ background. Then in [36], it was shown that the deformed sigma model constructed in [92] as a deformation of a generic string sigma model can also be obtained by the action of a coordinate dependent $O(d,d)$ transformation. Before we present the details, let us note that the approach adopted in the papers [93, 94, 95, 96], which regards YB deformation as an open-closed string map [97] also contributes to the understanding of the $O(d,d)$ structure of YB deformations, as the open-closed string map can be regarded as a special case of β-transformation [36].

The $O(d,d)$ transformation that generates the YB deformation is as follows[7]:

$$T_{\text{YB}} = L.\begin{pmatrix} 1 & 0 \\ \eta R_g^{IJ} & 1 \end{pmatrix}. \tag{89}$$

Here, the first factor L is as in (80) and the second factor is a β-transformation (61) with the bivector field R_g constructed from an R-matrix on the Lie algebra of the isometry group. An R-matrix on a Lie algebra g with a non-degenerate inner product is a skew-symmetric endomorphism on g, which satisfies the CYBE [98]:

$$[RX, RY] - R([RX, Y] + [X, RY]) = 0, \quad \forall X, Y \in g. \tag{90}$$

When this equation is satisfied, the skew-symmetric bracket

$$[X, Y]_R = [RX, Y] + [X, RY] \tag{91}$$

satisfies the Jacobi identity and hence also defines a Lie bracket. The dressed R-matrix R_g is constructed by extending R to the whole group manifold by the adjoint action:

$$R_g = Ad_{g^{-1}} R Ad_g. \tag{92}$$

Obviously, R_g satisfies CYBE if R does, since the adjoint action is an automorphism of the Lie bracket. An interesting fact is that the brackets $[,]$ and $[,]_R$ are compatible in the sense that it is possible to construct a Lie bracket (from the brackets $[,]$ and $[,]_R$) on the direct sum $g \oplus g^*$, that is (g, g^*) is a Lie bialgebra and $g \oplus g^*$ has a Drinfeld double structure (see footnote 5). Now, one can construct a bivector field from the dressed R-matrix R_g (which we will also call R_g) in the following way: first one defines a non-degenerate 2-form ω

$$\kappa(RX, Y) = \omega(X, Y), \quad X, Y \in g. \tag{93}$$

Here κ is the non-degenerate inner product on g. ω is indeed a 2-form as the skew-symmetry property of R implies that $\omega(X, Y) = -\omega(Y, X)$. Also, it is non-degenerate, at least on a subalgebra of g called the Frobenius subalgebra[8]. On this subalgebra one can define a bivector field β

$$\beta = \frac{1}{2} R_g^{IJ} T_I \wedge T_J, \tag{94}$$

with

$$\omega(T_I, T_J) = (R_g^{-1})_{IJ}. \tag{95}$$

In the above expressions, T_I, T_J are the generators of the Frobenius subalgebra. The fact that R_g satisfies the CYBE is then equivalent to the condition that the Schouten bracket (a natural

[7] The matrix (89) differs slightly both from the one in [32] and the one in [36]. This is because in [32] the $O(d,d)$ matrix (equal to $T_{\text{YB}} L^{-1}$) acts on the seed background directly, whereas here our formalism requires that the YB matrix should act on untwisted fields. On the other hand, [36] is interested in how untwisted fields are mapped to untwisted fields under YB deformation, and this mapping is generated by $L^{-1} T_{\text{YB}}$. This is the reason why the flux associated with the YB matrix found in [36] is Q-flux, whereas here we will find that the Q-flux is zero and the geometric flux is preserved. See Appendix B of [36] for more details.

[8] It is known that skew-symmetric solutions for CYBE on a finite dimensional Lie algebra g are in one-to-one correspondence with quasi-Frobenius subalgebras of g. On such a subalgebra the R-matrix acts as a non-degenerate operator. If this subalgebra is Abelian, the R-matrix is called Abelian; if this subalgebra is unimodular, the R-matrix is called unimodular. The rank of the R-matrix is equal to the dimension of the quasi-Frobenius subalgebra. For more detail, see [99] and Proposition 22.6 and Proposition 3.1.6 in the book [100].

extension of the Lie bracket to multivector fields) of β with itself (which is normally a trivector) vanishes:

$$[\beta, \beta]_S = 0. \tag{96}$$

Equivalently, the bivector field β defines a Poisson structure.

In order to find the supergravity background corresponding to the YB deformed sigma model, one has to form the untwisted DFT fields as we did in the previous section and transform them as in (81)-(83), where now the $O(d,d)$ and $Spin(d,d)$ matrices involved are T_{YB} and S_{YB}. The resulting fields solve equations of DFT (and hence of Type II supergravity since one remains in the supergravity frame) because the flux associated with the YB transformation is just geometric flux with $f_{ij}{}^k = C_{ij}{}^k$. In fact for an arbitrary bivector field, a twist matrix of the form (89) would also have R-flux. However, it is identically zero in our case, resulting directly from the fact that β defines a Poisson structure.

5. Outlook

Symmetry plays a fundamental role in physics. The dynamics of some of the most successful theories in physics is *dictated* by symmetry principles. This is the case with the standard model of particle physics, where the theory is based on gauge symmetries described by Lie algebras and also is the case with general relativity, where the structure of the theory is determined by demanding invariance under diffeomorphisms and Lorentz transformations. String theory possesses duality symmetries, which do not exist in conventional field theories. In recent years, remarkable progress has been made in constructing field theories which are *dictated* by these duality symmetries. This review paper has focused on some aspects of one of these theories, Double Field Theory, which is based on T-duality.

As we have seen in parts of the paper, construction of DFT requires using a mathematical framework that goes beyond semi-Riemannian geometry. Most of the mathematical structures are constructed (at least in the supergravity frame) on the direct sum bundle $TM \oplus T^*M$, a prototypical Courant algebroid. This Courant algebroid also plays a fundamental role for the generalized geometry program of Nigel Hitchin [45, 46], and indeed many constructions in DFT are closely related to those in generalized geometry.

Instead of discussing these interesting mathematical structures in detail, we have preferred in this paper to focus on a rather practical feature of DFT: its role in generating new, interesting supergravity solutions. Recently, it has been understood that certain (integrable) deformations of string sigma models can be described as coordinate dependent $O(d,d)$ transformations. On the other hand, when embedded in DFT, transformations of the supergravity fields under $O(d,d)$ (and its spinorial counterpart) can be described very easily. This makes DFT a very convenient setting for studying such deformations. We studied in detail two such (coordinate dependent) $O(d,d)$ transformations. The first one generates the non-Abelian T-dual of a given supergravity background with non-Abelian isometries, and the other one generates the supergravity fields in the target space of the Yang-Baxter deformation of a string sigma model. In both cases, the transformed fields can be expressed in a separated form as in a SS reduction, so the transformed theory can be easily analyzed by using tools from GDFT. This is a deformation of DFT determined by fluxes, and we have seen that invariance of fluxes under the associated $O(d,d)$ transformation is the key idea in showing that NATD and YB deformations are both solution generating transformations. It is also the key idea in showing that NATD preserves supersymmetry.

The approach we have discussed here makes it clear that classification of solution generating transformations for supergravity can be turned into a less difficult, algebraic problem: classification of fluxes. Indeed, this approach has been used very recently in the interesting papers [101, 102, 38]. An interesting future direction would be to try and understand solution generating transformations in terms of morphisms of Courant algebroids. Indeed, we have seen that fluxes, under the conditions imposed by the requirement of consistency of DFT, form structure functions of a Courant algebroid. Hence, understanding morphisms of Courant algebroids preserving structure functions might be a fruitful way to explore solution generating transformations.

T-duality covariant formulation of string theory has many remarkable features. It leads to new, interesting mathematical structures and also offers practical solutions to some interesting problems. We believe both these directions are worth further exploration.

Acknowledgements

This review has been prepared as a contribution to the Festschrift in honor of Prof. Tekin Dereli on the occasion of his retirement and his birthday. I am truly fortunate that Prof. Dereli was my supervisor and mentor as a graduate student and I would like to take this opportunity to thank him for his enduring guidance and support and for being an inspiration to me and to many other students and scientists in Turkey.

This work is supported by the Turkish Council of Research and Technology (TÜBİTAK) through the ARDEB 1001 project with grant number 121F123. I thank Emine Diriöz and Seçil Tunalı for collaboration on some parts of the work that has been reviewed here.

References

[1] A. Giveon, M. Porrati and E. Rabinovici, "Target space duality in string theory," Phys. Rept. **244** (1994), 77-202 doi:10.1016/0370-1573(94)90070-1 [arXiv:hep-th/9401139 [hep-th]].

[2] E. Cremmer, B. Julia, H. Lu and C. N. Pope, "Dualization of dualities. 1.," Nucl. Phys. B **523** (1998), 73-144 doi:10.1016/S0550-3213(98)00136-9 [arXiv:hep-th/9710119 [hep-th]].

[3] E. Cremmer and B. Julia, "The SO(8) Supergravity," Nucl. Phys. B **159** (1979), 141-212 doi:10.1016/0550-3213(79)90331-6

[4] E. Cremmer and B. Julia, "The N=8 Supergravity Theory. 1. The Lagrangian," Phys. Lett. B **80** (1978), 48 doi:10.1016/0370-2693(78)90303-9

[5] E. Cremmer, "Dimensional Reduction In Field Theory And Hidden Symmetries In Extended Supergravity," LPTENS 81/18 *Lectures given at ICTP Spring School Supergravity, Trieste, Italy, Apr 22 - May 6, 1981*

[6] B. Julia, "Group Disintegrations," LPTENS 80/16 *Invited paper presented at Nuffield Gravity Workshop, Cambridge, Eng., Jun 22 - Jul 12, 1980*

[7] M. J. Duff and J. X. Lu, "Duality Rotations In Membrane Theory," Nucl. Phys. B **347** (1990) 394. ,

[8] M. J. Duff, "Duality Rotations In String Theory," Nucl. Phys.

[9] W. Siegel, "Two vierbein formalism for string inspired axionic gravity," Phys. Rev. D **47** (1993) 5453 doi:10.1103/PhysRevD.47.5453 [hep-th/9302036].

[10] W. Siegel, "Superspace duality in low-energy superstrings," Phys. Rev. D **48** (1993) 2826 doi:10.1103/PhysRevD.48.2826 [hep-th/9305073].

[11] W. Siegel, "Manifest duality in low-energy superstrings," hep-th/9308133.

[12] O. Hohm, C. Hull and B. Zwiebach, "Background independent action for double field theory," JHEP **07** (2010), 016 doi:10.1007/JHEP07(2010)016 [arXiv:1003.5027 [hep-th]].

[13] C. Hull and B. Zwiebach, "The Gauge algebra of double field theory and Courant brackets," JHEP **09** (2009), 090 doi:10.1088/1126-6708/2009/09/090 [arXiv:0908.1792 [hep-th]].

[14] O. Hohm, C. Hull and B. Zwiebach, "Generalized metric formulation of double field theory," JHEP **1008** (2010) 008 doi:10.1007/JHEP08(2010)008 [arXiv:1006.4823 [hep-th]].

[15] O. Hohm, S. K. Kwak and B. Zwiebach, "Double Field Theory of Type II Strings," JHEP **09** (2011), 013 doi:10.1007/JHEP09(2011)013 [arXiv:1107.0008 [hep-th]].

[16] E. Bergshoeff, R. Kallosh, T. Ortin, D. Roest and A. Van Proeyen, "New formulations of D = 10 supersymmetry and D8 - O8 domain walls," Class. Quant. Grav. **18** (2001), 3359-3382 doi:10.1088/0264-9381/18/17/303 [arXiv:hep-th/0103233 [hep-th]].

[17] A. Çatal-Özer, "Duality Twisted Reductions of Double Field Theory of Type II Strings," JHEP **09** (2017), 044 doi:10.1007/JHEP09(2017)044 [arXiv:1705.08181 [hep-th]].

[18] J. Scherk and J. H. Schwarz, "How to get masses from extra dimensions", Nucl. Phys. B **153**, (1979) 61 .

[19] J. Scherk and J. H. Schwarz, "Spontaneous Breaking Of Supersymmetry Through Dimensional Reduction," Phys. Lett. B **82** (1979) 60.

[20] D. Geissbuhler, "Double Field Theory and N=4 Gauged Supergravity," JHEP **1111** (2011) 116 doi:10.1007/JHEP11(2011)116 [arXiv:1109.4280 [hep-th]].

[21] G. Aldazabal, W. Baron, D. Marqués and C. Nunez, "The effective action of Double Field Theory," JHEP **1111** (2011) 052 Erratum: [JHEP **1111** (2011) 109] doi:10.1007/JHEP11(2011)052, 10.1007/JHEP11(2011)109 [arXiv:1109.0290 [hep-th]].
[22] M. Graña and D. Marqués, "Gauged double field theory," JHEP **1204** (2012) 020 doi:10.1007/JHEP04(2012)020 [arXiv:1201.2924 [hep-th]].
[23] W. Cho, J. J. Fernández-Melgarejo, I. Jeon and J. H. Park, "Supersymmetric gauged double field theory: systematic derivation by virtue of twist," JHEP **1508** (2015) 084 doi:10.1007/JHEP08(2015)084 [arXiv:1505.01301 [hep-th]].
[24] D. S. Berman and K. Lee, "Supersymmetry for Gauged Double Field Theory and Generalised Scherk-Schwarz Reductions," Nucl. Phys. B **881** (2014) 369 doi:10.1016/j.nuclphysb.2014.02.015 [arXiv:1305.2747 [hep-th]].
[25] G. Dibitetto, J. J. Fernandez-Melgarejo, D. Marqués and D. Roest, "Duality orbits of non-geometric fluxes," Fortsch. Phys. **60** (2012), 1123-1149 doi:10.1002/prop.201200078 [arXiv:1203.6562 [hep-th]].
[26] J. Shelton, W. Taylor and B. Wecht, "Nongeometric flux compactifications," JHEP **10** (2005), 085 doi:10.1088/1126-6708/2005/10/085 [arXiv:hep-th/0508133 [hep-th]].
[27] D. Andriot, O. Hohm, M. Larfors, D. Lüst and P. Patalong, "Non-Geometric Fluxes in Supergravity and Double Field Theory," Fortsch. Phys. **60** (2012) 1150 doi:10.1002/prop.201200085 [arXiv:1204.1979 [hep-th]].
[28] E. Plauschinn, "Non-geometric backgrounds in string theory," Phys. Rept. **798** (2019), 1-122 doi:10.1016/j.physrep.2018.12.002 [arXiv:1811.11203 [hep-th]].
[29] F. Hassler, "Poisson-Lie T-Duality in Double Field Theory," Phys. Lett. B **807** (2020), 135455 doi:10.1016/j.physletb.2020.135455 [arXiv:1707.08624 [hep-th]].
[30] D. Lüst and D. Osten, "Generalised fluxes, Yang-Baxter deformations and the O(d,d) structure of non-abelian T-duality," JHEP **1805** (2018) 165 doi:10.1007/JHEP05(2018)165 [arXiv:1803.03971 [hep-th]].
[31] S. Demulder, F. Hassler and D. C. Thompson, "Doubled aspects of generalised dualities and integrable deformations," JHEP **02** (2019), 189 doi:10.1007/JHEP02(2019)189 [arXiv:1810.11446 [hep-th]].
[32] J. I. Sakamoto, Y. Sakatani and K. Yoshida, "Homogeneous Yang-Baxter deformations as generalized diffeomorphisms," J. Phys. A **50** (2017) no.41, 415401 doi:10.1088/1751-8121/aa8896 [arXiv:1705.07116 [hep-th]].
[33] J. I. Sakamoto and Y. Sakatani, "Local β-deformations and Yang-Baxter sigma model," JHEP **1806** (2018) 147 doi:10.1007/JHEP06(2018)147 [arXiv:1803.05903 [hep-th]].
[34] A. Çatal-Özer, "Non-Abelian T-duality as a transformation in double field theory," JHEP **08** (2019), 115 doi:10.1007/JHEP08(2019)115 [arXiv:1904.00362 [hep-th]].
[35] Y. Sakatani, "Type II DFT solutions from Poisson-Lie T-duality/plurality," doi:10.1093/ptep/ptz071 [arXiv:1903.12175 [hep-th]].
[36] A. Çatal-Özer and S. Tunalı, "Yang-Baxter deformation as an $O(d,d)$ transformation," Class. Quant. Grav. **37** (2020) no.7, 075003 doi:10.1088/1361-6382/ab6f7e [arXiv:1906.09053 [hep-th]].
[37] D. Orlando, S. Reffert, Y. Sekiguchi and K. Yoshida, "$O(d,d)$ transformations preserve classical integrability," Nucl. Phys. B **950** (2020), 114880 doi:10.1016/j.nuclphysb.2019.114880 [arXiv:1907.03759 [hep-th]].
[38] T. Codina and D. Marqués, "Generalized Dualities and Higher Derivatives," JHEP **10** (2020), 002 doi:10.1007/JHEP10(2020)002 [arXiv:2007.09494 [hep-th]].
[39] G. Arutyunov, S. Frolov, B. Hoare, R. Roiban and A. A. Tseytlin, "Scale invariance of the η-deformed $AdS_5 \times S^5$ superstring, T-duality and modified type II equations," Nucl. Phys. B **903** (2016) 262 doi:10.1016/j.nuclphysb.2015.12.012 [arXiv:1511.05795 [hep-th]].
[40] L. Wulff and A. A. Tseytlin, "Kappa-symmetry of superstring sigma model and generalized 10d supergravity equations," JHEP **1606** (2016) 174 doi:10.1007/JHEP06(2016)174 [arXiv:1605.04884 [hep-th]].
[41] Y. Sakatani, S. Uehara and K. Yoshida, "Generalized gravity from modified DFT," JHEP **04** (2017), 123 doi:10.1007/JHEP04(2017)123 [arXiv:1611.05856 [hep-th]].
[42] J. i. Sakamoto, Y. Sakatani and K. Yoshida, "Weyl invariance for generalized supergravity backgrounds from the doubled formalism," PTEP **2017** (2017) no.5, 053B07 doi:10.1093/ptep/ptx067 [arXiv:1703.09213 [hep-th]].
[43] O. Hohm and S. K. Kwak, "Frame-like Geometry of Double Field Theory," J. Phys. A **44** (2011), 085404 doi:10.1088/1751-8113/44/8/085404 [arXiv:1011.4101 [hep-th]].
[44] M. Gualtieri, "Generalized complex geometry," math/0401221 [math-dg].
[45] N. Hitchin, "Lectures on generalized geometry," arXiv:1008.0973 [math.DG];
[46] N. Hitchin, "Generalized Calabi-Yau manifolds," Quart. J. Math. **54** (2003) 281 doi:10.1093/qjmath/54.3.281 [math/0209099 [math-dg]].
[47] S. Mukai, "Symplectic Structure of the Moduli Space of Sheaves on an Abelian or K3 Surface," Invent. Math., 77:101116, 1984.
[48] Z. J. Liu, A. Weinstein and P. Xu, "Manin Triples for Lie Bialgebroids," J. Diff. Geom. **45** (1997) no.3,

547-574 [arXiv:dg-ga/9508013 [math.DG]].

[49] D. Roytenberg. "Courant algebroids, derived brackets and even symplectic supermanifolds" PhD thesis, UC Berkeley, 1999. math.DG/9910078.

[50] T. Courant. "Dirac manifolds" Trans. Amer. Math. Soc., 319:631661, 1990.

[51] C. M. Hull, "Massive string theories from M-theory and F-theory", JHEP **9811**, (1998) 027 [arXiv:hep-th/9811021].

[52] C. M. Hull and R. A. Reid-Edwards, "Flux compactifications of string theory on twisted tori," Fortsch. Phys. **57** (2009), 862-894 doi:10.1002/prop.200900076 [arXiv:hep-th/0503114 [hep-th]].

[53] A. Çatal-Özer, "Duality twists on a group manifold," JHEP **10** (2006), 072 doi:10.1088/1126-6708/2006/10/072 [arXiv:hep-th/0606278 [hep-th]].

[54] C. M. Hull and A. Çatal-Özer, "Compactifications with S duality twists," JHEP **10** (2003), 034 doi:10.1088/1126-6708/2003/10/034 [arXiv:hep-th/0308133 [hep-th]].

[55] A. Çatal-Özer, "Scherk-Schwarz Reductions of Effective String Theories in Even Dimensions, Ph.D. thesis, Middle East Technical U., 2003, https://tez.yok.gov.tr/UlusalTezMerkezi/tezSorguSonucYeni.jsp

[56] J. Schon and M. Weidner, "Gauged N=4 supergravities," JHEP **0605** (2006) 034 doi:10.1088/1126-6708/2006/05/034 [hep-th/0602024].

[57] D. Geissbuhler, D. Marqués, C. Nunez and V. Penas, "Exploring Double Field Theory," JHEP **1306** (2013) 101 doi:10.1007/JHEP06(2013)101 [arXiv:1304.1472 [hep-th]].

[58] G. Aldazabal, D. Marqués, C. Nunez and J. A. Rosabal, "On Type IIB moduli stabilization and N = 4, 8 supergravities," Nucl. Phys. B **849** (2011) 80 doi:10.1016/j.nuclphysb.2011.03.016 [arXiv:1101.5954 [hep-th]];

[59] A. Çatal-Özer, "Massive deformations of Type IIA theory within double field theory," JHEP **02** (2018), 179 doi:10.1007/JHEP02(2018)179 [arXiv:1706.08883 [hep-th]].

[60] N. Halmagyi, "Non-geometric backgrounds and the first order string sigma model," [arXiv:0906.2891 [hep-th]].

[61] R. Blumenhagen, A. Deser, E. Plauschinn and F. Rennecke, "Bianchi Identities for Non-Geometric Fluxes - From Quasi-Poisson Structures to Courant Algebroids," Fortsch. Phys. **60** (2012), 1217-1228 doi:10.1002/prop.201200099 [arXiv:1205.1522 [hep-th]].

[62] A. Chatzistavrakidis, L. Jonke, F. S. Khoo and R. J. Szabo, "Double Field Theory and Membrane Sigma-Models," JHEP **07** (2018), 015 doi:10.1007/JHEP07(2018)015 [arXiv:1802.07003 [hep-th]].

[63] K. Uchino, "Remarks on the Definition of a Courant Algebroid", Letters in Mathematical Physics (2002) 60(2) doi:10.1023/A:1016179410273 [arXiv:math/0204010[math.DG]]

[64] A. Chatzistavrakidis, L. Jonke, F. S. Khoo and R. J. Szabo, "The Algebroid Structure of Double Field Theory," PoS **CORFU2018** (2019), 132 doi:10.22323/1.347.0132 [arXiv:1903.01765 [hep-th]].

[65] N. Ikeda, "Lectures on AKSZ Sigma Models for Physicists," doi:10.1142/9789813144613_0003 [arXiv:1204.3714 [hep-th]].

[66] X. C. de la Ossa and F. Quevedo, "Duality symmetries from non-Abelian isometries in string theory," Nucl. Phys. B **403** (1993) 377 doi:10.1016/0550-3213(93)90041-M [hep-th/9210021].

[67] E. Alvarez, L. Alvarez-Gaume and Y. Lozano, "Non-Abelian duality in WZW models,"

[68] A. Giveon and M. Rocek, "On non-Abelian duality," Nucl. Phys. B **421** (1994) 173 doi:10.1016/0550-3213(94)90230-5 [hep-th/9308154].

[69] K. Sfetsos, "Gauged WZW models and non-Abelian duality," Phys. Rev. D **50** (1994) 2784 doi:10.1103/PhysRevD.50.2784 [hep-th/9402031].

[70] E. Alvarez, L. Alvarez-Gaume and Y. Lozano, "On non-Abelian duality," Nucl. Phys. B **424** (1994) 155 doi:10.1016/0550-3213(94)90093-0 [hep-th/9403155].

[71] K. Sfetsos and D. C. Thompson, "On non-abelian T-dual geometries with Ramond fluxes," Nucl. Phys. B **846** (2011) 21 doi:10.1016/j.nuclphysb.2010.12.013 [arXiv:1012.1320 [hep-th]].

[72] Y. Lozano, E. Ó Colgáin, K. Sfetsos and D. C. Thompson, "Non-abelian T-duality, Ramond fields and coset geometries," JHEP **1106** (2011) 106 doi:10.1007/JHEP06(2011)106 [arXiv:1104.5196 [hep-th]].

[73] M. Bugden, "Non-abelian T-folds," JHEP **03** (2019), 189 doi:10.1007/JHEP03(2019)189 [arXiv:1901.03782 [hep-th]].

[74] A. Giveon, M. Porrati and E. Rabinovici, "Target space duality in string theory," Phys. Rept. **244** (1994), 77-202 doi:10.1016/0370-1573(94)90070-1 [arXiv:hep-th/9401139 [hep-th]].

[75] A. Çatal-Özer and E. Diriöz, "$Pin(d,d)$ covariance of pure spinor equations for supersymmetric vacua and non-Abelian T-duality," JHEP **12** (2021), 071 doi:10.1007/JHEP12(2021)071 [arXiv:2109.14580 [hep-th]].

[76] M. Graña, R. Minasian, M. Petrini and A. Tomasiello, "Generalized structures of N=1 vacua," JHEP **11** (2005), 020 doi:10.1088/1126-6708/2005/11/020 [arXiv:hep-th/0505212 [hep-th]].

[77] M. Graña, R. Minasian, M. Petrini and A. Tomasiello, "A Scan for new N=1 vacua on twisted tori," JHEP **05** (2007), 031 doi:10.1088/1126-6708/2007/05/031 [arXiv:hep-th/0609124 [hep-th]].

[78] L. Martucci and P. Smyth, "Supersymmetric D-branes and calibrations on general N=1 backgrounds", JHEP, 11, (2005), doi = "10.1088/1126-6708/2005/11/048", hep-th/0507099.

[79] C. Klimcik, "Yang-Baxter sigma models and dS/AdS T duality," JHEP **0212** (2002) 051 doi:10.1088/1126-6708/2002/12/051 [hep-th/0210095].

[80] C. Klimcik, "On integrability of the Yang-Baxter sigma-model," J. Math. Phys. **50** (2009) 043508 doi:10.1063/1.3116242 [arXiv:0802.3518 [hep-th]].

[81] F. Delduc, M. Magro and B. Vicedo, "On classical q-deformations of integrable sigma-models," JHEP **1311** (2013) 192 doi:10.1007/JHEP11(2013)192 [arXiv:1308.3581 [hep-th]].

[82] F. Delduc, M. Magro and B. Vicedo, "An integrable deformation of the $AdS_5 \times S^5$ superstring action," Phys. Rev. Lett. **112** (2014) no.5, 051601 doi:10.1103/PhysRevLett.112.051601 [arXiv:1309.5850 [hep-th]].

[83] G. Arutyunov, R. Borsato and S. Frolov, "S-matrix for strings on η-deformed $AdS_5 \times S^5$," JHEP **1404** (2014) 002 doi:10.1007/JHEP04(2014)002 [arXiv:1312.3542 [hep-th]].

[84] I. Kawaguchi, T. Matsumoto and K. Yoshida, "Jordanian deformations of the $AdS_5 \times S^5$ superstring," JHEP **1404** (2014) 153 doi:10.1007/JHEP04(2014)153 [arXiv:1401.4855 [hep-th]].

[85] O. Lunin and J. M. Maldacena, "Deforming field theories with U(1) x U(1) global symmetry and their gravity duals," JHEP **0505** (2005) 033 doi:10.1088/1126-6708/2005/05/033 [hep-th/0502086].

[86] S. Frolov, "Lax pair for strings in Lunin-Maldacena background," JHEP **0505** (2005) 069 doi:10.1088/1126-6708/2005/05/069 [hep-th/0503201].

[87] A. Çatal-Özer, "Lunin-Maldacena deformations with three parameters," JHEP **02** (2006), 026 doi:10.1088/1126-6708/2006/02/026 [arXiv:hep-th/0512290 [hep-th]].

[88] A. Çatal-Özer and N. S. Değer, "Beta, dipole and noncommutative deformations of M-theory backgrounds with one or more parameters," Class. Quant. Grav. **26** (2009), 245015 doi:10.1088/0264-9381/26/24/245015 [arXiv:0904.0629 [hep-th]].

[89] B. Hoare and A. A. Tseytlin, "Homogeneous Yang-Baxter deformations as non-abelian duals of the AdS_5 sigma-model," J. Phys. A **49** (2016) no.49, 494001 doi:10.1088/1751-8113/49/49/494001 [arXiv:1609.02550 [hep-th]].

[90] R. Borsato and L. Wulff, "Integrable deformations of T-Dual σ models," Phys. Rev. Lett. **117** (2016) no.25, 251602 doi:10.1103/PhysRevLett.117.251602 [arXiv:1609.09834 [hep-th]].

[91] R. Borsato and L. Wulff, "On non-abelian T-duality and deformations of supercoset string sigma-models," JHEP **1710** (2017) 024 doi:10.1007/JHEP10(2017)024 [arXiv:1706.10169 [hep-th]].

[92] R. Borsato and L. Wulff, "Non-abelian T-duality and Yang-Baxter deformations of Green-Schwarz strings," JHEP **1808** (2018) 027 doi:10.1007/JHEP08(2018)027 [arXiv:1806.04083 [hep-th]].

[93] T. Araujo, I. Bakhmatov, E. Ó. Colgáin, J. I. Sakamoto, M. M. Sheikh-Jabbari and K. Yoshida, "Conformal twists, Yang-Baxter σ-models and holographic noncommutativity," J. Phys. A **51** (2018) no.23, 235401 doi:10.1088/1751-8121/aac195 [arXiv:1705.02063 [hep-th]].

[94] T. Araujo, E. Ó Colgáin, J. I. Sakamoto, M. M. Sheikh-Jabbari and K. Yoshida, "I in generalized supergravity," Eur. Phys. J. C **77** (2017) no.11, 739 doi:10.1140/epjc/s10052-017-5316-5 [arXiv:1708.03163 [hep-th]].

[95] T. Araujo, I. Bakhmatov, E. Ó. Colgáin, J. I. Sakamoto, M. M. Sheikh-Jabbari and K. Yoshida, "Yang-Baxter σ-models, conformal twists, and noncommutative Yang-Mills theory," Phys. Rev. D **95** (2017) no.10, 105006 doi:10.1103/PhysRevD.95.105006 [arXiv:1702.02861 [hep-th]].

[96] T. Araujo, E. Ó. Colgáin, Y. Sakatani, M. M. Sheikh-Jabbari and H. Yavartanoo, "Holographic integration of $T\bar{T}$ and $J\bar{T}$ via $O(d,d)$," JHEP **1903** (2019) 168 doi:10.1007/JHEP03(2019)168 [arXiv:1811.03050 [hep-th]].

[97] N. Seiberg and E. Witten, "String theory and noncommutative geometry," JHEP **9909** (1999) 032 doi:10.1088/1126-6708/1999/09/032 [hep-th/9908142].

[98] M. Jimbo, "Yang-Baxter equation in integrable systems," Adv. Ser. Math. Phys. **10** (1989).

[99] R. Borsato and L. Wulff, "Target space supergeometry of η and λ-deformed strings," JHEP **1610** (2016) 045 doi:10.1007/JHEP10(2016)045 [arXiv:1608.03570 [hep-th]].

[100] V. Chari and A. Pressley, "A guide to quantum groups," Cambridge, UK: Univ. Pr. (1994)

[101] R. Borsato and S. Driezen, "Supergravity solution-generating techniques and canonical transformations of σ-models from $O(D,D)$," JHEP **05** (2021), 180 doi:10.1007/JHEP05(2021)180 [arXiv:2102.04498 [hep-th]].

[102] R. Borsato, S. Driezen and F. Hassler, "An Algebraic Classification of Solution Generating Techniques," [arXiv:2109.06185 [hep-th]].

Journal of Physics: Conference Series **2191** (2022) 012008 doi:10.1088/1742-6596/2191/1/012008

A Review of Third Way Consistent Theories

Nihat Sadik Deger
Department of Mathematics, Bogazici University, Bebek, 34342, Istanbul, Turkey
E-mail: `sadik.deger@boun.edu.tr`

Abstract. We will give an overview of "third way consistent" theories. Field equations of such models do not come from the variation of a local action without auxiliary fields, yet their covariant divergences still vanish on-shell. First examples were discovered in three dimensions which were pure massive gravity and Yang-Mills theories. However, recently interacting p-form theories with this property in arbitrary dimensions were also constructed. After explaining construction of these theories and some of their general features, we will discuss some open problems and future directions.

1. Introduction

The action principle has undeniably been central in the development of modern physics. Equations that describe 4 fundamental forces of our universe, namely those of General Relativity and the Standard Model, can be derived using this principle. Having an action helps in understanding interactions, symmetries and quantizing the theory. However, there are some theories for which no simple action exists. A well-known example is the 10-dimensional Type IIB supergravity which contains a 4-form field with a self-dual field strength. As a result, its equations of motion can not be derived from a Lorentz invariant action without extra fields [1]. Similarly, there is no action for supergravities in which the global scaling symmetry is gauged [2]. These models are defined via their field equations whose consistency works as usual. In principle, there might also be some physical phenomena or theoretical questions whose explanations require inclusion of interaction terms in the field equations that do not come from an action without auxiliary fields and Bianchi identities are satisfied in an unusual way. Such a situation occurred in 3-dimensions where search for a pure higher curvature gravity theory which has both bulk and boundary unitarity led to the discovery of the Minimal Massive Gravity (MMG) model [3]. The novelty of its modified Einstein's equation is that, its covariant divergence vanishes only after using the Einstein's equation again. This rather surprising way of establishing consistency is called the "third way". Other such gravity [4, 5] and gauge theory [6] examples in 3-dimensions were found later. Recently, such p-form theories in arbitrary dimensions are built in [7]. Moreover, in [8] $N = 1$ supersymmetric version of [6] is obtained.

In this paper we will focus on the construction of this new type of theories and leave other important issues, such as their exact solutions, couplings etc. aside. One common feature they all share in their construction is the vital role played by a shift in the connection of the model. As will be seen, in principle it is possible to obtain more examples in three dimensions using the methods that we will describe. However, finding new models in higher dimensions is a challenge. We will start by defining the term "third way" in the next section which is also reviewed in [9].

Constructions of known gravity and gauge models with this property are explained in sections 3 and 4 respectively. We conclude in section 5 by indicating some future directions.

2. What is the "Third Way"?

To explain what is meant by the "third way" let us consider the vacuum Einstein's equation:

$$G_{\mu\nu} \equiv R_{\mu\nu} - \frac{1}{2} R g_{\mu\nu} = 0\,, \tag{1}$$

which can be derived from the Einstein-Hilbert action. Any addition to this equation should be compatible with the fact that the divergence of the Einstein tensor $G_{\mu\nu}$ vanishes identically. There are two familiar and one uncommon sorts of modifications:

i) Adding divergence free gravitational terms: Historically, it was Einstein himself who first considered such a modification by introducing a cosmological constant Λ which also follows from an action. In 3-dimensions a famous generalization is the (cosmological) topologically massive gravity (TMG) [10] whose field equation is:

$$G_{\mu\nu} = -\Lambda g_{\mu\nu} - \frac{1}{\mu} C_{\mu\nu}\,, \tag{2}$$

where μ is a mass parameter. The Cotton tensor $C_{\mu\nu}$ is defined as

$$C_{\mu\nu} \equiv \epsilon_{\mu\rho\sigma} \nabla^{\rho} S^{\sigma}{}_{\nu}\,, \tag{3}$$

where $S_{\mu\nu} \equiv G_{\mu\nu} + \frac{1}{4} R g_{\mu\nu}$ is the Schouten tensor. It is straightforward to show that the Cotton tensor is covariantly conserved. Indeed this is not a surprise, since any additional gravitational term to (1) that comes from a diffeomorphism invariant action will automatically be divergent free and the Cotton tensor can be obtained from the variation of the gravitational Chern-Simons action.

ii) Adding other fields: Let us now consider the following modification of (1):

$$G_{\mu\nu} = \kappa (F_{\mu\rho} F^{\rho}{}_{\nu} - \frac{1}{4} g_{\mu\nu} F_{\rho\sigma} F^{\rho\sigma})\,, \tag{4}$$

where $F_{\mu\nu} \equiv \partial_\mu A_\nu - \partial_\nu A_\mu$ is the field strength of an abelian vector field A_μ and κ is a coupling constant. If we now take the divergence of this equation we see that the r.h.s vanishes only when

$$\nabla_\alpha F^{\alpha\beta} = 0\,. \tag{5}$$

Actually (4) and (5) are nothing but equations of motion of the Einstein-Maxwell action. In general, any covariant coupling of a matter action to the Einstein-Hilbert action will change the gravitational field equation as $G_{\mu\nu} = \kappa T_{\mu\nu}$ where $T_{\mu\nu}$ is the matter energy-momentum tensor which is conserved when matter field equations are satisfied by the Noether's theorem.

iii) Adding gravitational terms à la third way: Now consider the following generalization of the TMG field equation (2):

$$G_{\mu\nu} = -\Lambda g_{\mu\nu} - \frac{1}{\mu} C_{\mu\nu} - \frac{\gamma}{\mu^2} J_{\mu\nu}\,, \tag{6}$$

where γ is a dimensionless constant and the new piece is given by [3]:

$$J_{\mu\nu} \equiv \frac{1}{2} \epsilon_{\mu\rho\sigma} \epsilon_{\nu\alpha\beta} S^{\rho\alpha} S^{\sigma\beta}\,. \tag{7}$$

Since the tensor J is built out of metric and curvature tensors only, for this to be a valid modification, one would expect its divergence to vanish identically as in the class i) that we considered above. Instead, one derives:

$$\nabla_\mu J^{\mu\nu} = \epsilon^{\nu\alpha\beta} S_\alpha{}^\lambda C_{\lambda\beta} \,. \tag{8}$$

This shows that the J-term can not come from variation of an action of the metric field alone and hence one may conclude that equation (6) is inconsistent. However, note in (8) that the divergence of J is proportional to the Cotton tensor which appears in the main field equation (6). Now replacing the Cotton tensor in (8) using (6) one gets:

$$\nabla_\mu J^{\mu\nu} = -\mu \epsilon^{\nu\alpha\beta} S_\alpha{}^\lambda \left(S_{\lambda\beta} - \frac{1}{4} R g_{\lambda\beta} + \Lambda g_{\lambda\beta} + \frac{\gamma}{\mu^2} J_{\lambda\beta} \right) , \tag{9}$$

which is zero since each term vanishes. Note in particular that, $SJ \sim S^3 = 0$ due to the way indices are arranged. Hence, we conclude that equation (6) makes sense on-shell and the model it describes is called the Minimal Massive Gravity (MMG) [3]. Actually, here we have a 'mixture' of the two cases that we discussed above; namely we have a purely gravitational modification (as in case i)) but we have to use the gravity equation itself again (as in case ii)) for consistency. This is what is meant by the "third way" and as we will see it is not something particular to gravity; such gauge theories also exist.

3. Gravity Examples

Coming up with a third way consistent gravity equation such as (6) is highly non-trivial. That is probably why this class of theories were not discovered until recently. Their origin in 3-dimensions becomes more transparent using the Chern-Simons-like description of gravity developed in [11–14] which we summarize below.[1]

In this formulation dynamical fields (X_K's with $K = 0, 1, 2, , ...$) are a set of Lorentz-vector valued 1-forms $X^a_K = X^a_{K\mu} dx^\mu$ where the generators of the SO(1,2) Lorentz algebra satisfy $[J_a, J_b] = \epsilon_{abc} J^c$. The index K indicates the weight of the field and we split those with even and odd weight as f_I (with weight $2I$) and h_I (with weight $(2I+1)$) respectively. Moreover, we define $f_0 \equiv e$ (with weight zero) as the dreibein and $h_0 \equiv \omega$ (with weight 1) as the (dual) spin-connection ($2\omega^a = \epsilon^{abc}\omega_{ab}$). So, the set of fields that will be used in this construction ordered with increasing weight are: $(e, \omega, f_1, h_1, f_2, h_2, ...)$. To build an action we have the following rules:

- Lagrangian should be a 3-form.
- Derivatives should be covariant using the spin connection: $D \equiv d + \frac{1}{2}[\omega,]$. Covariant derivative has weight 1.
- The spin connection ω should appear only in covariant derivatives and in the Chern-Simons combination: $< \omega \wedge d\omega + \frac{2}{3}\omega \wedge \omega \wedge \omega >$, where the bracket notation stands for the matrix trace.
- The vanishing of the torsion ($De = 0$) should be among the equations of motion. We also assume that the dreibein is invertible. These two together imply that ω is solvable in terms of e.
- Equations of motion should be integrable in the sense that each field should be solvable in terms of the lower weight fields or their covariant derivatives. In other words, it should be possible to order the equations with increasing weight starting with $De = 0$ (which is weight 1) such that the weight p equation has the form: $DX_{p-1} + e \wedge X_p = 0$ [2] which can easily be solved for X_p.

[1] Construction of such models directly in the metric formulation is described in [15].
[2] This equation may also include terms with weight less than p.

Journal of Physics: Conference Series **2191** (2022) 012008 doi:10.1088/1742-6596/2191/1/012008

This means that the whole equation system can be solved recursively until the last one. Finally, using solutions for the auxiliary fields in the highest weight equation will give the gravity field equation expressed in terms of the metric.

Note that the last item in the list restricts possible actions quite strongly. The allowed Lagrangian terms with the above rules up to weight 3 are as follows:

Weight 0: $e \wedge e \wedge e$
Weight 1: $e \wedge De$
Weight 2: $e \wedge e \wedge f_1$, $e \wedge R$ (where $R^a = D\omega^a$ is the dualized curvature 2-form.)
Weight 3: $< \omega \wedge d\omega + \frac{2}{3}\omega \wedge \omega \wedge \omega >$, $f_1 \wedge De$ (equivalent to $Df_1 \wedge e$), $e \wedge e \wedge h_1$

Actions with only even (odd) weight terms are called parity even (odd). A parity even (odd) action S_k has terms with weight $(k+2)$ or less and should be built out of fields with weight less than $(k+2)$. With this terminology the first parity even action S_0 includes 3-forms with weights 2 or 0 which are built out of e and ω:

$$S_0 = -\frac{1}{\kappa^2} \int < e \wedge R - \frac{\Lambda}{3} e \wedge e \wedge e >, \tag{10}$$

where κ is a constant. From this action we get $G_{\mu\nu} + \Lambda g_{\mu\nu} = 0$ as the metric field equation. Similarly, the first parity odd action which is compatible with our rules is:

$$S_1 = \frac{1}{2\kappa^2 \mu} \int < \omega \wedge d\omega + \frac{2}{3}\omega \wedge \omega \wedge \omega + 2f_1 \wedge De >, \tag{11}$$

which leads to the metric field equation $C_{\mu\nu} = 0$. As the weight of the actions get bigger the number of possible terms that can be included grows rapidly and figuring out which ones should be kept gets involved. However, in [13] a systematic construction of S_{k+2} from S_k is given and since S_0 and S_1 are known this gives an infinite number of massive gravity models which are free of scalar ghosts. Of course, it is also possible to consider parity violating actions by mixing parity odd and even ones. The first such action that leads to an integrable system of equations is

$$S_{\text{TMG}} = S_0 + S_1 , \tag{12}$$

which gives rise to the TMG field equation (2).

Clearly MMG model (6) can not be obtained from such a construction since in the above all auxiliary fields are solvable in terms of the dreibein and after solving, we can back substitute them in the auxiliary action to get a metric action. Hence, to get the MMG equation (6) we need to relax some of our rules. Let us now deform the TMG action (12) as follows [8]:

$$S_{\text{MMG}} = S_{\text{TMG}} + \frac{\alpha}{\kappa^2 \mu^2} \int \langle e \wedge f_1 \wedge f_1 \rangle , \tag{13}$$

where α is a dimensionless constant. Note that with this addition we are not increasing the number of auxiliary fields since f_1 was already in S_1 but the weight of this additional action is 4 which was not allowed by our rules. From (13) one obtains

$$\begin{aligned}
&\delta f_1 \qquad && De + \tfrac{\alpha}{\mu}\,[e, f_1] = 0\,, \\
&\delta \omega \qquad && \tfrac{1}{\mu}\,(R + [e, f_1]) - De = 0\,, \\
&\delta e \qquad && \tfrac{1}{\mu} Df_1 + \tfrac{\alpha}{2\mu^2}[f_1, f_1] - R + \tfrac{1}{2}\Lambda[e, e] = 0\,.
\end{aligned} \tag{14}$$

Journal of Physics: Conference Series **2191** (2022) 012008 doi:10.1088/1742-6596/2191/1/012008

Obviously here the torsion is non-zero but if we shift the spin connection as

$$\omega \to \omega - \frac{\alpha}{\mu} f_1 \,, \tag{15}$$

equations (14) become [3]:

$$\begin{aligned} &De = 0\,, \\ &R + (1+\alpha)^2[e, f_1] + \tfrac{\alpha}{2}\Lambda[e,e] = 0\,, \\ &\tfrac{1+\alpha}{\mu} D f_1 - \tfrac{\alpha(1+\alpha)}{2\mu^2}[f_1, f_1] - R + \tfrac{1}{2}\Lambda[e,e] = 0\,, \end{aligned} \tag{16}$$

where D and R are now defined with respect to the shifted connection. Note that, there is no torsion any more and assuming $\alpha \neq -1$ the second equation can be solved for f_1 as $(f_1)_{\mu\nu} = -(S_{\mu\nu} + \alpha\Lambda g_{\mu\nu}/2)/(1+\alpha)^2$. Substituting this to the last equation above we get the MMG field equation (6) with $\gamma = -\alpha/(1+\alpha)^2$. The extra term $[f_1, f_1]$ is the source of the J-tensor (7).

The above shows that we can relax the no torsion condition in this construction. It is enough to have an integrable system of equations after a linear shift of the connection with auxiliary fields. However, after this shift, back substitution of solved fields to the auxiliary action is not allowed any more and hence there is no metric action for such a gravity model.

Although this procedure gives a valuable insight about the origin of MMG, it is far from obvious how to construct other such models. The next example was found almost 4 years after MMG in [4] by systematically analyzing the most general Chern-Simons like action for the fields (e, ω, f_1, h_1) with no weight restriction for the action. Requiring integrability after the shift of the connection (which is with h_1) they found such a first order action for these fields which is parity odd. Its field equation is fourth order in derivatives of the metric. It is clear that this approach gets very complicated as the number of auxiliary fields increases. In [4] initially there are 20 possible terms in the action. In [5] it was shown that this model can be obtained with less effort starting from the S_3 action constructed in [13] and by truncating the highest weight field f_2 as $f_2 \to c_1 f_1 + c_2 e$ where c_i's are constant. In general

$$S_{2N+1}[f_{N+1} \to (f_N, \cdots, e)] \longrightarrow \tilde{S}_{2N}\,. \tag{17}$$

After the truncation the dynamical terms in $\tilde{S}_{2N}$ are fixed. Note that $\tilde{S}_{2N}$ is parity odd. For the integrability, one may also need to add some parity odd non-dynamical terms which might violate the weight constraint. However, they are easy to guess once all the dynamical terms are known. In [5] this top-down approach was used to construct another third-way consistent model whose field content is $(e, \omega, f_1, h_1, f_2, h_2)$ starting from S_5. Its metric field equation contains sixth order derivatives.

4. Gauge Theory Examples

We start with 3-dimensions and then continue with higher dimensional examples. Let A^I_μ be a Yang-Mills (YM) gauge field that transforms in the adjoint representation of an an arbitrary gauge group G with structure constants f^I_{JK}. Its field strength is defined as

$$F^I_{\mu\nu} \equiv 2\partial_{[\mu}A^I_{\nu]} + f^I_{JK}A^J_\mu A^K_\nu\,. \tag{18}$$

Following [6] let us consider the following YM field equation in 3-dimensions:

$$\epsilon_\mu{}^{\nu\rho} D_\nu \tilde{F}^I_\rho + \mu \tilde{F}^I_\mu = -\frac{1}{2m}\epsilon_\mu{}^{\nu\rho} f^I_{JK}\tilde{F}^J_\nu \tilde{F}^K_\rho\,, \tag{19}$$

where μ and m are mass parameters and $\tilde{F}^I_\mu \equiv \frac{1}{2}\epsilon_\mu{}^{\nu\rho}F^I_{\nu\rho}$ is the dual field strength[3]. The gauge covariant divergence of the l.h.s vanishes automatically thanks to the fact that $[D_\mu, D_\nu]\tilde{F}^I_\rho \sim f^I_{JK}F^J_{\mu\nu}\tilde{F}^K_\rho$ and due to the YM Bianchi identity $D^\mu\tilde{F}^I_\mu = 0$. This, of course is not a surprise since these terms come from the YM and Chern-Simons actions respectively and hence they should be conserved. Therefore, for equation (19) to be consistent r.h.s should also be conserved. However, it is easy to convince oneself that this term can not come from variation of an action for A^I_μ alone. Indeed, taking its divergence we see that it is not immediately zero but using (19) again we find

$$D_\mu(\epsilon^{\mu\nu\rho}f^I_{JK}\tilde{F}^J_\nu\tilde{F}^K_\rho) \sim f^I_{JK}(\epsilon^{\mu\nu\rho}D_\mu\tilde{F}^J_\nu)\tilde{F}^K_\rho \sim f^I_{JK}f^J_{MN}\epsilon^{\mu\nu\rho}\tilde{F}^M_\mu\tilde{F}^N_\nu\tilde{F}^K_\rho\,, \tag{20}$$

which vanishes due to the Jacobi identity $f^I_{J[K}f^J_{MN]} = 0$. In [6] also an auxiliary action was constructed for this model. However, it was not clear how to find other such examples as was done in the gravity case [4, 5] after MMG [3]. This issue was resolved in [7] which we explain now. To do this we switch to differential form notation for convenience. We begin with the Chern-Simons field equation:

$$F(A) \equiv dA + A^2 = 0\,, \tag{21}$$

and shift the connection with a Lie algebra valued 1-form C

$$A \to A + C\,. \tag{22}$$

We declare what we obtain after this to be our field equation. That gives,

$$F(A+C) = F(A) + D_AC + C^2 = 0\,, \tag{23}$$

where the covariant derivative of C is $D_AC \equiv dC + AC + CA$. Now (23) is our new field equation and for consistency $D_AF(A+C) = 0$ must hold. To see this we use the Bianchi identity $D_AF(A) = 0$ after which we get

$$D_AF(A+C) = D_A^2C + D_AC^2 = [F(A), C] + [D_AC, C] = -[C^2, C] = 0\,, \tag{24}$$

where we used the field equation (23) in the last step. Hence, we see that equation (23) is third way consistent for any C. Moreover, if C is conserved, that is $D_A * C = 0$, then we can generalize it as

$$F(A) + D_AC + C^2 + \tau * C = 0\,, \tag{25}$$

where τ is an arbitrary constant. Equation (25) is again third way consistent due to $[*C, C] = 0$. If we choose $C = \tilde{F}/m$ and $\tau = (m - \mu)$ then this equation becomes (19). But any other choice of C will be a generalization and hence this construction gives infinitely many third way consistent models. Moreover, if C has a conserved part M and non-conserved part N, that is $C = M + N, D * M = 0, D * N \neq 0$, then we can modify (23) as

$$F(A) + D_AC + C^2 + c_1 * M + c_2 * N = 0\,, \tag{26}$$

where c_1 and c_2 are constants, if

$$c_2D * N = (c_1 - c_2)[* N, M]\,. \tag{27}$$

Until now we were assuming the spacetime dimension to be three, however in the above construction this did not play any role except when we moved from (23) to (25) which requires the Hodge dual of C to be a 2-form. Moreover, the particular choice $C = *F$ is only possible in 3-dimensions. Giving up on these, it is straightforward to generalize (23) to higher dimensions.

[3] This equation with some special choices for the mass parameters appeared earlier in [16, 17].

For that purpose, let B^I's be Lie algebra valued $(d-2)$-forms in $d \geq 3$ dimensions whose YM covariant field strengths defined as $H^I \equiv DB^I \equiv dB^I + f^I_{JK} A^J B^K$. Note that the dual field strength, denoted by $\tilde{H}$, is a 1-form. Hence, we can choose $C = \kappa \tilde{H}$ in the shift (22) where κ is a constant and (23) becomes

$$F(A)^I + \kappa\, D_A \tilde{H}^I = -\frac{1}{2}\kappa^2 f^I_{JK}\, \tilde{H}^J \wedge \tilde{H}^K \,, \tag{28}$$

which is third way consistent by construction. However, compatibility of YM and $(d-2)$-form gauge symmetries require $F(A)^I = 0$ [7]. Hence, we set $A^I = 0$ in (28) and end up with

$$d\tilde{H}^I = -\tfrac{1}{2}\kappa\, \tilde{H}^J \wedge \tilde{H}^K\, f^I_{JK} \,, \tag{29}$$

with $H^I = dB^I$.[4] If we hit this equation with the exterior derivative d, the l.h.s is zero since it is exact and the r.h.s also vanishes on shell applying the Jacobi identity. Hence, this is a third way consistent interacting theory of $(d-2)$-forms. Also note that it has the standard p-form gauge symmetry $\delta B^I = d\xi^I$ where ξ^I is a $(d-3)$-form. We can generalize this equation by including more terms in the shift (22) that are constructed out of H^I and its derivatives. We may also add a matter current J^I as

$$d\tilde{H}^I + \tfrac{1}{2}\kappa\, \tilde{H}^J \wedge \tilde{H}^K\, f^I_{JK} = J^I \,, \tag{30}$$

provided that it satisfies

$$dJ^I = \kappa\, J^M \wedge \tilde{H}^N\, f^I_{MN} \,. \tag{31}$$

The current J^I can be expressed in terms of a 1-form y^I as

$$J^I = -dy^I - \kappa f^I_{MN} y^M \tilde{H}^N - \frac{1}{2}\kappa f^I_{MN} y^M \wedge y^N \,, \tag{32}$$

which can be seen by including y^I in the shift (22).

5. Future Directions

There are several interesting problems about this new class of theories some of which we discuss below under three headlines.

Constructing other examples: All the known gravity examples [3–5] are in three dimensions and the construction method of [5] works only for *exotic* gravity models which have parity odd actions in the Chern-Simons like formulation but parity even metric field equations. Adapting this method to parity violating models like the MMG [3] is desirable. Even more challenging is to find examples in higher dimensions. Recall that the consistency of (6) works thanks to some very particular properties of the tensors (S, C, J) which can schematically be listed as follows: i) $J \sim S^2$, ii) $\nabla J \sim CS$ since $C \sim \nabla S$, iii) C is divergence free and part of the main field equation (6), iv) $S.E_C = 0$ where E_C is the C-field equation that is obtained from (6). Finding analogues of these tensors in higher dimensions seems difficult. Adapting the first order formulation of 3-dimensional higher derivative gravities [11–14] that we reviewed in section 3 to other dimensions could be the key to achieve this. On the gauge theory side, it would be nice to find a YM covariant version of (29) where unlike (28) gauge symmetries are compatible with each other. Using tensor hierarchies [18] might be useful for this purpose. Finding an action with auxiliary fields for (29) is another open problem.

[4] In 3-dimensions B^I is a vector field and (29) can be thought as the ungauged version of (19).

Supersymmetric extensions: Until now only one example of a supersymmetric third way consistent model has been constructed [8] which is the $N = 1$ supersymmetric version of the 3-dimensional YM theory of [6] that we introduced in section 4. The starting point of [8] is the $N = 1$ off-shell supersymmetric topologically massive YM theory which contains a Majorana spinor field as the fermionic partner of the YM field. Its YM field equation looks exactly like (19) but instead of the F^2 term on the r.h.s. there is a fermionic current j^I_μ. This current is conserved using the spinor field equation. By deforming the YM equation of this system with the F^2 term and demanding that under supersymmetry field equations are transformed to each other, the necessary extra terms in both equations are found. It is remarkable that this is possible without adding further bosonic terms to (19). Moreover, it turns out that the end result can be obtained starting from the original system and performing a transformation of the form

$$A^I_\mu \to A^I_\mu + \alpha_1 \tilde{F}^I_\mu + \alpha_2 j^I_\mu , \tag{33}$$

in the spinor field equation. The constants α_1 and α_2 are uniquely fixed by supersymmetry. The final YM equation can be obtained by a similar shift starting from $\tilde{F}^I_\mu = 0$. Hence, the shift idea that already appeared in (15) and (22) again plays a crucial role. The construction of [8] that we outlined above shows existence of a new class of supersymmetric theories that were not realized before and hence encourages many interesting directions to pursue such as trying to couple this model with other $N = 1$ supersymmetric multiplets, constructing its extended supersymmetric versions, finding supersymmetric versions of gravity examples such as MMG (6) and p-form theories (29).

The third way and symmetry breaking: Another way of understanding terms that lead to third way consistency is via symmetry breaking. In [19] this was done for the MMG by extending the Poincaré algebra to Hietarinta/Maxwell algebra [20] which has one additional generator Z_a. Recall that the MMG auxiliary action (13) contains 3 fields (e, ω, f_1) and they are now associated with the generators of this group, namely translations P_a, rotations J_a and Z_a, respectively. Then, one can construct the Chern-Simons action

$$S = \int < \Omega \wedge d\Omega + \frac{2}{3} \Omega \wedge \Omega \wedge \Omega > , \tag{34}$$

for the Hietarinta/Maxwell algebra valued 1-form $\Omega = e^a P_a + \omega^a J_a + f_1^a Z_a$. Spontaneous breaking of the Hietarinta/Maxwell symmetry to its Poincaré sub-algebra gives rise to extra terms in (34) and the resulting action is equivalent to that of MMG (13) after some field redefinitions. It should be possible to generalize this method for exotic gravity [4, 5] and gauge theory examples [6, 7]. The YM model of [6] with $\mu = 0$ is already known to arise from a Higgs mechanism [16]. In [21] an alternative description of MMG based only on dreibein and spin connection is given where the Lorentz gauge symmetry is broken. Finally, in [22] it is shown that MMG is related with the break of the Weyl symmetry. It would be interesting to clarify connections between these different approaches.

Acknowledgements

It is a great pleasure to dedicate this work to Prof. Tekin Dereli on the occasion of his 72nd birthday. I was fortunate to have him as my master's supervisor, and I benefited from his support and guidance since then. With his distinguished scientific output and leadership, he has been an inspiring example to many of us. I am grateful to ENS de Lyon, Laboratoire de Physique for hospitality and Research Fellowship Program of Embassy of France in Turkey for financial support during the final phase of this study. I also would like to thank Henning Samtleben for helpful discussions.

References

[1] N. Marcus and J. H. Schwarz, "Field Theories That Have No Manifestly Lorentz Invariant Formulation," Phys. Lett. B **115** (1982), 111.

[2] A. Le Diffon and H. Samtleben, "Supergravities without an Action: Gauging the Trombone," Nucl. Phys. B **811** (2009), 1-35, [arXiv:0809.5180 [hep-th]].

[3] E. Bergshoeff, O. Hohm, W. Merbis, A. J. Routh and P. K. Townsend, "Minimal Massive 3D Gravity," Class. Quant. Grav. **31** (2014) 145008, [arXiv:1404.2867 [hep-th]].

[4] M. Özkan, Y. Pang and P. K. Townsend, "Exotic Massive 3D Gravity," JHEP **08** (2018) 035, [arXiv:1806.04179 [hep-th]].

[5] H. R. Afshar and N. S. Deger, "Exotic massive 3D gravities from truncation," JHEP **11** (2019) 145, [arXiv:1909.06305 [hep-th]].

[6] A. S. Arvanitakis, A. Sevrin and P. K. Townsend, "Yang-Mills as massive Chern-Simons theory: a third way to three-dimensional gauge theories," Phys. Rev. Lett. **114** (2015) no.18, 181603, [arXiv:1501.07548 [hep-th]].

[7] M. Broccoli, N. S. Deger and S. Theisen, "Third Way to Interacting p-Form Theories," Phys. Rev. Lett. **127** (2021) no.9, 091603, [arXiv:2103.13243 [hep-th]].

[8] N. S. Deger and J. Rosseel, "Novel 3D supersymmetric massive Yang-Mills theory," Phys. Rev. D **104** (2021) no.8, L081701, [arXiv:2105.13300 [hep-th]].

[9] E. Bergshoeff, W. Merbis, A. J. Routh and P. K. Townsend, "The Third Way to 3D Gravity," Int. J. Mod. Phys. D **24** (2015) no.12, 1544015, [arXiv:1506.05949 [gr-qc]].

[10] S. Deser, R. Jackiw and S. Templeton, "Topologically Massive Gauge Theories," Annals Phys. **140** (1982), 372-411 [erratum: Annals Phys. **185** (1988), 406].

[11] O. Hohm, A. Routh, P. K. Townsend and B. Zhang, "On the Hamiltonian form of 3D massive gravity," Phys. Rev. D **86** (2012), 084035, [arXiv:1208.0038 [hep-th]].

[12] E. A. Bergshoeff, O. Hohm, W. Merbis, A. J. Routh and P. K. Townsend, "Chern-Simons-like Gravity Theories," Lect. Notes Phys. **892** (2015), 181-201, [arXiv:1402.1688 [hep-th]].

[13] H. R. Afshar, E. A. Bergshoeff and W. Merbis, "Extended massive gravity in three dimensions," JHEP **08** (2014), 115, [arXiv:1405.6213 [hep-th]].

[14] W. Merbis, "Chern-Simons-like Theories of Gravity," [arXiv:1411.6888 [hep-th]].

[15] G. Alkac, M. Tek and B. Tekin, "Bachian Gravity in Three Dimensions," Phys. Rev. D **98** (2018) no.10, 104021, [arXiv:1810.03504 [hep-th]].

[16] S. Mukhi, "Unravelling the novel Higgs mechanism in (2+1)d Chern-Simons theories," JHEP **12** (2011) 083, [arXiv:1110.3048 [hep-th]].

[17] B. E. W. Nilsson, "Critical solutions of topologically gauged N = 8 CFTs in three dimensions," JHEP **04** (2014), 107, [arXiv:1304.2270 [hep-th]].

[18] B. de Wit, H. Nicolai and H. Samtleben, "Gauged Supergravities, Tensor Hierarchies, and M-Theory," JHEP **02** (2008), 044, [arXiv:0801.1294 [hep-th]].

[19] D. Chernyavsky, N. S. Deger and D. Sorokin, "Spontaneously broken *3d* Hietarinta/Maxwell Chern–Simons theory and minimal massive gravity," Eur. Phys. J. C **80** (2020) no.6, 556, [arXiv:2002.07592 [hep-th]].

[20] J. Hietarinta, "Supersymmetry Generators of Arbitrary Spin," Phys. Rev. D **13** (1976), 838.

[21] M. Geiller and K. Noui, "Metric formulation of the simple theory of 3d massive gravity," Phys. Rev. D **100** (2019) no.6, 064066, [arXiv:1905.04390 [gr-qc]].

[22] T. Dereli and C. Yetismisoglu, "Weyl Covariant Theories of Gravity in Riemann-Cartan-Weyl Space-times II. Minimal Massive Gravity," Class. Quantum Grav. **36** (2019), 215005, [arXiv:1904.11255 [gr-qc]].

A remarkable dynamical symmetry of the Landau problem

Tekin Dereli[1], **Philippe Nounahon**[2], **Todor Popov**[3,4,5]
[1]Department of physics, Koç University, 34450 Sariyer-İstanbul, Turkey
[2]Institut de mathématique et de sciences physiques, B.P 613, Porto-Novo, Bénin
[3]INRNE, Bulgarian Academy of Sciences, Tzarigradsko chaussée 72, Sofia 1784, Bulgaria
[4]American University in Bulgaria, Sv. Buchvarova Str. 8, Blagoevgrad 2700, Bulgaria
[5]African University of Science and Technology, P.M.B 681, Garki, Abuja F.C.T., Nigeria.

E-mail: `tdereli@ku.edu.tr, tpopov@aubg.edu`

Abstract. We show that the dynamical group of an electron in a constant magnetic field is the group of symplectomorphisms $Sp(4,\mathbb{R})$. It is generated by the spinorial realization of the conformal algebra $\mathfrak{so}(2,3)$ considered in Dirac's seminal paper "A Remarkable Representation of the 3 + 2 de Sitter Group". The symplectic group $Sp(4,\mathbb{R})$ is the double covering of the conformal group $SO(2,3)$ of 2+1 dimensional Minkowski spacetime which is in turn the dynamical group of a hydrogen atom in 2 space dimensions. The Newton-Hooke duality between the 2D hydrogen atom and the Landau problem is explained via the Tits-Kantor-Koecher construction of the conformal symmetries of the Jordan algebra of real symmetric 2×2 matrices. The connection between the Landau problem and the 3D hydrogen atom is elucidated by the reduction of a Dirac spinor to a Majorana one in the Kustaanheimo-Stiefel spinorial regularization.

1. Introduction

In the heart of the quantum theory is the passionate history of the hydrogen atom spectrum. By good fortune the Coulomb potential describing the "the action at a distance" is the twin brother of the potential of the Newton universal interaction. Therefore the quantum motion of electron in the hydrogen atom amounts to the quantization of the Kepler orbits. In some poetical sense, the heavenly spheres mirror the depth of the micro-cosmos.

In his famous paper, Vladimir Fock [13] explained the accidental degeneracies in the spectrum of the hydrogen atom by the existence of an additional integral of motion given by the Laplace-Runge-Lenz. Fock's result allows for a generalization. Namely, a n-dimensional hydrogen atom whose Schrödinger equation for the bounded states in momentum space is transformed by stereographic projection to the Laplace's equation on a n-dimensional sphere S^n[9].

In this work we study the duality between the 2-dimensional (2D) hydrogen atom on one hand and the Landau problem for the quantization of the electronic orbits in the uniform magnetic field on the other:

$$\text{2D } e^- \text{ in electric field} \quad \overset{Newton-Hooke}{\leftrightarrow} \quad \text{2D } e^- \text{ in magnetic field} . \tag{1}$$

The classical geodesic motion on the sphere, say on the equator, is seen as the periodic circular Larmour motion in the magnetic field directed between the poles. The Fock method applied for

2D hydrogen system with potential $-k/r$ sends orbital eigenfunctions to spherical harmonics in momentum space [6, 28, 4, 26, 31] which are the harmonic functions of the sphere S^2, *i.e.*, eigenfunctions of an isotropic harmonic oscillator with two degrees of freedom. The mapping from 2D harmonic oscillator to 2D Coulomb-Kepler problem is known to be the Levi-Civita transform [4]. In this paper we show that the dynamical group of the Landau problem is the double covering $Sp(4,\mathbb{R})$ of the conformal group $SO(2,3)$ of the 2D hydrogen atom and arises from the exponentiation of the remarkable spinorial representation of $SO(2,3)$ found by Dirac in his famous paper "A Remarkable Representation of the 3 + 2 de Sitter Group"[11].

Different dualities between Coulomb-Kepler motion and harmonic oscillators has been already studied by many autors[1] [4, 9, 1] in various dimensions, however to the best of our knowledge the implications about the Landau problem were not highlighted. A tremendous amount of work on the spectrum generating algebra of the hydrogen atom has been done by Asım Barut and his collaborators [2, 3, 5](See also [22]). Gradually it became clear that the spectrum of 3D hydrogen atom carries a minimal (massless) representation with helicity $\lambda = 0$ of the conformal group $SO(2,4)$. It allows for a dual description in terms of 4D harmonic oscillator via a ladder $U(2,2)$-representation that has been put forward by Mack and Todorov[21]. The correspondence with the massless $U(2,2)$-representation having arbitrary helicity λ (isomorphic to a $SO(2,4)$-representation) turns out to be the quantization of the Kustaanheimo-Stiefel regularization [20] in celestial mechanics. This correspondence holds true not only for the quantum Kepler motion of electron in the field of the positively charged nucleus but also for more general binary systems of coupled dyons [3, 32].

A powerful approach to the conformal dynamical symmetries of generalized quantum Kepler problems based on Jordan algebras was proposed by Guowu Meng [23]. To an Euclidean Jordan algebra one associates a symmetric null ray cone whose automorphisms form a conformal group. From the minimal data of the Jordan algebra $\mathfrak{J}_2^C$ of 2×2 complex Hermitian matrices one is able to build the whole conformal algebra $\mathfrak{so}(2,4)$ of 3D hydrogen atom[23, 27]. We show that the reduction from 3D to 2D hydrogen atom dynamical group is done by simply imposing the reality condition on the Jordan algebra of observables $\mathfrak{J}_2^C$:

$$\begin{array}{ccccc} \mathfrak{co}(\mathfrak{J}_2^{\mathbb{C}}) & = & \mathfrak{so}(2,4) & \overset{Kustaanheimo-Stiefel}{\longleftarrow} & \mathfrak{su}(2,2) \\ {\scriptstyle x=x^t}\uparrow & & & & \uparrow{\scriptstyle \psi=\psi^c} \\ \mathfrak{co}(\mathfrak{J}_2^{\mathbb{R}}) & = & \mathfrak{so}(2,3) & \overset{Levi-Civita}{\longleftarrow} & \mathfrak{sp}(4,\mathbb{R}) \end{array} \tag{2}$$

The projection $\pi : \mathfrak{J}_2^{\mathbb{C}} \to \mathfrak{J}_2^{\mathbb{R}}$ simply projects out the complex Pauli matrix σ_2. On the other hand the 3D hydrogen atom is in duality with the 4D harmonic oscillator (the quantization of the harmonic motion on the sphere S^3) the duality mapping being the Kustaanheimo-Stiefel transformation [20], the so called spinorial regularization removing collision Kepler orbits from the phase space. Similarly the Levi-Civita transformation connects the 2D harmonic oscillator with the 2D hydrogen atom (1). We found out that the spinorial reduction from 4D harmonic oscillator (regularized motion on S^3) to 2D harmonic oscillator (Landau problem, *i.e.*, motion on S^2) is the reduction from Dirac to the Majorana spinor. In other words on the spinorial side of the diagram we also apply the reality condition $\psi = \psi^c$.

2. Landau Problem and Harmonic oscillator

An electron in an uniform magnetic field propagates on circular orbits with Larmour frequency $\omega = \frac{eB}{mc}$. The Landau quantization of the circular orbits leads to an isotropic quantum harmonic

[1] For comprehensive lecture notes and further references we send the reader to [30].

oscillator with two modes. The minimal coupling of the electron's charge density to the external magnetic field described by the electromagnetic vector potential $\boldsymbol{\mathcal{A}}$ leads to the Hamiltonian

$$H = \frac{1}{2m}\left(\boldsymbol{p} - \frac{e}{c}\boldsymbol{\mathcal{A}}\right)^2 =: \frac{1}{2m}\boldsymbol{P}^2 . \tag{3}$$

The constant uniform magnetic field $\boldsymbol{B} = B\hat{z}$ along the z-direction can be obtained from different potentials $\boldsymbol{\mathcal{A}} = (\mathcal{A}_x, \mathcal{A}_y)$ in the plane. We choose the symmetric gauge

$$\boldsymbol{\mathcal{A}} = (\mathcal{A}_x, \mathcal{A}_y) = \frac{B}{2}(-y, x) , \qquad \mathcal{A}_i = -\frac{B}{2}\epsilon_{ij}x^j \tag{4}$$

but most of our conclusions are gauge independent. The gauge independent kinetical momenta $\boldsymbol{P}$ and the coordinates $\boldsymbol{X}$ of the center of the cyclotron motion are related to the phase space canonical coordinates (x, y, p_x, p_y) by

$$P_i = m\dot{x}_i = m\frac{\partial H}{\partial p_i} = p_i - \frac{e}{c}\mathcal{A}_i, \qquad X_i = x_i + \frac{1}{m\omega}\epsilon_{ij}P^j . \tag{5}$$

The center of mass coordinates $\boldsymbol{X} = (X, Y)$ are integrals of motion $\dot{X} = 0 = \dot{Y}$ and decouple from the system. One has two independent Heisenberg algebras $[\boldsymbol{P}, \boldsymbol{X}] = 0$

$$[P_x, P_y] = i\frac{\hbar e}{c}B = i\hbar m\omega = \frac{i\hbar^2}{\ell^2} , \qquad [X, Y] = -i\frac{\hbar}{m\omega} = -i\ell^2 .$$

Here ℓ stands for the magnetic length $\ell^2 = \frac{\hbar}{m\omega} = \frac{\hbar c}{eB}$. We introduce also dimensionless canonical coordinates of the phase space

$$x = \sqrt{2}\xi\ell \qquad y = \sqrt{2}\eta\ell \qquad p_x = p_\xi\sqrt{m\omega\hbar}/\sqrt{2} \qquad p_y = p_\eta\sqrt{m\omega\hbar}/\sqrt{2}$$

satisfying the canonical commutation relations

$$[\xi, p_\xi] = i = [\eta, p_\eta] \qquad [\xi, p_\eta] = 0 = [\eta, p_\xi] .$$

The Hamiltonian H in term of the new variables boils down to an isotropic harmonic oscillator Hamiltonian plus a "magnetic" term proportional to the angular momentum:

$$H = \frac{P_x^2 + P_y^2}{2m} = \frac{\hbar\omega}{4}\left\{p_\xi^2 + p_\eta^2 + (\xi^2 + \eta^2)\right\} - \frac{\hbar\omega}{2}(\xi p_\eta - \eta p_\xi) .$$

The kinetic momenta $\boldsymbol{P} = \boldsymbol{p} - \frac{e}{c}\boldsymbol{\mathcal{A}}$ are quantized by the energy creation and annihilation operators $a^\pm$. The guiding center coordinates X and Y are integrals of motion, they are quantized by the *magnetic translation operators*[2] $b^\pm$:

$$\begin{aligned} a^\pm &= \tfrac{a_x^\pm \mp i a_y^\pm}{\sqrt{2}} = \tfrac{-P_y \mp i P_x}{\sqrt{2m\omega\hbar}} , \quad & P_x &= \tfrac{p_\xi + \eta}{\sqrt{2}}\sqrt{m\omega\hbar} , \quad & P_y &= \tfrac{p_\eta - \xi}{\sqrt{2}}\sqrt{m\omega\hbar} , \\ b^\pm &= \tfrac{a_x^\pm \pm i a_y^\pm}{\sqrt{2}} = \tfrac{X \pm iY}{\ell\sqrt{2}} , \quad & X &= \left(\tfrac{\xi + p_\eta}{\sqrt{2}}\right)\ell , \quad & Y &= \left(\tfrac{\eta - p_\xi}{\sqrt{2}}\right)\ell \end{aligned} \tag{6}$$

where we have used two commuting Heisenberg algebras with generators

$$a_x^\pm = (\xi \mp i p_\xi)/\sqrt{2} , \qquad a_y^\pm = (\eta \mp i p_\eta)/\sqrt{2} \qquad \text{such that} \qquad [a_i^-, a_i^+] = 1 . \tag{7}$$

[2] The geometrical meaning of the Zak's magnetic translations in the Landau problem has been clarified in the work [10].

Journal of Physics: Conference Series **2191** (2022) 012009 doi:10.1088/1742-6596/2191/1/012009

The operators $a^{\pm}$ shift between different energy levels of the Hamiltonian

$$H = \frac{\hbar\omega}{2}\{a^+, a^-\} , \qquad [H, a^{\pm}] = \pm a^{\pm} , \qquad [H, b^{\pm}] = 0 . \tag{8}$$

The "zero mode" generators $b^{\pm}$ commute with $a^{\pm}$ since the momenta $\boldsymbol{P} = (P_x, P_y)$ commute with the guiding center coordinates $\boldsymbol{X} = (X, Y)$. Hence the magnetic translations $b^{\pm}$ are responsible for the degeneracy of the Landau levels. The angular momentum operator $L_z = xp_y - yp_x$ is the generator of the rotational symmetry, the operators $b^{\pm}$ increase (decrease) the angular momentum eigenvalue

$$\frac{L_z}{\hbar} = \frac{1}{2}\{b^-, b^+\} - \frac{1}{2}\{a^-, a^+\} , \qquad [\frac{L_z}{\hbar}, b^{\pm}] = \pm b^{\pm} , \qquad [L_z, H] = 0 .$$

The angular momentum commutes with the energy operator H, however, there is a larger accidental group $SO(3)$ of transformations preserving the energy. Its origin is rooted in the analog of the Laplace-Runge-Lenz vector in the Landau problem, the so called magnetic LRL vector [33] which is an integral of motion beside the angular momentum. We proceed by exploring the dynamical group of the Landau problem in a more systematic way.

3. Dirac dynamical algebra $\mathfrak{so}(2,3)$

In his seminal paper "A Remarkable Representation of the 3 + 2 de Sitter Group"[11] Dirac came out with a realization of $\mathfrak{so}(2,3)$ which is quadratic in the phase space coordinates on a plane: p_ξ, p_η, ξ, η. The symplectic group $Sp(4,\mathbb{R})$ is a natural symmetry for the Landau problem since the magnetic field is encoded into the symplectic form of the phase space.

We now show that the Dirac's algebra $\mathfrak{so}(2,3)$ of the De Sitter group $SO(2,3)$ can be thought of as the group of symplectomorphisms, in view of the isomorphism

$$SO(2,3) \cong Sp(4,\mathbb{R})/\mathbb{Z}_2$$

thus yielding a dynamical group of the Landau's problem. The group $SO(2,3)$ has a homogeneous space dS_4 defined by the Dirac quadric in 5-dimensional flat ambient space with coordinates y^A

$$y_{-1}^2 + y_0^2 - y_1^2 - y_2^2 - y_3^2 = R^2 = \eta_{ab}y^a y^b , \qquad \eta_{ab} = diag(+1, +1, -1, -1, -1) .$$

The group of motions of the 4-dimensional hyperboloid is the orthogonal de Sitter group $SO(2,3)$, the conformal group of the Minkowski space $\mathbb{R}^{1,2}$. The generators m_{ab} of the Lie algebra $\mathfrak{so}(2,3)$ satisfy the commutations relations

$$[m_{ab}, m_{cd}] = 0 , \qquad [m_{ab}, m_{bc}] = -i\eta_{bb}m_{ac} \tag{9}$$

where the indices a, b, c, d are assumed to be all distinct from the set $\{-1, 0, 1, 2, 3\}$.

The Dirac's remarkable representation [11] is given by the following quadratic generators[3]

[3] We adopt a different convention for the indices of the matrix m_{ab}, the mapping between our convention and the Dirac's one reads $\{1, 2, 3, -1, 0\} \to \{1, 2, 3, 4, 5\}$.

alternatively written with the Heisenberg algebras generators $a_i^{\pm}$ (7):

$$\begin{array}{llcll}
m_{12} & = & \frac{1}{2}(\xi p_\eta - \eta p_\xi) & = & \frac{1}{2i}\left(a_x^+ a_y^- - a_y^+ a_x^-\right) , \\
m_{23} & = & \frac{1}{4}\left(p_\xi^2 - p_\eta^2 + \xi^2 - \eta^2\right) & = & -\frac{1}{2}\left(a_x^+ a_x^- - a_y^+ a_y^-\right) , \\
m_{31} & = & -\frac{1}{2}\left(\xi\eta + p_\xi p_\eta\right) & = & -\frac{1}{2}\left(a_x^+ a_y^- + a_y^+ a_x^-\right) , \\
m_{1-1} & = & \frac{1}{2}\left(\xi\eta - p_\xi p_\eta\right) & = & \frac{1}{2}\left(a_x^- a_y^- + a_x^+ a_y^+\right) , \\
m_{2-1} & = & \frac{1}{4}\left(\xi^2 - \eta^2 + p_\eta^2 - p_\xi^2\right) & = & \frac{1}{4}\left(a_x^- a_x^- + a_x^+ a_x^+ - a_y^- a_y^- - a_y^+ a_y^+\right) \\
m_{3-1} & = & \frac{1}{2}\left(\xi p_\xi + \eta p_\eta\right) - \frac{i}{2} & = & \frac{i}{4}\left(a_x^+ a_x^+ - a_x^- a_x^- + a_y^+ a_y^+ - a_y^- a_y^-\right) , \\
m_{01} & = & \frac{i}{2}\left(\xi p_\eta + \eta p_\xi\right) & = & -\frac{1}{2}\left(a_x^+ a_y^+ - a_x^- a_y^-\right) , \\
m_{02} & = & \frac{1}{2}\left(\xi p_\xi - \eta p_\eta\right) & = & \frac{i}{4}\left(a_x^+ a_x^+ - a_x^- a_x^- - a_y^+ a_y^+ + a_y^- a_y^-\right) , \\
m_{03} & = & \frac{1}{4}\left(p_\xi^2 + p_\eta^2 - \xi^2 - \eta^2\right) & = & \frac{1}{4}\left(a_x^+ a_x^+ + a_x^- a_x^- + a_y^+ a_y^+ + a_y^- a_y^-\right) , \\
m_{-10} & = & \frac{1}{4}\left(p_\xi^2 + p_\eta^2 + \xi^2 + \eta^2\right) & = & \frac{1}{2}\left(a_x^+ a_x^- + a_y^- a_y^+\right) .
\end{array} \quad (10)$$

Therefore the Dirac $\mathfrak{so}(2,3)$-representation is rooted in the oscillator algebra with two modes $a_x^{\pm}$ and $a_y^{\pm}$ [7]. We will further related these two modes $a_x^{\pm}$ and $a_y^{\pm}$ through eqs (6) to the energy and magnetic translation creation and annihilation operators $a^{\pm}$ and $b^{\pm}$. We will get the Dirac conformal algebra $\mathfrak{so}(2,3)$ playing the role of infinitesimal symplectomorphisms of the Landau problem.

It is convenient to introduce (anti)holomorphic coordinates z $(\bar{z})$ on the phase space such that

$$\begin{array}{lll}
z & = & (x+iy)/2\ell = \frac{1}{\sqrt{2}}(\xi + i\eta) \qquad \partial = \frac{\partial}{\partial z} = \ell\left(\frac{\partial}{\partial x} - i\frac{\partial}{\partial y}\right) = \frac{1}{\sqrt{2}}\left(\partial_\xi - i\partial_\eta\right) , \\
\bar{z} & = & (x-iy)/2\ell = \frac{1}{\sqrt{2}}(\xi - i\eta) \qquad \bar{\partial} = \frac{\partial}{\partial \bar{z}} = \ell\left(\frac{\partial}{\partial x} + i\frac{\partial}{\partial y}\right) = \frac{1}{\sqrt{2}}\left(\partial_\xi + i\partial_\eta\right) ,
\end{array}$$

where the momenta are related to the (anti)holomorphic derivatives ∂ $(\bar{\partial})$ through

$$\partial = \frac{i}{\sqrt{2}}(p_\xi - ip_\eta) , \quad \bar{\partial} = \frac{i}{\sqrt{2}}(p_\xi + ip_\eta) , \quad ip_\xi = \frac{1}{\sqrt{2}}(\partial + \bar{\partial}) , \quad p_\eta = \frac{1}{\sqrt{2}}(\partial - \bar{\partial}) .$$

The energy and magnetic translation creation and annihilation operators $a^{\pm}$ and $b^{\pm}$ in the Landau problem from equation (6) are then expressed in the holomorphic phase space coordinates as follows [4]

$$\begin{array}{lllllll}
a^- & = & \frac{1}{\sqrt{2}}(z + \bar{\partial}) & \quad b^- & = & \frac{1}{\sqrt{2}}(\bar{z} + \partial) , \\
a^+ & = & \frac{1}{\sqrt{2}}(\bar{z} - \partial) & \quad b^+ & = & \frac{1}{\sqrt{2}}(z - \bar{\partial}) .
\end{array} \quad (11)$$

We are now able to recast the Dirac's generators m_{ab} from eq (10) in the form of quadratic

[4] The operators $a^{\pm}$ and $b^{\pm}$ yields one more parametrization of the phase space $\mathbb{R}^4$. The inverse transformation reads

$$\begin{array}{lllllll}
z & = & \frac{1}{\sqrt{2}}(a^- + b^+) & \quad \bar{\partial} & = & \frac{1}{\sqrt{2}}(a^- - b^+) \\
\bar{z} & = & \frac{1}{\sqrt{2}}(a^+ + b^-) & \quad \partial & = & \frac{1}{\sqrt{2}}(b^- - a^+)
\end{array}$$

polynomials of the creation and annihilation operators $a^\pm$ and $b^\pm$ as follows

$$\begin{array}{rclcl}
m_{12} &=& \frac{1}{2}(z\partial - \bar{z}\bar{\partial}) &=& \frac{1}{4}\left(\{b^-,b^+\} - \{a^-,a^+\}\right) , \\
m_{23} &=& \frac{1}{4}(z^2+\bar{z}^2-\partial^2-\bar{\partial}^2) &=& \frac{1}{4}\left(\{a^-,b^+\} + \{a^+,b^-\}\right) , \\
m_{31} &=& \frac{i}{4}(z^2-\bar{z}^2+\partial^2-\bar{\partial}^2) &=& \frac{i}{4}\left(\{a^-,b^+\} - \{a^+,b^-\}\right) , \\
m_{1-1} &=& \frac{1}{4i}(z^2-\bar{z}^2-\partial^2+\bar{\partial}^2) &=& \frac{i}{4}\left(a^+a^+ - a^-a^- + b^-b^- - b^+b^+\right) , \\
m_{2-1} &=& \frac{1}{4}(z^2+\bar{z}^2+\partial^2+\bar{\partial}^2) &=& \frac{1}{4}\left(a^-a^- + a^+a^+ + b^-b^- + b^+b^+\right) , \\
m_{3-1} &=& -\frac{i}{2}\left(z\partial + \bar{z}\bar{\partial} - 1\right) &=& -\frac{i}{4}\left(\{a^-,b^-\} - \{a^+,b^+\}\right) , \\
m_{01} &=& -\frac{1}{2}\left(z\bar{\partial} - \bar{z}\partial\right) &=& -\frac{1}{4}\left(a^-a^- + a^+a^+ - b^-b^- - b^+b^+\right) , \\
m_{02} &=& -\frac{i}{2}\left(\bar{z}\partial + z\bar{\partial}\right) &=& -\frac{i}{4}\left(a^-a^- - a^+a^+ + b^-b^- - b^+b^+\right) , \\
m_{03} &=& \frac{1}{2}\left(z\bar{z} + \partial\bar{\partial}\right) &=& -\frac{1}{4}\left(\{a^+,b^+\} + \{a^-,b^-\}\right) , \\
m_{-10} &=& \frac{1}{2}\left(z\bar{z} - \partial\bar{\partial}\right) &=& \frac{1}{4}\left(\{a^-,a^+\} + \{b^-,b^+\}\right) .
\end{array} \tag{12}$$

Weyl Spinors and $Sp(4,\mathbb{R})$. The whole dynamic Dirac algebra $\mathfrak{so}(2,3) \cong \mathfrak{sp}(4,\mathbb{R})$ can be compactly represented if we pack the Landau's creation and annihilation operators $a^\pm$ and $b^\pm$ into a two-component Weyl spinor

$$\chi^\alpha = \begin{pmatrix} b^- \\ a^- \end{pmatrix} = \frac{1}{\sqrt{2}}\begin{pmatrix} \bar{z}+\partial \\ z+\bar{\partial} \end{pmatrix} , \qquad \chi^*_\alpha = (b^+ \ a^+) = \frac{1}{\sqrt{2}}(z-\bar{\partial} \ , \ \bar{z}-\partial) . \tag{13}$$

Barut and Duru [4] have found that the conformal algebra $\mathfrak{so}(2,3)$ generated by two oscillators (12) can be written by the spinorial operators

$$\begin{array}{rclrcl}
m_{ij} &=& \frac{1}{2}\epsilon_{ijk}\chi^*\sigma_k^T\chi , & m_{-1i} &=& \frac{i}{4}(\chi^*\sigma_i^T\epsilon^T(\chi^*)^T - \chi^T\epsilon\sigma_i^T\chi) , \\
m_{-10} &=& \frac{1}{2}(\chi^*\chi+1) , & m_{0i} &=& \frac{1}{4}(\chi^*\sigma_i^T\epsilon^T(\chi^*)^T + \chi^T\epsilon\sigma_i\chi)
\end{array} \tag{14}$$

with the help of the Pauli matrices $(\sigma_i)_{\alpha\beta}$ and the charge operator $\epsilon = i\sigma_2$. They have obtained the isomorphism of the spinorial $\mathfrak{so}(2,3)$ representation (14) with the spectrum generating algebra of the 2D hydrogen atom. The latter spectrum has been derived as a solution of infinite-component Majorana equation (see also the work of Stoyanov and Todorov [29]).

The subalgebra generated by m_{-10}, m_{-13} and m_{03} is the radial subalgebra $\mathfrak{so}(1,2)$. The generator m_{-10} is the conformal Hamiltonian attached to the harmonic oscillator while the true hamiltonian is $\frac{H}{\hbar\omega} = m_{-10} - m_{12}$. As the motion of a harmonic oscillator is periodic the evolution parameter τ attached to the operator m_{-10} will be compactified to S^1 and will be referred to as *conformal time* τ .

The $\mathfrak{so}(3)$-algebra spanned by the operators m_{12}, m_{23}, m_{31} is commuting with the conformal Hamiltonian m_{-10}

$$[m_{-10}, m_{ij}] = 0 \qquad i,j = 1,2,3 \ .$$

The $\mathfrak{so}(3)$-algebra is the dynamical "accidental" symmetry extending the rotational symmetry generated by the angular momentum operator $2m_{12} = L_z/\hbar$. The rotational group element $R(\phi) = \exp im_{12}\phi$ would actually live in a spinor representation of $\mathfrak{so}(3)$.

4. Jordan algebra toolbox

The tight connection between the two quantum problems sharing the same dynamical conformal symmetry $\mathfrak{so}(2,3)$: Landau problem and 2D hydrogen atom find their natural formulations in the setting of Jordan algebras. We shall follow the ideas of Murat Günaydin [14] to employ a Jordan algebra in the construction of a regular linear representation of the conformal group. For the sake of completeness we first introduce the main notions related to Jordan algebra in general and only then we specialize to the important examples of the Jordan algebras of the 2×2 real Hermitian matrices $\mathfrak{J}_2^{\mathbb{R}}$ and complex Hermitian matrices $\mathfrak{J}_2^{\mathbb{C}}$ yielding the linear representation of $SO(2,3)$ and $SO(2,4)$, respectively.

Jordan algebras. A commutative multiplication law $\circ : \mathfrak{J}\times\mathfrak{J}\to\mathfrak{J}$, satisfying the Jordan identity

$$a\circ b = b\circ a \qquad a\circ\left(a^2\circ b\right) = a^2\circ(a\circ b) \tag{15}$$

defines a Jordan algebra $(\mathfrak{J},\circ)$. Any associative matrix algebra (over $\mathbb{R},\mathbb{C},\mathbb{H}$) can be converted into a *special* Jordan algebra via the product $a\circ b = \frac{1}{2}(ab+ba)$, where ab is the standard associative matrix multiplication.

Jordan triple product and conformal group $Co(\mathfrak{J})$ representations. Any Jordan algebra $\mathfrak{J}$ has a Jordan triple product $(\bullet,\bullet,\bullet) : \mathfrak{J}\times\mathfrak{J}\times\mathfrak{J}\to\mathfrak{J}$

$$(abc) = a\circ(b\circ c) - b\circ(a\circ c) + (a\circ b)\circ c\,. \tag{16}$$

The Jordan triple product defines a Jordan triple system with identities

$$(abc) = (cba) \tag{17}$$

$$(ab(cdx)) - (cd(abx)) = (a(dcb)x) - ((cda)bx)\,. \tag{18}$$

For any pair of elements $(x,y)\in\mathfrak{J}\times\mathfrak{J}$ one has a linear map $S_x^y : \mathfrak{J}\to\mathfrak{J}$ with a matrix fixed by the structure constants $(\Sigma_\mu^\nu)_\rho^\sigma$ of the Jordan triple product

$$S_x^y(z) = (xyz) \qquad (e_\mu, e_\nu, e_\rho) = \Sigma_{\mu\rho}^{\nu\sigma} e_\sigma\,. \tag{19}$$

Tits, Kantor and Koecher (TKK) construction of $\mathfrak{co}(\mathfrak{J})$. The conformal group $Co(\mathfrak{J})$ is generated by vector fields closing a *conformal* Lie algebra $\mathfrak{co}(\mathfrak{J})$. The general construction of a conformal Lie algebra $\mathfrak{co}(\mathfrak{J})$ from a given Jordan algebra $\mathfrak{J}$ is due to Tits, Kantor and Koecher.

Any Jordan triple system generated in $x\in\mathfrak{J}$ gives rise to a 3-graded Lie algebra $\mathfrak{co}(\mathfrak{J})$, endowed with an involution $\dagger$ via

$$(x,y,z) := [[x,y^\dagger],z]\,.$$

Conversely, any 3-graded Lie algebra $[\mathfrak{g}_i,\mathfrak{g}_j]\subset\mathfrak{g}_{i+j}$ (with $\mathfrak{g}_i = 0$ when $i\neq 0,\pm1$) endowed with a graded involution $\dagger$, $\mathfrak{g}_k^\dagger = \mathfrak{g}_{-k}$ determines a Jordan triple system. The 3-graded Lie algebra $\mathfrak{co}(\mathfrak{J})$ has the graded decomposition

$$\mathfrak{co}(\mathfrak{J}) = \mathfrak{g}_{+1}\oplus\mathfrak{g}_0\oplus\mathfrak{g}_{-1} := \mathfrak{J}^*\oplus\mathfrak{str}(\mathfrak{J})\oplus\mathfrak{J} \tag{20}$$

where the abelian subalgebra $\mathfrak{g}_{-1}(\mathfrak{g}_{+1})$ is generated in the space $\mathfrak{J}(\mathfrak{J}^*)$. The grading operator $D\in\mathfrak{g}_0$ is the dilatation $[D,g]=kg$, for any $g\in\mathfrak{g}_k$. The grading alone implies that $\mathfrak{g}_{+1}$ and $\mathfrak{g}_{-1}$ are abelian Lie subalgebras of $\mathfrak{co}(\mathfrak{J})$ and their mutual commutators belong to $\mathfrak{g}_0$, $[\mathfrak{g}_{-1},\mathfrak{g}_{+1}]\subset\mathfrak{g}_0$. It turns out that all elements in $\mathfrak{g}_0$ can be represented as commutators, the structure algebra of $\mathfrak{J}$, $\mathfrak{str}(\mathfrak{J}) := \mathfrak{g}_0 = [\mathfrak{g}_{-1},\mathfrak{g}_{+1}]$ and $\mathfrak{g}_{\pm1}$ are its fundamental and antifundamental representations in view of

$$[\mathfrak{g}_0,\mathfrak{g}_{\pm1}] = \mathfrak{g}_{\pm1}\,.$$

Coordinate-free definition of $\mathfrak{co}(\mathfrak{J})$. The relations[5] of the conformal algebra $\mathfrak{co}(\mathfrak{J})$ are compactly written with the help of the Jordan triple product conveying the essence of the

[5] The normalization $[U_a,U^b] = -2S_a^b$ is adopted for future convenience following [23].

TKK construction [14]

$$\begin{array}{rclrclr} [U_a, U^b] &=& -2S_a^b\,, & [U_a, U_b] &=& 0\,, & U_a \in \mathfrak{g}_{-1} \\ [S_a^b, U_c] &=& U_{(abc)}\,, & [S_a^b, S_c^d] &=& S^d_{(abc)} - S_c^{(bad)}\,, & S_a^b \in \mathfrak{g}_0 \\ [S_a^b, U^c] &=& -U^{(bac)}\,, & [U^a, U^b] &=& 0\,, & U^b \in \mathfrak{g}_{+1}\,. \end{array} \tag{21}$$

The involution $\dagger$ acts as $U_a^\dagger = U^a$ and consequently $(S_a^b)^\dagger = S_b^a$.

TKK for Jordan algebra $\mathfrak{J}_2^{\mathbb{C}} = H_2(\mathbb{C})$. The associative algebra of 2×2 Hermitian matrices with complex elements $H_2(\mathbb{C})$ has a basis of Pauli matrices σ_μ. The symmetric product $\circ$ gives rise to the (special) Jordan algebra $\mathfrak{J}_2^C$

$$\sigma_i \circ \sigma_j := \frac{1}{2}\{\sigma_i, \sigma_j\} = \delta_{ij}\,, \qquad \sigma_0 \circ \sigma_i = \sigma_i\,, \qquad \sigma_0 \circ \sigma_0 = \sigma_0\,.$$

An element $x \in H_2(\mathbb{C})$ is parametrized by its Minkowski coordinates

$$x = x^\mu \sigma_\mu = x^0 \mathbb{1} + x^i \sigma_i$$

where σ_0 is the unit matrix $\mathbb{1}$ and σ_i are the Pauli matrices. Hence

$$x = x^\mu \sigma_\mu = \begin{pmatrix} x_0 + x_3 & x_1 - ix_2 \\ x_1 + ix_2 & x_0 - x_3 \end{pmatrix}, \qquad x^\mu := \frac{1}{2} tr(x\sigma_\mu) \tag{22}$$

and the Minkowski metric g with a signature $(+,-,-,-)$ is given by the determinant

$$\det x = x_\mu x^\mu = (x^0)^2 - \delta_{ij} x^i x^j = g_{\mu\nu} x^\mu x^\nu.$$

The evaluation of the Jordan triple structure constants $\Sigma^{\beta\rho}_{\alpha\gamma}$ in the $\mathfrak{J}_2^{\mathbb{C}}$ basis

$$(\sigma_\alpha, \sigma_\beta, \sigma_\gamma) = \Sigma^{\beta\rho}_{\alpha\gamma} \sigma_\rho = \sigma_\alpha \circ (\sigma_\beta \circ \sigma_\gamma) - \sigma_\beta \circ (\sigma_\gamma \circ \sigma_\alpha) + (\sigma_\alpha \circ \sigma_\beta) \circ \sigma_\gamma$$

gives back the concise formula,

$$\Sigma^{\beta\rho}_{\alpha\gamma} = \delta^\rho_\gamma \delta^\beta_\alpha + \delta^\rho_\alpha \delta^\beta_\gamma - g^{\beta\rho} g_{\alpha\gamma}\,. \tag{23}$$

The latter formula establishes via TKK construction a dichotomy between Minkowski's spacetime Jordan algebra $\mathfrak{J}_2^{\mathbb{C}}$ and the 3-graded Lie algebra $\mathfrak{so}(2,4)$.

Indeed the Jordan triple product induces via the TKK construction (Eq. (21)) a $\mathfrak{co}(\mathfrak{J}_2^{\mathbb{C}})$-representation [14, 25] by the following vector fields:

$\mathfrak{co}(\mathfrak{J})$	operator $\in \mathfrak{co}(\mathfrak{J})$	mapping	x−rep basis $\mathfrak{so}(2,4)$	deg(x)
$\mathfrak{J}$	$U_a = -ia^\mu P_\mu \quad \in \mathfrak{g}_{-1}$	$x \mapsto a$	$P_\nu = i\partial_\nu$	0
$\mathfrak{str}(\mathfrak{J})$	$S_a^b = ia^\nu b_\mu S^\mu_\nu \quad \in \mathfrak{g}_0$	$x \mapsto (a, b, x)$	$S^\mu_\nu = -i\Sigma^{\mu\beta}_{\nu\alpha} x^\alpha \partial_\beta$	1
$\mathfrak{J}^*$	$U^b = ib_\mu K^\mu \quad \in \mathfrak{g}_{+1}$	$x \mapsto -(x, b, x)$	$K^\mu = i\Sigma^{\mu\beta}_{\nu\alpha} x^\nu x^\alpha \partial_\beta$	2

The 15 generators of the conformal group $Co(\mathfrak{J}_2^{\mathbb{C}}) = SO(2,4)$ are obtained after the evaluation of the Jordan structure constants $\Sigma^{\mu\beta}_{\nu\alpha}$ (23)

$$\begin{array}{llcll} \mathfrak{g}_{-1} & -iP_\nu &=& \partial_\nu & \text{translations} \\ \mathfrak{g}_0 & iM^\mu{}_\nu &=& -x^\mu \partial_\nu + x_\nu \partial^\mu & \text{Lorentz transformations} \\ \mathfrak{g}_0 & iD &=& x^\mu \partial_\mu & \text{dilatation} \\ \mathfrak{g}_{+1} & iK^\mu &=& -2x^\mu x^\nu \partial_\nu + x^\nu x_\nu \partial^\mu & \text{special conformal transformations} \end{array} \tag{24}$$

yielding a minimal representation the Lie algebra $\mathfrak{so}(2,4)$. The Lorentz generators $M^{\mu}{}_{\nu}$ are the basis of the reduced structure algebra $\mathfrak{str}_0(\mathfrak{J}_2^{\mathbb{C}}) = \mathfrak{so}(1,3)$ (generating the Lorentz norm $\det x$ preserving group). The structure algebra $\mathfrak{str}(\mathfrak{J}_2^{\mathbb{C}})$ is generated by S^a_b. When we choose coordinates as in the table above $\mathfrak{str}(\mathfrak{J}_2^{\mathbb{C}})$ is the span of the generators $S^{\mu}_{\nu} = \frac{i}{2}[K^{\mu}, P_{\nu}] = M^{\mu}{}_{\nu} - \delta^{\mu}_{\nu} D$. Then the TKK construction (21) for $\mathfrak{J}_2^{\mathbb{C}}$ yields the commutation relations of the conformal algebra $\mathfrak{so}(2,4)$ of the Minkowski space $\mathbb{R}^{1,3}$ $(\mu, \nu, \lambda = 0,1,2,3)$

$$\begin{aligned} [K_{\mu}, P_{\nu}] &= 2i(g_{\mu\nu} D - M_{\mu\nu}) , & [D, P_{\mu}] &= iP_{\mu} , & [D, K_{\mu}] &= -iK_{\mu} , \\ [K_{\lambda}, M_{\mu\nu}] &= i(g_{\lambda\mu} K_{\nu} - g_{\lambda\nu} K_{\mu}) , & [D, M_{\mu\nu}] &= 0 , & [P_{\mu}, P_{\nu}] &= 0 = [K_{\mu}, K_{\nu}] , \\ [P_{\lambda}, M_{\mu\nu}] &= i(g_{\lambda\mu} P_{\nu} - g_{\lambda\nu} P_{\mu}) , & [M_{\mu\alpha}, M_{\beta\nu}] &= 0 , & [M_{\mu\alpha}, M_{\alpha\nu}] &= i g_{\alpha\alpha} M_{\mu\nu} . \end{aligned} \tag{25}$$

In the last two relations the indices are assumed distinct.

The conformal inversion $I(x^0, \boldsymbol{x}) = \left(\frac{x^0}{x^2}, -\frac{\boldsymbol{x}}{x^2}\right)$ is an involution, $I^2 = \mathbb{1}$. It anticommutes with the grading operator, $ID = -DI$ and induces the involution $\dagger$ through

$$K^{\mu} = IP_{\mu}I = P_{\mu}^{\dagger} .$$

The conformal inversion changes dimensions; namely, the length to inverse length.

TKK for Jordan algebra $\mathfrak{J}_2^{\mathbb{R}} = H_2(\mathbb{R})$. The real symmetric matrices are spanned by the subset $\{\sigma_0, \sigma_1, \sigma_3\}$ of real Pauli matrices with coordinates $\{y_0, y_1, y_2\}$

$$y = \sum_{\mu=0,1,3} x^{\mu}\sigma_{\mu} = \begin{pmatrix} y_0 + y_2 & y_1 \\ y_1 & y_0 - y_2 \end{pmatrix} , \quad y^T = y . \tag{26}$$

We simply skip the Pauli matrix σ_2 (which is not symmetric) from the $3+1$ Minkowski spinor $x = x^{\mu}\sigma_{\mu}$. Renaming the components $x^0 = y^0$, $x^1 = y^1$and $x^3 = y^2$ we end up with a real spinorial representation of the Minkowski spacetime $\mathbb{R}^{1,2}$ such that the deteminant yields the metric

$$\det y = y_{\tilde{\mu}} y^{\tilde{\mu}} = (y^0)^2 - (y^1)^2 - (y^2)^2 = g_{\tilde{\mu}\tilde{\nu}} y^{\tilde{\mu}} y^{\tilde{\nu}} \qquad \tilde{\mu}, \tilde{\nu} = 0,1,2 .$$

Tits-Kantor-Koecher construction applied to the Jordan algebra $\mathfrak{J}_2^{\mathbb{C}}$ and $\mathfrak{J}_2^{\mathbb{R}}$ yields, respectively, the conformal algebra of the Minkowski spacetime $\mathbb{R}^{1,3}$, eq. (25):

$$\mathfrak{co}(\mathfrak{J}_2^{\mathbb{C}}) = \mathfrak{so}(2,4) = \underbrace{(\mathfrak{J}_2^{\mathbb{C}})^*}_{K^{\mu}} \oplus \overbrace{(\underbrace{\mathfrak{so}(1,3)}_{M^{\mu}_{\nu}} \oplus \underbrace{\mathbb{R}}_{D})}^{\mathfrak{str}(\mathfrak{J}_2^{\mathbb{C}})} \oplus \underbrace{\mathfrak{J}_2^{\mathbb{C}}}_{P_{\nu}} \tag{27}$$

and the conformal algebra of the Minkowski space $\mathbb{R}^{1,2}$, eq. (25):

$$\mathfrak{co}(\mathfrak{J}_2^{\mathbb{R}}) = \mathfrak{so}(2,3) = \underbrace{(\mathfrak{J}_2^{\mathbb{R}})^*}_{K^{\tilde{\mu}}} \oplus \overbrace{(\underbrace{\mathfrak{so}(1,2)}_{M^{\tilde{\mu}}_{\tilde{\nu}}} \oplus \underbrace{\mathbb{R}}_{D})}^{\mathfrak{str}(\mathfrak{J}_2^{\mathbb{R}})} \oplus \underbrace{\mathfrak{J}_2^{\mathbb{R}}}_{P_{\tilde{\nu}}} \qquad \mu, \nu = 0,1,2. \tag{28}$$

Geometry of null cones. In his famous Erlangen program Félix Klein associated any geometric space with its group of motion, *i.e.*, its underlying group of symmetries.

Proposition 4.1 *Let the compactified Minkowski space $\mathcal{M}_{1,3} = \mathcal{N}/\mathbb{R}^*$ be the space of the isotropic rays in $\mathbb{R}^{2,4}$ (proportional isotropic vectors are identified)*

$$\mathcal{N} = \{\vec{x} \in \mathbb{R}^{2,4} | x_{-1}^2 + x_0^2 - x_1^2 - x_2^2 - x_3^2 - x_5^2 = 0; x \neq 0\}.$$

The compactified null cone $\mathcal{M}_{1,3} = \mathcal{N}/\mathbb{R}^* \cong (S^1 \times S^3)/\mathbb{Z}_2$ *is a homogeneous space for the conformal group* $SO(2,4)$ *generated in the Lie algebra* $\mathfrak{co}(\mathfrak{J}_2^{\mathbb{C}}) = \mathfrak{so}(2,4)$, *eqs (31).*

The intersection of $\mathcal{M}_{1,3}$ *by hyperplane* $x_2 = 0$ *yields the space* $\mathcal{M}_{1,2} = \tilde{\mathcal{N}}/\mathbb{R}^*$ *of the isotropic rays in* $\mathbb{R}^{2,3}$

$$\tilde{\mathcal{N}} = \{\vec{y} \in \mathbb{R}^{2,3} | y_{-1}^2 + y_0^2 - y_1^2 - y_2^2 - y_3^2 = 0; y \neq 0\}$$

where the remaining coordinates after the reduction are renamed according to

$$\{x_{-1}, x_0, x_1, x_3, x_5\} \leftrightarrow \{y_{-1}, y_0, y_1, y_2, y_3\}\ .$$

The compactified null cone $\mathcal{M}_{1,2} = \tilde{\mathcal{N}}/\mathbb{R}^* \cong (S^1 \times S^2)/\mathbb{Z}_2$ *of the isotropic rays in* $\mathbb{R}^{2,3}$ *is a homogeneous space for the conformal group* $SO(2,3)$ *generated in the Lie algebra* $\mathfrak{co}(\mathfrak{J}_2^{\mathbb{R}}) = \mathfrak{so}(2,3)$ *(see eqs (31))* .

In this sense, the geometry of the compactified Minkowski spaces $\mathcal{M}_{1,2} = (S^1 \times S^2)/\mathbb{Z}_2$ and $\mathcal{M}_{1,3} = (S^1 \times S^3)/\mathbb{Z}_2$ are associated with the conformal symmetry algebras $\mathfrak{so}(2,3)$ and $\mathfrak{so}(2,4)$ of the Jordan algebras of real and complex hermitian 2×2 matrices. The intersection with the plane $x_2 = 0$ corresponds to the reduction of the complex Jordan algebra to real $\mathfrak{J}_2^{\mathbb{R}}$ Jordan algebra $\mathfrak{J}_2^{\mathbb{C}}$

$$\mathfrak{J}_2^{\mathbb{R}} = \{y \in \mathfrak{J}_2^{\mathbb{C}} | y^T = y\}\ ,$$

thus reducing the conformal symmetry from $\mathfrak{so}(2,4)$ to $\mathfrak{so}(2,3)$.

The advantage of the TKK construction of the Jordan algebra symmetries is that it yields explicitly a conformal group representationof the conformal spacetime symmetries as linear transformations of a null-ray cone.

5. Hydrogen atom from 3D to 2D

The non-relativistic hydrogen atom in three space dimensions is central in the development of the quantum mechanics. Barut and collaborators (see e.g. [3, 5]) have shown that the states in the 3D hydrogen atom spectrum transform in a helicity zero massless irreducible representation [21] of the dynamical group $SO(2,4)$ with generators

$$\begin{array}{rclrclrcl} \boldsymbol{L} & = & \boldsymbol{r} \times \boldsymbol{p} & B_0 - A_0 & = & r & B_0 + A_0 & = & r\boldsymbol{p}^2 \\ \boldsymbol{\Gamma} & = & r\boldsymbol{p} & \boldsymbol{B} - \boldsymbol{A} & = & \boldsymbol{r} & \boldsymbol{B} + \boldsymbol{A} & = & \boldsymbol{r}\boldsymbol{p}^2 - 2\boldsymbol{p}(\boldsymbol{r} \cdot \boldsymbol{p}) \\ D & = & \boldsymbol{r} \cdot \boldsymbol{p} - i & & & & & & \end{array} \quad . \tag{29}$$

On the other hand the conformal symmetry $SO(2,4)$ of the hydrogen spectrum is also the group of causal space-time automorphisms of the Minkowski space $\mathbb{R}^{1,3}$. The dictionary between the two is given by the table

$\mathfrak{co}(\mathfrak{J})$	space-time cone	hydrogen atom
$\mathfrak{g}_0$	$M_{\mu\nu}$; D	L_{ij}, Γ_i; D
$\mathfrak{g}_{-1}$	K_μ	$B_\mu + A_\mu$
$\mathfrak{g}_{+1}$	P_μ	$B_\mu - A_\mu$

.

The dynamical group $SO(1,2)$ generated by $\{A_0, D, B_0\}$ is the "radial" group of the conformal transformations of the time coordinate $\mathbb{R}^{0,1}$ [17]

$$\begin{array}{lllll} t \to t' & = & t + t_0 & P_0 = B_0 - A_0 & \text{time translations} \\ t \to t' & = & \lambda t & D & \text{time dilation} \\ t \to \frac{1}{t'} & = & \frac{1}{t} + \frac{1}{t_0} & K_0 = A_0 + B_0 & \text{special conformal} \end{array} \quad .$$

The conformal transformation group $SO(1,2)$ arises from the TKK construction of the Jordan algebra $\mathfrak{J}_1^{\mathbb{R}} = \mathbb{R}$ yielding

$$\mathfrak{co}(\mathbb{R}) = \mathfrak{so}(1,2) = (\mathfrak{J}_1^{\mathbb{R}})^* \oplus \mathfrak{str}(\mathfrak{J}_1^{\mathbb{R}}) \oplus \mathfrak{J}_1^{\mathbb{R}} = \mathbb{R}K_0 \oplus \mathbb{R}D \oplus \mathbb{R}P_0 \ .$$

The sign of energy chooses different $SO(1,2)$-generator to be a conformal hamiltonian:

$$\begin{array}{rclll} B_0 & = & r(p^2+1)/2 & E<0 & \text{bound states ,} \\ A_0 & = & r(p^2-1)/2 & E>0 & \text{scattering states ,} \\ A_0+B_0 & = & rp^2 & E=0 & \text{free motion .} \end{array}$$

The conformal $SO(1,2)$ group can be seen also as the even generators of the Schrödinger group in the context of non-relativistic conformal symmetries (for the correspondence between Schrödinger and conformal group see [16]).

The 15 generators of the 3D hydrogen dynamical group $SO(2,4)$ act by linear transformations of the null-ray cone. They have been identified in [27] (see the table above) with the causal automorphisms of the light cone $\mathfrak{co}(\mathfrak{J}_2^{\mathbb{C}})$, see eqns (27, 31)

$$\begin{pmatrix} 0 & B_0 & B_1 & B_2 & B_3 & D \\ & 0 & \Gamma_1 & \Gamma_2 & \Gamma_3 & A_0 \\ & & 0 & L_3 & -L_2 & A_1 \\ & & & 0 & L_1 & A_2 \\ & & & & 0 & A_3 \\ & & & & & 0 \end{pmatrix} = \begin{pmatrix} 0 & L_{-10} & L_{-11} & L_{-12} & L_{-13} & L_{-15} \\ & 0 & L_{01} & L_{02} & L_{03} & L_{05} \\ & & 0 & L_{12} & L_{13} & L_{15} \\ & & & 0 & L_{23} & L_{25} \\ & & & & 0 & L_{35} \\ & & & & & 0 \end{pmatrix} . \qquad (30)$$

Stated differently the compactified Minkowski space $\mathcal{M}_{1,3} = (S^1 \times S^3)/\mathbb{Z}_2$ carries a minimal representation of the conformal group $SO(2,4)$ stemming from the TKK construction of the Jordan algebra $\mathfrak{J}_2^{\mathbb{C}}$.

The conformal algebra generators L_{AB} satisfy the commutation relations

$$[L_{AB}, L_{CD}] = -i(\eta_{AC}L_{BD} + \eta_{BD}L_{AC} - \eta_{AD}L_{BC} - \eta_{BC}L_{AD}) \qquad (31)$$

where the set of indices contains the auxiliary indices -1 and 5 in addition to the spacetime indices $\mu = 0,1,2,3$

$$L_{AB} \in \mathfrak{so}(2,4)\ , \qquad \eta_{AB} = diag(1,1,-1,-1,-1,-1)\ , \qquad A,B \in \{-1,0,1,2,3,5\}\ .$$

While reducing to the spacetime $\mathbb{R}^{1,2}$ with indices $\mu = 0,1,2$ we get an algebra representation equivalent to the Dirac's remarkable $\mathfrak{so}(2,3)$-representation (9)

$$L_{ab} \in \mathfrak{so}(2,3)\ , \qquad \eta_{ab} = diag(1,1,-1,-1,-1)\ , \qquad a,b \in \{-1,0,1,2,3=5\}^6\ .$$

When the motion of the electron is constrained to a plane we obtain a system which we will refer to as 2D hydrogen atom, the reduction of the usual 3D atom to two space dimensions. The dynamical algebras of the 2D hydrogen atom and the Landau problem are isomorphic, the isomorphism is simply the transposition

$$L_{ab} \quad \leftrightarrow \qquad m_{ba} = -m_{ab}\ . \qquad (32)$$

[6] The auxiliary index 5 comes with the Dirac matrix $\gamma_5 = \gamma_0\gamma_1\gamma_2\gamma_3$. In $\mathbb{R}^{1,2}$ we have $\gamma_5 = \gamma_0\gamma_1\gamma_2$ and similarly $i\sigma_3 = \sigma_0\sigma_1\sigma_2$ so we adopt a double notation $3 = 5$.

The isotropic rays in the five dimensional space $\mathbb{R}^{2,3}$ carry a linear representation of the conformal group $SO(2,3)$ with generators :

$$L_{ab} = \begin{pmatrix} 0 & B_0 & B_1 & B_2 & D \\ & 0 & \Gamma_1 & \Gamma_2 & A_0 \\ & & 0 & L_3 & A_1 \\ & & & 0 & A_2 \\ & & & & 0 \end{pmatrix} = \begin{pmatrix} 0 & L_{-10} & L_{-11} & L_{-12} & L_{-15} \\ & 0 & L_{01} & L_{02} & L_{05} \\ & & 0 & L_{12} & L_{15} \\ & & & 0 & L_{25} \\ & & & & 0 \end{pmatrix} .$$

The dynamical symmetry group of the $2D$ hydrogen atom is the conformal group $SO(2,3)$ stemming from the TKK algebra $\mathfrak{co}(\mathfrak{J}_2^{\mathbb{R}})$ (28). The compactified Minkowski space $\mathcal{M}_{1,2} \cong (S^1 \times S^2)/\mathbb{Z}_2$ is a homogeneous space for the dynamical $SO(2,3)$ group of the 2D hydrogen atom. We conclude that in the Jordan algebra language the reduction from $SO(2,4)$ to $SO(2,3)$ is projecting the complex Jordan algebras $\mathfrak{J}_2^{\mathbb{C}}$ to the real $\mathfrak{J}_2^{\mathbb{R}}$.

The sphere S^3 in $\mathcal{M}_{1,3}$ arises as compactification of the flat momentum space via the Cayley transform whereas S^1 stays for the compatified time coordinate. The maximally compact subgroup $SO(4) \subset SO(2,4)$ stabilizes the sphere S^3 interpolating between states with equal energy. A dual point of view is advocated in a recent paper [19] where S^3 in $\mathcal{M}_{1,3}$ is thought as a configuration space of the quark-antiquark system. In that way the compactified Minkowski space $\mathcal{M}_{1,3}$ becomes a toy model for the simplest QCD system: a meson. In other words one obtains the "QCD hydrogen atom" where the interaction potential is the curved analogue of the Coulomb potential, inherently introduced by the Green function of the Laplace-Beltrami operator on S^3. On closed surfaces charges appear in pairs thus the charge neutrality is naturally leading to a confinement. The phenomenologically observed degeneracies of the spectrum of meson masses can be then attributed to the conformal symmetries [18] of the compactified Minkowski space $(S^1 \times S^3)/\mathbb{Z}_2$ seen as a configuration space for one color charge degree of freedom. From that perspective the mesons are the QCD cousins of the system "electron-constant magnetic field" living in the space $\mathcal{M}_{1,2} \cong (S^1 \times S^2)/\mathbb{Z}_2$.

6. Reduction of Kustaanheimo-Steifel transform

The Kustaanheimo-Stiefel transform [20] in celestial mechanics removes the singular trajectories due to binary collisions from the phase space of the 3D Kepler motion and estabilishes a correspondence between the 4D isotropic harmonic oscillator 3D Coulomb-Kepler problem. The spectrum of bounded states for the 3D hydrogen atom (*i.e.* the quantum Kepler problem) is then symplectically equivalent to the spectrum of the harmonic oscillator with 4 bosonic modes. One has an inclusion $SU(2,2) \subset Sp(8,\mathbb{R})$ [21].

The elliptic Kepler orbits for negative energies $E < 0$ correspond to geodesic motion on 3-dimensional sphere S^3. The sphere S^3 arises through the stereographic projection of the momenta. A great circle in S^3 is the hodograph[7] of an elliptic orbit [24]. The regularized Kepler orbits live on the cotangent bundle T^*S^3 to the sphere S^3 with the zero section removed [9]

$$T^+S^3 = T^*S^3 - \{0_{sec}\} .$$

Definition 6.1 *The Kustaanheimo-Stiefel (KS) transform is the mapping between the cotangent bundles (with north poles deleted)* [8]

$$KS: \quad T^+S^3 \to T^+S^2 \qquad \subset (\mathbb{R}^*)^4 \times \mathbb{R}^4 \to (\mathbb{R}^*)^3 \times \mathbb{R}^3$$

[7] The hodograph is a curve drawn by the velocity vector, that is, the trajectory in the momentum space. It turns out the the hodographs of Kepler orbits for hyperbolic ($E > 0$) and parabolic $E = 0$ motions are segments of circles.

[8] We adopt the notation $\mathbb{R}^* = \mathbb{R} - \{0\}$.

where the 4D harmonic oscillator phase space coordinates $(u, w) \in (\mathbb{R}^*)^4 \times \mathbb{R}^4$ are related with the 3D Kepler phase space $(x, p) \in (\mathbb{R}^*)^3 \times \mathbb{R}^3$ coordinates through the Hopf fibration map (see eq. (35))

$$\begin{aligned} x_1 &= u_1u_3 + u_2u_4 \ , \\ x_2 &= u_2u_3 - u_1u_4 \ , \\ x_3 &= -u_1^2 - u_2^2 + u_3^2 + u_4^2 \ , \end{aligned} \tag{33}$$

extended with the derived momentum relations ($|\boldsymbol{z}|^2 = u_1^2 + u_2^2 + u_3^2 + u_4^2$)

$$\begin{aligned} p_1 &= -(u_1w_3 + w_1u_3 + u_2w_4 + w_2u_4)/|\boldsymbol{z}|^2 \ , \\ p_2 &= -(u_2w_3 + w_2u_3 - w_1u_4 - u_1w_4)/|\boldsymbol{z}|^2 \ , \ . \\ p_3 &= (u_1w_1 + u_2w_2 - u_3w_3 - u_4w_4)/|\boldsymbol{z}|^2 \ . \end{aligned} \tag{34}$$

The coordinates on T^+S^3 are subject to the constraint

$$K = u_1w_2 - u_2w_1 + u_3w_4 - u_4w_3 = 0 \ .$$

This constraint is in fact a constant of motion as a sum of two angular momentum components. The non-vanishing of the constraint K is playing a crucial role in the generalizations of the magnetised problem of a electric charge in the field of a Dirac monopole and systems of two dyonic particles [32].

The Hopf fibration is essentially representing a 3D vector $\boldsymbol{x} \in \mathbb{R}^3$ as a "square root" of a spinor $\boldsymbol{z} \in \mathbb{C}^2$ in view of

$$|\boldsymbol{x}| = |\boldsymbol{z}|^2 \ , \qquad |\boldsymbol{x}| = \sqrt{x_1^2 + x_2^2 + x_3^2} \ , \qquad \boldsymbol{z} = \begin{pmatrix} u_1 + iu_2 \\ u_3 + iu_4 \end{pmatrix} \ , \qquad |\boldsymbol{z}|^2 = \boldsymbol{z}^\dagger \boldsymbol{z}. \tag{35}$$

The KS transformation can be seen as a phase space extension of the Hopf fibration

$$0 \to S^1 \hookrightarrow S^3 \to S^2 \to 0 \ ,$$

the kernel consists of the spinors $e^{i\theta}\boldsymbol{z} \in S^1$.

Reduction of KS transform to the Levi-Civita transform is done by the choice

$$u_2 = u_4 = 0 \ , \qquad w_2 = w_4 = 0$$

which amounts to taking a square root of a vector in the 2D plane $x_1 = \eta$ and $x_3 = \xi$.

The Levi-Civita transform stems from the change to parabolic coorindates u_1 and u_3 in the complex plane

$$\xi + i\eta = Z^2 \ , \qquad \begin{aligned} \xi &= u_1^2 - u_3^2 \ , \\ \eta &= 2u_1u_3 \ , \end{aligned} \qquad \begin{aligned} p_\xi &= (u_1w_1 - w_3u_3)/|Z|^2 \ , \\ p_\eta &= -(u_1w_3 + w_1u_3)/|Z|^2 \end{aligned} \tag{36}$$

where the real spinor $\psi = \begin{pmatrix} u_1 \\ u_3 \end{pmatrix}$ is written with one complex number $Z = u_1 + iu_3$. Levi-Civita mapping is then a sympletic extension of the the trivial Hopf fibration

$$0 \to S^0 \hookrightarrow S^1 \to S^1 \to 0$$

where the dimension zero sphere $S^0 = \mathbb{Z}_2$ is in the kernel thus reflecting the fact that any pair of parabolic (spinor) coordinates (u_1, u_3) and $(-u_1, -u_3)$ parametrize one and the same point (ξ, η) in the complex plane.

The Levi-Civita transform (36) is then a 2-to-1 mapping between 2D harmonic oscillator phase-space (u_1, u_3, w_1, w_3) and the phase space of the 2D Kepler problem $(\xi, \eta, p_\xi, p_\eta)$. Upon quantization it yields the Newton-Hooke duality between the Landau problem and the 2D hydrogen atom (1). We are now going to show that the phase space (u_1, u_3, w_1, w_3) is naturally parametrized by a Majorana spinor in dimension 4.

7. Ladder $U(2,2)$ representation

Let $\psi = (\psi^\alpha)$ be operator-valued Dirac spinor with 4 components satisfying the canonical commutation relations

$$[\psi^\alpha, \bar{\psi}_\beta] = \delta^\alpha_\beta \ , \qquad [\psi^\alpha, \psi^\beta] = 0 \ .$$

A canonical representation of a pair of 2D harmonic oscillators with two complex variables z^α and the holomorphic derivatives $iw_\alpha = \frac{\partial}{\partial z^\alpha}$ is given by

$$\boldsymbol{z} = \begin{pmatrix} z^1 \\ z^2 \end{pmatrix} \qquad \boldsymbol{\partial} = \begin{pmatrix} \frac{\partial}{\partial z^1} \\ \frac{\partial}{\partial z^2} \end{pmatrix} \qquad \psi = \begin{pmatrix} \bar{z} \\ \partial \end{pmatrix} \qquad \bar{\psi} = (-\bar{\partial}, z) \ . \tag{37}$$

The ladder representation of $U(2,2)$ [21] is a spinorial representation realized by the operators[9]

$$J^{AB} = \bar{\psi}\sigma^{AB}\psi \ , \qquad C_1 = \bar{\psi}\psi \tag{38}$$

where the 4×4 matrices σ^{AB} close a defining representation of $\mathfrak{su}(2,2)$: these are the matrices in $\mathfrak{sl}(4,\mathbb{C})$ preserving a pseudo-Hermitian form β with signature $(++--)$

$$(\phi, \psi) = \phi_1^*\psi_1 + \phi_2^*\psi_2 - \phi_3^*\psi_3 - \phi_4^*\psi_4 = \phi^\dagger\beta\psi = \bar{\phi}\psi \ .$$

The Hermitian matrix $\beta = \beta^\dagger$ depends on the basis, it fixes the choice of the Dirac matrix $\gamma_0 := \beta$ and the invariance of the form implies

$$\beta\sigma^{AB}\beta^{-1} = (\sigma^{AB})^\dagger \ .$$

The linear Casimir operator C_1 is the center of $U(2,2)$. It is represented by an integer multiple of the unit $\mathbb{1}$:

$$C_1 + 2 = -2\lambda = z^\alpha \frac{\partial}{\partial z^\alpha} - \bar{z}^\alpha \frac{\partial}{\partial \bar{z}^\alpha} \qquad \lambda = 0, \pm\frac{1}{2}, \pm 1, \ldots$$

where the half-integer helicities λ are labelling the zero-mass representations of $U(2,2)$. The hydrogen atom is described by helicity $\lambda = 0$ representation [21].

A finite-dimensional representation of $\mathfrak{su}(2,2)$ is generated by the 4×4 Dirac gamma matrices γ^μ,

$$\{\gamma^\mu, \gamma^\nu\} = 2\eta^{\mu\nu} \qquad \mu, \nu = 0, 1, 2, 3 \ .$$

The 15 $\mathfrak{su}(2,2)$-generators σ^{AB} can be concisely written[10](see e.g. [21]) :

$$\begin{aligned} \sigma^{\mu\nu} &= \tfrac{i}{4}[\gamma^\mu, \gamma^\nu] \ , & \sigma^{-15} &= -\tfrac{1}{2}\gamma^5 \ , \\ \sigma^{\mu 5} &= \tfrac{i}{4}[\gamma^\mu, \gamma^5] \ , & \sigma^{-1\mu} &= -\tfrac{1}{2}\gamma^\mu \ . \end{aligned} \tag{39}$$

Here we denoted $\gamma^5 := \gamma^0\gamma^1\gamma^2\gamma^3$. In view of the isomorphism $\mathfrak{su}(2,2) \cong \mathfrak{so}(2,4)$ the operators σ^{AB} satisfy the $\mathfrak{so}(2,4)$ commutation relations (31). They give rise to a spinorial representation of the conformal algebra $\mathfrak{so}(2,4)$, eq. (30) through the Kustaanheimo-Stiefel correspondence between the quantum Coulomb-Kepler problem and the 4D harmonic oscillator

$$L_{AB} \quad \leftrightarrow \quad J^{AB} = \bar{\psi}\sigma^{AB}\psi \ , \qquad A, B \in \{-1, 0, 1, 2, 3, 5\} \ .$$

[9] Here the summation on spinorial indices is implicit.

[10] We get another concise expression $\sigma^{AB} = \frac{i}{2}\gamma^A\gamma^B$ for $A < B$ in $\{-1, 0, 1, 2, 3, 5\}$ if we set the unit matrix multiplier $\gamma^{-1} := i\mathbb{1}$.

Proposition 7.1 *The Majorana condition $\psi = \psi^c$ is reducing the Dirac $SU(2,2)$-spinor ψ_α to a real $Sp(4,\mathbb{R})$-spinor. The reality condition is essentially reducing the 4D harmonic oscillator (Dirac spinor) to the 2D oscillator (Weyl spinor).*

Proof. The Majorana condition is the invariance under the charge-conjugation involution $C = i\gamma^0\gamma^2$. We choose the Dirac representation for the Clifford algebra generators γ^μ

$$\gamma^0 = \begin{pmatrix} \sigma_0 & 0 \\ 0 & -\sigma_0 \end{pmatrix}, \qquad \gamma^i = \begin{pmatrix} 0 & \sigma_i \\ -\sigma_i & 0 \end{pmatrix}, \qquad C = i\gamma^0\gamma^2 = \begin{pmatrix} 0 & \epsilon \\ \epsilon & 0 \end{pmatrix}.$$

A Majorana spinor is a Weyl spinor written in 4-dimensional form. The Majorana condition $\psi = \psi^c$ yields

$$\psi = \begin{pmatrix} \chi \\ \epsilon^T(\chi^*)^T \end{pmatrix}, \qquad \bar{\psi} = \psi^*\gamma^0 = \begin{pmatrix} \chi^* & -\chi^T\epsilon \end{pmatrix}. \tag{40}$$

The Dirac's remarkable spinorial $\mathfrak{so}(2,3)$-algebra represention (14) is nothing else than the ladder representation for the Majorana spinor $\psi = \psi^c$

$$m_{ba} \quad \leftrightarrow \quad J^{ab} = \bar{\psi}^c \sigma^{ab} \psi^c \qquad\qquad a, b \in \{-1, 0, 1, 2, 3\}.$$

The Weyl spinor $\chi_\alpha^{(*)}$ has components identified in eq. (13) with the Heisenberg algebra generators $a^\pm$ and $b^\pm$ in Landau problem. Hence the dynamical algebra $\mathfrak{so}(2,3)$ of the Landau problem (12) is the span of the infinitesimal symplectomorphisms $\mathfrak{sp}(4,\mathbb{R})$ of the phase space $(\xi, \eta, p_\xi, p_\eta) \in \mathbb{R}^4$ obtained from the Majorana reduction of the ladder representation (38) of $\mathfrak{su}(2,2) \cong \mathfrak{sp}(8,\mathbb{R}) \cap \mathfrak{so}(4,4)$.

8. Conclusion and outlook

In celestial mechanics the regularization is the transformation of the singular equations of the Kepler motion to the regular equations of the harmonic oscillator. The duality between the Newton's universal law and Hooke's law of harmonic motion has been know since the time of Newton and Hooke [8]. Its reincarnation in quantum mechanics is the duality between the hydrogen atom and the harmonic oscillator via the Fock's method of quantization. At the quantum level, the Levi-Civita regularization induces the duality transform between the 2D hydrogen atom and 2D harmonic oscillator, whereas the Kustaanheimo-Stiefel regularization bridges between the 3D hydrogen atom and the 4D harmonic oscillator. We have given an unified picture (2) of the dynamical groups in duality derived from the conformal symmetries on Jordan algebras and the TKK construction. We point out that the dynamical group $Sp(4,\mathbb{R})$ of the quantum motion of an electron in a constant magnetic field (Landau problem) is mapped via the Levi-Civita regularization to the conformal $SO(2,3)$ spectrum generating algebra of the 2D hydrogen atom. The symplectic group $Sp(4,\mathbb{R})$ is a natural symmetry for the Landau problem since the magnetic field is encoded into the symplectic form on the phase space of the planar motion. On the other hand the carrier of the electromagnetic interaction is the zero-mass photon propagating on the light cone in Minkowski spacetime $\mathbb{R}^{1,2}$ whose causal automorphisms form the group $SO(2,3)$. In a similar fashion, the Kustaanheimo-Stiefel regularization connects the zero-mass spinorial representation of $SU(2,2)$ with the minimal $SO(2,4)$-representation of light cone automorphisms in Minkowski spacetime $\mathbb{R}^{1,3}$.

The merit of the Jordan algebra formalism is that the reduction from 3D to 2D hydrogen atom is done by the reduction of the Jordan algebra $\mathfrak{J}_2^{\mathbb{C}}$ to $\mathfrak{J}_2^{\mathbb{R}}$. We have shown that on the spinorial side of the Newton-Hooke correspondence (1), a 4D Dirac spinor is reduced to the Majorana spinor with 4 real independent components (or alternatively Weyl spinor with 2 complex), that is, we impose the natural reality condition for spinors.

We have also noted that the Levi-Civita transform connecting the Landau problem and the 2D hydrogen atom is an extension of the Hopf fibration $S^1 \to S^1$. We have drawn the parallel with the Kustaanheimo-Stiefel transform thought as an extension of the Hopf fibration $S^3 \to S^2$. We believe that phase space extentions of the remaining Hopf fibrations

$$\begin{array}{cccccccccc} \mathbb{H} & 0 & \to & S^3 & \hookrightarrow & S^7 & \to & S^4 & \to & 0\,, \\ \mathbb{O} & 0 & \to & S^7 & \hookrightarrow & S^{15} & \to & S^8 & \to & 0\,, \end{array}$$

are of potential interest for high energy physics in view of the exceptional role that the octonions could play in the quark-lepton symmetry of the Standard Model [12, 15].

Acknowledgments
It is a pleasure to thank Michel Dubois-Violette, Ludmil Hadjiivanov, Malte Henkel, Richard Kerner, Mariana Kirchbach, Petko Nikolov, Stoimen Stoimenov and Ivan Todorov for many inspiring discussions. One of us (T.P.) is supported by the Bulgarian National Science Fund Research Grant DN 18/3.

References

[1] Bars, Itzhak, and Jonathan L. Rosner. "Duality between hydrogen atom and oscillator systems via hidden $SO(d, 2)$ symmetry and 2T-physics." *Journal of Physics A: Mathematical and Theoretical* **53** (2020): 234001.

[2] Barut, A. O. and Hagen Kleinert. "Transition probabilities of the hydrogen atom from noncompact dynamical groups." *Physical Review* **156** (1967): 1541.

[3] Barut, A.O. and G.L. Bornzin. "$SO(4, 2)$-Formulation of the Symmetry Breaking in Relativistic Kepler Problems with or without Magnetic Charges", *Journal of Mathematical Physics* **12** (1971) 841-846.

[4] Barut, A. O., and I. H. Duru. "Introduction of internal coordinates into the infinite-component Majorana equation." *Proceedings of the Royal Society of London. A. Mathematical and Physical Sciences* **333** (1973): 217-224.

[5] Barut, A. O., C. K. E. Schneider, and Raj Wilson. "Quantum theory of infinite component fields." *Journal of Mathematical Physics* **20** (1979): 2244-2256.

[6] Cisneros, Arturo, and Harold V. McIntosh. "Symmetry of the two?dimensional hydrogen atom." *Journal of Mathematical Physics* **10**, no. 2 (1969): 277-286.

[7] Başkal, Sibel, Young S. Kim, and Marilyn E. Noz. "Poincaré Symmetry from Heisenbergs Uncertainty Relations." Symmetry 11 (2019): 409.

[8] Chandrasekhar, Subrahmanyan. Newton's Principia for the common reader. Oxford University Press, 2003.

[9] Cordani, Bruno. The Kepler Problem: Group Theotretical Aspects, Regularization and Quantization, with Application to the Study of Perturbations. Vol. 29. Springer Science Business Media, 2003.

[10] Dereli, Tekin and Popov, Todor. Bloch waves and non-commutative tori of magnetic translations. *Journal of Mathematical Physics* **62** (2021), 101701.

[11] Dirac, P. A. M., A remarkable representation of the 3+ 2 de Sitter group. *Journal of Mathematical Physics* (1963), 901-909.

[12] Dubois-Violette, Michel. "Exceptional quantum geometry and particle physics." *Nuclear Physics B***912** (2016), 426-449.

[13] Fock, Vladimir. *Z. Phys.* **98**(1935), 145.

[14] Günaydin, Murat. "Generalized conformal and superconformal group actions and Jordan algebras." *Modern Physics Letters A* **8**(1993), 1407-1416.

[15] Günaydin, Murat and Feza Gürsey. "Quark structure and octonions" *J. Math. Phys.* **14**, 1651-1667.

[16] Henkel, Malte, and Jeremie Unterberger. "Schrödinger invariance and spacetime symmetries." *Nuclear Physics* **B 660** (2003): 407-435.

[17] Jackiw, R. "Dynamical symmetry of the magnetic vortex." *Annals of Physics* 201.1 (1990): 83-116.

[18] Kirchbach, M., and C. B. Compean. "Modelling duality between bound and resonant meson spectra by means of free quantum motions on the de Sitter space-time dS_4." *The European Physical Journal A***52** (2016): 1-21.

[19] Kirchbach, M., Popov, T. Vallejo, J.A. "Color confinement at the boundary of the conformally compactified AdS_5" *J. High Energ. Phys.* (2021), 1-35.

[20] Kustaanheimo, Paul and E. Stiefel. "Perturbation theory of Kepler motion based on spinor regularization." *J. Reine Angew. Math.* **218** (1965), 204-219.

[21] Mack, G. and Ivan Todorov, "Irreducibility of the ladder representations of $U(2,2)$ when restricted to the Poincaré subgroup", *Journal of Mathematical Physics* **10**(1969), 2078-2085.
[22] Malkin, I.A. and V.I. Manko, Symmetry of the Hydrogen Atom, *JETP Letters* (1966), 146148.
[23] Meng, Guowu. "Euclidean Jordan algebras, hidden actions, and J-Kepler problems." *Journal of mathematical physics* **52** (2011): 112104.
[24] Milnor, John. "On the geometry of the Kepler problem." *The American Mathematical Monthly* **90**(1983): 353-365.
[25] Palmkvist, Jakob. "A generalization of the Kantor-Koecher-Tits construction." arXiv preprint arXiv:1308.3761 (2013).
[26] Parfitt, D. G. W., and M. E. Portnoi. "The two-dimensional hydrogen atom revisited." *Journal of Mathematical Physics* **43**, (2002): 4681-4691.
[27] Popov, Todor. "Jordan Algebra and Hydrogen Atom." In Quantum Theory And Symmetries, pp. 231-244. Springer, Singapore, 2017.
[28] Shibuya, Tai-ichi, and Carl E. Wulfman. "The Kepler problem in two-dimensional momentum space." *American Journal of Physics* **33**, (1965): 570-574.
[29] Stoyanov, D. Tz, and I. T. Todorov. "Majorana Representations of the Lorentz Group and Infinite-Component Fields." *Journal of Mathematical Physics* **9**,(1968): 2146-2167.
[30] Ter-Antonyan, V. "Dyon-oscillator duality." arXiv preprint quant-ph/0003106 (2000).
[31] Torres Del Castillo, G. F. "On the connection between the Kepler problem and the isotropic harmonic oscillator in classical mechanics." *Revista Mexicana de Fisica* **44**, (1998): 333-338.
[32] Torres del Castillo, Gerardo Francisco, and Ortega A. Lopez. "The three-dimensional Kepler problem amd the four-dimensional isotropic harmonic oscillator" *Revista Mexicana De Fisica* **45**, (1999): 1-8.
[33] Velasco-Martinez, D., V. G. Ibarra-Sierra, J. C. Sandoval-Santana, A. Kunold, and J. L. Cardoso. "Constants of motion for the planar orbit of a charged particle in a static and uniform magnetic field: the magnetic LaplaceRungeLenz vector." *European Journal of Physics* **35**, (2014): 055018.

Journal of Physics: Conference Series **2191** (2022) 012010 doi:10.1088/1742-6596/2191/1/012010

Dark Range of ω In Brans-Dicke Gravity

Tekin Dereli[1,2] **and Yorgo Şenikoğlu**[2]
[1] Department of Physics, Koç University, 34450 Sarıyer,İstanbul,Turkey
[2] Department of Basic Sciences, Faculty of Engineering and Natural Sciences, Maltepe University, 34857 Maltepe,İstanbul, Turkey
E-mail: `tdereli@ku.edu.tr`, `yorgosenikoglu@maltepe.edu.tr`

Abstract.
The variational field equations of Brans-Dicke scalar-tensor theory of gravitation that is coupled to a mass-varying sterile neutrino field are presented in Riemannian and non-Riemannian setting in the language of exterior differential forms over four-dimensional spacetime. In a gravitational plane wave geometry in Rosen coordinates, with the scalar field set equal to a constant, the field equations admit ghost neutrino solutions. These correspond to Majorana spinor configurations for which the total neutrino stress-energy-momentum tensor vanishes. We show here that in the absence of a sterile neutrino in the gravitational wave geometry, there are scalar field configurations for which the total scalar field stress-energy-momentum tensor vanishes for negative values of the Brans-Dicke parameter $-3/2 < \omega < 0$.

1. Introduction

General relativity is a geometrical theory of gravitation that is determined by the metric of spacetime. Brans-Dicke theory of gravity[1, 2, 3], a scalar-tensor theory, introduces an adjustment to the initial framework of Einstein in order to incorporate Mach's Principle by including a scalar field ϕ. The astrophysical observations supplied to scientist with the construct of low energy string theories with scalar fields are very efficient in depicting the large scale structure of spacetime and subatomic physics; so it is very reasonable to recognize them also at large scales.

Brans-Dicke theory covers the geodesic postulate as well; the motion of a test mass is determined by the geodesics with the gravitational effects of all particles embedded in the metric and the associated Levi-Civita connection. The assumption that the matter lagrangian is independent of the scalar field ϕ secures the equivalence principle. The string unification of gravity and quantum corrections to higher orders make researchers think on the property and reasonableness of the theory of gravitational interaction being depicted, by geometry alone. There are several hints that a non-Riemannian setting of spacetime may contribute a refined and coherent way to describe gravitational fields[4, 5]. A first order variation of the Brans-Dicke Lagrangian implies that the spacetime connection can admit a non-vanishing torsion. The field equations obtained are equivalent to up to a shift in the coupling constant [4].

Under the influence of gravity alone, non-spinning test masses follow geodesic equations of motion in a Riemannian spacetime and autoparallels of a connection with torsion [6] in a non-Riemannian geometry. The existence of other matter fields will reshape the geometry of spacetime, and in the presence spinorial matter, torsion will be picked up by differential equations [7, 8]. In this article, the gradient of the scalar field determines the space-time torsion, and a particular choice of the scalar field applied to parallely propagating gravitational plane waves will reduce the field equations to the vacuum Einstein equations. In particular, we are motivated on the interpretation in a Riemannian and non-Riemannian geometry. Explicitly, the Brans-Dicke parameter and the autoparallels of the connection with torsion which differ from the geodesic equations of motion will help us understand the coherence of the explanation.

The article is organized as follows: in Section 2, we present the Brans-Dicke theory of gravity with the gravitational field equations and the equations of motion of a non-spinning test mass in a non-Riemannian setting in the language of exterior differential forms. Section 3 details the plane wave solutions in Rosen coordinates and a particular choice of scalar field that reduce the field equations to source-free Einstein equations. In Section 4 we consider a mass-varying sterile neutrino field coupled to Brans-Dicke gravity. In particular two special sub-cases of solutions that describe ghost neutrinos on the one hand and ghost scalars on the other are discussed. Section 5 is devoted to concluding remarks.

2. Brans-Dicke scalar-tensor gravity

The Brans-Dicke field equations will be determined from the action $I = \int_M \mathcal{L}$, where M denotes the 4-dimensional spacetime manifold and the Brans-Dicke Lagrange density 4-form

$$\mathcal{L} = \frac{\alpha^2}{2} R_{ab} \wedge *(e^a \wedge e^b) - \frac{c}{2} d\alpha \wedge *d\alpha. \tag{1}$$

Here $\{e^a\}$'s are the co-frame 1-forms in terms of which the space-time metric is given by $g = \eta_{ab} e^a \otimes e^b$ with $\eta_{ab} = diag(- + ++)$. $*$ stands for the Hodge map. The volume form is $*1 = e^0 \wedge e^1 \wedge e^2 \wedge e^3$. We wrote the Brans-Dicke scalar field as $\phi = \alpha^2$ for convenience. Then $\{\omega^a{}_b\}$ denote the connection 1-forms that satisfy the Cartan structure equations

$$de^a + \omega^a{}_b \wedge e^b = T^a \tag{2}$$

with the torsion 2-forms T^a. On the other hand the curvature 2-forms $R^a{}_b$ of spacetime are given by

$$d\omega^a{}_b + \omega^a{}_c \wedge \omega^c{}_b = R^a{}_b. \tag{3}$$

We vary the action independently with respect to the co-frame 1-forms e^a, connection 1-forms $\omega^a{}_b$ and the scalar field α. The corresponding coupled field equations in the non-Riemannian description turn out to be $(c \neq 0)$:

$$\begin{aligned} -\frac{\alpha^2}{2} R^{bc} \wedge *(e_a \wedge e_b \wedge e_c) &= c\, \tau_a[\alpha], \\ T^a &= e^a \wedge \frac{d\alpha}{\alpha}, \\ cd * d\alpha^2 &= 0; \end{aligned} \tag{4}$$

where

$$\tau_a[\alpha] = \frac{1}{2} \left(\iota_a d\alpha * d\alpha + d\alpha \wedge \iota_a * d\alpha \right) \equiv T_{ab} * e^b \tag{5}$$

are the energy-momentum 3-forms of the scalar field and ι_a denote the interior products that satisfy $\iota_a e^b = \delta^b{}_a$. We can now introduce the Levi-Civita connection 1-forms $\{\hat{\omega}^a{}_b\}$, fixed in a unique way by the metric tensor through the Cartan structure equations

$$de^a + \hat{\omega}^a{}_b \wedge e^b = 0. \tag{6}$$

Then in terms of them, the equivalent form of the Brans-Dicke field equations (4) in the Riemannian description turn our to be:

$$
\begin{aligned}
-\frac{\alpha^2}{2}\hat{R}^{bc}\wedge *(e_a\wedge e_b\wedge e_c) &= \frac{(c-6)}{2}(\iota_a d\alpha * d\alpha + d\alpha\wedge\iota_a * d\alpha) \\
&\quad +\hat{D}(\iota_a * d\alpha^2), \\
cd * d\alpha^2 &= 0,
\end{aligned} \tag{7}
$$

where $\hat{D}$ denotes the covariant exterior derivative with respect to the Levi-Civita connection. Thus our coupling constant c can be related with the classical Brans-Dicke coupling constant ω as

$$c = 4\omega + 6. \tag{8}$$

We note that the right hand side of the Einstein field equations in a Riemannian setting is composed of two terms[9]. The first term involves the Brans-Dicke coupling constant as an over-all factor in front of the scalar field stress-energy-momentum tensor. This term comes from the kinetic part of the scalar field in the action. While the second term does not contain any parameter. In fact it comes from the geometry of the spacetime itself, however, it is only implicit in the non-Riemannian description. We wish to point out here that there exists such parameter values, that is, there is a *dark range of* ω, where these two terms on the right hand side cancel each other out and the field equations reduce to the ordinary source-free Einstein equations as we have demonstrated above. Thus in certain spacetime geometries the geodesic equations of motion are not affected by the scalar field at all in the conventional Riemannian approach. There is a non-trivial scalar field but the motion of non-spinning test masses are not influenced by it. We have a conceptual problem in the conventional Riemannian interpretation whereas in the non-Riemannian description there is none.

3. Gravitational plane waves

In order to demonstrate our point, we wish to study below a specific family of solutions of the Brans-Dicke field equations. Let us consider the plane fronted gravitational wave metric in Rosen coordinates (u, v, x, y) to solve the Brans-Dicke field equations[10, 11]:

$$g = 2\mathrm{d}u\otimes\mathrm{d}v + \frac{\mathrm{d}x\otimes\mathrm{d}x}{f(u)^2} + \frac{\mathrm{d}y\otimes\mathrm{d}y}{h(u)^2}. \tag{9}$$

We further take a scalar field $\alpha = \alpha(u)$ and note that the scalar field equation to the system is identically satisfied. Working out the expressions for the curvature, torsion and the scalar field stress-energy-momentum tensors, the Einstein field equations to be solved reduce to the following second order differential equation:

$$\frac{f_{uu}}{f} - \frac{2f_u^2}{f^2} + \frac{h_{uu}}{h} - \frac{2h_u^2}{h} - \frac{2\alpha_{uu}}{\alpha} + \frac{4\alpha_u^2}{\alpha^2} = \frac{c\alpha_u^2}{\alpha^2}. \tag{10}$$

Therefore for a particular choice of α such that

$$\alpha(u) = \Big((\omega+1)(C_1 u + C_2)\Big)^{\frac{1}{2(\omega+1)}}, \tag{11}$$

where C_1 and C_2 are arbitrary integration constants, all the α-dependent terms in (10) drop out and we recover the source-free Einstein field equation:

$$\Big(\frac{f_{uu}}{f} - \frac{2f_u^2}{f^2} + \frac{h_{uu}}{h} - \frac{2h_u^2}{h}\Big) = 0. \tag{12}$$

Let us further decorate this observation by explicitly giving the geodesic equations of motion as functions of proper time τ and constants of motion p_u, p_x, p_y neatly as [11]

$$\begin{aligned} u(\tau) &= u(0) + \frac{p_u \tau}{m}, \\ v(\tau) &= v(0) - \frac{m\tau}{2p_u} - \frac{p_x^2}{2p_u^2}\int_0^\tau f(u)^2 du - \frac{p_y^2}{2p_u^2}\int_0^\tau h(u)^2 du, \\ x(\tau) &= x(0) + \frac{p_x}{p_u}\int_0^\tau f(u)^2 du, \quad y(\tau) = y(0) + \frac{p_y}{p_u}\int_0^\tau h(u)^2 du. \end{aligned} \tag{13}$$

In a similar way the auto-parallel equations would be given as

$$\begin{aligned} \tau &= \frac{m}{p_u}\int_0^{\frac{p_u\tau}{m\alpha}} \alpha(u) du, \\ v(\tau) &= v(0) - \frac{m^2}{2p_u^2}\int_0^{\frac{p_u\tau}{m\alpha}} \alpha(u)^2 du - \frac{p_x^2}{2p_u^2}\int_0^{\frac{p_u\tau}{m\alpha}} f^2(u) du - \frac{p_y^2}{2p_u^2}\int_0^{\frac{p_u\tau}{m\alpha}} h^2(u) du, \\ x(\tau) &= x(0) + \frac{p_x}{p_u}\int_0^{\frac{p_u\tau}{m\alpha}} f^2(u) du, \quad y(\tau) = y(0) + \frac{p_y}{p_u}\int_0^{\frac{p_u\tau}{m\alpha}} h^2(u) du, \end{aligned} \tag{14}$$

In the former case a non-spinning test mass does not interact with the scalar field whereas in the latter case it clearly does.
Taking the Brans-Dicke field equations with torsion and looking at solutions, we have found a particular gravitational plane wave solution. The metric is the source-free (vacuum) Einstein metric and the terms coming from torsion and the stress-energy momentum tensor cancel out and the scalar field equation is satisfied. In the non-Riemannian description, we do not have any issues, on the right hand side, the energy momentum tensor is different than zero and positive definite. The left hand side of the field equations, the geometry is affected by torsion, given by the gradient of the scalar field. When we look at the autoparallel equations they are not the source-free Einstein geodesic equations. We find different equations of motion as expected. Had we insisted on the classical Brans-Dicke interpretation, this solution would be a solution again. But what kind of a solution? If one insists upon the conventional interpretation of the Brans-Dicke gravity, then the above class of gravitational wave solutions with $-3/2 < \omega < 0$ should be left aside as unphysical[6, 12, 13]. On the other we wish to argue below for a broader perspective in which such solutions may play a role, for instance, in a cosmological context.

4. A mass-varying sterile neutrino field coupled to scalar-tensor gravity

The starting point of our argument concerns the Einstein-massless scalar field theory coupled to a mass-varying sterile neutrino field. We discussed this model some years ago[14] in a cosmological context by a variational principle from the following action density 4-form in the Einstein picture: written in the Einstein picture:

$$\mathcal{L} = \frac{1}{2\kappa^2}\mathcal{R} * 1 - \frac{1}{2}d\Phi \wedge *d\Phi + V(\Phi) * 1 + Her\left\{\frac{i}{2}\bar{\Psi}(\gamma \wedge *\nabla)\Psi\right\} - \frac{i}{2}M(\Phi)(\bar{\Psi}\Psi) * 1. \tag{15}$$

Here $\kappa^2 = 8\pi G$ is Newton's gravitational constant. Φ is a 0-form representing an acceleron scalar while Ψ is a Majorana 4-spinor that represents a sterile neutrino with a varying mass $M(\Phi)$. One may also introduce a non-trivial acceleron potential function $V(\Phi)$. Some classes of exact cosmological solutions of the corresponding variational field equation, for the cases $V(\Phi) = \pm V_0$, were constructed and shown to be consistent asymptotically as $t \to 0$ with the observed expansion of the flat 3-universe in ΛCDM models[14]. Let us consider the same model

Journal of Physics: Conference Series **2191** (2022) 012010 doi:10.1088/1742-6596/2191/1/012010

as above instead in the Jordan-Brans-Dicke picture. In order to pass from the Einstein picture to the Jordan-Brans-Dicke picture we employ the following field re-definitions:

$$\tilde{g} = e^{2\sigma} g, \quad \tilde{\Psi} \equiv \psi = e^{-\frac{3}{2}\sigma} \Psi, \quad \tilde{\Phi} \equiv \phi = e^{-\sigma} \Phi \tag{16}$$

where we choose a scale function

$$\sigma = \omega \ln \Phi. \tag{17}$$

A real parameter ω is introduced which at the end turns out to be nothing but the Brans-Dicke coupling constant. Then the action density 4-form in the Jordan-Brans-Dicke picture, droppping the $\tilde{}$s for notational ease, reads

$$\begin{aligned} \mathcal{L} &= \frac{\alpha^2}{2} \mathcal{R} * 1 - \omega d\alpha \wedge * d\alpha + \bar{V}(\alpha) * 1 \\ &+ Her \left\{ \frac{i}{2} \bar{\psi} (\gamma \wedge * \nabla) \psi \right\} - \frac{i}{2} \bar{M}(\alpha) (\bar{\psi} \psi) * 1 + 3\omega (\bar{\psi} * \gamma \psi) \wedge \frac{d\alpha}{\alpha}, \end{aligned} \tag{18}$$

where $\alpha^2 = \phi$. The scalar potential $\bar{V}(\alpha)$ and the varying-neutrino mass $\bar{M}(\alpha)$ are suitably re-defined. It is remarkable that the gradient of the Brans-Dicke scalar couples to the vector current associated with a sterile neutrino field with a particular coupling strength proportional to the Brans-Dicke coupling constant. An analogous vector coupling between the acceleron scalar and the sterile neutrino field was absent in our earlier work[14]. The above expression in the Jordan-Brans-Dicke picture suggests such a coupling of the Brans-Dicke scalar to the vector current of the sterile neutrino might have been introduced in the Einstein picture in the first place. Now, we are ready to discuss two sub-classes of special types of solutions in the Jordan-Brans-Dicke picture. We consider only the cases $\bar{V}(\alpha) = 0 = \bar{M}(\alpha)$. This is a simplifying assumption that may be relaxed.

(i) Suppose $\alpha = \frac{1}{\kappa}$ is a constant. Our coupled field equations above reduce to the system of Einstein-massless neutrino field equations. It had already been noted many years ago that the Einstein-massless neutrino field equations admit exact plane wave solutions for which the contribution of the neutrino stress-energy-momentum tensor on the right hand side of the Einstein field equations identically vanish[15, 16, 17, 18, 19]. Such field configurations are called *ghost neutrinos* because it seems as if the propagation of a non-vanishing neutrino wave has no effect on the spacetime metric. A consistent physical interpretation can be provided within the framework of Einstein-Cartan gravity with torsion coupled to a massless neutrino field[20, 21, 22, 23]. In what follows, it is our wish to point out a similar configuration that we call a *ghost acceleron* exists in the Einstein-Majorana fermion-scalar field theory expressed, however, in the Jordan-Brans-Dicke picture.

(ii) Suppose now the sterile neutrino field identically vanishes, $\psi = 0$. Then our coupled field equations reduce simply to the Brans-Dicke field equations. The left hand side ensures the Einstein vacuum equations while at the right hand side, we have already shown above that there are scalar field configurations for which the improvement term and the scalar stress-energy-momentum tensor cancel each other out. We might interpret these solutions as *ghost accelerons*. The fundamental problem arises when we consider the geodesic hypothesis. The presence of the scalar field is assumed not to affect the geodesics in this case. We have a ghost scalar in the same sense analogous to the ghost neutrino problem[18, 20]. The theory is not scale invariant, the metric is the vacuum Einstein metric and the test particle equations of motion are the geodesics that are implied by the source-free Einstein field equations. Thus test masses wouldn't be able to feel the scalar field. In the conventional interpretation of the Brans-Dicke gravity this is an issue. But in a non-Riemannian interpretation with torsion it is not. This observation we believe gives one more support for a non-Riemannian interpretation of Brans-Dicke gravity.

5. Concluding Remarks

In this article, we studied a gravitational plane wave solution of the Brans-Dicke field equations to point the consistency interpretation of the non-Riemannian setting. The spacetime geometry is relaxed to admit torsion, depending on the gradient of the scalar field. A particular choice of the scalar field α reduces the field equations to the source-free Einstein field equations. The clarification on the sign is of utmost importance. There are two terms that cancel out in the right hand side, one is the shifted energy-momentum tensor and another term that comes from the geometry. What we explicitly want to show is the existence of such a configuration for $0 < c < 6$ i.e. for a negative ω, specifically $-\frac{3}{2} < \omega < 0$. This may well be a dark energy candidate. On the other hand, we have detailed that if we try to explain the theory in a Riemannian setting, we come to a pathological case where there is a scalar field present that cancels, in a way, the right hand side of the Einstein equations and that in the same geodesic hypothesis the scalar test mass does not interact with this scalar field. If we adopt a more consistent hypothesis, that scalar test masses should follow auto-parallels of the connection with torsion, then the identification of the energy-momentum tensor and the equations of motion are coherent. Torsion explicitly affects as it should, the equations of motion of the scalar test masses. Brans-Dicke field equations and the equivalence principle should be given in a non-Riemannian setting to be viable, modest and coherent. That is why we have discussed this in a particular plane wave example with a negative Brans-Dicke parameter ω, dark ω, which can depict some form of dark energy.

6. Acknowledgement

Y.Ş. is grateful to Koç University for its hospitality and partial support.

7. References

[1] C.H.Brans,R.H.Dicke,*Mach's principle and a relativistic theory of gravitation*,Phys.Rev.**124**,925(1961)
[2] R.H.Dicke,*Mach's principle and invariance under transformation of units*,Phys.Rev.**125**,2163 (1962)
[3] C.H. Brans,*Mach's principle and a relativistic theory of gravitation II*,Phys.Rev.**125**,2194 (1962)
[4] T.Dereli,R.W.Tucker,*Weyl scalings and spinor matter interactions in scalar-tensor theories of gravitation*,Phys.Lett.**B110**(1982)206
[5] P.Teyssandier,R.W.Tucker,C.Wang,*On an interpretation of non-Riemannian gravitation*,Acta Phys.Polon.**B29**(1998)987
[6] T.Dereli,R.W.Tucker,*On the detection of scalar field induced space-time torsion*,Mod.Phys.Lett.**A17**(2002)421
[7] D.Burton,T.Dereli,R.W.Tucker,*On the motion of matter in gravitational fields*, in **Symmetries in Gravity and Field Theory**, V.Aldaya, J.M.Cervero,Y Pilar Garcia (Editors) (Ediciones Universidad Salamanca, 2004) pp.237-249 [arXiv:gr-qc/0107017]
[8] D.A.Burton,R.W.Tucker,C.H.Wang,*Spinning particles in scalar-tensor gravity*,Phys.Lett.**A372**(2008)3141
[9] C.G.Callan Jr.,S.Coleman,R.Jackiw *A new improved energy momentum tensor*,Ann.Phys. **59**(1970)42-73
[10] M.Adak, T.Dereli, Y.Şenikoğlu, *Non-Riemannian description of Robinson-Trautman spacetimes in Brans-Dicke gravity*, Int. J. Mod. Phys. **D 28**, 1950070 (2019)
[11] T.Dereli,Y.Şenikoğlu, *Gravitational Plane Waves in a Non-Riemannian Description of Brans-Dicke Gravity*, Phys.Scr.**95**(2020)045223
[12] V.Faraoni,*The limit $\omega \to \infty$ limit of Brans-Dicke theory*,Phys.Lett.**A245**(1998)26
[13] V.Faraoni,J.Côté,*Two new approaches to the anomalous limit of Brans-Dicke to Einstein gravity*, Phys.Rev.**D99**(2019)064013
[14] B.Balantekin,T.Dereli,*Exact cosmological solution of the coupled Einstein-Majorana fermion-scalar field equations* , Phys.Rev.**D75**(2007) 024039
[15] J.Madore,*On the neutrino in general relativity*,Lett.Nuo.Cim.**5**(1972)48
[16] J.B.Griffiths,*Gravitational radiation and neutrinos*,Comm.Math.Phys.**28**(1972)295
[17] C.D.Collinson,P.B.Morris,*Spacetimes admitting neutrino fields with zero energy and momentum*,J.Phys.**A6**(1973)915
[18] T.M.Davis,J.R.Ray *Ghost neutrinos in general relativity*, Phys.Rev.**D9**(1974)334
[19] J.B.Griffiths,*On the Dirac fields in a curved spacetime*, J.Phys.**A12**(1979)2429

Journal of Physics: Conference Series **2191** (2022) 012010 doi:10.1088/1742-6596/2191/1/012010

[20] T.Dereli,R.W.Tucker *Exact neutrino solutions in the presence of torsion*,Phys.Lett.**A 82**,(1981)229
[21] J.B.Griffiths,*Neutrino fields in Einstein-Cartan theory*, Gen.Rel.Grav.**13**(1981) 227
[22] A.Dimakis,F.Müller-Hoissen,*Massive ghost Dirac equations in Einstein-Cartan theory*,Phys.Lett.**A92**(1982)431
[23] A.Dimakis,F.Müller-Hoissen,*Solutions of the Einstein-Cartan-Dirac equations with vanishing energy-momentum tensor*,J.Math.Phys.**26**(1985)1040

Journal of Physics: Conference Series **2191** (2022) 012011 doi:10.1088/1742-6596/2191/1/012011

Statistical geometry and Hessian structures on pre-Leibniz algebroids

Keremcan Doğan[1, 2]
[1]Department of Physics, Koç University, İstanbul, Turkey
[2]Department of Mathematics, İstanbul Technical University, İstanbul, Turkey
E-mail: kedogan[at]ku.edu.tr

Abstract. We introduce statistical, conjugate connection and Hessian structures on anti-commutable pre-Leibniz algebroids. Anti-commutable pre-Leibniz algebroids are special cases of local pre-Leibniz algebroids, which are still general enough to include many physically motivated algebroids such as Lie, Courant, metric and higher-Courant algebroids. They create a natural framework for generalizations of differential geometric structures on a smooth manifold. The symmetrization of the bracket on an anti-commutable pre-Leibniz algebroid satisfies a certain property depending on a choice of an equivalence class of connections which are called admissible. These admissible connections are shown to be necessary to generalize aforementioned structures on pre-Leibniz algebroids. Consequently, we prove that, provided certain conditions are met, statistical and conjugate connection structures are equivalent when defined for admissible connections. Moreover, we also show that for 'projected-torsion-free' connections, one can generalize Hessian metrics and Hessian structures. We prove that any Hessian structure yields a statistical structure, where these results are completely parallel to the ones in the manifold setting. We also prove a mild generalization of the fundamental theorem of statistical geometry. Moreover, we generalize α-connections, strongly conjugate connections and relative torsion operator, and prove some analogous results.

Keywords: Statistical Geometry, Conjugate Connection Structures, Hessian Structures, Pre-Leibniz Algebroids, Admissible Connections

1. Introduction

Algebroids are mathematical frameworks that are suitable to generalize differential geometric structures. Many different notions of algebroids are studied extensively in the literature. For example, metric-affine geometries can be constructed on the tangent bundle of a smooth manifold. Tangent bundles are special cases for Lie algebroids, and geometric structures on the tangent bundle can be easily generalized to an arbitrary Lie algebroid [1]. On the other hand, generalized geometry on the direct sum of the tangent and cotangent bundles can be written on an arbitrary exact Courant algebroid [2]. Generalized geometry is closely related to the local double field theory formulation of the bosonic sector of string theories [3]. Local double field theory geometrizes the Kalb-Ramond B-field and T-duality, whereas exceptional field theories do the same thing for other sectors and U-duality by making use of different algebroids [4]. There are many other algebroid structures such as metric algebroids [5], dull alberoids [6], conformal Courant algebroids [7] that are used in the string or M theory literature. All of these are special cases for local pre-Leibniz algebroids as the latter is defined by using just a few axioms. Hence, the construction of differential geometric objects on local pre-Leibniz algebroids creates a chance to study all of these examples in a more comprehensive scheme. This was done in one of our previous papers [8] by extending the ideas of [9], where we constructed metric-connection geometries on local pre-Leibniz algebroids.

Even though these algebroids yield fruitful generalizations, and they satisfy many analogous properties to the usual metric-affine geometries; many features, and in particular the geometric

Journal of Physics: Conference Series **2191** (2022) 012011 doi:10.1088/1742-6596/2191/1/012011

intuition, are lost. In order to solve this issue, we defined anti-commutable pre-Leibniz algebroids in [10]. These algebroids are defined as pre-Leibniz algebroids whose bracket's symmetrization satisfies a certain property which depends on a choice of an equivalence class of connections that we called admissible. Even though anti-commutable ones are a special case for local pre-Leibniz algebroids, they are still general enough to include many other cases such that Lie, Courant, metric, higher-Courant and conformal Courant algebroids. We proved that for Lie algebroids, any connection is admissible; whereas for Courant and metric algebroids, the admissibility condition is equivalent to metric-compatibility. For other cases, we also proved an equivalence with a metric-compatibility condition in a generalized sense. Our claim in [10] was that while working on the geometry of an algebroid structure, one should consider only admissible connections. We tried to justify this claim by proving many important properties including Bianchi identity, Cartan structure equations and the decomposition of the connection in terms of its torsion and non-metricity tensors for admissible connections. Moreover, we proved that any torsion-free, and in particular Levi-Civita, connection has to be admissible. In other words, there are no Levi-Civita connections on local pre-Leibniz algebroids which are not anti-commutable. In this paper, we present another application of anti-commutable pre-Leibniz algebroids and admissible connections. We construct the generalizations of conjugate connection, statistical and Hessian structures on anti-commutable pre-Leibniz algebroids. We prove that many analogous properties of the usual formulations of these structures hold only for admissible connections.

Statistical, conjugate connection and Hessian structures on a smooth manifold are important ingredients in statistical geometry. The relevance of these geometric structures to statistics comes from the fact that every statistical structure has a corresponding statistical model [11] which is defined through a function that represents how a parameter is related to the potential outcomes of a random variable. The correspondence is of the form of the following dictionary [12]. Points of the manifold correspond to probability density functions, whereas the metric is identified with the Fisher information matrix which is related to the amount of information that a random variable carries about an unknown parameter. Tangent vectors of the manifold correspond to score functions that are gradients of the log-likelihood functions which measure the fit of the model to the sample data. Two conjugate affine connections correspond to the two special connections $\nabla^{(1)}$ and $\nabla^{(-1)}$ in a statistical model. Moreover, the non-metricty tensor is identified with the skewness tensor which measures the third cumulants of the model. Without getting into the details of this correspondence, the focus will be on the differential geometric framework.

In Section (2), the constructions of metric-connection geometries on pre-Leibniz algebroids are outlined. Tensors, vector fields, connections, metrics and related structures such as torsion, curvature and non-metricity tensors are defined in this framework. Anti-commutable pre-Leibniz algebroids and admissible connections on them are explained whereas some useful properties are pointed out. In Section (3), statistical geometry on a smooth manifold is summarized where the main references are [12], [13] and [14]. Conjugate connection and statistical structures are defined, and their equivalence is explained. In Section (4), the algebroid versions of conjugate connection and statistical structures are introduced, and their equivalence under certain conditions is proven. These conditions make use of the fact that for admissible connections, torsion operator is anti-symmetric. In Section (5), Hessian metrics and Hessian structures on manifolds are briefly summarized, and their relation to statistical structures with flat connections is explained. Generalizations of Hessian structures are defined for pre-Leibniz algebroids in Section (6), where it is shown that any Hessian structure in this setting yields a statistical structure. Moreover, a mild generalization of the fundamental theorem of statistical geometry is proven. Similarly, these proofs depend on the admissibility of the connection.

Throughout the paper, Einstein's summation convention of repeated indicies is used. Every construction is assumed to be in the smooth category. Moreover, M always denotes an orientable, paracompact, Hausdorff, smooth manifold. The tangent bundle of M is denoted by $T(M)$, whereas $\mathfrak{X}(M)$ denotes the set of vector fields. Vector fields are derivations over the ring of smooth functions $C^\infty(M, \mathbb{R})$, and the action of a vector field V on a smooth function f will be denoted by $V(f)$.

Given an affine connection ∇, the torsion and curvature operators are defined as

$$\begin{aligned} T(\nabla)(U,V) &:= \nabla_U V - \nabla_V U - [U,V], \\ R(\nabla)(U,V)W &:= \nabla_U \nabla_V W - \nabla_V \nabla_U W - \nabla_{[U,V]} W, \end{aligned} \tag{1.1}$$

for $U, V, W \in \mathfrak{X}(M)$, where $[.,.]$ is the Lie bracket. For an affine connection ∇ and a metric g, the non-metricity tensor is defined as $Q(\nabla, g) := \nabla g$. While constructing metric-connection geometries on pre-Leibniz algebroids in Section (2), even though we use different notations for most of the structures, some of them will be identical including connections ∇, local frames (X_a), and their dual local co-frames (e^a) with the ones in the manifold setting. For the other objects, we usually use the notations with reversed lower/upper-case letters. For instance, V, ω denote a vector field and a p-form in the usual differential geometry, whereas v, Ω denote a vector field and a p-form on an algebroid. We hope that this notation is clear enough to notice the analogies between the two settings.

2. Anti-commutable pre-Leibniz algebroids

On an arbitrary (real) vector bundle E over M, one can introduce (q, r)-type E-tensors as:

$$Tens^{(q,r)}(E) := \Gamma\left(\bigotimes_{i=1}^{q} E \otimes \bigotimes_{j=1}^{r} E^*\right). \tag{2.1}$$

The sections of the bundle E itself are called E-vector fields, and their set is denoted by $\mathfrak{X}(E) := \Gamma(E)$. A local basis (X_a) of E is called a local E-frame, and its dual $(e^a \mid e^a(X_b) = \delta^a{}_b)$ is called a local E-coframe. An anti-symmetric $(0, p)$-type E-tensor is called an E-p-form, and their set is denoted by $\Omega^p(E)$. A $(0, 2)$-type E-tensor is called an E-metric if it is symmetric and non-degenerate.

In order to introduce E-connections, one should work with anchored vector bundles. An anchored vector bundle is a doublet (E, ρ) such that E is a vector bundle over M, and $\rho : E \to T(M)$ is a vector bundle morphism. A linear E-connection on the anchored vector bundle (E, ρ) is an $\mathbb{R}$-bilinear map $\nabla : \mathfrak{X}(E) \times \mathfrak{X}(E) \to \mathfrak{X}(E)$ satisfying

$$\begin{aligned} \nabla_u(fv) &= \rho(u)(f)v + f\nabla_u v, \\ \nabla_{fu} v &= f\nabla_u v, \end{aligned} \tag{2.2}$$

for all $u, v \in \mathfrak{X}(E), f \in C^\infty(M, \mathbb{R})$ [15]. E-connection coefficients are defined as

$$\Gamma(\nabla)^a{}_{bc} := e^a(\nabla_{X_b} X_c), . \tag{2.3}$$

One can extend the action of a linear E-connection to the whole E-tensor algebra by the Leibniz rule. The E-non-metricity tensor corresponding to a linear E-connection ∇ and an E-metric g is defined as

$$Q(\nabla, g) := \nabla g. \tag{2.4}$$

If it vanishes, then ∇ is called E-metric-g-compatible.

For the generalizations of torsion and curvature operator, one should introduce a bracket. The bracket is not necessarily anti-symmetic as opposed to the Lie bracket of usual vector fields, so one should work with two Leibniz rules. A triplet $(E, \rho, [.,.]_E)$ is called an almost-Leibniz algebroid if (E, ρ) is an anchored vector bundle, and $[.,.]_E : \mathfrak{X}(E) \times \mathfrak{X}(E) \to \mathfrak{X}(E)$ is an $\mathbb{R}$-bilinear map which satisfies the right-Leibniz rule

$$[u, fv]_E = \rho(u)(f)v + f[u, v]_E, \tag{2.5}$$

for all $u, v \in \mathfrak{X}(E), f \in C^\infty(M, \mathbb{R})$. Moreover, a quartet $(E, \rho, [.,.]_E, L)$ is called a local almost-Leibniz algebroid [16], [9] if $(E, \rho, [.,.]_E)$ is an almost-Leibniz algebroid which satisfies the following left-Leibniz rule for some $C^\infty(M,\mathbb{R})$-multilinear map $L : \Omega^1(E) \times \mathfrak{X}(E) \times \mathfrak{X}(E) \to \mathfrak{X}(E)$ called "*the locality operator*"

$$[fu, v]_E = -\rho(v)(f)u + f[u, v]_E + L(Df, u, v), \tag{2.6}$$

for all $u, v \in \mathfrak{X}(E), f \in C^\infty(M,\mathbb{R})$, where $D : C^\infty(M,\mathbb{R}) \to \Omega^1(E)$ is the coboundary map defined by $(Df)(u) := \rho(u)(f)$ [17].

The naive generalizations of torsion and curvature operators are not $C^\infty(M,\mathbb{R})$-linear, so one needs to modify their definitions to get tensorial quantities. The E-torsion operator of a linear E-connection ∇ is defined as a map $T(\nabla) : \mathfrak{X}(E) \times \mathfrak{X}(E) \to \mathfrak{X}(E)$ such that

$$T(\nabla)(u, v) := \nabla_u v - \nabla_v - [u, v]_E + L(e^a, \nabla_{X_a} u, v). \tag{2.7}$$

If $T(\nabla) = 0$, then the linear E-connection ∇ is called E-torsion-free.

For the curvature operator, one needs to work on a pre-Leibniz algebroid which is defined as an almost-Leibniz algebroid $(E, \rho, [.,.]_E)$ satisfying

$$\rho([u, v]_E) = [\rho(u), \rho(v)], \tag{2.8}$$

for all $u, v \in \mathfrak{X}(E)$, where the bracket on the right-hand side is the Lie bracket of usual vector fields. Moreover, one needs to assume that the anchored vector bundle (E, ρ) is regular, i. e. the anchor ρ is of locally constant rank. On a regular local pre-Leibniz algebroid, a "*locality projector*" is defined as a $C^\infty(M,\mathbb{R})$-linear map $\mathcal{P} : \mathfrak{X}(E) \to \mathfrak{X}(E)$ satisfying [8]

- The image of the composition map $\widehat{L} := \mathcal{P}L$ is in the kernel of the anchor,
- When restricted on the kernel of the anchor, $\mathcal{P}$ should be the identity map.

Note that these properties imply that $\mathcal{P}^2 = \mathcal{P}$. Given a locality projector on a regular local pre-Leibniz algebroid, one can define the E-curvature operator of a linear E-connection ∇ as a map $R(\nabla) : \mathfrak{X}(E) \times \mathfrak{X}(E) \times \mathfrak{X}(E) \to \mathfrak{X}(E)$ such that

$$R(\nabla)(u, v)w := \nabla_u \nabla_v w - \nabla_v \nabla_u w - \nabla_{[u,v]_E} + \nabla_{\widehat{L}(e^a, \nabla_{X_a} u, v)} w. \tag{2.9}$$

If $R(\nabla) = 0$, then the linear E-connection ∇ is called E-flat.

In [10], motivated from the modifications of E-torsion and E-curvature operators, we introduced the concept of "anti-commutable" almost-Leibniz algebroids as a special case of local almost-Leibniz algebroids. An anti-commutable almost-Leibniz algebroid is defined as an almost-Leibniz algebroid $(E, \rho, [.,.]_E)$ which satisfies

$$[u, v]_E + [v, u]_E = L(e^a, \nabla_{X_a} u, v) + L(e^a, \nabla_{X_a} v, u), \tag{2.10}$$

for some $C^\infty(M,\mathbb{R})$-multilinear map $L : \Omega^1(E) \times \mathfrak{X}(E) \times \mathfrak{X}(E) \to \mathfrak{X}(E)$ and some linear E-connection ∇, where (X_a) is a local E-frame with its dual (e^a). One can show that $(E, \rho, [.,.]_E, L)$ is a local almost-Leibniz algebroids. Given a fixed L, the choice of the linear E-connection ∇ is not unique, and any linear E-connection satisfying Equation (2.10) is called admissible. We proved that many algebroid structures in the literature are anti-commutable including Lie, Courant, metric, higher-Courant and conformal Courant algebroids [10]. For Lie algebroids, every linear E-connection is admissible because of the anti-symmetry of the bracket the locality operator can be chosen as 0. For metric and Courant algebroids, a linear E-connection is admissible if and only if it is compatible with the E-metric in the definition of the algebroid. For the remaining two cases, we proved a similar characterization for a generalization of the metric-compatibility condition.

Equation (2.10) motivates one to introduce "*the modified bracket*":

$$[u, v]_E^\nabla := [u, v]_E - L(e^a, \nabla_{X_a} u, v). \tag{2.11}$$

It is indeed still a local almost-Leibniz bracket; in particular it is an almost-dull bracket meaning that the locality operator is 0. In terms of this modified bracket, the E-torsion operator (2.7) becomes

$$T(\nabla)(u, v) = \nabla_u v - \nabla_v u - [u, v]_E^\nabla. \tag{2.12}$$

The modified bracket $[.,.]_E^{\nabla}$ is anti-symmetric if and only if the linear E-connection ∇ is admissible. This trivially implies that any E-torsion-free linear E-connection has to be admissible. Moreover, it implies that the E-torsion operator of an admissible linear E-connection is anti-symmetric. Similarly, for the E-curvature operator, one can introduce "*the projected modified bracket*":

$$[u,v]_E^{\widehat{\nabla}} := [u,v]_E - \widehat{L}(e^a, \nabla_{X_a}u, v). \tag{2.13}$$

It is a pre-dull bracket, and it is anti-symmetric for an admissible linear E-connection. Evidently, the E-curvature operator for an admissible linear E-connection ∇ satisfies

$$R(\nabla)(u,v)w = -R(\nabla)(v,u)w, \tag{2.14}$$

for all $u,v,w \in \mathfrak{X}(E)$. As the E-torsion and E-curvature operators are defined by using two different brackets, in [10] we introduced the $C^\infty(M,\mathbb{R})$-bilinear "*projected E-torsion operator*":

$$\widehat{T}(u,v) := \nabla_u v - \nabla_v u - [u,v]_E^{\widehat{\nabla}}. \tag{2.15}$$

The E-curvature and projected E-torsion operators of any linear E-connection ∇ satisfy the following Ricci identity:

$$\nabla^2_{u,v}w - \nabla^2_{v,u}w = R(\nabla)(u,v)w - \nabla_{\widehat{T}(\nabla)(u,v)}w, \tag{2.16}$$

for all $u,v,w \in \mathfrak{X}(E)$, where the second order E-covariant derivative is defined as

$$\nabla^2_{u,v}w := \nabla_u\nabla_v w - \nabla_{\nabla_u v}w. \tag{2.17}$$

In terms of the projected modified brackets, one can introduce the E-version of the exterior derivative, which we called the projected E-exterior derivative $\widehat{d} : \Omega^p(E) \to \Omega^{p+1}(E)$:

$$\begin{aligned}\widehat{d}(\nabla)\Omega(v_1,\ldots,v_{p+1}) := &\sum_{1\le i\le p+1} (-1)^{i+1}\rho(v_i)\left(\Omega(v_1,\ldots,\check{v}_i,\ldots,v_{p+1})\right)\\ &+\sum_{1\le i<j\le p+1} (-1)^{i+j}\Omega\left([v_i,v_j]_E^{\widehat{\nabla}}, v_1,\ldots,\check{v}_i,\ldots,\check{v}_j,\ldots,v_{p+1}\right),\end{aligned} \tag{2.18}$$

where $\check{v}$ indicates that v is excluded [10]. This map does not square to 0 in general, but when acting on smooth functions it squares to 0, i. e. $\widehat{d}(\nabla)^2 f = 0$ for all $f \in C^\infty(M,\mathbb{R})$. Similarly one can define the E-exterior derivative $d(\nabla)$ by replacing the projected modified brackets by the modified brackets in Equation (2.18), which does not square to 0 even for smooth functions.

A linear E-connection is called an E-Levi-Civita connection if it is E-torsion-free and E-metric-g-compatible. In [10], we proved that a linear E-connection is E-Levi-Civita if and only if it is an admissible E-Koszul connection. An E-Koszul connection is defined as a linear E-connection that satisfies the modification of the Koszul formula [9]:

$$\begin{aligned}2g(\nabla_u v, w) = &\rho(u)(g(v,w)) + \rho(v)(g(u,w)) - \rho(w)(g(u,v))\\ &- g([v,w]_E^{\nabla}, u) - g([u,w]_E^{\nabla}, v) + g([u,v]_E^{\nabla}, w),\end{aligned} \tag{2.19}$$

for all $u,v,w \in \mathfrak{X}(E)$.

3. Statistical structures on manifolds

By the fundamental theorem of semi-Riemannian geometry, the Levi-Civita connection ${}^g\nabla$ is the only torsion-free affine connection that preserves the metric g under parallel transport. On the other hand, by using two affine connections, one can also preserve the metric. This can be done via the conjugate affine connections. For an affine connection ∇ on a smooth manifold M, its conjugate (or dual) with respect to the metric g is defined as another affine connection ∇^* satisfying

$$U(g(V,W)) = g(\nabla_U V, W) + g(V, \nabla^*_U W), \tag{3.1}$$

for all $U, V, W \in \mathfrak{X}(M)$. On a local frame (X_a), this can be written as

$$X_a(g_{bc}) = \Gamma(\nabla)^d{}_{ab} g_{dc} + \Gamma(\nabla^*)^d{}_{ac} g_{bd}, \tag{3.2}$$

where the connection coefficients are defined by $\Gamma(\nabla)^a{}_{bc} := e^a(\nabla_{X_b} X_c)$. The triplet (g, ∇, ∇^*) is called a conjugate connection structure on M. The invertibility of the metric g implies that for every affine connection ∇, there exists a unique conjugate affine connection with respect to the metric g. One can compare the definition of the conjugate affine connections with the metric-g-compatibility condition, where the latter implies

$$U(g(V, W)) = g(\nabla_U V, W) + g(V, \nabla_U W), \tag{3.3}$$

for all $U, V, W \in \mathfrak{X}(M)$. Hence, one can prove that an affine connection ∇ is metric-g-compatible if and only if it is equal to its conjugate with respect to g. Moreover, as the conjugation is an involution, i.e. $(\nabla^*)^* = \nabla$ for all affine connections ∇, the mean affine connection defined by

$$\nabla^0 := \frac{1}{2}(\nabla + \nabla^*) \tag{3.4}$$

is metric-g-compatible for every ∇.

Non-metricity tensors of two conjugate affine connections are related by

$$Q(\nabla, g)(U, V, W) = -Q(\nabla^*, g)(U, V, W) = g\left(\Delta(\nabla^*, \nabla)(U, V), W\right), \tag{3.5}$$

for all $U, V, W \in \mathfrak{X}(M)$, where the difference operator between two affine connections is defined by

$$\Delta(\nabla, \nabla')(U, V) := \nabla_U V - \nabla'_U V. \tag{3.6}$$

A symmetric $(0, p)$-type tensor Z with $p \geq 2$ is called a ∇-Codazzi tensor of order p if ∇Z is totally symmetric. In particular, the metric g is a ∇-Codazzi tensor of order 2 when the non-metricity tensor $Q(\nabla, g)$ is totally symmetric. If one considers a torsion-free affine connection ∇, then the following are equivalent:

(i) The conjugate affine connection is torsion-free, i.e. $T(\nabla^*) = 0$.

(ii) The metric g is ∇-Codazzi.

(iii) The metric g is ∇^*-Codazzi.

(iv) The mean connection is the unique Levi-Civita connection, i.e. $\nabla^0 = {}^g\nabla$.

The doublet (g, C) is called a statistical structure if g is a metric, and C is a totally symmetric $(0, 3)$-type tensor. By the equivalence above, it has been shown that if one starts with a conjugate connection structure with torsion-free conjugate connections, then one can construct a statistical structure by choosing $C = Q(\nabla, g)$. In this context, C is known as the Amari-Chentsov or skewness tensor.

One can also construct a conjugate connection structure starting from a statistical structure (g, C). By using the total symmetry of C, the affine connections ∇ and ∇^* can be defined as

$$\begin{aligned} 2g(\nabla_U V, W) &= 2g({}^g\nabla_U V, W) - C(U, V, W), \\ 2g(\nabla^*_U V, W) &= 2g({}^g\nabla_U V, W) + C(U, V, W), \end{aligned} \tag{3.7}$$

for all $U, V, W \in \mathfrak{X}(M)$. One can show that both ∇ and ∇^* are torsion-free and their non-metricity tensors are given by

$$Q(\nabla, g) = -Q(\nabla^*, g) = -C, \tag{3.8}$$

which follows from the general fact

$$\begin{aligned} 2g(\nabla'_U V, W) = 2g({}^g\nabla'_U V, W) & \\ - Q(\nabla', g)(U, V, W) - Q(\nabla', g)(V, U, W) + Q(\nabla', g)(W, U, V) & \\ - g(T(\nabla')(V, W), U) - g(T(\nabla')(U, W), V) + g(T(\nabla')(U, V), W), & \end{aligned} \tag{3.9}$$

for any affine connection ∇' by using the total symmetry of C. One should note that the total symmetry is necessary and sufficient for the simplifications that lead to Equation (3.7). Because of the equivalence between statistical structures and conjugate connections structures, the former is often defined as a doublet (g, ∇) such that g is ∇-Codazzi. In this sense, the doublet $(g, {}^g\nabla)$ is a statistical structure, and it is often called the trivial statistical structure.

A possible generalization would be introducing two objects C and B which compensate the non-metricity and torsion tensors, respectively. One can introduce

$$\begin{aligned} 2g(\nabla_U V, W) &= 2g({}^g\nabla_U V, W) - C(U, V, W) + g(B(V, W), U) - g(B(U, W), V) + g(B(U, V), W), \\ 2g(\nabla^*_U V, W) &= 2g({}^g\nabla_U V, W) + C(U, V, W) + g(B(V, W), U) - g(B(U, W), V) + g(B(U, V), W), \end{aligned} \tag{3.10}$$

for all $U, V, W \in \mathfrak{X}(M)$. By Equation (3.9), it can be proven that ∇ and ∇^* are conjugate affine connections whose non-metricity and torsion satisfy

$$\begin{aligned} Q(\nabla, g) &= -Q(\nabla^*, g) = -C, \\ T(\nabla) &= T(\nabla^*) = B. \end{aligned} \tag{3.11}$$

Here, the crucial point is that having the same torsion for two affine connections is necessary and sufficient to have a symmetric difference operator. As the non-metricity tensors of two conjugate connections depend on the difference operator as in (3.5), having the same torsion guarantees that the non-metricity tensor is totally symmetric.

There is a closely related notion called quasi-statistical structure which includes torsionful affine connections. A doublet (g, ∇) is called a quasi-statistical structure if

$$(\nabla_U g)(V, W) = (\nabla_V g)(U, W) - g(T(\nabla)(U, V), W), \tag{3.12}$$

for all $U, V, W \in \mathfrak{X}(M)$. One can show that if (g, ∇) is a quasi-statistical structure, then the conjugate affine connection is torsion-free, i.e. $T(\nabla^*) = 0$.

Two conjugate affine connections ∇ and ∇^* are said to be conjugate in the strong sense if

$$\nabla_U V - \nabla^*_V U = [U, V], \tag{3.13}$$

for all $U, V \in \mathfrak{X}(M)$. For strongly conjugate affine connections, one has

$$T(\nabla) = -T(\nabla^*). \tag{3.14}$$

In particular, they have to be torsion-free at the same time. Furthermore, there are even stronger facts. For example, if ∇ has a strong conjugate, then it has to be torsion-free. On the other hand, one can show that the Levi-Civita connection is strongly conjugate to itself. Combining these facts, one can conclude that if ∇ and ∇^* are strongly conjugate, then $\nabla = \nabla^*$ so that ∇ is the unique Levi-Civita connection.

As there are two affine connections in the framework of statistical geometry, it is natural to define a relative torsion operator of two affine connections ∇ and ∇'

$$T(\nabla, \nabla')(U, V) := \nabla_U V - \nabla'_V U - [U, V], \tag{3.15}$$

One can explicitly show that it is $C^\infty(M, \mathbb{R})$-bilinear. The relative torsion operator for two conjugate affine connections ∇ and ∇^* satisfies the following properties

$$T(\nabla, \nabla^*)(U, V) = -T(\nabla^*, \nabla)(V, U), \tag{3.16}$$

for all $U, V \in \mathfrak{X}(M)$, and

$$T(\nabla, \nabla^*) + T(\nabla^*, \nabla) = T(\nabla) + T(\nabla^*). \tag{3.17}$$

Both of these equalities are actually valid for any two affine connections. Note that if $T(\nabla, \nabla^*) = 0$, then $\nabla = \nabla^*$ and ∇ is the Levi-Civita connection.

One can consider the one-parameter family of affine connections of the form a convex combination with respect to $\alpha \in \mathbb{R}$:

$$\nabla^{(\alpha)} := \frac{1+\alpha}{2}\nabla^{*} + \frac{1-\alpha}{2}\nabla. \tag{3.18}$$

These α-dependent affine connections are called α-connections. In particular one has

$$\nabla^{(1)} = \nabla^{*}, \qquad \nabla^{(-1)} = \nabla, \qquad \nabla^{(0)} = \nabla^{0}. \tag{3.19}$$

One can show that $(\nabla^{(\alpha)})^{*} = \nabla^{(-\alpha)}$. Moreover, torsion and non-metricity tensors of an α-connection can be expressed as

$$\begin{aligned} T(\nabla^{(\alpha)}) &= \frac{1+\alpha}{2}T(\nabla^{*}) + \frac{1}{2}T(\nabla), \\ Q(\nabla^{(\alpha)}, g) &= -\alpha Q(\nabla, g) = \alpha Q(\nabla^{*}, g). \end{aligned} \tag{3.20}$$

Similarly, one can evaluate the curvature of α-connections as follows:

$$\begin{aligned} R(\nabla^{(\alpha)})(U,V)W &= \frac{1+\alpha}{2}R(\nabla^{*})(U,V)W + \frac{1-\alpha}{2}R(\nabla)(U,V)W \\ &+ \frac{1-\alpha^{2}}{4}\left[\Delta(\nabla,\nabla^{*})\left(V, \Delta(\nabla,\nabla^{*})(U,W)\right) - \Delta(\nabla,\nabla^{*})\left(U, \Delta(\nabla,\nabla^{*})(V,W)\right)\right], \end{aligned} \tag{3.21}$$

for all $U, V, W \in \mathfrak{X}(M)$. This expression implies that for a pair of flat conjugate affine connections ∇ and ∇^{*},

$$R(\nabla^{(\alpha)}) = R(\nabla^{(-\alpha)}). \tag{3.22}$$

4. Statistical structures on almost-Leibniz algebroids

In this section, the statistical structures on manifolds will be generalized to the almost-Leibniz (and pre-Leibniz for curvature related topics) algebroid framework. The notions will be almost identical to the ones in the previous section with small differences. We will prove the generalizations of some of the basic results from the statistical geometry literature.

As an affine connection on a manifold can be generalized to a linear E-connection on anchored vector bundle (E, ρ), it is also possible to define conjugate linear E-connections. For a linear E-connection, its conjugate with respect to an E-metric g is defined as another linear E-connection ∇^{*} which satisfies

$$\rho(u)(g(v,w)) = g(\nabla_{u}v, w) + g(v, \nabla^{*}_{u}w), \tag{4.1}$$

for all $u, v, w \in \mathfrak{X}(E)$. This notion and its relation to contrast functions was studied in the Lie algebroid framework in [18].

Definition 4.1. *On an anchored vector bundle, the triplet* (g, ∇, ∇^{*}) *is called a conjugate E-connection structure.*

On a local E-frame (X_a), the conjugation condition (4.1) becomes

$$\rho(X_a)(g_{bc}) = \Gamma(\nabla)^{d}{}_{ab}g_{dc} + \Gamma(\nabla^{*})^{d}{}_{ac}g_{bd}. \tag{4.2}$$

Similarly to the usual case, the invertibility of the E-metric g implies that for every linear E-connection, there exists a unique conjugate linear E-connection with respect to the E-metric g. As the E-metric-g-compatibility condition for a linear E-connection ∇ can be expressed as

$$\rho(u)(g(v,w)) = g(\nabla_{u}v, w) + g(v, \nabla_{u}w), \tag{4.3}$$

for all $u, v, w \in \mathfrak{X}(E)$, one can see that a linear E-connection ∇ is E-metric-g-compatible if and only if $\nabla = \nabla^*$. Moreover, as the conjugation is involutive, the mean linear E-connection defined by

$$\nabla^0 := \frac{1}{2}(\nabla + \nabla^*) \tag{4.4}$$

is an E-metric-g-compatible connection.

In order to continue with other geometric concepts, we need to introduce a bracket. Hence, we should assume that we are working on a local almost-Leibniz algebroid $(E, \rho, [.,.]_E, L)$. The naive definition of strongly conjugate connections does not work here as replacing the Lie bracket in Equation (3.13) with the bracket $[.,.]_E$ creates some problems. For instance, the left-hand side of Equation (3.13) is anti-symmetric, so the bracket that will replace the Lie bracket should also be anti-symmetric. This is possible when one considers the modified bracket $[.,.]_E^\nabla$ with respect to a linear E-connection ∇ by the locality operator L. Therefore, we often should focus on only anti-commutable almost-Leibniz algebroids and consider admissible linear E-connections because in this case the modified bracket is anti-symmetric. Yet, as there are two connections in the left-hand side of Equation (3.13), it is natural to introduce the following term

$$\frac{1}{2}\left([u,v]_E^\nabla + [u,v]_E^{\nabla^*}\right). \tag{4.5}$$

This term is anti-symmetric as needed, and when $L = 0$, as in the usual geometry, it reduces to the original bracket. This leads us to the following definition.

Definition 4.2. *A pair of conjugate linear E-connections ∇ and ∇^* are said to be conjugate in the strong sense if they satisfy*

$$\nabla_u v - \nabla_v^* u = \frac{1}{2}\left([u,v]_E^\nabla + [u,v]_E^{\nabla^*}\right), \tag{4.6}$$

for all $u, v \in \mathfrak{X}(E)$.

With this definition, we can reproduce the results about strong conjugate connections from the manifold framework.

Proposition 4.1. *If ∇ and ∇^* are two admissible strongly conjugate linear E-connections, then*

$$T(\nabla) = -T(\nabla^*). \tag{4.7}$$

Proof. It can be proven by direct calculation:

$$\begin{aligned}
T(\nabla)(u,v) &= \nabla_u v - \nabla_v u - [u,v]_E^\nabla \\
&= \left\{\nabla_v^* u + \frac{1}{2}\left([u,v]_E^\nabla + [u,v]_E^{\nabla^*}\right)\right\} - \left\{\nabla_v^* u + \frac{1}{2}\left([v,u]_E^\nabla + [v,u]_E^{\nabla^*}\right)\right\} - [u,v]_E^\nabla \\
&= \nabla_v^* u - \nabla_u^* v - [u,v]_E^{\nabla^*} \\
&= -T(\nabla^*)(u,v).
\end{aligned}$$

The admissibility of both ∇ and ∇^* is crucial as it implies the anti-symmetry of the modified brackets which is used in the proof. □

This proposition is implied by the stronger one:

Proposition 4.2. *If an admissible linear E-connection ∇ has an admissible strong conjugate, then it is E-torsion-free.*

Proof. The strong conjugate condition (4.6) implies that

$$\rho(g(u,v)) = g(\nabla_w u, v) + g\left(u, \nabla_v w - \frac{1}{2}\left([v,w]_E^{\nabla} + [v,w]_E^{\nabla^*}\right)\right).$$

Using the symmetry of the E-metric g and the anti-symmetry of the modified brackets for admissible linear E-connections, one gets

$$g(T(\nabla)(w,u),v) + g(T(\nabla)(v,w),u) = \frac{1}{2}\left\{g\left([v,w]_E^{\nabla} - [v,w]_E^{\nabla^*}, u\right) - g\left([u,w]_E^{\nabla} - [u,w]_E^{\nabla^*}, v\right)\right\}.$$

Subtracting the cyclic permutations from this equation yields

$$-2g(T(\nabla)(u,v),w) = 0,$$

which gives the E-torsion-freeness due to the non-degeneracy of the E-metric g. □

As we see, the admissibility of two conjugate linear E-connections are sometimes necessary for proving analogous results. Being admissible for both ∇ and ∇^* necessitates the following condition:

$$L(e^a, \Delta(\nabla,\nabla^*)(X_a,u),v) = -L(e^a, \Delta(\nabla,\nabla^*)(X_a,v),u), \tag{4.8}$$

for all $u, v \in \mathfrak{X}(E)$.

As in the usual case, one can evaluate the E-non-metricity tensors of two conjugate linear E-connections ∇ and ∇^* in terms of their difference operator.

Proposition 4.3. *For a linear E-connection ∇ and its conjugate ∇^*, the E-non-metricity tensors satisfy*

$$Q(\nabla,g)(u,v,w) = -Q(\nabla^*,g)(u,v,w) = g\left(\Delta(\nabla^*,\nabla)(u,v),w\right), \tag{4.9}$$

for all $u, v, w \in \mathfrak{X}(E)$.

Proof. The E-non-metricity tensor $Q(\nabla,g)$ is given by

$$Q(\nabla,g)(u,v,w) = \rho(u)(g(v,w)) - g(\nabla_u v, w) - g(v, \nabla_u w).$$

The conjugation relation (4.1) yields

$$\rho(u)(g(v,w)) - g(v,\nabla_u w) = g(\nabla^*_u v, w),$$

so that the E-non-metricity tensor becomes

$$Q(\nabla,g)(u,v,w) = g(\nabla^*_u v, w) - g(\nabla_u v, w) = g(\Delta(\nabla^*,\nabla)(u,v),w),$$

which is the desired result. □

Proposition 4.4. *If an admissible linear E-connection ∇ has an admissible strong conjugate, then it is an E-Levi-Civita connection.*

Proof. The symmetric part of the difference operator of strongly conjugate linear E-connections ∇ and ∇^* is equal to

$$\Delta(\nabla,\nabla^*)(u,v) + \Delta(\nabla,\nabla^*)(v,u) = \frac{1}{2}\left([u,v]_E^{\nabla} + [u,v]_E^{\nabla^*}\right) + \frac{1}{2}\left([v,u]_E^{\nabla} + [v,u]_E^{\nabla^*}\right).$$

Hence, if ∇ and ∇^* are admissible then the difference operator is anti-symmetric. By Proposition (4.3), the E-non-metricity tensor is also anti-symmetric in the first two components:

$$Q(\nabla,g)(u,v,w) = -Q(\nabla,g)(v,u,w),$$

for all $u, v, w \in \mathfrak{X}(E)$. Since $Q(\nabla,g)$ is symmetric in the last two components, this anti-symmetry forces $Q(\nabla,g)$ to be 0. Hence it is E-metric-g-compatible. By Proposition (4.2), it is also E-torsion-free. Hence, $\nabla = \nabla^*$ is an E-Levi-Civita connection. □

In the manifold setting, Proposition (4.3) implies the total symmetry of $Q(\nabla, g)$ when ∇ is torsion-free. This is not valid for the algebroid case as the difference operator of any two E-torsion-free linear E-connections ∇ and ∇' satisfies

$$\Delta(\nabla, \nabla')(u, v) - \Delta(\nabla, \nabla')(v, u) = [u, v]_E^{\nabla} - [u, v]_E^{\nabla'} = L(e^a, \Delta(\nabla, \nabla')(X_a, u), v), \tag{4.10}$$

for all $u, v \in \mathfrak{X}(E)$. One can instead show that the difference operator of any two linear E-connections ∇ and ∇' is symmetric if and only if

$$T(\nabla)(u, v) - T(\nabla')(u, v) = [u, v]_E^{\nabla'} - [u, v]_E^{\nabla}. \tag{4.11}$$

This fact forces one to generalize statistical structures by using two geometrical objects that correspond to both E-non-metricity and E-torsion tensors because one cannot have conjugate linear E-connections with the same E-torsion while their E-non-metricity tensors are totally symmetric.

Definition 4.3. *On an almost-Leibniz algebroid $(E, \rho, [.,.]_E)$, the triplet (g, C, B) is called an E-statistical structure if g is an E-metric, C is a totally symmetric $(0,3)$-type E-tensor and $B : \mathfrak{X}(E) \times \mathfrak{X}(E) \to \mathfrak{X}(E)$ is an anti-symmetric $C^\infty(M, \mathbb{R})$-bilinear map.*

One can prove the equivalence of conjugate E-connection structures and E-statistical structures provided they satisfy the necessary conditions which will be explained shortly. Given a conjugate E-connection structure (g, ∇, ∇^*), if

$$T(\nabla)(u, v) = T(\nabla^*)(u, v) + [u, v]_E^{\nabla^*} - [u, v]_E^{\nabla}, \tag{4.12}$$

then the E-non-metricity tensors are totally symmetric. Hence, one can choose

$$C := -Q(\nabla, g). \tag{4.13}$$

Moreover, if one assumes that ∇ is E-torsion-free, then one can choose

$$B := T(\nabla^*). \tag{4.14}$$

If ∇^* is admissible, then $T(\nabla^*)$ is anti-symmetric. Therefore, we prove that given a conjugate E-connection structure (g, ∇, ∇^*), with admissible linear E-connections such that

$$T(\nabla^*)(u, v) = [u, v]_E^{\nabla} - [u, v]_E^{\nabla^*}, \tag{4.15}$$

one can construct an E-statistical structure (g, C, B) by the identifications (4.13) and (4.14). Conversely, given an E-statistical structure (g, C, B), one can look for two admissible linear E-connections ∇ and ∇^* satisfying

$$\begin{aligned} 2g(\nabla_u v, w) &= 2g({}^g\widetilde{\nabla}_u v, w) - C(u, v, w), \\ 2g(\nabla^*_u v, w) &= 2g({}^g\widetilde{\nabla^*}_u v, w) + C(u, v, w) - g(B(v, w), u) - g(B(u, w), v) + g(B(u, v), w), \end{aligned} \tag{4.16}$$

for all $u, v, w \in \mathfrak{X}(E)$. Here, ${}^g\widetilde{\nabla}$ and ${}^g\widetilde{\nabla^*}$ are the unique E-Levi-Civita connections on the almost-Lie algebroids $(E, \rho, [.,.]_E^{\nabla})$ and $(E, \rho, [.,.]_E^{\nabla^*})$, respectively. Note that both left- and right-hand sides of these equations depend on ∇ and ∇^*, so they are not constructive. Instead, one has to solve these equations for the linear E-connections. We assume that such solutions exist. For almost-Lie, pre-Lie and Lie algebroids, such solutions trivially exist. Moreover, for metric and Courant algebroids, existence of E-Levi-Civita connections indicates that these solutions exist. For these algebroids, E-Levi-Civita connections are not unqiue, see for example [19] for the proof for Courant algebroids.

Proposition 4.5. *The linear E-connections ∇ and ∇^* satisfying (4.16) are conjugates with respect to g provided that*

$$B(u, v) = [u, v]_E^{\nabla} - [u, v]_E^{\nabla^*}. \tag{4.17}$$

Moreover, they satisfy

$$\begin{aligned} Q(\nabla, g) &= -Q(\nabla^*, g) = -C, \\ T(\nabla) &= 0, \\ T(\nabla^*) &= B. \end{aligned} \tag{4.18}$$

Proof. In Ref. [10], we proved that for any admissible linear E-connection ∇, we have the decomposition

$$\begin{aligned} 2g(\nabla_u v, w) &= 2g({}^g\widetilde{\nabla}_u v, w) \\ &\quad - Q(\nabla, g)(u, v, w) - Q(\nabla, g)(v, u, w) + Q(\nabla, g)(w, u, v) \\ &\quad - g(T(\nabla)(v, w), u) - g(T(\nabla)(u, w), v) + g(T(\nabla)(u, v), w). \end{aligned}$$

This yields

$$C(u, v, w) = -Q(\nabla, g)(u, v, w) - Q(\nabla, g)(v, u, w) + Q(\nabla, g)(w, u, v),$$

which by the total symmetry of C forces us to have

$$C = -Q(\nabla, g) = Q(\nabla^*, g).$$

By comparing Equations (4.16) with the above decomposition, we can immediately conclude that

$$T(\nabla) = 0, \qquad T(\nabla^*) = B.$$

Now, we need to prove that they are indeed conjugate. In order to do that first we focus on the E-Levi-Civita connections ${}^g\widetilde{\nabla}$ and ${}^g\widetilde{\nabla^*}$. They are given by the Koszul formula with their corresponding modified brackets. Hence, we have

$$\begin{aligned} 2g({}^g\widetilde{\nabla}_u v, w) &= \rho(u)(g(v, w)) + \rho(v)(g(u, w)) - \rho(w)(g(u, v)) \\ &\quad - g([v, w]_E^{\nabla}, u) - g([u, w]_E^{\nabla}, v) + g([u, v]_E^{\nabla}, w), \end{aligned}$$

and

$$\begin{aligned} 2g({}^g\widetilde{\nabla^*}_u v, w) &= \rho(u)(g(v, w)) + \rho(v)(g(u, w)) - \rho(w)(g(u, v)) \\ &\quad - g([v, w]_E^{\nabla^*}, u) - g([u, w]_E^{\nabla^*}, v) + g([u, v]_E^{\nabla^*}, w). \end{aligned}$$

Adding these two equations together, we get

$$\begin{aligned} 2g({}^g\widetilde{\nabla}_u v, w) + 2g({}^g\widetilde{\nabla^*}_u v, w) &= 2\rho(u)(g(v, w)) - g([v, w]^{\nabla} - [v, w]^{\nabla^*}, u) \\ &\quad - g([u, w]^{\nabla} - [u, w]^{\nabla^*}, v) + g([u, v]^{\nabla} - [u, v]^{\nabla^*}, w). \end{aligned}$$

Note that the extra terms next to $2\rho(u)(g(v, w))$ cancel the terms coming from the operator B because of the assumption (4.17). Hence, we conclude

$$2g(\nabla_u v, w) + 2g(v, \nabla_u^* w) = 2\rho(u)(g(v, w)),$$

which implies that ∇ and ∇^* are conjugate. □

Note that the crucial point here is the difference of E-torsion tensors. Instead of E-torsion-free ∇, we could have introduced an E-torsionful linear E-connection if the difference of their E-torsion tensors is given by the difference of their modified bracket as in (4.17). Moreover, the difference between these E-torsion tensors guarantees that the difference E-tensor $\Delta(\nabla, \nabla^*)$ is symmetric, so that the E-non-metricity tensor $Q(\nabla, g)$ is totally symmetric by Proposition (4.3), which is consistent with the fact that C is totally symmetric.

Another way to incorporate E-torsion into the picture is to generalize quasi-statistical structures.

Definition 4.4. *On an almost-Leibniz algebroid $(E, \rho, [.,.]_E)$, a doublet (g, ∇) is called an E-quasi-statistical structure if g is an E-metric, ∇ is a linear E-connection satisfying*

$$Q(\nabla, g)(u, v, w) = Q(\nabla, g)(v, u, w) - g(T(\nabla)(u, v), w), \tag{4.19}$$

for all $u, v, w \in \mathfrak{X}(E)$.

As in the usual case, the property (4.19) fixes the E-torsion of the conjugate ∇^*:

Proposition 4.6. *If (g, ∇) is an E-quasi-statistical structure, then*

$$T(\nabla^*)(u,v) = [u,v]_E^{\nabla} - [u,v]_E^{\nabla^*}, \tag{4.20}$$

for all $u, v \in \mathfrak{X}(E)$.

Proof. The conjugation relation (4.1) implies that

$$g(\nabla^*_u v, w) = \rho(u)(g(v,w)) - g(\nabla_u w, v),$$

which yields

$$\begin{aligned} g(T(\nabla^*)(u,v) + [u,v]_E^{\nabla^*}, w) &= \rho(u)(g(v,w)) - \rho(v)(g(u,w)) - g(\nabla_u w, v) + g(\nabla_v w, u) \\ &= g([u,v]_E^{\nabla}, w), \end{aligned}$$

where for the last equality, the defining equation (4.19) is used. The non-degeneracy of the E-metric g gives the desired result. □

Quasi-statistical structures in the generalized geometry framework with small differences can be found in [20].

Similarly to the strong conjugation condition, one needs to modify the naive generalization of the relative torsion (3.15).

Definition 4.5. *The relative E-torsion operator of two linear E-connections ∇ and ∇' is a map $T(\nabla, \nabla') : \mathfrak{X}(E) \times \mathfrak{X}(E) \to \mathfrak{X}(E)$ defined as*

$$T(\nabla, \nabla')(u,v) := \nabla_u v - \nabla'_v u - \frac{1}{2}\left([u,v]_E^{\nabla} + [u,v]_E^{\nabla'}\right). \tag{4.21}$$

One can explicitly show that this is $C^\infty(M, \mathbb{R})$-bilinear, so that one can define an E-tensor. For two conjugate linear E-connections ∇ and ∇^*, being strongly conjugate is equivalent to having a vanishing relative E-torsion operator.

Proposition 4.7. *If ∇ and ∇' are two admissible linear E-connections, then the relative E-torsion operator satisfies*

$$T(\nabla, \nabla')(u,v) = -T(\nabla', \nabla)(v,u), \tag{4.22}$$

for all $u, v \in \mathfrak{X}(E)$.

Proof. It can be proven by direct calculation:

$$\begin{aligned} T(\nabla, \nabla')(u,v) &= \nabla_u v - \nabla'_v u - \frac{1}{2}\left([u,v]_E^{\nabla} + [u,v]_E^{\nabla'}\right) \\ &= -\left\{\nabla'_v u - \nabla_u v - \frac{1}{2}\left([v,u]_E^{\nabla} + [v,u]_E^{\nabla'}\right)\right\} \\ &= -T(\nabla', \nabla)(v,u), \end{aligned}$$

where the anti-symmetry of the both modified brackets is used. □

Proposition 4.8. *If ∇ and ∇' are two linear E-connections (not necessarily admissible), then*

$$T(\nabla, \nabla') + T(\nabla', \nabla) = T(\nabla) + T(\nabla'). \tag{4.23}$$

Proof. Similarly to the previous proposition, it can be proven by a simple direct computation:

$$\begin{aligned} T(\nabla,\nabla')(u,v)+T(\nabla',\nabla)(v,u) &= \left\{\nabla_u v-\nabla'_v u-\frac{1}{2}\left([u,v]_E^{\nabla}+[u,v]_E^{\nabla'}\right)\right\} \\ &\quad+\left\{\nabla'_u v-\nabla_v u-\frac{1}{2}\left([u,v]_E^{\nabla'}+[u,v]_E^{\nabla}\right)\right\} \\ &= \left\{\nabla_u v-\nabla_v u-[u,v]_E^{\nabla}\right\}+\left\{\nabla'_u v-\nabla'_v u-[u,v]_E^{\nabla'}\right\} \\ &= T(\nabla)(u,v)+T(\nabla')(u,v), \end{aligned}$$

which is the desired result. □

Note that Proposition (4.7) and Proposition (4.8) hold in particular for conjugate linear E-connections ∇ and ∇^*.

Proposition 4.9. *If ∇ and ∇^* are E-torsion-free conjugate linear E-connections, then their mean linear E-connection ∇^0 is an E-Levi-Civita connection.*

Proof. The mean linear E-connection ∇^0 is defined as $\frac{1}{2}(\nabla+\nabla^*)$ by Equation (4.4). Its E-torsion can be evaluated as

$$\begin{aligned} T(\nabla^0)(u,v) &= \nabla^0_u v-\nabla^0_v u-[u,v]_E^{\nabla^0} \\ &= \frac{1}{2}\left\{(\nabla_u v+\nabla^*_u v)-(\nabla_v u+\nabla^*_v u)-\left([u,v]_E^{\nabla}+[u,v]_E^{\nabla^*}\right)\right\} \\ &= \frac{1}{2}\left[T(\nabla)(u,v)+T(\nabla^*)(u,v)\right]=0. \end{aligned}$$

Since the mean linear E-connection ∇^0 is also E-metric-g-compatible, it is an E-Levi-Civita connection. □

In [10], we proved that all E-torsion-free and in particular E-Levi-Civita connections have to admissible. Even though the admissibility is not used in the previous proposition, it necessarily holds.

Similarly to the usual case, one can consider the one-parameter family of linear E-connections for $\alpha\in\mathbb{R}$:

$$\nabla^{(\alpha)} := \frac{1+\alpha}{2}\nabla^*+\frac{1-\alpha}{2}\nabla. \tag{4.24}$$

These α-dependent linear E-connections are called α-E-connections. In particular one has

$$\nabla^{(1)}=\nabla^*,\qquad \nabla^{(-1)}=\nabla,\qquad \nabla^{(0)}=\nabla^0. \tag{4.25}$$

α-E-connections in the generalized geometry setting can be found in [21]. Identical to the manifold framework, for α-E-connections, the following proposition holds.

Proposition 4.10. *For any $\alpha\in\mathbb{R}$, the following relations hold*

$$\begin{aligned} (\nabla^{(\alpha)})^* &= \nabla^{(-\alpha)}, \\ T(\nabla^{(\alpha)}) &= \frac{1+\alpha}{2}T(\nabla^*)+\frac{1-\alpha}{2}T(\nabla), \\ Q(\nabla^{(\alpha)},g) &= -\alpha Q(\nabla,g)=\alpha Q(\nabla^*,g), \end{aligned} \tag{4.26}$$

for all $u,v,w\in\mathfrak{X}(E)$.

Proof. The first identity follows from the fact that the conjugation is an involution. The second one can be proven by using the decomposition of the modified bracket corresponding to some $\nabla^{(\alpha)}$:

$$[u,v]_E^{\nabla^{(\alpha)}} = \frac{1+\alpha}{2}[u,v]_E^{\nabla^*}+\frac{1-\alpha}{2}[u,v]_E^{\nabla}.$$

Moreover, the last one is a consequence of Proposition (4.3) where

$$Q(\nabla,g)=-Q(\nabla^*,g)$$

is proven. □

Proposition 4.11. *On a local pre-Leibniz algebroid with a locality projector, the curvature of α-E-connections can be expressed as*

$$
\begin{aligned}
R(\nabla^{(\alpha)})(u,v)w &= \frac{1+\alpha}{2}R(\nabla^*)(u,v)w + \frac{1-\alpha}{2}R(\nabla)(u,v)w \\
&+ \frac{1-\alpha^2}{4}\left[\Delta(\nabla,\nabla^*)\left(v,\Delta(\nabla,\nabla^*)(u,w)\right) - \Delta(\nabla,\nabla^*)\left(u,\Delta(\nabla,\nabla^*)(v,w)\right)\right] \\
&+ \frac{1-\alpha^2}{4}\Delta(\nabla,\nabla^*)\left([u,v]_E^{\widehat{\nabla}} - [u,v]_E^{\widehat{\nabla^*}}, w\right),
\end{aligned} \tag{4.27}
$$

for all $u,v,w \in \mathfrak{X}(E)$.

Proof. This follows from the fact that the terms with two E-covariant derivatives can be expressed as

$$
\begin{aligned}
\nabla^{(\alpha)}_u \nabla^{(\alpha)}_v w &= \left(\frac{1+\alpha}{2}\right)^2 \nabla^*_u \nabla^*_v w + \frac{1-\alpha^2}{4}\nabla^*_u \nabla_v w \\
&+ \frac{1-\alpha^2}{4}\nabla_u \nabla^*_v w + \left(\frac{1-\alpha}{2}\right)^2 \nabla_u \nabla_v w,
\end{aligned}
$$

and the term with the projected modified can be written as

$$
\begin{aligned}
\nabla^{(\alpha)}_{[u,v]_E^{\widehat{\nabla^{(\alpha)}}}} w &= \left(\frac{1+\alpha}{2}\right)^2 \nabla^*_{[u,v]_E^{\widehat{\nabla^*}}} w + \frac{1-\alpha^2}{4}\nabla^*_{[u,v]_E^{\widehat{\nabla^*}}} w \\
&+ \left(\frac{1-\alpha}{2}\right)^2 \nabla^*_{[u,v]_E^{\widehat{\nabla}}} w + \frac{1-\alpha^2}{4}\nabla_{[u,v]_E^{\widehat{\nabla}}} w,
\end{aligned}
$$

for all $u,v,w \in \mathfrak{X}(E)$. Combining these equations, one gets the desired result. □

Note that the expression (4.27) reduces to Equation (3.21) when the locality operator vanishes. Moreover, it implies that for a pair of flat conjugate linear E-connections ∇ and ∇,

$$
R(\nabla^{(\alpha)}) = R(\nabla^{(-\alpha)}), \tag{4.28}
$$

which is also valid for the usual case (3.22).

5. Hessian structures on manifolds

The action of the second order covariant derivative of a smooth function is called the Hessian of the function:

$$
H(\nabla)(f)(U,V) := \nabla^2_{U,V} f := U(V(f)) - df(\nabla_U V) = (\nabla df)(U,V). \tag{5.1}
$$

It is $C^\infty(M,\mathbb{R})$-bilinear, so that it defines a $(0,2)$-type tensor. The Hessian $H(\nabla)(f)$ is symmetric if and only if ∇ is torsion-free. Hence, for a torsion-free affine connection, the Hessian defines a metric if it is also non-degenerate. This leads one to the definition of Hessian metrics. For a flat affine connection ∇, a metric g is called a Hessian metric if g can be locally expressed by the Hessian of some smooth function $f \in C^\infty(M,\mathbb{R})$:

$$
g = H(\nabla)(f). \tag{5.2}
$$

In this case, the doublet (g,∇) is called a Hessian structure, and f is called the potential. For a Hessian structure (g,∇), the affine connection ∇ is necessarily torsion-free, and it is flat by definition. Hence, (M,g,∇) defines a symmetric teleparallel geometry. For a detailed exposition of Hessian structures see [22].

Hessian structures are directly related to statistical structures. The doublet (g,∇) with flat ∇ is a Hessian structure if and only if g is ∇-Codazzi so that (g,∇) is a statistical structure. Hence, a

Hessian structure can be equivalently defined as a statistical structure with a flat affine connection [23]. The following fundamental theorem of statistical geometry guarantees that if ∇ is flat, so is ∇^*:

$$g(R(\nabla)(U,V)W,Z) = -g(R(\nabla^*)(U,V)Z,W), \tag{5.3}$$

for all $U,V,W,Z \in \mathfrak{X}(M)$. An affine connection ∇ is said to have a constant curvature $\kappa \in \mathbb{R}$ if

$$R(\nabla)(U,V)W = \kappa\left[g(V,W)U - g(U,W)V\right], \tag{5.4}$$

for all $U,V,W \in \mathfrak{X}(M)$. The fundamental theorem of statistical geometry also implies that if ∇ has a constant curvature κ so does ∇^* [14].

6. Hessian structures on pre-Leibniz algebroids

Completely analogous to the definition of the Hessian of a function, one can define the E-Hessian on an anchored vector bundle (E,ρ).

Definition 6.1. *The action of the second order E-covariant derivative on a smooth function is called the E-Hessian of the function:*

$$H(\nabla)(f)(u,v) := \nabla^2_{u,v}f := \rho(u)(\rho(v)(f)) - Df(\nabla_u v) = (\nabla Df)(u,v). \tag{6.1}$$

One can explicitly show that, $H(\nabla)(f)$ is indeed a $(0,2)$-type E-tensor. In order to define Hessian E-metrics, one needs to figure out when the E-Hessian is symmetric. Similarly to the usual case for torsion-free connections, the symmetry is equivalent to projected E-torsion-freeness provided certain assumptions are satisfied.

Proposition 6.1. *Let $(E,\rho,[\cdot,\cdot],L)$ be a local pre-Leibniz algebroid. For an admissible linear E-connection ∇ whose projected E-torsion operator's image is not entirely in the kernel of the anchor ρ, the following are equivalent:*

(i) The E-Hessian $H(\nabla)(f)$ of any smooth function f is symmetric.

(ii) ∇ is projected E-torsion-free, i.e. $\widehat{T}(\nabla) = 0$.

(iii) For any $\Omega \in \Omega^1(E), u,v \in \mathfrak{X}(E)$

$$\widehat{d}(\nabla)\Omega(u,v) = \left(\nabla_u\Omega\right)(v) - \left(\nabla_v\Omega\right)(u). \tag{6.2}$$

Proof. We will prove this statement by showing $(1 \iff 2)$ & $(2 \iff 3)$. For the first equivalence, we start with the observation

$$\begin{aligned} H(\nabla)(f)(u,v) - H(\nabla)(f)(v,u) &= \{\rho(u)(\rho(v)(f)) - Df(\nabla_u v)\} - \{\rho(v)(\rho(u)(f)) - Df(\nabla_v u)\} \\ &= [\rho(u),\rho(v)](f) - \{Df(\nabla_u v) - Df(\nabla_v u)\} \\ &= [\rho(u),\rho(v)](f) - \rho(\widehat{T}(\nabla)(u,v))(f) - \rho([u,v]_E^{\widehat{\nabla}})(f) \\ &= -\rho(\widehat{T}(\nabla)(u,v))(f). \end{aligned}$$

Hence, $H(\nabla)(f)$ is symmetric if and only if the projected E-torsion $\widehat{T}(\nabla)$ vanishes as it is assumed that the image of the projected E-torsion operator is not a subset of the kernel of the anchor. Note that, here we used the fact that the projected modified bracket is a pre-Leibniz bracket. We did not need the anti-symmetry of the bracket, so this equivalence is actually valid for any linear E-connection. This is not the case for the other equivalence as one needs the admissibility for the definition of the projected E-exterior derivative. For the equivalence $(2 \iff 3)$, we start with

$$\begin{aligned} \widehat{d}(\nabla)\Omega(u,v) &= \rho(u)(\Omega(v)) - \rho(v)(\Omega(u)) - \Omega([u,v]_E^{\widehat{\nabla}}) \\ &= \{(\nabla_u\Omega)(v) + \Omega(\nabla_u v)\} - \{(\nabla_v\Omega)(u) + \Omega(\nabla_v u)\} - \Omega([u,v]_E^{\widehat{\nabla}}) \\ &= \{(\nabla_u\Omega)(v) - (\nabla_v\Omega)(u)\} + \{\Omega(\nabla_u v) - \Omega(\nabla_v u)\} - \Omega([u,v]_E^{\widehat{\nabla}}) \\ &= \{(\nabla_u\Omega)(v) - (\nabla_v\Omega)(u)\} + \Omega\left(\widehat{T}(\nabla)(u,v) + [u,v]_E^{\widehat{\nabla}}\right) - \Omega([u,v]_E^{\widehat{\nabla}}) \\ &= \{(\nabla_u\Omega)(v) - (\nabla_v\Omega)(u)\} + \Omega\left(\widehat{T}(\nabla)(u,v)\right). \end{aligned}$$

Journal of Physics: Conference Series **2191** (2022) 012011 doi:10.1088/1742-6596/2191/1/012011

Hence, 3 holds if and only if the projected E-torsion vanishes. Therefore, we proved all the equivalences. □

Note that the equivalence of 2 & 3 is also valid for E-torsion-free admissible linear E-connections if the term $\widehat{d}(\nabla)$ in Equation (6.2) is replaced by $d(\nabla)$. They are not equivalent to 1 as one uses the pre-Leibniz property of the projected modified bracket. Moreover, if the image of the projected E-torsion operator is in the kernel of the anchor, then the E-Hessian is automatically symmetric without any condition on the linear E-connection. For example, if the anchor is trivial, i. e. $\rho = 0$, then for every linear E-connection, the E-Hessian is symmetric.

Corollary 6.1. *If the image of the projected E-torsion operator of a linear E-connection ∇ is a subset of the kernel of the anchor, then the E-Hessian $H(\nabla)(f)$ is symmetric.*

Proof. In the proof of the previous proposition (6.1), we proved that

$$H(\nabla)(f)(u,v) - H(\nabla)(f)(v,u) = -\rho(\widehat{T}(\nabla)(u,v))(f).$$

Hence if the image of $\widehat{T}(\nabla)$ is in the kernel of the anchor ρ, the right-hand side is zero so that the E-Hessian is symmetric. □

Corollary 6.2. *Let ∇ be a projected E-torsion-free linear E-connection with the E-conjugate ∇^* satisfying*

$$T(\nabla)(u,v) - T(\nabla^*)(u,v) = [u,v]_E^{\nabla} - [u,v]_E^{\nabla^*}, \tag{6.3}$$

then the E-Hessian of ∇^* is also symmetric.

Proof. If Equation (6.3) holds, then the difference between the projected E-torsion operators of ∇ and ∇^* reads

$$\widehat{T}(\nabla)(u,v) - \widehat{T}(\nabla^*)(u,v) = [u,v]_E^{\widehat{\nabla}} - [u,v]_E^{\widehat{\nabla^*}},$$

which implies by the projected E-torsion-free assumption for ∇

$$\widehat{T}(\nabla^*)(u,v) = -\widehat{L}(e^a, \Delta(\nabla,\nabla^*)(X_a,u),v).$$

As the image of $\widehat{L}$ is in the kernel of the anchor, by the previous corollary (6.1), we get the desired result. □

Note that the assumption (6.3) in the previous corollary (6.2) is valid for conjugate E-connections (4.16) constructed from an E-statistical structure.

For projected E-torsion-free linear E-connections (or for linear E-connections whose projected E-torsion operator's image is in the kernel of the anchor), one can define Hessian E-metrics.

Definition 6.2. *For an E-flat, projected E-torsion-free linear E-connection ∇, an E-metric g is called a Hessian E-metric if it is locally of the form*

$$g = H(\nabla)(f), \tag{6.4}$$

for some $f \in C^\infty(M,\mathbb{R})$. In this case, the doublet (g,∇) is called an E-Hessian structure.

Similarly to the usual case, there is a relation between E-Hessian structures and E-statistical structures.

Proposition 6.2. *If (g,∇) is an E-Hessian structure, then g is ∇-Codazzi.*

Proof. If g is an E-Hessian metric, it is of the form

$$g(u,v) = H(\nabla)(f)(u,v) = \rho(u)(\rho(v)(f)) - Df(\nabla_u v),$$

for some $f \in C^\infty(M, \mathbb{R})$ so that the E-non-metricity tensor of ∇ and g is given by

$$\begin{aligned} Q(\nabla, g)(u, v, w) &= \rho(u)(g(v, w)) - g(\nabla_u, w) - g(v, \nabla_u w) \\ &= \rho(u)(\rho(v)(\rho(w)(f)) - Df(\nabla_v w)) \\ &\quad - \rho(\nabla_u v)(\rho(w)(f)) - Df(\nabla_{\nabla_u v} w) \\ &\quad - \rho(v)(\rho(\nabla_u w)(f)) - Df(\nabla_v \nabla_u w) \\ &= \rho(u)(\rho(v)(\rho(w)(f))) - \rho(\nabla_u v)(\rho(w)(f)) \\ &\quad - Df(\nabla_{\nabla_u v} w) - Df(\nabla_v \nabla_u w). \end{aligned}$$

By the projected E-torsion-freeness and E-flatness, the Ricci identity (2.16) yields

$$\nabla_{\nabla_u v} w + \nabla_v \nabla_u w = \nabla_{\nabla_v u} w + \nabla_u \nabla_v w,$$

which implies that the following term is symmetric in u and v

$$Df'(\nabla_{\nabla_u v} w) + Df'(\nabla_v \nabla_u w),$$

for all $f' \in C^\infty(M, \mathbb{R})$. The projected E-torsion-freeness also implies the symmetry of the following term in u and v due to the fact that the projected modified bracket is a pre-Leibniz bracket:

$$\rho(\nabla_u v)(\rho(w)(f')) - \rho(u)(\rho(v)(\rho(w)(f'))),$$

for all $f' \in C^\infty(M, \mathbb{R})$. These two observations together imply that the E-non-metricity tensor $Q(\nabla, g)(u, v, w)$ is symmetric in u and v. Hence, g is ∇-Codazzi. □

Corollary 6.3. *Let (g, ∇) be an E-Hessian structure. Then, the triplet $(g, Q(\nabla, g), T(\nabla))$ is an E-statistical structure if ∇ is admissible.*

Proof. By the previous proposition (6.2), g is ∇-Codazzi so that the E-non-metricity tensor $Q(\nabla, g)$ is totally symmetric for any ∇. Moreover, for an admissible linear E-connection ∇, the E-torsion operator $T(\nabla)$ is anti-symmetric. Hence, $(g, Q(\nabla, g), T(\nabla))$ is an E-statistical structure. □

In the manifold setting, the other implication is also valid. In other words, for an statistical structure (g, ∇) with a flat connection ∇, the metric g is a Hessian metric. Because of the local condition in the definition of Hessian metrics, this proof makes use of Poincaré lemma which states that every closed form is locally exact. On the algebroid case, when acting on smooth functions, the projected E-exterior derivative squares to 0, and the action of the coboundary map coincides with the action of the projected exterior derivative, i. e. $Df = \widehat{d}(\nabla)f$. Yet, the former does not hold for E-p-forms with $p \geq 1$, and there is no analogue of the Poincaré lemma. Therefore, we were not able to produce a proof of the other implication for the algebroid setting.

Under certain assumptions, one can generalize the fundamental theorem (5.3) of statistical geometry.

Proposition 6.3. *Let (g, ∇, ∇^*) be a conjugate E-connection structure on a local pre-Leibniz algebroid $(E, \rho, [.,.]_E)$ such that both projected modified brackets $[.,.]_E^{\widehat{\nabla}}$ and $[.,.]_E^{\widehat{\nabla}^*}$ share a holonomic local E-frame. Then,*

$$g(R(\nabla)(u, v)w, z) + g(R(\nabla^*)(u, v)z, w) = 0, \tag{6.5}$$

for all $u, v, w, z \in \mathfrak{X}(E)$.

Proof. We start with the left-hand side LHS of Equation (6.5) on a local E-frame (X_a) which is holonomic for both projected modified brackets, so that the terms with these brackets vanish:

$$\begin{aligned} LHS(X_a, X_b, X_c, X_d) &= g(R(\nabla)(X_a, X_b)X_c, X_d) + g(R(\nabla^*)(X_a, X_b)X_d, X_c) \\ &= g(\nabla_{X_a}\nabla_{X_b}X_c - \nabla_{X_b}\nabla_{X_a}X_c, X_d) + g(\nabla^*_{X_a}\nabla^*_{X_b}X_d - \nabla^*_{X_b}\nabla^*_{X_a}X_d, X_c). \end{aligned}$$

Next, we observe that the conjugation condition (4.1) implies that

$$g(\nabla_{X_a}\nabla_{X_b}X_c, X_d) = \rho(X_a)(g(\nabla_{X_b}X_c, X_d)) - g(\nabla_{X_b}X_c, \nabla^*_{X_a}X_d),$$
$$g(\nabla^*_{X_a}\nabla^*_{X_b}X_c, X_d) = \rho(X_a)(g(\nabla^*_{X_b}X_c, X_d)) - g(\nabla^*_{X_b}X_c, \nabla_{X_a}X_d).$$

Inserting these equations in LHS and canceling out, one gets:

$$LHS(X_a, X_b, X_c, X_d) = [\rho(X_a), \rho(X_b)](g(X_c, X_d)).$$

As both projected modified brackets are pre-Lie brackets, one has

$$[\rho(X_a), \rho(X_b)] = \rho([X_a, X_b]_E^{\widehat{\nabla}}) = \rho(0) = 0,$$

which implies the desired result by the $C^\infty(M,\mathbb{R})$-linearity of g and $R(\nabla)$.

□

Having a mutual holonomic local E-frame (X_a) for the projected modified brackets $[.,.]_E^{\widehat{\nabla}}$ and $[.,.]_E^{\widehat{\nabla}^*}$ necessitates the following conditions:

$$\widehat{\gamma}(\nabla)^a{}_{bc} = \gamma^a{}_{bc} - \Gamma(\nabla)^e{}_{db}\widehat{L}^{ad}{}_{ec} = 0,$$
$$\widehat{\gamma}(\nabla^*)^a{}_{bc} = \gamma^a{}_{bc} - \Gamma(\nabla^*)^e{}_{db}\widehat{L}^{ad}{}_{ec} = 0.$$

These together imply that

$$\Gamma(\nabla)^e{}_{db}\widehat{L}^{ad}{}_{ec} = \Gamma(\nabla^*)^e{}_{db}\widehat{L}^{ad}{}_{ec} = -\gamma^a{}_{bc}. \tag{6.6}$$

This mutual holonomic local E-frame assumption becomes trivial when one restricts to the usual manifold setting. As $L = 0$ for the tangent bundle, the projected modified anholonomy coefficients are identical with the usual anholonomy coefficients. The coordinate frames are by definition holonomic, so the assumption of Proposition (6.3) is automatically satisfied. It seems possible to weaken this assumption, yet we could not find an alternative proof that does not depend on the choice of a local E-frame. For example, on a g-orthonormal local E-frame, which always exists, the term $[\rho(X_a), \rho(X_b)](g(X_c, X_d))$ automatically vanishes. Similarly, g-orthonormality also cancels out other terms, yet the following term does not vanish:

$$- g\left(\nabla_{L(e^e, \Delta(\nabla,\nabla^*)(X_e, X_a), X_b)}X_c, X_d\right). \tag{6.7}$$

Completely identical to the usual case, one can define linear E-connections with constant E-curvature.

Definition 6.3. *A linear E-connection ∇ is said to have constant E-curvature $\kappa \in \mathbb{R}$ if*

$$R(\nabla)(u, v)w = \kappa\,[g(v, w)u - g(u, w)v], \tag{6.8}$$

for all $u, v, w \in \mathfrak{X}(E)$.

Note that the right-hand side of Equation (6.8) is anti-symmetric in u and v, so that having constant E-curvature necessitates the anti-symmetry of the E-curvature operator $R(\nabla)(u, v)w$ in u and v. This is the case when the linear E-connection is admissible because when ∇ is admissible, the projected modified bracket $[.,.]_E^{\widehat{\nabla}}$ is anti-symmetric, so that the E-curvature operator is anti-symmetric.

Corollary 6.4. *Under the same assumptions as Proposition (6.3), a linear E-connection ∇ has a constant curvature κ if and only if its conjugate ∇^* has a constant curvature κ.*

Proof. By the previous proposition (6.3), one has

$$g(R(\nabla)(u,v)w,z) = -g(R(\nabla^*)(u,v)z,w),$$

which implies for linear E-connection ∇ with a constant E-curvature

$$\begin{aligned} g(R(\nabla^*)(u,v)z,w) &= -g\left(\kappa\left[g(v,w)u - g(u,w)v\right], z\right) \\ &= -\kappa\left[g(v,w)g(u,z) - g(u,w)g(v,z)\right] \\ &= -g\left(\kappa\left[g(u,z)v - g(v,z)u\right], w\right) \\ &= g\left(\kappa\left[g(v,z)u - g(u,z)v\right], w\right). \end{aligned}$$

As the E-metric g is non-degenerate, this yields

$$R(\nabla^*)(u,v)z = \kappa\left[g(v,z)u - g(u,z)v\right],$$

which is the desired result. □

7. Concluding remarks

In this paper, we introduced statistical, conjugate connection and Hessian structures on anti-commutable pre-Leibniz algebroids. These algebroids are defined with a property in which the symmetrization of the bracket satisfies a certain condition depending on the choice of an equivalence class of connections. Such 'admissible' connections turn out to be necessary for appropriate generalizations of these structures which are normally defined in the manifold setting. The admissibility condition for a connection results in the anti-symmetry of torsion and curvature operators, which is a necessary property for many features of metric-affine geometries. In our previous paper [10], we proved Bianchi identity, Cartan structure equations and the decomposition of the connection in terms of its torsion and non-metricity tensors for admissible connections. In this paper, we extended the theory of admissible connections, and showed that the admissibility condition is crucial for the generalizations of aforementioned structures. We proved that statistical and conjugate connection structures are equivalent provided certain assumptions hold. We also introduced generalizations of strongly conjugate connections, α-connections, relative torsion tensors and proved many results completely analogous to the usual manifold setting. We defined the generalizations Hessian metrics and figured out the necessary conditions for a connections to yield Hessian metrics. Moreover, we proved that the existence of Hessian structures yields a statistical structure. We also proved a mild generalization of the fundamental theorem of statistical geometry concerning the curvature operators of two conjugate connections provided certain conditions about anholonomy coefficients are met.

In the near future, we plan to work on other applications of admissible connections and study the generalizations of complex structures. As a special case, we wish to work on Kähler structures and their relations to statistical and Hessian structures. One of our aims will be to find out how one can generalize the fact that the tangent bundle of a smooth manifold endowed with a Hessian metric carries a natural Kähler structure [23] and work on the details of this relation. There is also another closely related construction called a locally equiaffine structure which is defined by the existence of a local volume form whose covariant derivative vanishes [24]. These structures are closely tied with the Ricci symmetric connections, whereas the Ricci symmetry is equivalent to the existence of trivial Weyl structures [25]. Currently, we are working on the generalizations of conformal and projective structures in the algebroid framework. Weyl structures on manifolds naturally arises from the compatibility condition of these structures [26], and we wish to relate all these constructions together in a completely parallel way to the manifold setting. We leave out the possible applications of statistical structures on pre-Leibniz algebroids to statistical models to the experts in the field.

Acknowledgment
This work is dedicated to Tekin Dereli in honor of his 72nd birthday. He is one of the most renowned figures in Turkish mathematical physics community. I was quite lucky to be one of his PhD students and a member of a big academic family-tree (it has loops though!) created by him. I am indebted for his invaluable contribution to my knowledge and wish him a life full of science.

References

[1] C. Marle, Calculus on Lie algebroids, Lie groupoids and Poisson manifolds. *Dissertationes Mathematicae 457*, 1-57 (2008); arXiv:0806.0919 [math.DG].

[2] Z. Liu, A. Weinstein, & P. Xu, Manin triples for Lie bialgebroids. *Journal of Differential Geometry, 45*, 547-574 (1997); arXiv:dg-ga/9508013.

[3] O. Hohm, C. Hull, & B. Zwiebach, Generalized metric formulation of double field theory. *Journal of High Energy Physics, 8* (2010); arXiv:1006.4823 [hep-th].

[4] C. Hull, Generalised geometry for M-theory. *Journal of High Energy Physics 07* (2007); arXiv:hep-th/0701203.

[5] I. Vaisman, Towards a double field theory on para-Hermitian manifolds. *Journal of Mathematical Physics 54* (2013); arXiv:1209.0152 [math.DG].

[6] M. J. Lean, Dorfman connections and Courant algebroids. *Journal de Mathématiques Pures et Appliquées 116*, 1-39 (2018); arXiv:1209.6077 [math.DG].

[7] D. Baraglia, Conformal Courant algebroids and orientifold T-duality. *International Journal of Geometric Methods in Modern Physics 10*, 1-35 (2013); arXiv:1109.0875 [math.DG].

[8] T. Dereli, & K. Doğan, Metric-connection geometries on pre-Leibniz algebroids: A search for geometrical structure in string models. *Journal of Mathematical Physics 62*, (2021); arXiv:2006.05957 [hep-th].

[9] B. Jurčo, & J. Vysoký, Leibniz algebroids, generalized Bismut connections and Einstein-Hilbert actions. *Journal of Geometry and Physics 97*, 25-33 (2015); arXiv:1503.03069 [hep-th].

[10] T. Dereli, & K. Doğan, 'Anti-commutable' pre-Leibniz algebroids and admissible connections, (submitted), (2021), arXiv:2108.10199 [math.DG].

[11] H. V. Le, Statistical manifolds are statistical models. *Journal of Geometry 84*, 83-93 (2006); arXiv:0805.3405 [math.DG].

[12] O. Calin & C. Udriste, *Geometric modeling in probability and statistics*, (Springer, 2014).

[13] S. L. Lauritzen, Statistical manifolds. *Differential Geometry in Statistical Inference 10*, 163-216 (1987). 1987. *Journal of Symplectic Geometry 7*, 311-335 (2009).

[14] F. Nielsen, An elementary introduction to information geometry. *Entropy 22*, 1-62 (2020); arXiv:1808.08271 [cs.LG].

[15] R. J. Fernandes, Lie algebroids, holonomy and characteristic classes. *Advances in Mathematics, 170* , 119-179 (2002); arXiv:math/0007132[math.DG].

[16] J. Grabowski, D. Khudaverdyan, & N. Poncin, The supergeometry of Loday algebroids. *The Journal of Geometric Mechanics 5* , 185-21 (2011); arXiv:1103.5852 [math.DG].

[17] A. Wade, On some properties of Leibniz algebroids. In T. P. Robart, J. A. Leslie, & A. Banyaga (Eds.) *Infinite dimensional Lie groups in geometry and representation theory* (1st ed. 65-78), (Washington, DC: World Scientific, 2002).

[18] K. Grabowska, J. Grabowski, M. Kuś, & G. Marmo, Lie groupoids in information geometry. *Journal of Physics A: Mathematical and Theoretical 52*, (2019); arXiv:1904.00709 [math-ph].

[19] B. Jurčo, & J. Vysoký, Courant algebroid connections and string effective action. In *Proceedings of Tohoku Forum for Creativity, Special Volume: Noncommutative Geometry and Physics IV* (2016); arXiv:1612.01540 [math-ph].

[20] A. M. Blaga, & A Nannicini, Generalized quasi-statistical structures. *Bulletin of the Belgian Mathematical Society 27*, (2019); 731-754 arXiv:1809.04784 [math.DG].

[21] A. M. Blaga, & A Nannicini, α-connections in generalized geometry. *Journal of Geometry and Physics 219*, (2021); arXiv:2004.05036 [math.DG].

[22] H. Shima, *The geometry of Hessian structures.* (World Scientific, 2007).

[23] H. Furuhata, Hypersurfaces in statistical manifolds. *Differential Geometry and its Applications 27*, (2009); 420–429.

[24] H. Matsuzoe, J. Takeuchi, & S. Amari, Equiaffine structures on statistical manifolds and Bayesian statistics. *Differential Geometry and Its Applications 24*, (2006); 567–578.

[25] E. García-Río, P. Gilkey, S. Nikčević, & R. Váazquez-Lorenzo, *Applications of affine and Weyl geometry*, (Morgan & Claypool Publishers, 2013).

[26] V. S. Matveev, & E. Scholz, Light cone and Weyl compatibility of conformal and projective structures. *General Relativity and Gravitation 52*, 52-66 (2020); arXiv:2001.01494 [math.DG].

Journal of Physics: Conference Series **2191** (2022) 012012 doi:10.1088/1742-6596/2191/1/012012

From Feynman graphs to Witten diagrams

A. Domingo Gallegos, Umut Gürsoy, Natale Zinnato

Institute for Theoretical Physics, Utrecht University, Leuvenlaan 4, 3584 CE Utrecht, The Netherlands

E-mail: a.d.gallegospazos@uu.nl, u.gursoy@uu.nl, n.zinnato@uu.nl

Abstract. We investigate the possibility of generalizing Gopakumar's microscopic derivation of Witten diagrams in large N free quantum field theory [1] to interacting theories. For simplicity we consider a massless, matrix valued real scalar field with Φ^h interaction in d-dimensions. Using Schwinger's proper time formulation and organizing the sum over Feynman graphs by the number of loops ℓ, we show that the two-point function can be expressed as a sum over boundary-to-boundary propagators of bulk scalars in AdS_{d+1} with mass determined by ℓ.

This manuscript is intended as a contribution to the festschrift of prof. Tekin Dereli's on the occasion of his 72nd birthday[1].

1. Introduction

Gauge-string correspondence [2, 3, 4], despite all the successful checks it goes through and the plethora of work that applied it to study, again successfully, both strongly interacting QFTs and quantum gravity, lacks a satisfactory microscopic derivation and a deeper understanding thereof. A basic question is how to reformulate holographic QFT correlation functions, in particular Feynman graphs, such that emergence of gravitational dynamics becomes manifest, for example how the gravitational field propagators in the dual curved space-time arise from field theory amplitudes. Another question is, how to determine which QFTs are holographic which are not. Given a holographic QFT and assuming a limit where the dual geometry is well defined, is there an algorithm to determine this dual background precisely?

Among all the different approaches that have been suggested in the literature, entanglement entropy [5], geometrization of RG flows [7, 6, 8], bulk reconstruction [9], quantum error correction [10], tensor networks [11], etc., there is one which stands out as the most fundamental: deriving dual gravity propagators directly from the QFT Feynman graphs. As far as we know, for field theories in $d > 2$ this approach was first proposed by R. Gopakumar in the case of free field theory [1]. Gopakumar considered matrix valued free field theories and studied n-point function of composite single-trace operators in Schwinger's proper time formulation [13]. In the particular case of the three-point function he noticed a change of variables involving the moduli (Feynman parameter's of a given graph) that is called the *star-triangle duality*[2]. The name derives from an analogous relation that involves electric circuits[3] which relates the total effective impedance of a triangle shaped electric circuit to that of a tri-star circuit, see fig. 2. In Schwinger's formulation the total proper time of the graph is in one-to-one correspondence with

[1] I congratulate prof. Dereli on his birthday for his prolific career and for setting up an example with high standards for the younger generations of mathematical physicists.

[2] Earlier work relating matrix quantum mechanics and 2D non-critical string theory [12] relies on a similar type of duality.

[3] See for example, [14] for a precise account of the map between Feynman graphs and electric circuits.

the holographic direction of the dual gravity theory[4] [1] and the star-triangle relation becomes a clear manifestation of the gauge-string duality, or open-closed duality in string theory where the gauge theory 3-point function is represented by the triangle and the corresponding Witten diagram [3] in the dual theory is represented by the tri-star. See [18] for a more recent work, based on a different approach, that also derives dual gravity theory directly from field theory, in the case of vector models [19].

In this paper we suggest that a generalization of the star-triangle type duality of Feynman graphs into interacting field theories might be a fundamental manifestation of the gauge-string duality and a key to generalize it beyond the known specific cases[5]. In particular, we generalize Gopakumar's derivation of Witten's diagrams from free field theory to interacting theories. As a prototype, we take a real, massless N×N matrix-valued scalar field Φ with Φ^h interaction, for integer $h > 2$, in d-dimensions, and consider the two-point function $\langle\Phi(x_1)\Phi(x_2)\rangle$. The two-point function in the large N limit is given by Feynman graphs summed over the number of independent quantum loops ℓ which, in the large N limit, can further be classified in terms of 2D Euclidean Riemann surfaces embedded in d-dimensions. We show that each term in the sum over ℓ can be mapped onto a boundary-to-boundary propagator of a scalar field with mass m related to ℓ, in $d+1$ dimensional AdS space. This provides a dual "closed string" picture of the two-point function in terms of a *generalized Witten diagram* given by sum over AdS Witten diagrams. Even though we perform our calculations in a simple scalar ungauged theory, we will be assuming that our findings generalize to theories like $\mathcal{N} = 4$ super-Yang-Mills without conceptual difficulties.

In the next section we review Gopakumar's construction in free field theory. In section 3 we first review Schwinger's proper time formulation for interacting field theories and then generalize Gopakumar's computation to finite coupling. We end in section 4 by discussing open issues, generalization to higher point functions and an overlook.

2. Open-closed and star-triangle

The AdS/CFT correspondence originates from an equivalence between open and closed string descriptions of a set of D3 branes in IIB string theory [2]. Loosely speaking, and in the simplest case, this can be understood geometrically as in fig. 1 which depicts an equivalence between one-loop partition function of open strings in d-dimensions and propagation of a closed string in $d+1$ dimensions [23]. In the low energy limit where the massive string states decouple, open strings on N coincident D3 branes are effectively described by 4D $U(N)$ Yang-Mills gauge theory theory. On the other hand the closed string in fig. 1 turns out to propagate in the $\mathrm{AdS}_5 \times S^5$ geometry which is generated by the backreaction of the brane system. More precisely, the n-point function of gauge invariant operators in the Yang-Mills theory is given in terms of the closed string world-sheet path integral

$$\langle\mathcal{O}_1(k_1)\cdots\mathcal{O}_n(k_n)\rangle_g = \int_{M_{g,n}} \langle\mathcal{V}_1(k_1,z_1)\cdots\mathcal{V}_n(k_n,z_n)\rangle_{w.s.}\,, \tag{1}$$

where the subscript g on the RHS denotes the genus-g contribution to the Feynman graphs and $\mathcal{V}$s are the closed string vertex operators which correspond to gauge theory operators on the LHS. The integral is over the moduli of Riemann surfaces with genus g and n punctures.

To demonstrate this equivalence at the level of Feynman graphs, one must show, how the holes on the open string (gauge theory) side are glued together and generate closed string world-sheets with n-punctures. This mechanism was first proposed by 't Hooft [24] in the double scaling limit,

$$g_{YM} \to 0\,, \qquad N \to \infty\,, \qquad g^2_{YM}N = \lambda\,, \tag{2}$$

[4] See also [15, 16, 17].
[5] e.g. based on D-brane descriptions [20], lower dimensional examples [21, 22] and vector models [18, 19].

Figure 1. Equivalence between one-loop diagram of an open string and tree level propagation of a closed string.

where g_{YM} is the Yang-Mills coupling constant. Emergence of a dual description in this limit can be made explicit in 2D string theory, where the quantum mechanics of $N \times N$ hermitean matrices become dual to 2D non-critical string theory, see for example [21].

A strong indication that the same "gluing" happens in higher dimensional *free* field theories was noted in [1] utilising the proper time formulation of n-point functions, which we review below.

Schwinger's proper time formulation makes the point-like feature of QFT manifest. In particular, correlators of a quantum field are represented by propagation of quantum mechanical particle in proper time τ embedded in space-time as world-line $x^\mu(\tau)$. To see this one exponentiates the denominator in the two-point function

$$\langle \Phi(x_1)\Phi(x_2)\rangle = i\int d^dk \frac{e^{ik(x_1-x_2)}}{k^2+m^2-i\epsilon} = \int_0^\infty d\tau \langle x_1|e^{-i\tau(-\partial^2+m^2)}|x_2\rangle\,. \tag{3}$$

The RHS is nothing else but the path integral of a particle propagating in τ with hamiltonian $H_{pp} = k^2+m^2$. The integral over τ is moduli — a consequence of the reparametization invariance of the worldsheet [6]. A generalization of this representation to n-point functions in a free field theory involves introduction of vertex operators inside the path integral

$$\langle \phi(x_1)\cdots\phi(x_n)\rangle = \int_0^\infty \frac{d\tau}{\tau}\prod_{i=1}^n d\tau_i \langle e^{ik_1\hat{X}(\tau_1)}\dots e^{ik_n\hat{X}(\tau_n)}\rangle_{q.m.} \tag{4}$$

where the RHS is the path integral with the point particle hamiltonian $H_{pp} = k^2+m^2$. The integral is over the moduli of the Feynman graph given by the total proper time for the process and proper times at insertions of the vertex operators. Note the structural similarity between (1) and (4) which already implies the utility of the Schwinger's formulation to explore the basic mechanism behind the gauge-string duality.

As the path integral in (4) is Gaussian for free field theory, one can compute it explicitly [25] and express the result solely in terms of moduli integrals. More interestingly, one can find a judicious change of variables of moduli to reformulate the result in terms of propagators of scalar fields in AdS$_{d+1}$ [1, 17]. Consider $\mathcal{N}=4$ super-Yang-Mills at large N and in the free limit $\lambda = 0$, see (2). For the purpose of demonstration let us consider the simplest non-trivial case of the 3-point function and the operator $\mathrm{tr}\,\Phi^2$ where Φ is one of the 6 scalars in the theory. There is a single diagram that contributes to the connected 3-point function $\langle \mathrm{tr}\,\Phi^2(k_1)\Phi^2(k_2)\Phi^2(k_3)\rangle$ that is shown on the left figure in fig. 2.

Introducing a change of variables [1] $\alpha_i = \epsilon_{ijk}|\tau_j-\tau_k|/\tau$ from the moduli τ_i to Schwinger parameters one can rewrite the connected 3-point function as follows

$$\Omega(k_1,k_2,k_3) \propto \delta^d(\sum k_i)\int_0^\infty d\tau \int_0^1 \prod_{i=1}^3 d\alpha_i\,\delta(\sum \alpha_i - 1)\,e^{-\tau(k_1^2\alpha_2\alpha_3+k_2^2\alpha_3\alpha_1+k_3^2\alpha_1\alpha_2)} \tag{5}$$

[6] which can be removed by introducing an auxiallry worldsheet einbein $g^{\tau\tau}$ in the path integral.

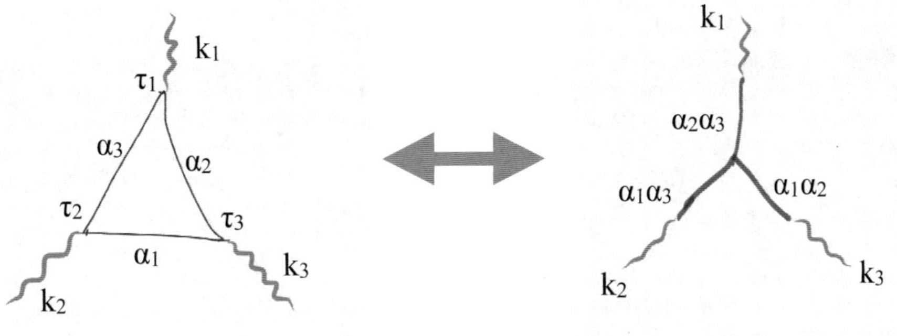

Figure 2. Star-triangle duality in free field theory. LHS shows the only Feynman graph that contributes to the 3-point function of $\mathrm{tr}\,\Phi^2$ where k_i are external momenta, τ_i are the moduli and α_i are the Schwinger parameters. RHS shows its equivalent under the duality.

This is precisely in the form given by product of three propagators with dual Schwinger parameters $\alpha_1\alpha_2$ etc. as shown on the RHS of fig. 2. This procedure explicitly achieves the "gluing" mentioned above in the sense that the hole on the "open string side" i.e. the LHS of fig. 2 is closed up on the "closed string side" i.e. the RHS of fig. 2. The RHS also resembles the Witten diagram for the 3-point function in AdS and this resemblance can be made precise by another change of variables $\alpha_i = \rho_i / \sum_{j=1}^{3} \rho_j$ [1] and defining the radial coordinate of the AdS space z_0 in terms of these Schwinger moduli as

$$z_0^2 = 4\tau \left(\sum_{i=1}^{3} \rho_i \right) \prod_{i=1}^{3} \alpha_i \,. \tag{6}$$

This results in the final expression after Fourier transforming to space-time as

$$\Omega(x_1, x_2, x_3) \propto \int_0^\infty \frac{dz_0}{z_0^{d+1}} \int d^d z \prod_{i=1}^{3} K_{\Delta_i}(x_i; z, z_0) \tag{7}$$

where $K_\Delta(x_i, y; z)$ are the boundary-to-bulk propagator for a scalar with mass $m^2 = \Delta(d-\Delta)$, with $\Delta = 2$, corresponding to $\mathrm{tr}\,\Phi^2$ operator in AdS_{d+1} on the Poincaré patch

$$ds^2 = \frac{1}{z_0^2} \left(dz_0^2 + \eta_{ab}\, dz^a dz^b \right) . \tag{8}$$

This computation can be generalized to an arbitrary string of Φ fields [15], presumably to other $\mathcal{N} = 4$ super-Yang-Mills operators and higher point functions [17].

3. Generalization to interacting theories

The derivation of AdS propagators from Feynman graphs in the free case, presented above, carries over to the interacting QFTs [26] to a large extent. For simplicity, we will discuss two-point functions in a massless matrix valued scalar field Φ in d-dimensions with an interaction potential Φ^h. The action is

$$\mathcal{S} = \int d^d x \,\mathrm{Tr} \left(-\frac{1}{2} (\partial\Phi)^2 + \mathfrak{g}\Phi^h \right) , \tag{9}$$

where $h > 2$ is the coordination number of the vertex associated with the interaction term Φ^h and $\mathfrak{g}$ is the coupling constant analogous to g_{YM}^2 in (2). After the rescaling $\Phi \to \sqrt{N}\Phi$ one obtains

$$\mathcal{S} = N \int d^d x \, \mathrm{Tr} \left(-\frac{1}{2}(\partial\Phi)^2 + \lambda \Phi^h \right) , \tag{10}$$

where $\lambda \equiv N^{(h-2)/2}\mathfrak{g}$ is the 't Hooft coupling.

We are interested in computing correlation functions of scalar fields in (10). We consider a Feynman diagram F of genus g_0, with I internal lines, V vertices and $\ell = I - V + 1 - 2g_0$ independent loops[7].

Schwinger's proper time formulation can be generalized [14] to express the value, Ω_F, of this generic graph F in terms of integrals of Schwinger parameters a_r, $r = 1, \cdots I$ introduced for each internal line[8] (analogous to τ in (3)):

$$\Omega_F(\vec{k}) = \delta^{(d)} (k_1 + \cdots + k_{n_e}) \int_0^\infty \left(\prod_{r=1}^{I} da_r \right) \mathcal{U}(a_r)^{-d/2} \, e^{-P\left(a_r;\vec{k}\right)} . \tag{11}$$

This is the expression for the amputated graph in Euclidean time with n_e external momenta which we collectively denote as $\vec{k} = (k_1, \ldots, k_{n_e})$. The non-trivial ingredients here are the *Symanzik polynomials*

$$\mathcal{U}(a) \equiv \sum_{T_1 \in \mathcal{T}_1} \prod_{r \notin T_1}^{\ell} a_r , \qquad P(a_r; \vec{k}) \equiv \frac{1}{\mathcal{U}(a_r)} \sum_{T_2 \in \mathcal{T}_2} \left(\prod_{r \notin T_2}^{\ell+1} a_r \right) \left(\sum_{b \in \mathcal{J}} k_b \right)^2 , \tag{12}$$

with $\mathcal{T}_1, \mathcal{T}_2$ being the sets of trees and 2-trees respectively[9], and $\mathcal{J}$ is one of the two disconnected components of a 2-tree.

Applying this to the connected two-point function and after a series of change of variables [26] and Fourier transform to position space one arrives at the following compact expression

$$\Omega_F(x,y) = 4^{\Delta-2} \pi^{d/2} \Gamma(\Delta) \frac{\mathcal{V}_F}{|x-y|^{2\Delta}} . \tag{13}$$

This has precisely the same form as a CFT$_d$ two-point function $\langle \mathcal{O}_\Delta \mathcal{O}_\Delta \rangle$ of a conformal field $\mathcal{O}_\Delta$ with the scaling dimension

$$\Delta = \left(\frac{d}{2} - \frac{h}{h-2} \right) \ell + \frac{d}{2} + 1 , \tag{14}$$

where we expressed the variables I and V in terms of ℓ using Euler's theorem for genus-0 graphs and an additional relation between I, V and the coordination number of the vertex h. It is tempting to call this quantity as the "conformal dimension of the Feynman graph" ℓ. Note that it does not depend on the detailed structure of the graph but only on the number of independent loops ℓ.

Moreover, the entire dependence on the moduli a_r is contained in the overall coefficient $\mathcal{V}_F$ in (13) that is given by

$$\mathcal{V}_F \equiv \ell \int_0^\infty \left(\prod_{r=1}^{I} da_r \right) \delta\left(1 - \mathcal{U}_F(a)\right) \mathcal{A}_F(a)^{\Delta - d/2} , \qquad \mathcal{A}_F(a) \equiv \sum_{T_2 \in \mathcal{T}_2} \prod_{r \notin T_2}^{\ell+1} a_r . \tag{15}$$

[7] We use the double line notation of 't Hooft implicitly. A "line" actually refers to a double-line in the discussion below.

[8] We suppress the matrix indices in what follows.

[9] If one removes ℓ internal lines from F such that there are no loops left, the remaining graph can be shown to be a simply-connected subgraph of F, which is called a *tree*, T_1. If $\ell + 1$ lines are removed then one is left with two disconnected components (trees) with no loops, called a *2-tree*, T_2.

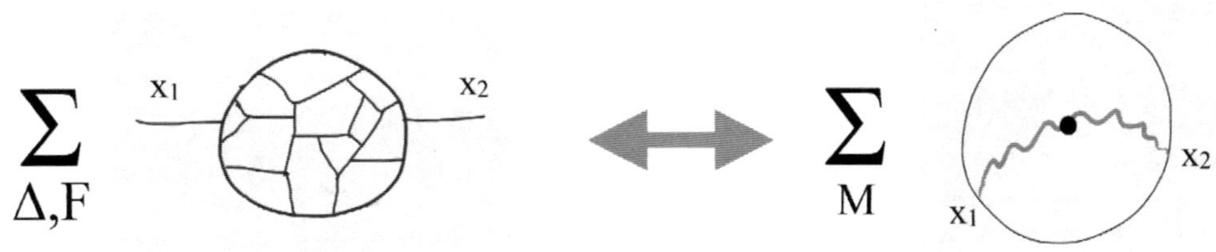

Figure 3. A schematic description of the duality between Feynman graphs and Witten diagrams. On the LHS the sum is over the scale dimension of a graph Δ, equivalently over the number of loops ℓ, see equation (14), and over all Feynman graphs with the same Δ. On the RHS the sum is over two-point Witten diagrams that invovle a bulk scalar with mass $M = \Delta(d-\Delta)$.

This coefficient depends on the particular Feynman graph through the sets $\mathcal{T}_1, \mathcal{T}_2$ in the Symanzik polynomials specific to a given graph F.

The fact that the contribution to the two-point function is in the CFT form (13) immediately suggests, through the standard AdS/CFT prescription, that the full two-point function can be written as a sum over two-point Witten diagrams as shown in fig. 3. This can indeed be shown explicitly [26] again following a judicious change of variables similar to section 2. The Witten diagram corresponding to the graph F, equation (13) reads[10]

$$\Omega_F(x_1, x_2) = \mathcal{V}'_F \lim_{\epsilon\to 0} \epsilon \int \frac{dz_0 d^d z}{z_0^{1+d}} \, z^{2\epsilon} \, K_{\Delta+\epsilon}(x_1; z, z_0) K_{\Delta+\epsilon}(x_2; z, z_0) \, , \tag{16}$$

where we defined a new coefficient $\mathcal{V}'_F = \mathcal{V}_F \pi^d 4^\Delta \Gamma(\Delta - \frac{d}{2})$ and introduced the AdS bulk-to-boundary propagator

$$K_\Delta(x; z, z_0) = \frac{\Gamma(\Delta)}{\pi^{d/2}\Gamma(\Delta - \frac{d}{2})} \frac{z_0^\Delta}{\left(z_0^2 + (x-z)^2\right)^\Delta} \, , \tag{17}$$

of a bulk field with mass $M = \Delta(d-\Delta)$. It is straightforward to check that the limit in (16) is finite.

The full perturbative two-point function is then given by the sum over the number of independent quantum loops ℓ of the contributing connected Feynman graphs

$$\Omega(x_1, x_2) = N \lim_{\epsilon\to 0} \sum_{\ell=0}^{\infty} \lambda^{2\ell/(h-2)} \, \mathcal{C}_\ell \, \epsilon \int_{AdS} z_0^{2\epsilon} K_{\Delta+\epsilon}(x_1; z, z_0) K_{\Delta+\epsilon}(x_2; z, z_0) \;\, , \tag{18}$$

where the coefficient

$$\mathcal{C}_\ell \equiv \sum_{F\in\mathcal{F}_\ell} \frac{\mathcal{V}'_F}{\sigma_F} \tag{19}$$

is given by the sum over all connected Feynman graphs with ℓ independent loops and the symmetry factor σ_F. This is our final expression for the two-point function expressed in terms of AdS propagators. It is a strikingly simple expression where all the complication from distinct Feynman graphs is absorbed in a single coefficient $\mathcal{C}_\ell$.

[10] The proportionality constant involves a UV cut-off, for the more precise expression we refer to [26].

4. Discussion

Our final result (18) contains divergences and should be understood as a formal expression. Apart from the standard issue of Borel summability of perturbative expansion in QFT, summands in (18) contain multiple UV and IR divergences contained in the coefficients $\mathcal{V}'_F$. The UV divergences all arise from independent loops in the Feynman graphs, in the limit $a_r \to 0$ of Schwinger parameters. These can be regulated in the standard manner by renormalizing the two-point function. The IR divergences, on the other hand, arise from the other boundary limits $a_r \to \infty$ which can also be regularized by introducing a small mass term in (10). These issues will be detailed in [26].

Independent of its holographic interpretation, (18) is interesting in the sense that it corresponds to an alternative form of the Källén-Lehmann representation of the two-point function in massless QFT. Indeed the sum over ℓ can easily be turned into an integral over Δ which labels excitations in the Hilbert space. Then the density of these states are determined by the coefficient in (19). The results we obtained here for the two-point function seem to generalize to higher point-functions barring some technical difficulties that will be addressed in [26]. In case of the three-point function this construction implies an interacting field theory analog of the star-triangle duality, fig. 2, which could be playing a central role in derivation of the gauge-string correspondence for generic holographic QFTs.

It is also tempting to speculate on the holographic interpretation of this formula. Assuming that the QFT we started with is holographic — for example $\mathcal{N} = 4$ super-Yang-Mills or its marginal or relevant deformations — then the RHS of (18) would correspond to a closed string propagator in a curved space-time that is asymptotically AdS_{d+1}. In the large N limit we are considering here, this is given by the genus-0 contribution to the world-sheet path integral in (1). Is it then possible to interpret the sum, (18), over ℓ as contribution of different string states to the string propagator in this curved spacetime? Is this space-time a solution to $(d+1)$ dimensional non-critical string theory? Is there a saddle point in this sum (which can be expressed as an integral over Δ) that corresponds to the gravity limit analogous to $\lambda \to \infty$ in $\mathcal{N} = 4$ super-Yang-Mills? Can we then read off the metric in this space-time from the coefficients $\mathcal{C}_\ell$? Would it then be possible to use our approach as an operational definition of the holographic dual to a given QFT e.g. QCD?

These are fascinating questions answers to which lie beyond our reach today. We hope that the approach we outlined in this paper will be instrumental for a deeper understanding of holographic duality and will shed new light on these questions.

References

[1] R. Gopakumar, Phys. Rev. D **70** (2004), 025009 doi:10.1103/PhysRevD.70.025009 [arXiv:hep-th/0308184 [hep-th]].
[2] J. M. Maldacena, Adv. Theor. Math. Phys. **2** (1998), 231-252 doi:10.1023/A:1026654312961 [arXiv:hep-th/9711200 [hep-th]].
[3] E. Witten, Adv. Theor. Math. Phys. **2** (1998), 253-291 doi:10.4310/ATMP.1998.v2.n2.a2 [arXiv:hep-th/9802150 [hep-th]].
[4] S. S. Gubser, I. R. Klebanov and A. M. Polyakov, Phys. Lett. B **428** (1998), 105-114 doi:10.1016/S0370-2693(98)00377-3 [arXiv:hep-th/9802109 [hep-th]].
[5] S. Ryu and T. Takayanagi, Phys. Rev. Lett. **96** (2006), 181602 doi:10.1103/PhysRevLett.96.181602 [arXiv:hep-th/0603001 [hep-th]].
[6] J. de Boer, E. P. Verlinde and H. L. Verlinde, JHEP **08** (2000), 003 doi:10.1088/1126-6708/2000/08/003 [arXiv:hep-th/9912012 [hep-th]].
[7] I. Heemskerk and J. Polchinski, JHEP **06** (2011), 031 doi:10.1007/JHEP06(2011)031 [arXiv:1010.1264 [hep-th]].
[8] S. S. Lee, JHEP **01** (2014), 076 doi:10.1007/JHEP01(2014)076 [arXiv:1305.3908 [hep-th]].
[9] D. Harlow, PoS **TASI2017** (2018), 002 doi:10.22323/1.305.0002 [arXiv:1802.01040 [hep-th]].
[10] A. Almheiri, X. Dong and D. Harlow, JHEP **04** (2015), 163 doi:10.1007/JHEP04(2015)163 [arXiv:1411.7041 [hep-th]].
[11] P. Hayden, S. Nezami, X. L. Qi, N. Thomas, M. Walter and Z. Yang, JHEP **11** (2016), 009 doi:10.1007/JHEP11(2016)009 [arXiv:1601.01694 [hep-th]].

[12] V. A. Kazakov, A. A. Migdal and I. K. Kostov, Phys. Lett. B **157** (1985), 295-300 doi:10.1016/0370-2693(85)90669-0
[13] J. S. Schwinger, Phys. Rev. **82** (1951), 664-679 doi:10.1103/PhysRev.82.664
[14] C. S. Lam and J. P. Lebrun, Nuovo Cim. A **59** (1969), 397-421 doi:10.1007/BF02753153
[15] R. Gopakumar, Phys. Rev. D **70** (2004), 025010 doi:10.1103/PhysRevD.70.025010 [arXiv:hep-th/0402063 [hep-th]].
[16] R. Gopakumar, Phys. Rev. D **72** (2005), 066008 doi:10.1103/PhysRevD.72.066008 [arXiv:hep-th/0504229 [hep-th]].
[17] R. Gopakumar, Comptes Rendus Physique **5** (2004), 1111-1119 doi:10.1016/j.crhy.2004.10.004 [arXiv:hep-th/0409233 [hep-th]].
[18] O. Aharony, S. M. Chester and E. Y. Urbach, JHEP **03** (2021), 208 doi:10.1007/JHEP03(2021)208 [arXiv:2011.06328 [hep-th]].
[19] S. S. Gubser, I. R. Klebanov and A. M. Polyakov, Nucl. Phys. B **636** (2002), 99-114 doi:10.1016/S0550-3213(02)00373-5 [arXiv:hep-th/0204051 [hep-th]].
[20] O. Aharony, S. S. Gubser, J. M. Maldacena, H. Ooguri and Y. Oz, Phys. Rept. **323** (2000), 183-386 doi:10.1016/S0370-1573(99)00083-6 [arXiv:hep-th/9905111 [hep-th]].
[21] I. R. Klebanov, [arXiv:hep-th/9108019 [hep-th]].
[22] J. Maldacena and D. Stanford, Phys. Rev. D **94** (2016) no.10, 106002 doi:10.1103/PhysRevD.94.106002 [arXiv:1604.07818 [hep-th]].
[23] A. M. Polyakov, Contemp. Concepts Phys. **3** (1987), 1-301
[24] G. 't Hooft, Nucl. Phys. B **72** (1974), 461 doi:10.1016/0550-3213(74)90154-0
[25] M. J. Strassler, Nucl. Phys. B **385** (1992), 145-184 doi:10.1016/0550-3213(92)90098-V [arXiv:hep-ph/9205205 [hep-ph]].
[26] D. Gallegos, U. Gürsoy, N. Zinnato, to appear.

Journal of Physics: Conference Series **2191** (2022) 012013 doi:10.1088/1742-6596/2191/1/012013

Integrability of Three Dimensional Gravity Field Equations

Metin Gürses

Department of Mathematics, Faculty of Sciences, Bilkent University, 06800 Ankara, Turkey

E-mail: `gurses@fen.bilkent.edu.tr`

Abstract. We show that the tree dimensional Einstein vacuum field equations with cosmological constant are integrable. Using the $sl(2, R)$ valued soliton connections we obtain the metric of the spacetime in terms of the dynamical variables of the integrable nonlinear partial differential equations.

1. Introduction

Gravity in three dimensions is highly simple. It has no degrees of freedom and the constraint equations can be explicitly solved. The main motivation to study three dimensional gravity was quantization as a toy model. For this purpose, the Einstein-Hilbert action was written as a Chern-Simons action with a suitable gauge group [1], [2]. Very recently [3], this property has been used to construct highly nontrivial exact solutions of the Einstein field equations with cosmological constant. The method presented here uses the zero curvature formalism of the integrable systems. The soliton connection is the Chern-Simons gauge potential with $SL(2, R)$ group as the gauge group. The soliton equations corresponding to this gauge connection are known as the AKNS equations containing the nonlinear Schrodinger and modified Korteweg de Vries equations.

In this work we shall show that the field equations

$$R_{\mu\nu} - \frac{1}{2} R g_{\mu\nu} = \Lambda g_{\mu\nu},$$

where Λ is the cosmological constant and $\mu, \nu = 1, 2, \cdots D$ are integrable when $D = 3$, by using the Newmann-Penrose null tetrad formalism. We find that SL(2,R) valued tetrad one form is the difference of two zero curvature connections without referring to the Chern-Simons theory.

Before starting to write the gravitational field equations in null tetrad formalism we give a brief introduction to the sl(2,R) valued soliton connections and AKNS sytem. Let $x^\mu = (t, x, y)$ be the local coordinates of (M, g) and the soliton connection $\mathbf{a}$ depends on the coordinates t and x. Hence

$$\mathbf{a} = P\, dx + Q\, dt,$$

where $P(t,x)$ and $Q(t,x)$ are some sl(2,R) matrices. They are given by

$$P = \begin{pmatrix} 2\xi & p(t,x) \\ q(t,x) & -2\xi \end{pmatrix}, Q = \begin{pmatrix} A & B \\ C & -A \end{pmatrix}$$

Here ξ is the eigenvalue parameter and p and q are dynamical variables which satisfy nonlinear integrable system of equations of evolutionary type. The functions A,B, and C depend on the dynamical variables p and q and their partial derivatives with respect to x.

Since the soliton connection **a** has no curvature then matrices P and Q satisfy the zero curvature condition

$$P_{,t} - Q_{,x} + [P,Q] = 0,$$

which gives the well known AKNS equations [4]

$$A_x - p\,C + q\,B = 0,$$
$$q_t + C_x - 2qA - 2\xi C = 0,$$
$$p_t + B_x + 2pA + 2\xi B = 0,$$

We expand A, B and C as polynomials of the eigenvalue parameter ξ, i.e.,

$$A = \sum_{n=0}^{N} A_n\,\xi^{N-n}, \quad B = \sum_{n=0}^{N} B_n\,\xi^{N-n}, \quad C = \sum_{n=0}^{N} C_n\,\xi^{N-n},$$

where the functions A_n. B_n and C_n depend on the dynamical variables p and q and their partial derivatives withe respect to x.

The AKNS System: Taking $N = 2,3$ we obtain the Nonlinear Schrodinger and modified Korteweg de Vries systems respectively. This means that when we begin with the Lax pair in $sl(2,R)$ algebra and assume them as a polynomial of the spectral parameter of degree less or equal to three then we obtain the following system of evolution equations :

$$q_t = a_2\,(-\frac{1}{2}\,q_{xx} + q^2\,p) + a_3\,(-\frac{1}{4}\,q_{xxx} + \frac{3}{2}\,qpq_x), \tag{1}$$

$$p_t = a_2\,(\frac{1}{2}\,p_{xx} - q\,p^2) + a_3\,(-\frac{1}{4}\,p_{xxx} + \frac{3}{2}\,qpp_x). \tag{2}$$

Here a_2 and a_3 are arbitrary constants. For general N we have a recursion relation $u_t = \mathcal{R}^N u_x$ ($N = 2,3\cdots$) where $u = (p,q)^T$ and $\mathcal{R}$ is the recursion operator of the AKNS system. All the members of the AKNS system have infinitely many conserved quantities, infinitely many symmetries, bi-Hamiltonian structure, and admit soliton solution generating techniques etc.

2. Null tetrad approach

Newmann-Penrose formalism [5] in four dimensions is much elegant when written in sl(2, C) algebra. It is known that the Einstein field equations admit prolonged Cartan frames [6] in this formalism. We start first with SL(2,C) valued tetrad one form in four dimensions

$$\sigma = \begin{pmatrix} \mathbf{l} & \mathbf{m} \\ \bar{\mathbf{m}} & \mathbf{n} \end{pmatrix}, \tag{3}$$

where $\mathbf{l}, \mathbf{n}, \mathbf{m}, \bar{\mathbf{m}}$ are null tetrad one forms. The metric takes the form

$$ds^2 = \mathbf{l} \oplus \mathbf{n} + \mathbf{n} \oplus \mathbf{l} + \mathbf{m} \oplus \bar{\mathbf{m}} + \bar{\mathbf{m}} \oplus \mathbf{m}. \tag{4}$$

Here a bar over a letter denotes complex conjugation. In our approach we will use the following SL(2,C) tetrad one form

$$\tilde{\sigma} = \begin{pmatrix} \mathbf{n} & -\bar{\mathbf{m}} \\ -\mathbf{m} & \mathbf{l} \end{pmatrix} = \varepsilon \sigma \varepsilon^{-1}, \tag{5}$$

where

$$\varepsilon = \begin{pmatrix} 0 & 1 \\ -1 & 0 \end{pmatrix}. \tag{6}$$

Spin Connection: w

$$d\tilde{\sigma} = -w\,\tilde{\sigma} + \tilde{\sigma}\,w^{\dagger}, \tag{7}$$

where w is the sl(2,C) valued connection one form given as

$$w = \begin{pmatrix} w_0 & w_2 \\ w_1 & -w_0 \end{pmatrix}, \tag{8}$$

where

$$w_0 = \gamma\mathbf{l} + \epsilon\mathbf{n} - \alpha\mathbf{m} - \beta\bar{\mathbf{m}}, \tag{9}$$
$$w_1 = -\tau\mathbf{l} + -\kappa\mathbf{n} + \rho\mathbf{m} + \sigma\bar{\mathbf{m}}, \tag{10}$$
$$w_2 = \nu\mathbf{l} + \pi\mathbf{n} - \lambda\mathbf{m} - \mu\bar{\mathbf{m}}, \tag{11}$$

where $\gamma, \epsilon, \alpha, \beta, \cdots$ are the Newman-Penrose spin coefficients. In the construction of the metric tensor we don't need these spin coefficients. In the sequel of the text we may use these Greek letters for different purposes.

Curvature two form: $\mathbf{R} = dw + w\,w$

$$\mathbf{R} = \begin{pmatrix} R_0 & R_2 \\ R_1 & -R_0 \end{pmatrix}, \tag{12}$$

where

$$R_0 = (\Lambda - \Phi_{11} + \psi_2)\mathbf{l}\,\mathbf{n} + \psi_3\mathbf{lm} + \Phi_{12}\mathbf{n}\bar{\mathbf{m}} - \psi_1\mathbf{n}\bar{\mathbf{m}}$$
$$+(\psi_2 - \Phi_{11} - \Lambda)\mathbf{m}\bar{\mathbf{m}}, \tag{13}$$
$$R_1 = -(\psi_1 + \Phi_{01})\mathbf{l}\,\mathbf{n} - (\psi_2 + 2\Lambda)\mathbf{l}\,\mathbf{m} - \Phi_{02}\mathbf{l}\bar{\mathbf{m}} + \Phi_{00}\mathbf{nm} + \psi_0\mathbf{n}\bar{\mathbf{m}}$$
$$+(\Phi_{01} - \psi_1)\mathbf{m}\bar{\mathbf{m}} \tag{14}$$
$$R_2 = (\psi_3 + \Phi_{21})\mathbf{l}\,\mathbf{n} - (\psi_2 + 2\Lambda)\mathbf{n}\bar{\mathbf{m}} - \Phi_{20}\mathbf{nm} + \Phi_{22}\mathbf{l}\,\bar{\mathbf{m}} + \psi_4\mathbf{lm}$$
$$-(\Phi_{21} - \psi_3)\mathbf{m}\bar{\mathbf{m}}, \tag{15}$$

where Φ_{ij} $(i,j = 0,1,2)$ are the trace free Ricci spin coefficients, ψ_a $(a = 0,1,2,3,4)$ are the Weyl spinors and $\Lambda = -R/6$, where R is the scalar curvature. Let $\Phi_{ij} = 0$ and $\psi_a = 0$ for all $(i,j = 0,1,2)$ and for all $(a = 0,1,2,3,4)$,then

$$\mathbf{R} = \Lambda \begin{pmatrix} \mathbf{ln} - \mathbf{m}\bar{\mathbf{m}} & -2\mathbf{n}\bar{\mathbf{m}} \\ -2\mathbf{lm} & -\mathbf{ln} + \mathbf{m}\bar{\mathbf{m}} \end{pmatrix}. \tag{16}$$

Now let

$$\Gamma_{\pm} = w \pm \ell_0\tilde{\sigma}\varepsilon, \tag{17}$$

Journal of Physics: Conference Series **2191** (2022) 012013 doi:10.1088/1742-6596/2191/1/012013

where ℓ_0 is any constant. Curvatures $\Omega_\pm$ of the connection $\Gamma_\pm$ are given by

$$\Omega_\pm = \mathbf{R} + \ell_0^2 \tilde{\sigma}\varepsilon\,\tilde{\sigma}\varepsilon \pm \ell_0 \tilde{\sigma}\left(w^\dagger \varepsilon + \varepsilon\, w\right). \tag{18}$$

To have $\Omega_\pm = 0$ we must have $\bar{w} = w$ and $\bar{\mathbf{m}} = \mathbf{m}$. In this case the group SL(2,C) reduces to SL(2,R) and dimension reduces to three and

$$\ell_0^2 = -\Lambda. \tag{19}$$

As a result we end up with a theorem saying that, the integrability of 4 dimensional gravity implies that null tetrad must be real and the spacetime dimension falls down to three and hence SL(2,C) group is contracted to SL(2,R). This is an Inönü-Wigner contraction [7].

3. Newman-Penrose Formalism in Three Dimensions

Following [8] SL(2,R) valued tetrad one form is given by

$$\tilde{\sigma} = \begin{pmatrix} \mathbf{n} & -\frac{1}{\sqrt{2}}\mathbf{m} \\ -\frac{1}{\sqrt{2}}\mathbf{m} & \mathbf{l} \end{pmatrix} = \varepsilon\sigma\varepsilon^{-1} \tag{20}$$

Connection one form: w

$$d\tilde{\sigma} = -w\,\tilde{\sigma} + \tilde{\sigma}\,w^t, \tag{21}$$

where w is the sl(2,R) valued connection one form given as

$$w = \begin{pmatrix} w_0 & w_2 \\ w_1 & -w_0 \end{pmatrix}, \tag{22}$$

where

$$w_0 = \frac{1}{2}(-\epsilon'\mathbf{l} + \epsilon\mathbf{n} - \alpha\mathbf{m}), \tag{23}$$

$$w_1 = \frac{1}{\sqrt{2}}(-\tau\mathbf{l} - \kappa\mathbf{n} + \sigma\mathbf{m}), \tag{24}$$

$$w_2 = \frac{1}{\sqrt{2}}(-\kappa'\mathbf{l} - \tau'\mathbf{n} + \sigma'\mathbf{m}), \tag{25}$$

where $\gamma, \epsilon, \alpha, \beta, \cdots$ are the Newman-Penrose spin coefficients. Then

$$d\ell = -\epsilon\mathbf{ln} + (\alpha - \tau)\mathbf{lm} + \kappa'\mathbf{nm}, \tag{26}$$

$$dn = \epsilon'\mathbf{ln} - \kappa'\mathbf{lm} - (\alpha + \tau')\mathbf{nm}, \tag{27}$$

$$dm = (\tau' - \tau)\mathbf{ln} - \sigma'\mathbf{lm} - \sigma\mathbf{nm}. \tag{28}$$

Curvature two form: $\mathbf{R} = dw + w\,w$

$$\mathbf{R} = \begin{pmatrix} R_0 & R_2 \\ R_1 & -R_0 \end{pmatrix}, \tag{29}$$

where

$$R_0 = (2\Phi_{11} + \Lambda)\,\mathbf{l}\,\mathbf{n} - \Phi_{12}\,\mathbf{lm} + \Phi_{10}\,\mathbf{nm},$$

$$R_1 = -\sqrt{2}\Phi_{01}\,\mathbf{l}\,\mathbf{n} + \frac{1}{\sqrt{2}}(\Phi_{02} - 2\,\Lambda)\,\mathbf{l}\,\mathbf{m} - \sqrt{2}\Phi_{00}\,\mathbf{nm},$$

$$R_2 = \sqrt{2}\Phi_{12}\,\mathbf{l}\,\mathbf{n} - \sqrt{2}\Phi_{22}\,\mathbf{l}\,\mathbf{m} + \frac{1}{\sqrt{2}}(\Phi_{02} - 2\,\Lambda)\,\mathbf{nm},$$

Journal of Physics: Conference Series **2191** (2022) 012013 doi:10.1088/1742-6596/2191/1/012013

where Φ_{ij} $(i,j=0,1,2)$ are the trace free Ricci spin coefficients and $\Lambda=-R/6$. Let $\Phi_{ij}=0$ for all $(i,j=0,1,2)$, then

$$\mathbf{R}=\Lambda\begin{pmatrix}\mathbf{ln} & -\sqrt{2}\,\mathbf{nm}\\ -\sqrt{2}\,\mathbf{lm} & -\mathbf{ln}\end{pmatrix}, \tag{30}$$

and

$$\tilde{\sigma}\,\varepsilon=\begin{pmatrix}\frac{1}{\sqrt{2}}\mathbf{m} & \mathbf{n}\\ -\mathbf{l} & -\frac{1}{\sqrt{2}}\mathbf{m}\end{pmatrix}. \tag{31}$$

Hence

$$\tilde{\sigma}\,\varepsilon\,\tilde{\sigma}\,\varepsilon=\begin{pmatrix}\mathbf{ln} & -\sqrt{2}\,\mathbf{nm}\\ -\sqrt{2}\,\mathbf{lm} & -\mathbf{ln}\end{pmatrix} \tag{32}$$

Now let

$$\Gamma_{\pm}=w\pm\lambda\tilde{\sigma}\varepsilon, \tag{33}$$

where λ is any constant. Curvatures $\Omega_{\pm}=d\Gamma_{\pm}+\Gamma_{\pm}\Gamma_{\pm}$ of the connections $\Gamma_{\pm}$ are found as

$$\Omega_{\pm}=\mathbf{R}+\lambda^2\tilde{\sigma}\varepsilon\,\tilde{\sigma}\varepsilon\pm\lambda\tilde{\sigma}\left(w^t\,\varepsilon+\varepsilon\,w\right) \tag{34}$$

Since w is an sl(2,R) connection then $w^t\varepsilon+\varepsilon w=0$. Taking $\lambda^2=-\Lambda$. then

$$\mathbf{R}+\lambda^2\tilde{\sigma}\varepsilon\,\tilde{\sigma}\varepsilon=0$$

Hence both Γ_+ and Γ_- are zero curvature sl(2,R) connections. Then three dimensional gravity is integrable and as a result we have the soliton sl(2,R) connections

$$\Gamma_{\pm}=w\pm\lambda\tilde{\sigma}\varepsilon$$

are the zero curvature connections for three dimensional AdS spacetimes. Writing $\Gamma_{\pm}$ in a suitable form for the integrable systems we have the tetrad one form

$$\tilde{\sigma}\,\varepsilon=\frac{1}{2\lambda}(\Gamma_+-\Gamma_-) \tag{35}$$

Here in our approach the connections $\Gamma_{\pm}$ are any zero curvature soliton connections, not coming from a Chern-Simons theory. Hence they do not depend on any boundary conditions as assumed by Cardenas et al [3]. Now we shall relate the flat gauge potential one forms $\Gamma_{\pm}$ to the well known AKNS soliton connection. Let $\Gamma_+=b_+^{-1}\,db_++b_+^{-1}\,\mathbf{a}^+\,b_+$ where b_+ is nonsingular matrix and $\mathbf{a}^+$ is a soliton connection one form. Then $\Omega_{\pm}=d\Gamma_{\pm}+\Gamma_{\pm}\Gamma_{\pm}=0$ implies that $\mathbf{a}^+$ is also a flat connection

$$d\mathbf{a}^++\mathbf{a}^+\,\mathbf{a}^+=0$$

Similarly let $\Gamma_-=b_-^{-1}\,db_-+b_-^{-1}\,\mathbf{a}^-\,b_-$ where b_- is another nonsingular matrix and

$$d\mathbf{a}^-+\mathbf{a}^-\,\mathbf{a}^-=0.$$

Choosing

$$\Gamma_{\pm}=\begin{pmatrix}X_{\pm} & Y_{\pm}\\ Z_{\pm} & -X_{\pm}\end{pmatrix} \tag{36}$$

where $X_{\pm}$, $Y_{\pm}$ and $Z_{\pm}$ are one forms. Then from (35) we find that

$$\mathbf{m} = \frac{1}{\sqrt{2}\lambda}\left(X_{+} - X_{-}\right), \tag{37}$$

$$\mathbf{n} = \frac{1}{2\lambda}\left(Y_{+} - Y_{-}\right), \tag{38}$$

$$\mathbf{l} = -\frac{1}{2\lambda}\left(Z_{+} - Z_{-}\right), \tag{39}$$

4. Soliton Connection and the Metric

Let $x^{\mu} = (t, x, y)$ be the local coordinates of (M, g) and the matrices $b_{\pm}$ depend on all coordinates and the soliton connections $a^{\pm}$ depend on the coordinates t and x. Hence

$$\mathbf{a}^{\pm} = P^{\pm}\, dx + Q^{\pm}\, dt$$

where $P(t,x)$ and $Q(t,x)$ are some $sL(2,R)$ matrices. They are given by

$$P^{\pm} = \begin{pmatrix} 2\xi^{\pm} & p^{\pm}(t,x) \\ q^{\pm}(t,x) & -2\xi^{\pm} \end{pmatrix}, Q^{\pm} = \begin{pmatrix} A^{\pm} & B^{\pm} \\ C^{\pm} & -A^{\pm} \end{pmatrix}.$$

Here $\xi^{\pm}$ are the eigenvalue parameters and $p^{\pm}$ and $q^{\pm}$ are dynamical variables which satisfy nonlinear integrable system of equations of evolutionary type. The functions $A^{\pm}$, $B^{\pm}$ and $C^{\pm}$ depend on the dynamical variables $p^{\pm}$ and $q^{\pm}$ and their partial derivatives withe respect to x. Choosing the b matrices as

$$b = \begin{pmatrix} \alpha & \beta \\ \gamma & \sigma \end{pmatrix}, \tag{40}$$

with unit determinant $\alpha\sigma - \beta\gamma = 1$. Functions α, β, γ and σ depend on all coordinates t, x, y. For each zero curvature connection Γ_{+} and Γ_{-} we use different gauge matrices as $b_{\pm}$. Then we find the one forms X, Y and Z (ignoring $\pm$ subscripts) as

$$\begin{aligned} X &= [2(\alpha\sigma + \beta\gamma)\xi + \sigma\gamma p - \alpha\beta q]\, dx + [(\alpha\sigma + \beta\gamma)A + \sigma\gamma B - \alpha\beta C]\, dt + \sigma d\alpha - \beta d\gamma, \\ Y &= \left[4\sigma\beta\xi + \sigma^2 p - \beta^2 q\right] dx + \left[2\sigma\beta A + \sigma^2 B - \beta^2 C\right] dt + \sigma d\beta - \beta d\sigma, \\ Z &= \left[-4\xi\gamma\alpha - \gamma^2 p + \alpha^2 q\right] dx + \left[-2\alpha\gamma A - \gamma^2 B + \alpha^2 C\right] dt - \gamma d\alpha + \alpha d\gamma. \end{aligned}$$

Finally by using the above expressions we obtain the tetrad one forms $\mathbf{l}$, $\mathbf{n}$, $\mathbf{m}$ from (37)-(39) and the metric tensor is given by

$$g_{\mu\nu} = l_{\mu}\, n_{\nu} + l_{\nu}\, n_{\mu} - m_{\mu}\, m_{\nu}. \tag{41}$$

The only gravitational field equations are the following six AKNS equations

$$A_x^{\pm} - p^{\pm}\, C^{\pm} + q^{\pm}\, B^{\pm} = 0, \tag{42}$$

$$q_t^{\pm} + C_x^{\pm} - 2q^{\pm} A^{\pm} - 2\xi^{\pm} C^{\pm} = 0, \tag{43}$$

$$p_t^{\pm} + B_x^{\pm} + 2p^{\pm} A^{\pm} + 2\xi^{\pm} B^{\pm} = 0. \tag{44}$$

Journal of Physics: Conference Series **2191** (2022) 012013 doi:10.1088/1742-6596/2191/1/012013

5. Connections with Non-vanishing Curvatures

Let Γ_+ and Γ_- be two sl(2,R) valued connection one forms with curvatures Ω_+ and Ω_- respectively. The three dimensional spacetime geometry has an SL(2,R) valued null tetrad one form $\tilde{\sigma}$ and sl(2,R) valued connection one form w given in terms of the connections Γ_+ and Γ_- as follows

$$\tilde{\sigma} = \frac{1}{2\lambda}(\Gamma_+ - \Gamma_-)\,\varepsilon, \quad w = \frac{1}{2}\,(\Gamma_+ + \Gamma_-), \tag{45}$$

where the matrix ε is defined in (6) and λ is a constant. Then we find that

$$d\tilde{\sigma} + w\tilde{\sigma} - \tilde{\sigma}w^t = T, \tag{46}$$

where T is the sl(2,R) valued torsion two form and found as $T = \Omega_+ - \Omega_-$. The curvature two form $\mathbf{R}$ of the spacetime geometry is found as

$$\mathbf{R} + \lambda^2\,\tilde{\sigma}\varepsilon\tilde{\sigma}\varepsilon = \frac{1}{2}(\Omega_+ + \Omega_-). \tag{47}$$

The last two equations (46) and (47) imlpy a 3 dimensional gravity with torsion and nonzero matter. When we let torsion to vanish then the connections Γ_+ and Γ_- have the same curvature, i.e., $\Omega_+ = \Omega_-$. Let us call this curvature as Ω. Hence the associated three dimensional gravity field equations are

$$d\tilde{\sigma} + w\tilde{\sigma} - \tilde{\sigma}w^t = 0, \tag{48}$$

$$\mathbf{R} + \lambda^2\,\tilde{\sigma}\varepsilon\tilde{\sigma}\varepsilon = \Omega. \tag{49}$$

By choosing connections Γ_+ and Γ_- properly the above field equations may correspond to a well defined non-vacuum case. We may consider the equations (45) as Backlund transformations from vacuum to non-vacuum solutions of the Einstein field equations.

6. Conclusion

Using the null tetrad formalism in three dimensions we showed the the vacuum equations with cosmological constant are integrable. This means that we determine the tetrad one forms in terms of the curvature free connection one forms. Choosing the zero curvature connection one forms (soliton connections) properly we find the metric of the spacetime in terms of the variables of the well known AKNS system including the nonlinear Schrodinger, KdV and Modified KdV equations. In our formalism the gauge matrices $b_\pm$ are taken independent and depend on all coordinates t, x, y. In [3] these matrices are assumed to be the inverses of each other and depend only on one coordinate. Some of these functions can be eliminated by using the SL(2,R) gauge transformation but we keep them for having the metric with maximum number of free functions. Since we have derived the tetrad one form $\tilde{\sigma}$ in terms of the zero curvature connections $\Gamma_\pm$ without referring to the Chern-Simons theory we do not necessarily need any boundary conditions to be satisfied by the tetrad functions.

We have some number of comments and questions to be answered. One of them is the infinite number of symmetries and conservation laws of the AKNS system. How these be perceived in the three dimensional gravity is unclear. Another point is the recursion operator of the AKNS system. By using the recursion operator of AKNS system one can go from one system to another one. How the corresponding three dimensional metrics are related is another puzzle. Finally, how we should interpret solitons and soliton collisions in three dimensional gravity needs further efforts.

In the last part of this paper we have considered the connections with non-vanishing curvatures and discussed possibility of obtaining non-vacuum solutions.

Journal of Physics: Conference Series **2191** (2022) 012013 doi:10.1088/1742-6596/2191/1/012013

7. Acknowledgment
I would like to thank Bayram Tekin and Tahsin Çağrı Şişman for their kind interest in this work.

References
[1] Witten E, $2+1$ *dimensional gravity as an exactly soluble system*, Nucl. Phys. **B 311**,46-78 (1988)
[2] Achocarro A and Townsend P.K,*A Chern-Simons Action For Three-Dimensional Anti-de Sitter Supergravity Theories*, Phyics Letters **180B** (1986).
[3] Cardenas M, Correa F, Lara K, and Pinol M, *Integrable Systems and Spacetime Dynamic*, Physical Review Letters **127**, 161601 (2021). (arXiv:2104.09676v2)
[4] Ablowitz M.J, Kaup D.J, Newell A.C, and Segur H, *Nonlinear-Evolution Equations of Physical Significance*, Phys. Rev. Lett. 31, 125 (1973).
[5] Newman E.T and Penrose R, *An approach to gravitational radiation by a method of spin coefficients* , Journ. Math. Phys. **3**, 566 (1962); ibid **4**, 998 (1963).
[6] Gürses M, *Prolongation Structure and a Backlund Transformation for Vacuum Field Equations*, Physics Letters **101A** , (1984) 388.
[7] Inönü E, Wigner E.P, *on the Contraction of Groups and Their Representations.* Proc. Natl. Acad. Sci. **39**, 510-24 (1953).
[8] Aliev A.N and Nutku Y, *Spinor Formulation of Topologically Massive Gravity*, Class. Quantum Gravity **12**,2914-2925 (1995)

Journal of Physics: Conference Series **2191** (2022) 012014 doi:10.1088/1742-6596/2191/1/012014

Echoes from the event horizon of a superfluid vortex

K Güven[1] and B Demirkaya[2]
[1]Department of Physics, Koç University, Istanbul, Turkey
[2]Department of Physics, Izmir Institute of Technology, Izmir, Turkey
E-mail: kguven@ku.edu.tr

Abstract. A vortex formed in the superfluid state of a Bose-Einstein condensate may exhibit superradiance à la blackhole for radially propagating acoustic fluctuations. The analogy is usually based on the so-called draining bathtub model of the vortex, in which an event horizon and ergosphere emerges when the radial velocity of the superfluid exceeds the propagation speed of sound in the condensate. The acoustic fluctuations mimic a massless scalar field in the curved Lorentzian space-time of the vortex and are governed by the Klein-Gordon wave equation. One common main approximation is the constant background density of the superfluid even in the presence of the vortex. This sets a constant relativistic sound speed. However, the vortex state solution of the Gross-Pitaevskii equation clearly shows that both the density and the speed of sound vary radially near the vortex core, where the event horizon and thus the superradiance will take place. What changes would this complex interdependence bring to the formulation and to the outcomes of the superradiance based on constant density approximation? Here, we recount this question posed under the guidance of Prof. Tekin Dereli and present recent results. We show that the self-consistent density modifies the amplification dynamics near the event horizon significantly, thereby altering the temporal and spectral fingerprint of the superradiance of the vortex.

1. Introduction

The superfluid Bose-Einstein condensates demonstrated great potential to investigate physical phenomena arising from black-holes analogously in the laboratory environment [1, 2, 3, 4]. The similarity of the formulation for the propagation of electromagnetic waves in gravitational fields and for the acoustic waves in moving fluids provides a mapping of the dynamics [5]. The experimental realization of vortex states in rotating trapped atomic BECs [6, 7], brought the acoustic black-hole analogy into the BEC superfluid [8, 2]. The acoustic superradiance and even the Penrose process in both BEC superfluids and classical fluids have been suggested in the literature [9, 10, 11], and quickly gained momentum as an active research subject. Eventually, an acoustic black hole in a needle-shaped BEC of 87Rb is experimentally demonstrated and recently spontaneous Hawking radiation, stimulated by quantum vacuum fluctuations, emanating from an analogue black hole in an atomic BEC is reported [12, 13, 14].

Theoretical/numerical studies concerning acoustic superradiance mostly investigated linear acoustic perturbations superimposed on a *constant* background density of the superfluid [9, 11, 15, 16, 17]. On one hand, the constant background density approximation (const-bg-dens) combined with the draining bathtub model (DBT) [18], elegantly demonstrates the acoustic superradiance. On the other hand, the BEC theory indicates that the self-consistent background density profile of the superfluid (self-bg-dens) in the vicinity of the vortex is not constant [19].

Journal of Physics: Conference Series **2191** (2022) 012014 doi:10.1088/1742-6596/2191/1/012014

This begs the question how good the const-bg-dens approximation captures the superradiance dynamics. Here, we present a comparative numerical study of the acoustic superradiance in the const-bg-dens and in the self-bg-dens models to address this question. First and foremost, the self-bg-dens formulation introduces a *radially varying relativistic speed* near the vortex, as opposed to it being a *constant*. This fact changes the acoustic black-hole portrait entirely. In the absence of a constant relativistic speed, scale-independence is lost. Indeed, the formulation reveals interdependencies among the rotating vortex parameters and the black-hole parameters, that do not appear under the const-bg-dens approximation. For multi-quantized vortex states, this would imply further complications as the self-bg-dens changes with the vortex quantization number. We should however note that, at present, the stability and experimental realization of multiquantized vortices is challenging and still under progress.

There are two shortcomings of the present work we should mention at the onset, and point the reader to the backing literature where available: (1) The dispersion of acoustic perturbations is not included, although they are inherent. We can probably get away (but not too far!) by arguing that it restricts the applicability of the model to weak-dispersive systems. A recent study [20] investigated acoustic superradiance with Bogoliubov dispersion and with the const-bg-dens approximation and concluded the following: (i) For weak dissipations the superradiance occurs at all frequencies of the acoustic fluctuations below the angular frequency of the vortex. This condition is the same as in the const-bg-dens and self-bg-dens cases. (ii) For strong dispersion, the superradiance may cease to exist below a critical rotation frequency of the vortex. We can speculate that this condition would emerge similarly in the present work when dispersion is included. (2) The BEC superfluid is assumed to be unconfined, which is not the case under experimental conditions. Again, one may assume that the variation of the confinement potential within the characteristic length scales of the acoustic scaterring from the vortex is negligible, and that the boundaries of the confinement are sufficiently far away from the vortex to interfere.

With this setting, the rest of the paper is organized as follows: In section 2 we recapitulate the vortex state formulation of the BEC superfluid. This is followed by the Klein-Gordon equation for the propagation of acoustic linear perturbations as section 3. The numerical implementation is described in section 4 and the results are given in section 5. Last section concludes the study.

2. The vortex state of an unconfined BEC

The evolution of the wavefunction of the bosons is governed by the Gross-Pitaevskii equation

$$i\hbar\frac{\partial\Psi(\mathbf{x},t)}{\partial t} = \left(-\frac{\hbar^2}{2m}\nabla^2 + V_{ext} + U_0\left|\Psi\right|^2\right)\Psi(\mathbf{x},t) \tag{1}$$

where V_{ext} is the external confining potential (set to zero in this study) and $V_H = \frac{4a\pi\hbar^2}{m}\left|\Psi\right|^2 \equiv U_0\left|\Psi\right|^2$ is the repulsive Hartree potential, with a scattering length, a. Steady state solutions can be expressed as $\Psi(\mathbf{x},t) = \psi(\mathbf{x})e^{-i\mu t/\hbar}$, where $\mu = \frac{\hbar^2}{2m\xi^2}$ is the chemical potential expressed in terms of the healing length $\xi = \sqrt{\frac{\hbar^2}{2\rho_\infty mU_0}}$. This is a characteristic length scale in which the condensate density tends to its bulk value $\rho_\infty = |\psi(r\to\infty)|^2$.

2.1. Density profile of the vortex

For the vortex state, we seek a steady state axisymmetric solution of Eq.1 with the spatial part expressed in polar coordinates in the form

$$\psi(\mathbf{x}) = f_q(r)e^{iq\theta}, \tag{2}$$

Here, $f_q(r)$ is the radial profile and the integer q denotes the winding (quantization) number of the vortex. Introducing dimensionless radial coordinate $\tilde{r} = r/\xi$ and the scaled function

$\chi_q = f_q/\sqrt{\rho_\infty}$ Eq.1 yields

$$-\frac{1}{\tilde{r}}\frac{d}{d\tilde{r}}\left(\tilde{r}\frac{d\chi_q}{d\tilde{r}}\right) + \frac{q^2\chi_q}{\tilde{r}^2} + \chi_q^3 - \chi_q = 0. \tag{3}$$

Equation 3 is subject to the boundary conditions $\chi_q(0) = 0, \chi_q(\infty) = 1$.

In the absence of the confinement of the condensate, multiquantized ($q > 1$) vortex states are unstable and decay into single quantized vortices[21, 22]. Experimentally obtained cigar-shaped condensates did not exhibit multiquantized vortices, either[23, 24]. Thus, while there has been theoretical[25] and experimental[26] effort to realize multiquantized vortices in different confinement potentials, the dynamical stability is an issue. The present work will consider the single-quantized ($q = 1$) case only.

The density profile dictated by Eq.3 has no closed form solution. In order to proceed with the analytical formulation of the metric and the solution of the Klein-Gordon equation, one has to resort to an approximate functional form. These were reported in the literature [27, 28, 29, 30], from which we pick the following functional form[29].

$$\chi_1 = \sqrt{\frac{\tilde{r}^2}{2+\tilde{r}^2}}, \tag{4}$$

It should be stressed that this approximation underlies the entire subsequent formulation and its results. Figure1 compares the numerical solution of Eq.3 with the function4. The background

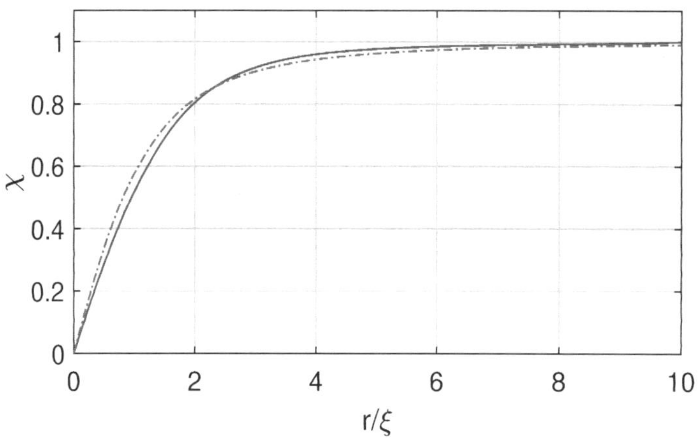

Figure 1: The numerical (solid) and approximate functional form (dash-dotted) of the radial wave function $\chi(r/\xi)$.

density profile of the vortex is then found from $\rho(r) = |\psi(r,\theta)|^2$

$$\rho_0(r) = \rho_\infty \frac{(r-r_0)^2}{(r-r_0)^2 + 2\xi^2} \tag{5}$$

where r_0 is the vortex center. The propagation speed of sound in the BEC is defined by the density $c = \sqrt{U_0\rho/m}$ and becomes

$$c(r) = c_\infty \sqrt{\frac{(r-r_0)^2}{(r-r_0)^2 + 2\xi^2}}, \tag{6}$$

where $c_\infty = \frac{\hbar}{m}\sqrt{4\pi a \rho_\infty}$ denotes the bulk value for the speed of sound.

Equations 5 and 6 comprise the key elements in the formulation of the vortex as an acoustic black hole. Their radial dependence becomes inherent in the superradiance dynamics.

2.2. Velocity profile of the vortex

At the vortex axis, the velocity of the condensate becomes singular. The consequence is that the circulation around a contour enclosing the vortex becomes nonzero and the velocity acquires a tangential component, described by a winding number.

$$\Theta = \oint \vec{v} \cdot d\vec{l} = 2\pi q \frac{\hbar}{m}, \qquad q = \pm 1, \pm 2... \tag{7}$$

$$v_{\hat{\phi}} = \frac{\Theta}{2\pi r}\hat{\theta} \equiv \frac{B}{r}\hat{\theta}. \tag{8}$$

Since the continuity equation $\frac{\partial \rho}{\partial t} + \nabla \cdot (\rho \vec{v}) = 0$, must be satisfied, the radial component of the velocity becomes

$$\nabla \cdot (\rho_0 \vec{v}) = 0 \Rightarrow v_{\hat{r}} = -\frac{A}{\rho(r) r} \tag{9}$$

where A is an empirical parameter to be set from experimental data (see section3.1). When contrasted with the velocity components of the draining bathtub model with a constant background density[31, 18], $\vec{v} = \frac{-A}{r}\hat{r} + \frac{B}{r}\hat{\theta}$ it is evident that the density modifies the radial velocity rather than being a mere scaling factor.

3. Acoustic Perturbations and the Klein-Gordon equation

For the superradiance, we consider small fluctuations superimposed on the background wavefunction $\psi = \psi_0 + \psi_1 = \sqrt{\rho} e^{i\Phi}$. With the linearized fluctuations in the density $\rho = \rho_0 + \rho_1$, and the phase $\Phi = \Phi_0 + \Phi_1$, substituting into Eq.1 yields the following set of equations

$$\frac{\partial \rho_1}{\partial t} + \frac{\hbar}{m}\nabla \cdot (\rho_0 \nabla \Phi_1) + \nabla \cdot (\rho_1 \vec{v}) = 0, \tag{10}$$

$$\partial_t \Phi_1 = -v \cdot \nabla \Phi_1 - U_0 \rho_1 + \frac{\hbar^2}{2m} D_2 \rho_1, \tag{11}$$

The velocity is expressed in terms of a "velocity potential" $\vec{v} = (\hbar/m)\nabla\Phi$. The last term in Eq.11 is known as the quantum pressure term[27] given explicitly below

$$D_2 \rho_1 = \frac{1}{2\sqrt{\rho_0}} \nabla^2 \frac{\rho_1}{\sqrt{\rho_0}} - \frac{\rho_1}{2\rho_0^{3/2}} \nabla^2 \sqrt{\rho_0}. \tag{12}$$

In the case of constant background density, this term can be neglected at all spatial ranges in comparison to other terms in Eq. 11, leading to the so-called hydrodynamic approximation. However, as the self-consistent density profile decays to zero at the vortex core, this simplification has to be justified in the vicinity of the event horizon. In the present study, we calculated and compared the respective magnitudes of the leading term in the quantum pressure and the Hartree potential $U_0 \rho_1$ near the event horizon and verified its validity.

Equations 10 and 11 yield the Klein-Gordon equation for the propagation of phase perturbations:

$$\frac{\partial}{\partial t}\left[\frac{\rho_0}{c^2}\left(\frac{\partial \Phi_1}{\partial t} + \vec{v} \cdot \nabla \Phi_1\right]\right) - \nabla \cdot (\rho_0 \nabla \Phi_1) + \nabla \cdot \left[\frac{\rho_0}{c^2}\left(\frac{\partial \Phi_1}{\partial t} + \vec{v} \cdot \nabla \Phi_1\right)\right] = 0. \tag{13}$$

By employing the techniques discussed in Ref.s [15], [32], one can reduce the Klein-Gordon equation into a set of coupled first-order partial differential equations. First, the following functions are defined

$$\Phi_1 = \phi_1(t,r) e^{im\phi} e^{ikz} \qquad \Pi = \pi_1(t,r) e^{im\phi} e^{ikz} \qquad \Gamma = \gamma_1(t,r) e^{im\phi} e^{ikz} \tag{14}$$

where(k, m) denote the axial and azimuthal wave numbers. We take translational symmetry along vortex axis (i.e $k = 0$) and single quantized vortex ($m = 1$). These functions are then used to define the conjugate fields

$$\Gamma = \frac{\partial \Phi_1}{\partial x^i} \qquad \Pi = -\frac{1}{c}\left(\frac{\partial \Phi_1}{\partial t} - \beta^i \Gamma_i\right), \tag{15}$$

where $x_i = (r, \theta, z)$ denote the cylindrical coordinates and $\beta^i = (-v_r, -v_\theta/r, 0)$ are the velocity components. Finally, the radial parts of these functions satisfy the following set of coupled first-order partial differential equations

$$\frac{\partial \phi_1}{\partial t} = -c\pi_1 - v_r\gamma_1 - \frac{imv_\theta}{r}\phi_1 \tag{16}$$

$$\begin{aligned}\frac{\partial \gamma_1}{\partial t} = &-\frac{\partial c}{\partial r}\pi_1 - c\frac{\partial \pi_1}{\partial r} - \frac{\partial v_r}{\partial r}\gamma_1 - v_r\frac{\partial \gamma_1}{\partial r}\\ &-\frac{\partial v_\theta}{\partial r}\frac{im}{r}\phi_1 + \frac{imv_\theta}{r^2}\phi_1 - \frac{imv_\theta}{r}\frac{\partial \phi_1}{\partial r}\end{aligned} \tag{17}$$

$$\begin{aligned}\frac{\partial \pi_1}{\partial t} =& \pi_1\left(-\frac{\partial c}{\partial r}\frac{v_r}{c} - \frac{imv_\theta}{r} - \frac{1}{r}\frac{\partial (rv_r)}{\partial r}\right) - v_r\frac{\partial \pi_1}{\partial r}\\ &+ \frac{\gamma_1}{c}\left(-\frac{c^2}{r} - \frac{c^2}{\rho}\frac{\partial \rho}{\partial r} + \frac{v_\theta^2}{r} + v_\theta\frac{\partial v_\theta}{\partial r}\right) - c\frac{\partial \gamma_1}{\partial r}\\ &+ \frac{\phi_1}{c}\left(\frac{m^2c^2}{r^2} + c^2k^2 - \frac{imv_r}{r}\frac{\partial v_\theta}{\partial r} - \frac{imv_rv_\theta}{r^2}\right)\end{aligned} \tag{18}$$

The numerical part of the study given in Sec.4 involves solving this set of equations.

3.1. The Black Hole profile of the vortex

The Klein-Gordon equation given in Eq. 13 renders an axisymmetric space-time with the metric

$$ds^2 = \frac{\rho}{c}\left[-cdt^2 + (dr - v_rdt)^2 + (rd\theta - v_\theta dt)^2\right]. \tag{19}$$

The off-diagonal elements in Eq. 19 can be reduced by the following coordinate transformation

$$dt = dt^* - v_r/\left(c^2 - v_r^2\right)dr, \tag{20}$$

$$d\theta = d\theta^* - v_rv_\theta/\left(r\left(c^2 - v_r^2\right)\right)dr, \tag{21}$$

$$dr = dr^*, \tag{22}$$

which yields,

$$ds^2 = \left[\left(v^2 - c^2\right)dt^2 + \frac{c^2}{c^2 - v_r^2}dr^2 + r^2d\theta^2 + 2v_\theta rdtd\theta\right]. \tag{23}$$

In the form above, the vanishing of the coefficient of dt^2 defines the ergosphere radius, r_e whereas the singularity of the metric determines the event horizon, r_h, in accordance with the Kerr black hole formulation in general relativity. In terms of the components of v given in Eq.s8, 9

$$v^2 - c^2 = 0 \Rightarrow \frac{A}{\rho_\infty c_\infty} - \frac{r_h^4}{\left(r_h^2 + 2\xi^2\right)^{3/2}} = 0 \tag{24}$$

Journal of Physics: Conference Series **2191** (2022) 012014 doi:10.1088/1742-6596/2191/1/012014

$$v_r^2 - c^2 = 0 \Rightarrow \frac{A^2\left(r_e^2 + 2\xi^2\right)^2}{\rho_\infty^2 r_e^6} + \frac{B^2}{r_e^2} - \frac{c_\infty^2 r_e^2}{r_e^2 + 2\xi^2} = 0 \tag{25}$$

The coefficient A may be determined by setting the event horizon at unit healing length $r_h = \xi$:

$$A = \frac{\xi \rho_\infty c_\infty}{3^{3/2}} \qquad B = \frac{\hbar}{m}. \tag{26}$$

This is because the vortex core size of cold-atom gases in experimental studies is estimated to be of the order of healing length [33] (i.e. $\sim$ 100nm). Note that B follows from Eq.8. For an experimental correspondence, Table 1 lists the parameters of a BEC of Rb-87 atoms taken from Ref.s[19, 34] and the resulting black hole parameters for a vortex state. We note that r_e is calculated as the only positive real root of Eq.25. Figure 2, shows the speed of sound and the

Table 1: The experimental (Ref.s [19, 34]) parameters of a BEC of Rb-87 atoms and the calculated parameters of the black-hole vortex.

a	m	ρ_∞	c_∞	ξ	r_h	r_e
5.77nm	$1.44 \times 10{-}25$kg	10^{21}m^{-3}	$6.2 \times 10^{-3}\text{m/s}$	83nm	ξ	1.84ξ

radial velocity of the condensate fluid near the vortex. Vertical dashed lines indicate the event horizon and the ergosphere radii, respectively. Noteworthy is the decrease of the speed of sound from the ergosphere boundary to the event horizon (roughly within one healing length range) by more than 25%. This significant variation is expected to have an observable effect on the superradiance dynamics.

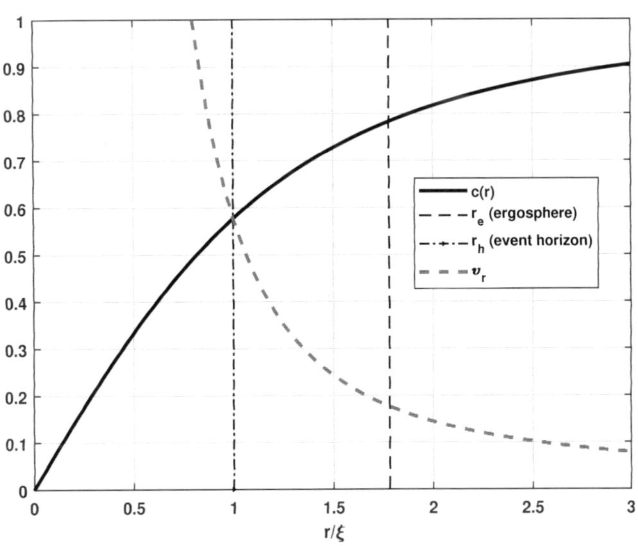

Figure 2: The speed of sound (Eq. 6), the radial velocity of the fluid (Eq. 9) along with the event horizon (r_h) and the ergoregion radius (r_e) calculated using the parameters given in Table 1.

4. Numerical Method

The propagation of the acoustic perturbations can be obtained by integrating the Eq.s 16, 17, 18. We employ Matlab's Partial Differential Equation Toolbox. The numerical calculations further implement the so-called "excision technique"[15, 35] to deal with the propagation through the event horizon and beyond numerically. From the calculated functions Π, Φ and Γ, we construct outgoing $(\Pi + \Phi)$ and incoming $(\Pi - \Phi)$ fields along the null ray. At large distances purely outgoing wave is implemented such that $\Gamma = \phi$. The spatiotemporal computational domain is defined as $0.5 < \tilde{r} < 110$, $0 < \tilde{t} < 110$, where $\tilde{r} = r/\xi$, $\tilde{t} = tc_\infty/\xi$ are dimensionless coordinates. The initial acoustic perturbation is defined in Eq. 14 as an axisymmetric, radially incoming Gaussian wave centered at r_0, with a width b as follows

$$\phi_1(0,r) = Nexp\left[-(r - r_0 + c(r)t)^2/b^2 - i\omega(r - r_0 + c(r)t)/c(r)\right]. \tag{27}$$

We set $r_0 = 50\xi$ and $b = 10\xi$. The initial π_1 and γ_1 are calculated by Eq.15. The main quantity we are interested in is the temporal variation of the energy of propagating acoustic perturbations,

$$E(t) = \int d^2r \frac{1}{2} M\rho \left|\vec{v}\right|^2 = (\hbar^2/2M) \int_0^{2\pi} d\theta \int_\xi^{r_{max}} \rho(r)(\nabla\Phi)^2 r dr, \tag{28}$$

with the scaled velocity

$$\vec{\tilde{v}} = -\frac{1}{3\sqrt{3}}\left(\frac{1}{\tilde{r}} + \frac{2}{\tilde{r}^3}\right)\hat{r} + \frac{\sqrt{2}}{\tilde{r}}\hat{\theta}. \tag{29}$$

5. Results: Temporal and spectral features of superradiance

We begin with the demonstration of the superradiance by contrasting the scattering of acoustic perturbations (density fluctuations ρ_1) from the event horizon of a vortex with (a) $m = 1$ and (b) $m = 0$ v(right panel) in Fig. 3 under const-bg-dens approximation. The self-bg-dens case is not shown as it is very similar to the const-bg-dens case in this plot type. At $t = 100 * c/a$ (the upper edge of the figures) the reflected wave returns to the launch point at $r_0 = 50\xi$. There is a clear difference between the superradiant and subradiant reflected profiles.

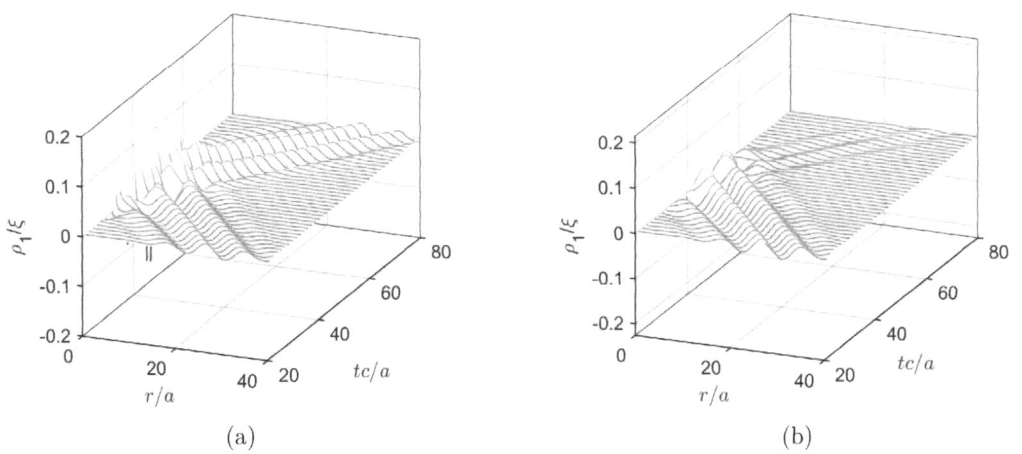

Figure 3: Propagation of the density fluctuations (ρ_1) in the scaled $r - t$ plane for (a) the superradiant ($m = 1$) and (b) the subradiant [$m = 0$)) settings of the vortex. The perturbation interacts with the event horizon between $t * c/a = 40$ and 60.

For a better representation in the self-bg-dens case, the bulk-normalized background density profile (Eq. 5) without and with the acoustic fluctuations at $t * c/a = 50$ superimposed are plotted in Fig. 4. For visibility, the profiles are cut out in an angular range and the profile with fluctuations is shifted vertically. The event horizon located at $\rho_0 = 0.5$ (see Fig. 3) remains just below the plotted range. The fluctuations against the rapidly decaying background density near the event horizon are hardly noticeable, but they are prominent near the ergosphere $r/a \sim 2$.

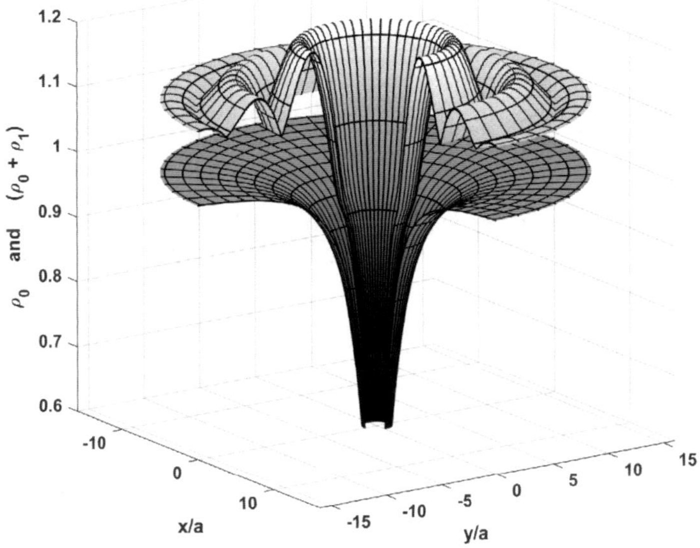

Figure 4: The self-consistent background density profile without (bottom) and with (top) superimposed density fluctuations near the vortex. The fluctuations are plotted at $t * c/a = 50$ while interacting with the event horizon. The top plot is displaced by 0.1 upward to enhance visibility.

Figure 5 is the main result of the paper. On the left, the variation of the relative energy (gain) for the superradiant and subradiant cases under const-bg-dens approximation are compared. In the subradiant case, the energy of the wave is completely absorbed by the vortex. The right panel shows the energy gain for the const-bg-dens and self-bg-dens formulations. The prominent difference is that the const-bg-dens results in an overshoot in the gain which *then* decreases as the wave is outgoing, whereas the non-constant density model reaches monotonically to a final gain value of the outgoing wave. In the self-bg-dens case, the fluctuations arrive slightly later, since the sound speed decreases near the vortex. Both formulations predict a gain of about 40%.

Finally, we investigate the spectral behavior of the superradiance. The energy gain factors achieved for acoustic fluctuations with different frequencies are plotted in Fig. 6, as obtained in const-bg-dens and self-bg-dens cases, respectively. The fluctuation wave frequency is scaled by the vortex rotation frequency, which is determined from $\vec{v_\theta} = \vec{\Omega} \times \vec{r}$ as

$$\Omega = \frac{B}{r_h^2} = q\frac{\hbar}{m}\frac{1}{r_h^2} = q\sqrt{2}\frac{c_\infty}{\xi} \tag{30}$$

The const-bg-dens and self-bg-dens profiles differ in the calculated superradiance mostly in the intermediate range of the scaled frequency, where the non-constant density model yielding a

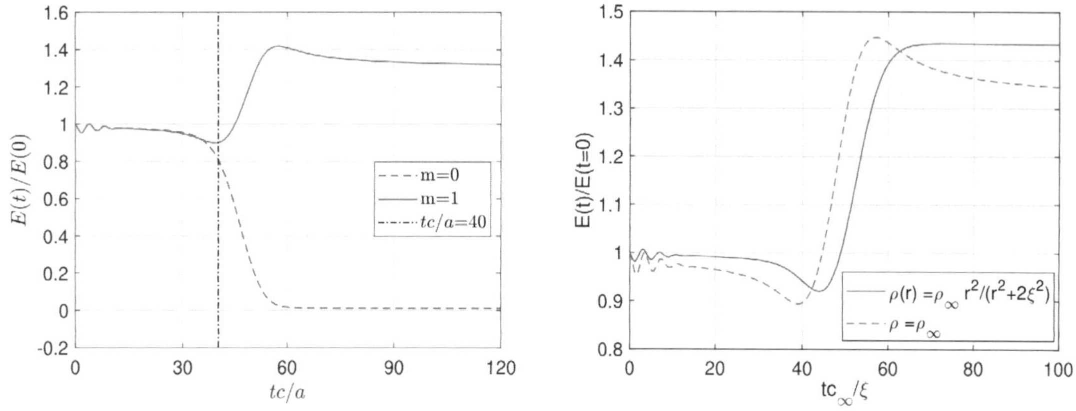

(a) (b)

Figure 5: [left] The energy gain for the superradiant ($m = 1$) (red) and the subradiant ($m = 0$) (blue) settings of the vortex. The vertical dashed line is the time the wave reaches the event horizon. [right] The energy gain under constant background density (red) and self-consistent background density (blue) formulations.

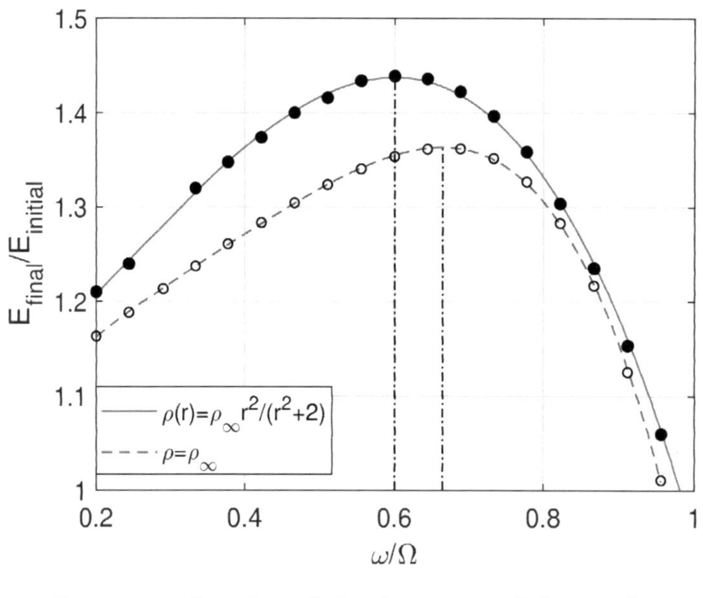

Figure 6: Superradiance as a function of the frequency of the incident acoustic perturbative wave, normalized by the vortex rotation frequency, and calculated with the constant density (circles) and self-consistent density (dots) background density profiles. Maximum gain for each case is marked by a vertical line.

slightly higher ($\sim 10\%$) gain. This is consistent with the result for single frequency shown in Fig. ??. For low frequency perturbations, $\omega/\Omega 1$ the difference decreases but remains finite. Both models exhibit a superradiance profile biased towards the upper midrange with respective maxima attained at $\tilde{\omega} = 0.60$ and $\tilde{\omega} = 0.67$.

6. Conclusion

We performed the formulation of the acoustic superradiance with the self-consistent background density model of the BEC vortex state. The resulting dynamics of superradiance is compared with that of the constant background density approximation. The main conclusion is that the self-consistent background density eliminates the overshoot behavior of the constant density approximation in the superradiance, and provides an overall enhanced gain for the superradiance effect for the single-quantized vortex state. Moreover, the self-consistent background density formulation implies that multiquantized vortices each would have a different background density profile and thus appear as a different black-hole, as the characteristic parameters (i.e. event horizon, ergosphere) changes. The vortex rotation frequency, which can be set independently as integer multiples of a fixed value in the constant density approximation, becomes an implicit function of the background density profile and changes nontrivially for multiquantized vortices. Admittedly, these implications may become relevant after the stability issues of multiquantized vortex states are resolved experimentally. We conclude by indicating two directions in which the present formulation may be advanced. The first one is to implement dispersion effects as it is inherent to any wave propagation phenomenon. The second is to include confinement potentials as the stability of multiquantized vortices may only become available in their presence.

Acknowledgments

The authors would like to express their sincere thanks to Professor Tekin Dereli for his "superradiant" guidance throughout this study and beyond.

References

[1] Barcelo C, Liberati S and Visser M 2001 *Classical and Quantum Gravity* **18** 1137
[2] Garay L J, Anglin J, Cirac J I and Zoller P 2002 *International Journal of Theoretical Physics* **41** 2073–2090
[3] Garay L J 2001 *Physical Review A* **63** 023611
[4] Lahav O, Itah A, Blumkin A, Gordon C, Rinott S, Zayats A and Steinhauer J 2010 *Phys. Rev. Lett.* **105**(24) 240401 URL `https://link.aps.org/doi/10.1103/PhysRevLett.105.240401`
[5] Unruh W G 1981 *Phys. Rev. Lett.* **46**(21) 1351–1353 URL `https://link.aps.org/doi/10.1103/PhysRevLett.46.1351`
[6] Carlson N N 1996 *Physica D: Nonlinear Phenomena* **98** 183 – 200 ISSN 0167-2789 URL `http://www.sciencedirect.com/science/article/pii/0167278996000528`
[7] Fetter A L and Svidzinsky A A 2001 *Journal of Physics: Condensed Matter* **13** R135
[8] Macher J and Parentani R 2009 *Physical Review A* **80** 043601
[9] Basak S and Majumdar P 2003 *Classical and Quantum Gravity* **20** 3907
[10] Torres T, Patrick S, Coutant A, Richartz M, Tedford E W and Weinfurtner S 2017 *Nature Physics*
[11] Ghazanfari N and Müstecaplıoğlu Ö E 2014 *Physical Review A* **89** 043619
[12] Steinhauer J 2014 *Nature Physics* **10** 864
[13] Steinhauer J 2016 *Nature Physics* **12** 959
[14] Kolobov V I, Golubkov K, de Nova J R M and Steinhauer J 2021 *Nature Physics* **17** 362–367
[15] Cherubini C, Federici F, Succi S and Tosi M 2005 *Physical Review D* **72** 084016
[16] Federici F, Cherubini C, Succi S and Tosi M 2006 *Physical Review A* **73** 033604
[17] Demirkaya B, Dereli T and Güven K 2018 *arXiv preprint arXiv:1806.02139*
[18] Visser M 1998 *Classical and Quantum Gravity* **15** 1767
[19] Dalfovo F, Giorgini S, Pitaevskii L P and Stringari S 1999 *Reviews of Modern Physics* **71** 463
[20] S P 2021 *Classical Quantum Gravity* **38** 095010
[21] Desyatnikov A S, Torner L and Kivshar Y S 2005 *arXiv preprint nlin/0501026*
[22] Zhang Y, Bao W and Du Q 2007 *European Journal of Applied Mathematics* **18** 607–630
[23] Leanhardt A E *et al.* 2003 *Phys. Rev. Lett.* **90** 140403
[24] Shin Y *et al.* 2004 *Phys. Rev. Lett.* **93** 160406
[25] Huhtamaki J A M, Möttönen M and Virtanen S M M 2006 *Phys. Rev. A.* **74** 063619
[26] Leanhardt A E *et al.* 2003 *Phys. Rev. Lett.* **89** 190403
[27] Pethick C J and Smith H 2002 *Bose-Einstein Condensation in Dilute Gases* (Cambridge University Press)
[28] Berloff N G 2004 *J. Phys. A: Math Gen.* **37** 1617–1632
[29] Slatyer T R and Savage C M 2005 *Classical and Quantum Gravity* **22** 3833 URL `http://stacks.iop.org/0264-9381/22/i=19/a=002`

[30] Chen W *et al.* 2021 *Commun. Theor. Phys.* **73** 085701
[31] Basak S and Majumdar P 2003 *Classical and Quantum Gravity* **20** 2929
[32] Scheel M A, Erickcek A L, Burko L M, Kidder L E, Pfeiffer H P and Teukolsky S A 2004 *Physical Review D* **69** 104006
[33] Verhelst N and Tempere J *Vortex Dynamics and Optical Vortices*
[34] Anderson M H, Ensher J R, Matthews M R, Wieman C E and Cornell E A 1995 *science* **269** 198–201
[35] Alcubierre M and Brügmann B 2001 *Phys. Rev. D* **63**(10) 104006 URL `https://link.aps.org/doi/10.1103/PhysRevD.63.104006`

Wahlquist metric revisited

M. Hortaçsu [1]
Mimar Sinan Güzel Sanatlar Üniversitesi,
Faculty of Arts and Lettres, Physics Department, Istanbul, Turkey

Abstract. Here we continue studying the Wahlquist metric. We know that the wave equation written for a zero mass scalar particle in the background of this metric gives Heun type solutions. To be able to use the existing literature on Heun functions, we try to put our wave equation to the standard form for these functions. Then we calculate the reflection coefficient of a wave coming from infinity and scattered at the center using this formalism.

1. Introduction

The work on Heun class solutions of wave equations written for particles in the background metric of different metrics is still popular. One of the most recent papers is given in [1]. Here we study a similar problem for a metric, which was given a while ago.

The Wahlquist metric [2, 3, 4] is an exact interior solution for a finite rotating body of perfect fluid to Einstein's field equations with equation of state corresponding to constant gravitational mass density. It is also given in the celebrated book *Exact Solutions to Einstein's Field Equations* [5]. This metric is axially symmetric, stationary and is type D. As stated in [6]", quoting Wahlquist [2], it can be "described as a superposition of a Kerr-NUT metric [7, 8] and is a rigidly rotation perfect fluid in the same space-time region". Equation of the state for the p perfect fluid is $\rho + 3p = \mu_0$, a constant.

The original metric written in [2], was slightly modified by Senovilla [9, 10], and put to new form by Mars [11], "to show that the Kerr-de Sitter and Kerr metrics are contained as sub cases". Mars states in [11] that Kramer [12] showed the vanishing of the Simon tensor [13] for this metric. Mars also states that the space-time admits a Killing tensor as shown in [14]".

[1] E-mail: hortacsu@itu.edu.tr

 2191 (2022) 012015 doi:10.1088/1742-6596/2191/1/012015

A negative point about this metric is the paper, [15], where it was shown that " the Wahlquist perfect fluid space-time can not be smoothly joined to an exterior asymptotically flat vacuum region since the conditions for matching the induced metrics and extrinsic curvatures are mutually contradictory". Thus, " the Wahlquist metric can not describe an isolated rotating body". Bradley et al., however, showed that this obstruction is removed for a limiting form of the Wahlquist solution [16], and also for every D-type metric [17] if the requirement of an asymptotically flatness is removed.
More recent work in this field exists, "where the existence of a rank-2 generalized closed conformal Killing-Yano tensor with a skew-symmetric torsion " [18], and "the separability of the Maxwell equation on the Wahlquist spacetime" are shown [19].
Here we start with the metric as given in [18] and try to write the wave equation, in the background of this metric, as given in [6], in the standard form given by [20]. How to perform this task is described in [21], as quoted by [22], and used meticulously by [23, 24]. The same method is recently used by Vieira et al in [25] . We need the standard form of the wave equation to be able to apply the information given in the existing literature to our work. We will first summarize our previous work, [6], then show how to convert the wave equation to the standard form. In the third section, we give the approximate reflection coefficient if a wave , coming from infinity is scattered at the origin. We conclude with a few remarks.

2. Summary of former work

Here, we will summarize the work in [6] below. First, using the metric given in [2], we try to calculate ϕ where it obeys the equation

$$\frac{1}{\sqrt{g}}\partial_\mu g^{\mu\nu}\sqrt{g}\partial_\nu\phi = 0. \tag{1}$$

Here g is the determinant of the metric coefficients $g_{\mu\nu}$.
We first write the metric as is given in [18], which is equivalent to the one giving in [11]. We follow some of the work in [18] in the first part, and [6] in the second part of this section.
Here "the comoving, pseudoconfocal, spatial coordinates are used, which are closely related to the oblate-spheroidal coordinates in Euclidean geometry" [2]. We set $4\pi G = c = 1$. Then, the metric reads

$$ds^2 = (v_1+v_2)(\frac{dz^2}{U}+\frac{dw^2}{V}) + \frac{U}{v_1+v_2}(d\tau+v_2 d\sigma)^2 - \frac{V}{v_1+v_2}(d\tau - v_1 d\sigma)^2 \tag{2}$$

where

$$U = Q_0 + a_1\frac{sinh(2\beta z)}{2\beta} - \frac{\nu_0}{\beta^2}\frac{cosh(2\beta z)-1}{2\beta^2}$$
$$-\frac{\mu_0}{2\beta^2}[\frac{cosh(2\beta z)-1}{2\beta^2} - z\frac{sinh(2\beta z)}{2\beta}], \tag{3}$$

$$V = Q_0 + a_2 \frac{sin(2\beta w)}{2\beta} + \frac{\nu_0}{\beta^2} \frac{-cos(2\beta w) + 1}{2\beta^2}$$
$$- \frac{\mu_0}{2\beta^2} [\frac{cos(2\beta w) - 1}{2\beta^2} + w \frac{sin(2\beta w)}{2\beta}] \tag{4}$$

and

$$v_1 = \frac{cosh(2\beta z) - 1}{2\beta^2}, v_2 = \frac{-cos(2\beta w) + 1}{2\beta^2}. \tag{5}$$

This metric has six real constants $Q_0, a_1, a_2, \nu_0, \mu_0, \beta$. Here a_1 is related to the NUT [26, 27] parameter; a_2 is related to the mass parameter. One writes the other variables to express the energy density, pressure and the fluid velocity of the perfect fluid. Here β is related to the scaling of z and w, both space coordinates, in the linear transformation of these variables given in the original paper by Wahlquist [2]. U and V are are related to h_2, h_1 in the original metric. All these parameters are scaled so that they do not vanish in the β going to zero limit.
When we use the ansatz

$$\phi = R(z)Y(w)T(\tau)S(\sigma) \tag{6}$$

for the solution, the wave equation written in this metric separates easily. We have two Killing vectors since the metric does not depend on τ and σ, related to t and θ in the original metric [2]. If we make a Wick rotation which changes w to $y = iw$, a_2 to $-ia_2$ [11] where i is the square root of minus unity, the metric becomes symmetrical

$$ds^2 = (v_1 + v_2)(\frac{dz^2}{U} - \frac{dw^2}{V}) + \frac{U}{v_1 + v_2}(d\tau - v_2 d\sigma)^2 - \frac{V}{v_1 + v_2}(d\tau + v_1 d\sigma)^2. \tag{7}$$

Then we have identical equations for z and y. The equation for z reads

$$\partial_z(U\partial_z)R(z) + (\frac{v_1^2}{U}\partial_\tau^2 + 2\frac{v_1}{U}\partial_\tau\partial_\sigma + \frac{1}{U}\partial_\sigma^2)R = 0. \tag{8}$$

We get exactly the same equation for the new variable y, with appropriate changes like U going to V and v_1 going to v_2.

$$\partial_y(V\partial_y)S(y) + (\frac{v_2^2}{V}\partial_\tau^2 + 2\frac{v_2}{V}\partial_\tau\partial_\sigma + \frac{1}{V}\partial_\sigma^2)S(y) = 0. \tag{9}$$

Since the functions v_1, U (similarly v_2, V) don't depend on τ and σ, the solutions for τ and σ are just exponential functions, giving us constants upon differentiation.

Here we will first try to write the differential equation for a zero mass scalar field for the radial coordinate z, in the background of this metric. Then, we make more independent variable changes to put our equation in the the standard form , as given in [20].
We find that this is not an easy task. The presence of hyperbolic sine and cosine functions in the wave equation prevents us from using standard analytical methods in solving differential equations. We change our variables to $x = exp(2\beta z)$ which makes it possible to write the hyperbolic sine and cosine functions in terms of powers. $sinh 2\beta z = 1/2[x - 1/x]$, $cosh 2\beta z - 1 = 1/2[x + 1/x - 2]$. Then, however, exists the relation $z = lnx/(2\beta)$.
The codes we have to solve differential equations analytically, does not recognize lnx. Thus, we can not give an exact solution for all values of the independent variable x. We can get solutions only in one patch, by using an algebraic expression approximating lnx in this region.
Upon the variable change from z to x, we get the new equation

$$\frac{d^2\psi(x)}{dx^2} + (\frac{\frac{dU}{dx}}{U} + \frac{1}{x})\frac{d}{dx}\psi(x) - \frac{1}{x^2U^2}(\omega^2 v_1^2 + 2v_1\omega s + s^2)\psi(x) = 0. \quad (10)$$

Here we used the solution to the τ equation in the form $exp(-i\omega\tau)$ and the solution for the $i\sigma$ equation a solution in the form $exp(-i\sigma s)$. Note that our metric had Killing vectors for both τ and σ. Here

$$U = Q_0 + a_1\frac{(x - 1/x)}{4\beta} - [\frac{\mu_0}{4\beta^4} + \frac{\nu_0}{4\beta^3}]\frac{(x-1)^2}{x} + \frac{\mu_0}{8\beta^4}\ln x\frac{(x^2-1)}{x}], \quad (11)$$

$$v_1 = \frac{(x-1)^2}{4\beta^2 x} \quad (12)$$

,

$$\frac{dU}{dx} = [\frac{a_1}{\beta}(1+\frac{1}{x^2}) - (\frac{\nu_0}{\beta^2} + \frac{\mu_0}{\beta^4}(1-\frac{1}{x^2}) + \frac{\mu_0}{8\beta^4 x}(x - \frac{1}{x}) + \frac{\mu_0}{8\beta^4}\ln x(1+\frac{1}{x^2})]. \quad (13)$$

From here on, we will define $\mu_0' = \frac{\mu_0}{(8\beta^4)}$ and $\nu_0' = \frac{\nu_0}{(4\beta^3)}$, $a_1' = a_1/\beta$ and use these new constants in our equations.
By studying the equation written in terms of the variable x, we know that there is a singularity near $x = 1$. To study the solution around this singularity, we expand lnx in the neighborhood of this point and use $x - 1$ instead of lnx in the wave equation we use in our calculation, after we differentiate U with respect to x.
We find that, if we keep all our constants non zero, we get [6] the exact solution in terms the general Heun function [20, 28, 29, 30] up to an exponential and terms (powers of polynomials) multiplying this function. The regular singularities are at zero and at three other finite points.

Journal of Physics: Conference Series **2191** (2022) 012015 doi:10.1088/1742-6596/2191/1/012015

Then we try the next possible choice. We try to find the solution when we keep all the terms in U aside from Q_0, which we set equal to zero. Then, our equation, again, has four regular singularities at $0, 1$ and the points

$$x_a = -\frac{1}{2\mu_0'}(M - [N]^{1/2}), \tag{14}$$

$$x_b = -\frac{1}{2\mu_0'}(M + [N]^{1/2}), \tag{15}$$

where

$$M = (a_1' - \nu_0' - 2\mu_0'), \tag{16}$$

$$N^2 = M^2 - 4\mu_0'(a_1' + \nu_0' + \mu_0'). \tag{17}$$

The $HeunG$, General Heun form solution is retained when we set Q_0 to zero. To further simplify our expressions, we define $a_1'' = \frac{a_1'}{\mu_0'}, \nu_0'' = \frac{\nu_0'}{\mu_0'}$.

3. Reduction to the standard form

How to reduce an equation with four regular singular points at finite values to an equation with singularities at zero, unity, a finite point and infinity is described in [22]. If the coefficients of the first derivatives satisfy a linear relation, one just has to make a homographic substitution to bring three of the original singularities b_1, b_2, b_4 to zero, unity and infinity. This transformation is

$$u = \frac{x - b_1}{x - b_4}\frac{b_2 - b_4}{b_2 - b_4} \tag{18}$$

In our case $b_1 = 0, b_1 = 1, b_3 = x_a, b_4 = x_b$, which gives $u = \frac{x(1-x_b)}{x-x_b}$. The singularity at x_b is moved to infinity, the singularity at x_a is moved to $\frac{-x_a(1-x_b)}{x_b-x_a}$. Then our equation reads

$$\frac{d^2\phi(u)}{du^2} + (\frac{1}{x_a x_b} - \frac{1}{u-1} + \frac{(1+x_b) - x_a x_b \frac{1}{x_a}}{(1-x_a)(u + \frac{x_a(1-x_b)}{x_b-x_a})})\frac{d\phi(u)}{du} + \frac{J}{2Tu}\phi(u)$$
$$+\frac{1}{4}(\frac{D}{T^2} + \frac{2s\omega}{EuT} + \frac{2s^2}{F(u-1)T} + \frac{4s^2}{H(u-1)^2} + \frac{\omega^2}{4(x_a x_b u)^2})\phi(u) = 0. \tag{19}$$

$$T = (u + \frac{x_a(1-x_b)}{x_b - x_a}), \tag{20}$$

$$D = \frac{1}{(x_b - x_a)^2}(\frac{\omega(1-x_a)}{2\beta^2 x_a} - \frac{2s}{1-x_a})^2, \tag{21}$$

$$E = 4x_a x_b(x_b - x_a), \tag{22}$$

$$F = (1-x_a)^2(1-x_b)(x_b - x_a) \tag{23}$$

$$H = (1 - x_a)^2(1 - x_b)^2, \tag{24}$$

$$J = -\frac{\omega^2 x_b(1 - x_a)}{\beta^4 x_a^2 x_b^2 (x_b - x_a)} \tag{25}$$

The new equation is not in the standard General Heun form, though. One has to remove the terms where the singularities are in quadratic form from the new differential equation. One makes a F-homotopic transformation on the dependent variable $\phi(u)$ in the form

$$\phi(u) = (u)^\lambda (u-1)^\xi (u + \frac{x_a(1 - x_b)}{x_b - x_a})^\kappa g(u), \tag{26}$$

which gives rather complicated expressions for these indices.

$$\lambda = \frac{1}{2}[1 \pm \sqrt{1 - \frac{w^2}{4\beta^4(x_a x_b)^2}}], \tag{27}$$

$$\xi = 1 \pm \sqrt{1 - \frac{4s^2}{(1-x_a)^2(1-x_b)2}}, \tag{28}$$

$$\kappa = -\frac{1}{2}[\frac{x_a}{x_a - x_b} - 1 \pm \sqrt{(\frac{x_a}{x_a - x_b})^2 - (\frac{w(x_a - 1)}{x_a \beta^2} - \frac{4s}{(x_a - 1)})^2}]. \tag{29}$$

Now we can write the new wave equation as

$$\frac{d^2 g(u)}{du^2} + (\frac{c}{u} + \frac{d}{u-1} + \frac{e}{(u + \frac{x_a(1-x_b)}{x_b - x_a})})\frac{d}{du}g(u) + \frac{ABu + Q}{u(u-1)(u + \frac{x_a(1-x_b)}{x_b - x_a})}g(u) = 0. \tag{30}$$

Here

$$c = [1 - \sqrt{1 - \frac{w^2}{4\beta^4(x_a x_b)^2}}], \tag{31}$$

$$d = 1 - 2\sqrt{1 - \frac{s^2}{(1-x_a)^2(1-x_b)2}}, \tag{32}$$

$$e = 1 - \sqrt{(\frac{x_a}{x_a - x_b})^2 - (\frac{w(x_a - 1)}{x_a \beta^2} - \frac{4s}{(x_a - 1)})^2} \tag{33}$$

Using the relation

$$c + d + e = A + B + 1, \tag{34}$$

which is valid when our equation has Heun form, we can also calculate A and B.

$$A = \frac{1}{2}(c + d + e - 1 + \sqrt{(c + d + e - 1)^2 - 4S}), \tag{35}$$

Journal of Physics: Conference Series **2191** (2022) 012015 doi:10.1088/1742-6596/2191/1/012015

$$B = \frac{1}{2}(c+d+e-1-\sqrt{(c+d+e-1)^2-4S}). \qquad (36)$$

The expression inside the square root above simplifies

$$(c+d+e-1)^2-4S = [\frac{2}{x_a x_b}+(-1\frac{x_a x_b-1}{x_b(x_b-x_a)})^2$$
$$-[\frac{\omega(1-x_b)}{2\beta^2 x_b(x_b-x_a)}-\frac{2s}{(1-x_b)(x_b-x_a)}]^2]. \qquad (37)$$

If we write $Q = Q_1 + Q_2$, we get

$$Q_1 = -\frac{1}{2}[ec+\frac{x_a(1-x_b)dc}{(x_b-x_a)}+\frac{x_a(1-x_b)}{x_a x_b(x_b-x_a)}+\frac{1+x_b-x_a x_b-\frac{1}{x_a}}{(x_b-x_a)(1-x_a)x_a x_b}], \qquad (38)$$

$$Q_2 = \frac{\omega^2 x_b(1-x_a)}{2\beta^4 x_a^2 x_b^2(x_b-x_a)}-\frac{s\omega}{2\beta^2 x_a x_b(x_b-x_a)}. \qquad (39)$$

4. Scattering at the origin

While changing our independent variable from z to x, we overlooked one point. Now x corresponds to minus infinity for z. To scatter from origin , we need to take $x = 1$. We, therefore, change our variables to $u - 1 = -y$. We know that $u = 1$ is the same point as $x = 1$. Then our equation reads

$$\frac{d^2g(y)}{dy^2}+(\frac{c}{y-1}+\frac{d}{y}+\frac{e}{(y-\frac{x_b(1-x_a)}{x_b-x_a})})\frac{d}{dy}g(y)+\frac{AB(y-1)-Q}{y(y-1)(y-\frac{x_b(1-x_a)}{x_b-x_a})}g(y)=0, \qquad (40)$$

which gives as solution

$$g(y) = H_G(\frac{x_b(1-x_a)}{x_b-x_a}, -Q-AB; A, B, d, c, e; y = 1-u), \qquad (41)$$

where H_G is the generalized Heun function, in the standard form[20]. We had a singular point at infinity. We want to have our wave to come from infinity and scatter at the origin. The other singular point at x_a can be arranged to be at a complex value.

For the scattering process, we will use the formula given by Dekar et al [31]. This formula is between two finite points. We bring the the point infinity to unity by using the transformation $t = \frac{y}{y-1}$. Now the point where y and t are equal to zero coincide, and y going to infinity is given by $t = 1$.

The new equation is not of the Heun form. To bring it back to the Heun form, we multiply the solution $g((t)$ by a power

$$g(t) = (t-1)^A h(t) \qquad (42)$$

and try to use $h(t)$ in our further calculations. Our new dependent variable $h(t)$ satisfies the equation

$$\frac{d^2h(t)}{dt^2}+(\frac{A+1-B}{t-1}+\frac{d}{t}+\frac{e}{t-\frac{x_b(1-x_a)}{x_a(1-x_b)}})\frac{d}{dt}h(t)+(\frac{(A(d+e-B)t+Q_1}{t(t-1)(t-\frac{x_b(1-x_a)}{x_a(1-x_b)})})h(t)=0. \quad (43)$$

At this point, we make a change in our notation. We want to use the standard Heun notation. This means we will redefine our parameters as $A+1-B=\delta, d=\gamma, d+e-B=\beta$. The other parameters are defined as $e=\epsilon, A=\alpha$ will remain as they are. Now, we can use the standard notation, since our equations will be of the Heun form in the further part of this work, and we will be able to use known relations for this function.
Now, the differential equation is written as

$$\frac{d^2h(t)}{dt^2}+(\frac{\gamma}{t}+\frac{\delta}{t-1}+\frac{\epsilon}{t-\frac{x_b(1-x_a)}{x_a(1-x_b)}})\frac{d}{dt}h(t)+(\frac{(\alpha\beta)t+Q_1}{t(t-1)(t-\frac{-x_b(1-x_a)}{x_a(1-x_b)})})h(t)=0. \quad (44)$$

$$Q_1=\frac{(x_b-x_a)(\alpha(\gamma+\epsilon-\beta)-Q)-\gamma\alpha(x_b(1-x_a)}{x_a(1-x_b)}. \quad (45)$$

From now on, we will rename $h(t)$ as y_1. Now we have to write y_1 in terms of the two linearly independent solutions, which we name $y_3(1-t)$ and $y_4(1-t)$, the two linearly independent solutions one gets when one expands around $1-t=v$.
In other words, we want to write

$$y_1(t)=C_1y_3(1-t)+C_2y_4(1-t). \quad (46)$$

The right hand side gives us the two Heun solutions when the same equation is expanded as a function of $1-t$. This expansion is easily done

$$y_3=H_G(\frac{(x_a-x_b)}{x_a(1-x_b)},Q_3;\alpha,\beta;\delta,\gamma,\epsilon;v=1-t), \quad (47)$$

where $Q_3=Q_1-\alpha\beta$, and which is the solution of the equation

$$\frac{d^2y_3(v)}{dv^2}+(\frac{\gamma}{v-1}+\frac{\delta}{v}+\frac{\epsilon}{v-\frac{x_a-x_a}{x_a(1-x_b)}})\frac{d}{dv}y_3(v)+(\frac{(\alpha\beta)v+Q_3}{v(v-1)(v-\frac{(x_a-x_b)}{x_a(1-x_b)})})y_3(v)=0. \quad (48)$$

Then, we make the transformation

$$y_4(v)=v^{(1-\delta)}j(v), \quad (49)$$

Journal of Physics: Conference Series **2191** (2022) 012015 doi:10.1088/1742-6596/2191/1/012015

where $j(v)$ satisfies the equation

$$\frac{d^2j(v)}{dv^2}+(\frac{\gamma}{v-1}+\frac{2-\delta}{v}+\frac{\epsilon}{v-\frac{x_a-x_a}{x_a(1-x_b)}})\frac{d}{dv}j(v)+(\frac{(\gamma+\epsilon-\alpha)(\gamma+\epsilon-\beta)v+Q_4}{v(v-1)(v-\frac{(x_a-x_b)}{x_a(1-x_b)})})j(v)=0. \tag{50}$$

to give

$$y_4=(1-t)^{(1-\delta)}(H_G(\frac{x_a-x_b}{x_a(1-x_b)},Q_4;\gamma+\epsilon-\alpha),\gamma+\epsilon-\beta;2-\delta,\gamma,\epsilon;v=1-t), \tag{51}$$

where $Q_4=Q_3-(1-\delta)(\gamma\frac{x_a-x_b}{x_a(1-x_b)}+\epsilon)$. Note that this last transformation is one of the transformations which does not change the Heun form of the differential equation. Furthermore,the relation $1+\alpha+\beta=\gamma+\delta+\epsilon$ is valid in this part of our calculations.
At this point we use the formula given in [31] for expanding $H_G(t)$ in terms of two linearly independent solutions of the equation written for $H_G(1-t=v)$, our equation (46).

$$y_1(t)=C_1y_3(1-t)+C_2y_4(1-t). \tag{52}$$

We find

$$C_1=H_G(\frac{x_b(1-x_a)}{x_a(1-x_b)},Q_3;\alpha,\beta;\gamma,\epsilon;1), \tag{53}$$

$$C_2=H_G(\frac{x_b(1-x_a)}{x_a(1-x_b)},Q_3-(\frac{x_b(1-x_a)}{x_a(1-x_b)})\gamma(1-\epsilon);\gamma+\epsilon-\alpha,\gamma+\epsilon-\beta;\gamma,\epsilon;1). \tag{54}$$

Note that the factor multiplying $j(1-t)$ in y_4 can be written as

$$\exp[(2i\sqrt{-1+\frac{s^2}{(1-x_a)^2(1-x_b)2}})\ln 1-t]. \tag{55}$$

Then we can write our eq. (46,52) as

$$\exp -i(V\ln(1-t))y_1(t)$$
$$=C_1\exp -i(V\ln(1-t))y_3(1-t)+C_2\exp(iV\ln(1-t))j(1-t), \tag{56}$$

where

$$V=\sqrt{-1+\frac{s^2}{(1-x_a)^2(1-x_b)2}}. \tag{57}$$

Then our reflection coefficient is given by

$$R=|\frac{C_2}{C_1}|^2. \tag{58}$$

This calculation is a formal one. We have to make sure that the Heun function, found by expansion around zero, is also finite at the second singular point for appropriate values of the parameters of our wave equation. A proper Leaver analysis [32] shows that for $\frac{x_a(1-x_b)}{x_b(1-x_a)}<1$, these terms are convergent at unity.

5. Conclusion

Here we put the wave equation for a zero mass scalar particle coupled to the Wahlquist metric to a form to give the Heun solutions in the standard notation and calculated the reflection coefficient for a particle coming from infinity at the origin.

Note that this calculation has weak points. We made an approximation by taking $lnx = x - 1$. This is valid only when x is close to unity. We made few transformations on the independent variable. We can not guarantee that this approximation is valid in the range we work. We have no choice, though, since we can not solve differential equations whose coefficients are transcendental functions. An interesting point is that we can satisfy an important algebraic constraint, given in [21, 20], in reducing the first equation to the standard form only if we make this approximation after we calculate $\frac{dU}{dx}$ exactly, and apply this approximation only after the differentiation is done.

6. Acknowledgement

This work was written to celebrate the 72nd birthday of my dear colleague Prof. Tekin Dereli, whom I met close to fifty years ago, in 1972, when he was still studying for his doctoral degree and I was a fresh Ph.D. while I was visiting M.E.T.U.in Ankara. We are close friends ever since.

In the preparation of this work, I am grateful to Prof. Nadir Ghazanfari for giving us the key paper by Dekar et. al. I am also grateful to Prof. Barış Yapışkan for important discussions and correcting a vital mistake in my calculations. I also thank Dr. Oğuzhan Kaşıkçı for technical assistance.

This work is morally supported by Science Academy, Istanbul, Turkey.

7. References

[1] Dariescu M A and Dariescu C 2021 *Astrophys.Space Sci.* **366** 44
[2] Wahlquist H D 1968 *Phys. Rev.* **172**, 01291
[3] Wahlquist H D 1992 *J.Math. Phys.,Erratum* **33** 3235
[4] Wahlquist H D 1993, in *Rotating Objects and Relativistic Physics*, ed F.J.Chinea and L.M. Gonzalez-Romero, Lecture Notes in Physics Vol. 423 (Berlin: Springer-Verlag) p.55
[5] Stephani H, Kramer D, Maccallum, M, Hoenselaer C , Herlt E 2003 *Exact Solutions to Einstein's Field Equations* 2nd Ed. , Cambridge Monographs on Mathematical Physics, Cambridge University Press , 334
[6] Birkandan T and Hortacsu M 2021, *European Phys.J.* C **81** 389
[7] Kerr R P 1963,*Phys. Rev. Letters* **9** 237
[8] Demianski M and Newman E T 1963, *Bull. Acad.Polon. Sci.* **14**, 653
[9] Senovilla J M M 1987,*Phys. Lett* A **123**, 211
[10] Senovilla J M M 1993, *Rotating Objects and Relativistic Physics*, edited by Chinea F J and Gonzalez-Romero L M , Lecture Notes in Physics Vol. 423 (Berlin: Springer-Verlag), p.73
[11] Mars M 2001,*Phys. Rev.* D **63**, 064022
[12] Kramer D 1985, *Class. Quantum Grav.* **2**, L135 ; 1986 *Astron. Nachr.* **307** 309 (1986)
[13] Simon W 1984 *Gen. Relativ. Grav.* **16**, 465

[14] Papakostas T 1987, *J. Math. Phys.* **29**, 1445
[15] Bradley M, Fodor G, Marklund M and Perjes Z 2000 *Class.Quant.Grav.* **17** 351
[16] Bradley M, Fodor G and Perjes Z 2000 *Class.Quant.Grav.* **17** 2635
[17] Bradley M, Eriksson D, Fodor G and Racz I 2007, *Phys.Rev.* D **75** 024013
[18] Hinoue K, Houri T, Rugina C and Yasui Y 2014, *Phys. Rev.* D **90**, 024037
[19] Houri T ,Tanahashi N and Yasui Y 2020, *Class. Quant. Grav.* **37**075005
[20] Arscott F M 1995 , *Heun's Differential Equations* ed A. Ronveaux , Oxford University Press , p.7.
[21] Forsyth A 1902 *Theory of Differential Equations, Part III*, Cambridge University Press
[22] Arscott F M 1995 , *Heun's Differential Equations* ed A. Ronveaux , Oxford University Press , p. 19
[23] Suzuki H, Takasugi E and Umetsu H 1998 *Prog. Theor. Phys* **100**, 491
[24] Suzuki H, Takasugi E and Umetsu H *Prog. Theor. Phys.* **103**, 723
[25] Vieira H S and Kokkotas, K D 2021 *Phys.Rev.* D **104** 024035
[26] Newman E, Tamburino L and Unti T 1963,*J. Math. Phys.* **4** , 915
[27] A-Badavi A and Halilsoy M 2006,*Gen. Relativ. Gravit.* **38** , 17629
[28] Slavyanov S Y and Lay W 2000 *Special Functions, A Unified Theory Based on Singularities* Oxford University Press, Oxford
[29] Fiziev P P 2010 *Class. Quantum Gravity* **27**, 135001
[30] Hortacsu M 2018, *Adv. High Energy Phys.* **2018** 8621573.
[31] Dekar L, Chetouani L, and Hammann T F 1998, *J. Math. Phys.* **39** 2551
[32] Leaver E W 1986, *J. Math. Phys.* **27**, 1238

Dynamics of a fermion with dipole moments in external wave fields

Yuri N. Obukhov
Theoretical Physics Laboratory, Nuclear Safety Institute, Russian Academy of Sciences, B.Tulskaya 52, 115191 Moscow, Russia
E-mail: obukhov@ibrae.ac.ru

Abstract. We discuss the motion of a massive fermion spin $\frac{1}{2}$ particle in external wave field configurations when the Pauli type non-minimal interaction of the gauge field strengths with particle's dipole moments is taken into account. The exact solution of the relativistic Dirac-Pauli equation is presented for the case of the plane electromagnetic wave.

1. Introduction

The questions which spacetime symmetry underlies the high energy physics (in particular, whether the Lorentz symmetry is valid on all scales) and what is the actual structure of spacetime at small distances (specifically, whether there are deviations from the Riemannian geometry) belong to the most fundamental problems in the modern physics. As it was noticed quite early by Einstein [1], these questions must be answered by experiment and are not a matter "of a convention to be chosen on grounds of mere expediency." These issues are extensively analyzed (both from the theoretical and experimental sides) in the current literature, see [2, 3, 4, 5, 6, 7, 8, 9, 10, 11] for example.

To probe the fundamental symmetries and the geometric structure of spacetime, one has to study the dynamics of matter (bodies, particles, fields) that necessarily has a microstructure. The spin represents the most familiar and the most important physical example of such a microstructure. As soon as the microstructural properties of matter are taken into account, they are treated as physical sources of the gravitational field in the framework of the natural gauge-theoretic Yang-Mills type models based on the Poincaré group [12, 13, 14, 15, 16, 17, 18]. In general, one then needs to go beyond the minimal coupling principle and to allow for the non-minimal coupling of matter to electromagnetic and gravitational fields via the phenomenological Pauli type terms in the Lagrangian.

Here we consider the Dirac fermion with dipole moments and non-minimal coupling moving under the action of the wave fields. Our basic notations and conventions are the same as in [10, 11, 12].

2. Dirac fermion with dipole moments in external classical fields

The relativistic Dirac theory of a fermion particle with spin $\frac{1}{2}$, electric charge q and mass m is based on the Dirac Lagrangian

$$L = \frac{i\hbar}{2}\left(\overline{\psi}\gamma^\alpha D_\alpha\psi - D_\alpha\overline{\psi}\gamma^\alpha\psi\right) - mc\overline{\psi}\psi, \qquad (1)$$

that accounts for the minimal coupling of the spinor field ψ to the external gauge fields – electromagnetic 4-vector potential A_i and the gauge gravitational potentials $e_i^\alpha, \Gamma_i{}^{\alpha\beta} = -\Gamma_i{}^{\beta\alpha}$ (the coframe and the local Lorentz connection)– via the covariant derivative

$$D_\alpha\psi = e^i_\alpha\left(\partial_i\psi - \frac{iq}{\hbar}A_i\psi + \frac{i}{4}\Gamma_i{}^{\beta\gamma}\sigma_{\beta\gamma}\psi\right), \tag{2}$$

plus the nonminimal Lagrangian with possible Pauli type terms [10]

$$\begin{aligned} L^{\rm non} &= \frac{\mu}{2c}\,F_{ij}\,\overline{\psi}\sigma^{ij}\psi \quad + \quad \frac{\delta}{2}\,\widetilde{F}_{ij}\,\overline{\psi}\sigma^{ij}\psi \\ +\frac{\tau}{2}\,T_{ij}{}^\alpha&\left(\overline{\psi}\sigma^{ij}D_\alpha\psi - D_\alpha\overline{\psi}\sigma^{ij}\psi\right) \quad + \quad \frac{\rho}{2}\,R_{ij}{}^{\alpha\beta}\overline{\psi}\left(\sigma^{ij}\sigma_{\alpha\beta} + \sigma_{\alpha\beta}\sigma^{ij}\right)\psi, \end{aligned} \tag{3}$$

which describe the coupling of the dipole moments to the corresponding gauge field strengths. Here $\sigma_{\alpha\beta} = \frac{i}{2}(\gamma_\alpha\gamma_\beta - \gamma_\beta\gamma_\alpha)$ are generators of the Lorentz group, and for the multiplet of the gauge fields $\{A_i, e_i^\alpha, \Gamma_i{}^{\alpha\beta}\}$ the corresponding multiplet of the gauge field strengths $\{F_{ij}, T_{ij}{}^\alpha, R_{ij}{}^{\alpha\beta}\}$ encompasses the electromagnetic Maxwell tensor $F_{ij} = \partial_i A_j - \partial_j A_i$, the gravitational translational $T_{ij}{}^\alpha = \partial_i e_j^\alpha - \partial_j e_i^\alpha + \Gamma_{i\beta}{}^\alpha e_j^\beta - \Gamma_{j\beta}{}^\alpha e_i^\beta$ (torsion) and the gravitational rotational $R_{ij}{}^{\alpha\beta} = \partial_i\Gamma_j{}^{\alpha\beta} - \partial_j\Gamma_i{}^{\alpha\beta} + \Gamma_{i\gamma}{}^\beta\Gamma_j{}^{\alpha\gamma} - \Gamma_{j\gamma}{}^\beta\Gamma_i{}^{\alpha\gamma}$ (curvature) Poincaré field strengths. We denote the dual tensor $\widetilde{F}_{ij} = \frac{1}{2}\eta_{ijkl}F^{kl}$, as usual, with the totally antisymmetric Levi-Civita tensor η_{ijkl}.

The phenomenological model (3) represents the general case of a fermion with an anomalous magnetic dipole moment, an electric dipole moment, and the gravitational dipole moments. The corresponding coupling parameters have the dimension $[\mu] = [\mu_B]$ of the magnetic dipole (Bohr's or nuclear magneton), $[\delta] = [q\,l]$ of the electric dipole (charge times length), and $[\rho] = [\tau] = [\hbar l]$ (spin times length). Explicitly, they read

$$\mu = a\,\frac{q\hbar}{2m}, \qquad \delta = b\,\frac{q\hbar}{2mc}, \qquad \tau = p_\tau\,\frac{\hbar}{2mc}, \qquad \rho = p_\rho\,\frac{\hbar}{2mc}. \tag{4}$$

The dimensionless constant parameters $a = (g-2)/2$ (with the gyromagnetic factor g), b, p_τ and p_ρ characterize the magnitude of the anomalous magnetic, electric dipole, and gravitational dipole moments, respectively.

The structure of the gravitational (translational and Lorentzian) dipole moments is fixed with the help of the Gordon decomposition of the Noether currents of the Dirac spinor field [19, 20, 21, 22, 23], generalizing the early analysis [24, 25, 26, 27]. In general, one can also expect the nonminimal interaction between the gauge fields [28, 29].

3. Wave configurations

The wave processes belong to the most fundamental phenomena in physics, and the interaction of a fermion with an external wave (electromagnetic, and/or gravitational) is a highly interesting physical problem.

From the geometric point of view, the key for the description of a wave on a spacetime manifold is the null shear-free geodetic vector field k^i

$$k_ik^i = 0, \qquad k^j\nabla_jk^i = 0. \tag{5}$$

The plane electromagnetic wave is then given by the field strength tensor F_{ij} that solves the Maxwell equations and satisfies

$$k^jF_{ij} = 0, \qquad k_{[i}F_{jk]} = 0, \qquad F_{ij}F^{ij} = 0. \tag{6}$$

Similarly, the plane gravitational wave [30, 31] is described by the corresponding gravitational field strengths, the torsion $T_{ij}{}^{\alpha}$ and the curvature $R_{ij}{}^{\alpha\beta}$, that satisfy

$$k^j T_{ij}{}^{\alpha} = 0, \qquad k_{[i}T_{jk]}{}^{\alpha} = 0, \qquad T_{ij}{}^{\alpha}T^{ij}{}_{\beta} = 0, \tag{7}$$

$$k^j R_{ij}{}^{\alpha\beta} = 0, \qquad k_{[i}R_{jk]}{}^{\alpha\beta} = 0, \qquad R_{ij}{}^{\alpha\beta}R^{ij}{}_{\mu\nu} = 0. \tag{8}$$

The actual form of the wave configurations depends on the Lagrangians of the electromagnetic and the gravitational fields.

After these general preliminaries, we now specialize to the case of the flat Minkowski spacetime with $e_i^{\alpha} = \delta_i^{\alpha}, \Gamma_i{}^{\alpha\beta} = 0$. Then $T_{ij}{}^{\alpha} = 0$ and $R_{ij}{}^{\alpha\beta} = 0$, and in Maxwell's theory the electromagnetic plane wave solution is given by

$$F_{ij} = k_i a_j - k_j a_i, \tag{9}$$

where the wave covector k_i is constant, and the polarization 4-vector a_j depends on the phase $\varphi = k_i x^i$ only

$$a_i = a_i(\varphi), \qquad k^i a_i = 0. \tag{10}$$

Then one can check that Maxwell's equations $\partial_j F^{ij} = 0$, $\partial_{[j}F_{kl]} = 0$ are satisfied, and the field invariants are trivial $F_{ij}F^{ij} = 0$, $\eta^{ijkl}F_{ij}F_{kl} = 0$.

Without a loss of generality, we assume that the wave with the frequency ω propagates along the z-axis. Hence

$$k_i = \Big(\omega, 0, 0, -\frac{\omega}{c}\Big), \qquad \varphi = k_i x^i = \omega\Big(t - \frac{z}{c}\Big). \tag{11}$$

We consider an elliptically polarized plane wave with

$$a_i = \widehat{a}\,(0, \cos\zeta\cos\varphi, \sin\zeta\sin\varphi, 0)\,. \tag{12}$$

The constant parameters $\widehat{a}$ and ζ specify the amplitude and the polarization of the wave. In particular, when $\zeta = 0$ or $\zeta = \frac{\pi}{2}$ we obtain a linear polarization, whereas $\zeta = \pm\frac{\pi}{4}$ yields the circular polarization.

4. Solution of the Dirac-Pauli equation

The Lagrangian $L + L^{\rm non}$, obtained by combining (1) with (3), yields the Dirac-Pauli wave equation for the electrically neutral fermion particle in the Minkowski spacetime

$$i\hbar\gamma^i\partial_i\psi - mc\psi + \frac{\mu}{2c}F_{ij}\sigma^{ij}\psi + \frac{\delta}{2}\widetilde{F}_{ij}\sigma^{ij}\psi = 0. \tag{13}$$

Noticing that $\widetilde{F}_{ij}\sigma^{ij} = -i\gamma_5 F_{ij}\sigma^{ij}$ and $\frac{1}{2}F_{ij}\sigma^{ij} = i(\gamma\cdot k)(\gamma\cdot a)$, we recast (13) into

$$i\hbar\gamma^i\partial_i\psi - mc\psi + \xi(\gamma\cdot k)(\gamma\cdot a)\psi = 0. \tag{14}$$

Here $(\gamma\cdot k) = \gamma^i k_i$ and we denoted $\xi = \frac{i\mu}{c} + \gamma_5\delta$. Obviously $(\gamma\cdot k)(\gamma\cdot a) = -(\gamma\cdot a)(\gamma\cdot k)$.

In order to construct an exact solution for the fermion, we use the plane-wave ansatz

$$\psi = e^{-\frac{i}{\hbar}p_i x^i}\,\phi(\varphi), \tag{15}$$

where $p_i = (\mathcal{E}, -\boldsymbol{p})$. By differentiation we find $i\hbar\gamma^i\partial_i\psi = e^{-\frac{i}{\hbar}p_i x^i}\left[(\gamma\cdot p)\,\phi + i\hbar(\gamma\, k)\phi'\right]$, where the prime denotes the derivative with respect to the phase. As a result, the Dirac-Pauli equation (14) reduces to

$$\left[(\gamma\cdot p) - mc\right]\phi + (\gamma\cdot k)\left[i\hbar\phi' + (\gamma\cdot a)\xi\phi\right] = 0. \tag{16}$$

Multiplying from left by $(\gamma \cdot k)$ and taking into account that $(\gamma \cdot k)(\gamma \cdot k) = k_i k^i = 0$, we derive

$$(\gamma \cdot k) \left[(\gamma \cdot p) - mc\right] \phi = 0. \tag{17}$$

Since $(\gamma \cdot k)(\gamma \cdot p) + (\gamma \cdot p)(\gamma \cdot k) = 2(p \cdot k)$, we have an equivalent form

$$\left[(\gamma \cdot p) + mc\right] (\gamma \cdot k)\phi = 2(p \cdot k)\phi. \tag{18}$$

By multiplying with $[(\gamma \cdot p) - mc]$ from the left, we obtain the algebraic equation

$$\left[(\gamma \cdot p) - mc\right]\left[(\gamma \cdot p) + mc\right] (\gamma \cdot k)\phi = 2(p \cdot k)\left[(\gamma \cdot p) - mc\right]\phi. \tag{19}$$

Choosing the 4-vector p_i such that

$$p_i p^i = m^2 c^2, \qquad \mathcal{E}^2 - m^2 c^4 - c^2 \boldsymbol{p}^2 = 0, \tag{20}$$

we have two classes of solutions with the two signs of the energy $\mathcal{E} = \pm\varepsilon$, where the positive quantity is introduced by

$$\varepsilon := +\sqrt{m^2 c^4 + c^2 \boldsymbol{p}^2}. \tag{21}$$

As a result, we find $[(\gamma \cdot p) - mc]\,\phi = 0$, and consequently from (16) we find

$$i\hbar \Phi' - \xi(\gamma \cdot a)\Phi = 0, \tag{22}$$

where we denoted $\Phi = (\gamma \cdot k)\phi$. After solving this equation, we obtain from (18) the spinor

$$\phi = \frac{(\gamma \cdot p) + mc}{2(p \cdot k)}\Phi, \tag{23}$$

and consequently, the explicit solution reads

$$\psi = e^{-\frac{i}{\hbar}p_i x^i}\,\frac{(\gamma \cdot p) + mc}{2(p \cdot k)}\Phi. \tag{24}$$

It remains thus to solve the ordinary differential equation (22) for the 4-spinor $\Phi(\varphi)$. The formal solution of (22) can be written as a T-ordered or as a Magnus exponential [32, 33, 34]. However, one can derive an exact solution in an explicit form as follows.

For the elliptically polarized plane wave (11), (12) we introduce the two auxiliary variables $\chi = \chi(\varphi)$ and $\alpha = \alpha(\varphi)$ by

$$\chi(\varphi) = \sqrt{\cos^2\zeta\cos^2\varphi + \sin^2\zeta\sin^2\varphi}, \qquad \cos\alpha = \frac{\cos\zeta\cos\varphi}{\chi(\varphi)}, \qquad \sin\alpha = \frac{\sin\zeta\sin\varphi}{\chi(\varphi)}. \tag{25}$$

These variables are obviously related via $\tan\alpha = \tan\zeta\,\tan\varphi$. As a result, (12) is recast into

$$a_i = \widehat{a}\chi\left(0, \cos\alpha, \sin\alpha, 0\right). \tag{26}$$

For $\zeta = \pm\frac{\pi}{4}$, we find $\chi = 1/\sqrt{2}$ and $\alpha = \pm\varphi$.

The crucial observation is as follows. The Lorentz transformation

$$\Lambda^j{}_i = \begin{pmatrix} 1 & 0 & 0 & 0 \\ 0 & \cos\alpha & -\sin\alpha & 0 \\ 0 & \sin\alpha & \cos\alpha & 0 \\ 0 & 0 & 0 & 1 \end{pmatrix} \tag{27}$$

brings the polarization vector to

$$\Lambda^j{}_i a_j = \overset{\circ}{a}_i, \qquad \overset{\circ}{a}_i = \widehat{a}\chi\,(0,1,0,0)\,. \tag{28}$$

The corresponding spinor transformation U is defined by $U^{-1}\gamma^i U = \Lambda^i{}_j\gamma^j$, and for (27) we find

$$U = \exp\left(-i\Sigma^3\frac{\alpha}{2}\right) = \cos\frac{\alpha}{2} - i\Sigma^3\sin\frac{\alpha}{2}, \qquad U^{-1} = \cos\frac{\alpha}{2} + i\Sigma^3\sin\frac{\alpha}{2}, \tag{29}$$

where the spin matrix is as usual

$$\Sigma^3 = \begin{pmatrix} \sigma^3 & 0 \\ 0 & \sigma^3 \end{pmatrix}. \tag{30}$$

Consequently, we have

$$(\gamma\cdot a) = U(\gamma\cdot\overset{\circ}{a})U^{-1}, \tag{31}$$

where, explicitly, $(\gamma\cdot\overset{\circ}{a}) = \widehat{a}\chi\gamma^1$.

It is now straightforward to verify that $(U^{-1})'U = \frac{i\alpha'}{2}\Sigma^3$, and thus substituting (31) into (22), and introducing $\Psi = U^{-1}\Phi$, we find

$$i\hbar\left[\Psi' - (U^{-1})'U\Psi\right] - \xi(\gamma\cdot\overset{\circ}{a})\Psi = 0. \tag{32}$$

With the help of the new auxiliary objects

$$\widehat{\lambda} = \frac{2\xi}{i\hbar}\frac{\widehat{a}\chi}{\alpha'} = \widetilde{\lambda}_m - i\gamma_5\widetilde{\lambda}_e, \qquad \widetilde{\lambda}_m = \frac{2\widehat{a}\mu}{c\hbar}\frac{\chi}{\alpha'}, \qquad \widetilde{\lambda}_e = \frac{2\widehat{a}\delta}{{}^-\hbar}\frac{\chi}{\alpha'}, \tag{33}$$

we recast (32) into

$$\frac{1}{\alpha'}\Psi' - \frac{1}{2}\left(i\Sigma^3 + \widehat{\lambda}\gamma^1\right)\Psi = 0. \tag{34}$$

We now specialize to the circularly polarized case with $\zeta = \pm\frac{\pi}{4}$. Then $\chi = 1/\sqrt{2}$ and $\alpha = \pm\varphi$ and thus $\alpha' = \pm 1$. Substituting this into (34), we find

$$\Psi' - \frac{1}{2}\left(\pm i\Sigma^3 + \widehat{\lambda}\gamma^1\right)\Psi = 0, \tag{35}$$

where $\widehat{\lambda} = \lambda_m - i\gamma_5\lambda_e$ with

$$\lambda_m = \frac{\widehat{a}\mu}{\sqrt{2}c\hbar}, \qquad \lambda_e = \frac{\widehat{a}\delta}{\sqrt{2}\hbar}. \tag{36}$$

Denote $M = \pm i\Sigma^3 + \widehat{\lambda}\gamma^1$. It is straightforward to verify that

$$M^\dagger = -M, \qquad M^2 = -1 - \lambda^2, \qquad \lambda^2 = \lambda_e^2 + \lambda_m^2, \tag{37}$$

and we define

$$\widehat{M} = \frac{M}{\sqrt{1+\lambda^2}}, \qquad \widehat{M}^\dagger = -\widehat{M}, \qquad \widehat{M}^2 = -1. \tag{38}$$

Then we obtain the solution of (35) in the form

$$\Psi = \exp\left(\frac{1}{2}M\varphi\right)\Psi_0 = \left(\cos\frac{\sqrt{1+\lambda^2}}{2}\varphi + \widehat{M}\sin\frac{\sqrt{1+\lambda^2}}{2}\varphi\right)\Psi_0. \tag{39}$$

Collecting together all intermediate results, we thus have the solution

$$\psi = e^{-\frac{i}{\hbar}p_i x^i}\,\frac{(\gamma\cdot p)+mc}{2(p\cdot k)}\exp\left(\mp i\Sigma^3\frac{\varphi}{2}\right)\exp\left([\pm i\Sigma^3+\widehat{\lambda}\gamma^1]\frac{\varphi}{2}\right)\Psi_0. \tag{40}$$

The form of the constant spinor Ψ_0 is fixed by the sign of the energy and the normalization condition. For each sign of the energy we have two linearly independent solutions. To describe them, we notice that in the absence of the dipole moments, $\xi = 0$, the Dirac-Pauli equation (14) reduces to the free Dirac equation that has the two ($A = 1, 2$) linearly independent solutions for the positive energy (denoting $\varepsilon' = \varepsilon + mc^2$)

$$\psi = e^{-\frac{i}{\hbar}\varepsilon t+\frac{i}{\hbar}\boldsymbol{p}\cdot\boldsymbol{x}}\,u^{(A)}(\boldsymbol{p}), \qquad u^{(A)}(\boldsymbol{p}) = \frac{1}{\sqrt{2\varepsilon}}\begin{pmatrix}\sqrt{\varepsilon'}\,\varphi^{(A)}\\ \frac{c\boldsymbol{p}\cdot\boldsymbol{\sigma}}{\sqrt{\varepsilon'}}\,\varphi^{(A)}\end{pmatrix}, \tag{41}$$

and the two ($A = 1, 2$) linearly independent solutions for the negative energy

$$\psi = e^{+\frac{i}{\hbar}\varepsilon t+\frac{i}{\hbar}\boldsymbol{p}\cdot\boldsymbol{x}}\,v^{(A)}(\boldsymbol{p}), \qquad v^{(A)}(\boldsymbol{p}) = \frac{1}{\sqrt{2\varepsilon}}\begin{pmatrix}-\frac{c\boldsymbol{p}\cdot\boldsymbol{\sigma}}{\sqrt{\varepsilon'}}\,\varphi^{(A)}\\ \sqrt{\varepsilon'}\,\varphi^{(A)}\end{pmatrix}, \tag{42}$$

with two possible choices of linearly independent 2-spinors $\varphi^{(1)} = \begin{pmatrix}1\\0\end{pmatrix}$, and $\varphi^{(2)} = \begin{pmatrix}0\\1\end{pmatrix}$.

By making use of these observations, for the positive energy we choose $\Psi_0 = N_{(+)}u^{(A)}$, and recast the solution (40) into

$$\psi = e^{-\frac{i}{\hbar}\varepsilon t+\frac{i}{\hbar}\boldsymbol{p}\cdot\boldsymbol{x}}\,w^{(A)}(\boldsymbol{p},\varphi), \qquad w^{(A)}(\boldsymbol{p},\varphi) = \sum_{B=1}^{2} u^{(B)}(\boldsymbol{p})\,\overset{(+)}{\kappa}{}^{BA}(\boldsymbol{p},\varphi), \tag{43}$$

where we introduced a 2×2 matrix

$$\overset{(+)}{\kappa}{}^{BA}(\boldsymbol{p},\varphi) = \frac{\varepsilon N_{(+)}}{c(k\cdot p)}\overline{u}^{(B)}\exp\left(\mp i\Sigma^3\frac{\varphi}{2}\right)\exp\left([\pm i\Sigma^3+\widehat{\lambda}\gamma^1]\frac{\varphi}{2}\right)u^{(A)}. \tag{44}$$

In a similar way, for the negative energy we choose $\Psi_0 = N_{(-)}v^{(A)}$, and recast the solution (40) into

$$\psi = e^{+\frac{i}{\hbar}\varepsilon t+\frac{i}{\hbar}\boldsymbol{p}\cdot\boldsymbol{x}}\,z^{(A)}(\boldsymbol{p},\varphi), \qquad z^{(A)}(\boldsymbol{p},\varphi) = \sum_{B=1}^{2} v^{(B)}(\boldsymbol{p})\,\overset{(-)}{\kappa}{}^{BA}(\boldsymbol{p},\varphi), \tag{45}$$

where we introduced a 2×2 matrix

$$\overset{(-)}{\kappa}{}^{BA}(\boldsymbol{p},\varphi) = \frac{\varepsilon N_{(-)}}{c(k\cdot p)}\overline{v}^{(B)}\exp\left(\mp i\Sigma^3\frac{\varphi}{2}\right)\exp\left([\pm i\Sigma^3+\widehat{\lambda}\gamma^1]\frac{\varphi}{2}\right)v^{(A)}. \tag{46}$$

The final step is to find the values of the normalization constants $N_{(\pm)}$. Since in the absence of the dipole moments, $\xi = 0$, hence $\widehat{\lambda} = 0$, the solutions (43) and (45) should obviously reduce to the free particle solutions (41) and (42), respectively, we fix $N_{(\pm)} = \pm(k\cdot p)/mc$.

The results obtained demonstrate that an external wave affects fermion's spin. The evolution of the spin state is described by the matrices $\overset{(\pm)}{\kappa}{}^{AB}$, and for the limiting case $\xi = 0$ the effect disappears $\overset{(\pm)}{\kappa}{}^{AB} = \delta^{AB}$.

5. Discussion

The first exact solution for the Dirac equation with the minimal coupling between a charged fermion and the plane electromagnetic field was obtained by Volkov [35, 36]. Our results extend Volkov's solution to the non-minimal coupling case (we presented here the solution for the neutral fermion, the extension to the charged case is straightforward but the corresponding formulas are too bulky and will be given elsewhere).

The dynamics of a neutron in the electromagnetic field, and in particular, in the plane wave was previously discussed in [37, 38, 39, 40, 41, 42, 43, 44, 45, 46, 47]. Our analysis now takes into account both the magnetic and the electric dipole moments. The crucial new technical point that underlies the current derivation is the use of the Lorentz transformation (27) and the corresponding spinor transformation (29) to simplify the polarization vector of the electromagnetic wave.

Among possible applications one can mention a number of the physically important problems such as the scattering and the radiation processes of a fermion particle in the external wave configurations, which are of interest both for the high-energy and the heavy-ion physics. These issues will be reported elsewhere.

The results obtained can be extended also to the case of the fermion with dipole moments in the plane gravitational wave (recall the early relevant work [11, 48, 49]), obtained as an exact solution of the Einstein field equation, as well as to the case of superposition of the electromagnetic and the gravitational waves.

Acknowledgments

It is a great pleasure for me to write this paper for the Festschrift of Tekin Dereli, who made important contributions to the various areas of the theoretical and mathematical physics, including the development of the gauge gravity theory, and the studies of the spin dynamics and the wave solutions in gravity and supergravity. My work was supported in part by the Russian Foundation for Basic Research (Grant No. 18-02-40056-mega).

References

[1] Einstein A 1921 *Sitzungsber. Preuss. Akad. Wiss. Phys.-math. Klasse* **1** 123
[2] Adak M, Dereli T, and Ryder L H 2001 *Class. Quantum Grav.* **18** 1503
[3] Adak M, Dereli T, and Ryder L H 2003 *Int. J. Mod. Phys. D* **12** 145
[4] Tasson J D 2012 *Phys. Rev. D* **86** 124021
[5] Kostelecký V A 2004 *Phys. Rev. D* **69** 105009
[6] Bonder Y 2013 *Phys. Rev. D* **88** 105011
[7] Puetzfeld D and Obukhov Y N 2008 *Phys. Lett. A* **372** 6711
[8] Hehl F W, Obukhov Y N, and Puetzfeld D 2013 *Phys. Lett. A* **377** 1775
[9] Puetzfeld D and Obukhov Y N 2013 *Phys. Lett. A* **377** 2447
[10] Obukhov Y N, Silenko A J, and Teryaev O V 2014 *Phys. Rev. D* **90** 124068
[11] Obukhov Y N, Silenko A J, and Teryaev O V 2017 *Phys. Rev. D* **96** 105005
[12] Hehl F W, McCrea J D, Mielke E W, and Ne'eman Y 1955 *Phys. Rept.* **258** 1
[13] Blagojević M 2002 *Gravitation and gauge symmetries* (Institute of Physics: Bristol)
[14] Blagojević M and Hehl F W (Eds) 2013 *Gauge Theories of Gravitation. A Reader with Commentaries* (Imperial College Press: London)
[15] Ponomarev V N, Barvinsky A, and Obukhov Y N 2017 *Geometrodynamical Methods and the Gauge Approach to the Gravitational Interaction* (Nauka: Moscow)
[16] Obukhov Y N 2006 *Int. J. Geom. Meth. Mod. Phys.* **3** 95
[17] Obukhov Y N 2018 *Int. J. Geom. Meth. Mod. Phys.* **15** (supp01) 1840005
[18] Hehl F W and Obukhov Y N 2020 in: *"One Hundred Years of Gauge Theory: Past, Present and Future Perspectives"*, Eds. S. De Bianchi and C. Kiefer, Fundamental Theories of Physics **199** (Springer: Cham) pp. 217-252
[19] Gordon W 1928 *Zeits. Physik* **50** 630
[20] Seitz M 1984 *Ann. Physik (Leipzig)* **41** 280
[21] Seitz M 1986 *Class. Quantum Grav.* **3** 175

[22] Obukhov Y N 1998 *Acta Phys. Polon. B* **29** 1131
[23] Hehl F W, Macías A, Mielke E W, and Obukhov Y N 1998 in: *"On Einstein's path", Essays in honor of E.Schucking*, Ed. A. Harvey (Springer: New York) 257
[24] Chou T T and Yang C N 1976 *Nucl. Phys. B* **107** 1
[25] Kobzarev I Y and Okun L B 1963 *Sov. Phys. JETP* **16** 1343
[26] Leitner J and Okubo S 1964 *Phys. Rev.* **136** B1542
[27] Hiida K and Yamaguchi Y 1965 *Prog. Theor. Phys. Suppl.* **E65** 261
[28] Baykal A and Dereli T 2015 *Phys. Rev. D* **92** 065018
[29] Dereli T and Sert Ö 2011 *Phys. Rev. D* **83** 065005
[30] Baykal A 2016 *Turk. J. Phys.* **40** 77
[31] Obukhov Y N 2017 *Phys. Rev. D* **95** 084028
[32] Feynman R P 1951 *Phys. Rev.* **84** 108
[33] Magnus W 1954 *Comm. Pure and Appl. Math.* **7** 649
[34] Blanes S, Casas F, Oteo J A, and Ros J 2009 *Phys. Rep.* **470** 151
[35] Wolkow D M 1935 *Zeits. Physik* **94** 250
[36] Volkov D M 1937 *Zh. Eksp. Teor. Fiz.* **7** 1286 (in Russian)
[37] Redmond P J 1965 *J. Math. Phys.* **6** 1163
[38] Baranov I G 1967 *Sov. Phys. J.* **10** 23
[39] Ternov I M, Bagrov V G, and Klimenko Y I 1968 *Sov. Phys. J.* **11** 29
[40] Charkrabarti A 1968 *Nuovo Cim. A* **56** 604
[41] Radyuk A F and Fedorov F I 1973 *Dokl. AN SSSR* **211** 1091 (in Russian)
[42] Bagrov V G, Bozrikov P V, Gitman D M, Klimenko Y I, and Khudomyasov A I 1974 *Sov. Phys. J.* **17** 1027
[43] Becker W and Mitter H 1974 *J. Phys. A: Math., Nucl. and Gen.* **7** 1266
[44] Salamin Y I 1993 *J. Phys. A: Math. Gen.* **26** 6067
[45] Simonovits R and Mitter H 1993 *Phys. Lett. A* **181** 109
[46] Sen Gupta N D 1995 *Pramana* **45** 327
[47] Savchenko O Y 1997 *JETP* **84** 656
[48] Dereli T and Tucker R W 1982 *J. Phys. A: Math. Gen.* **15** 1625
[49] Gonçalves B, Obukhov Y N, and Shapiro I S 2007 *Phys. Rev. D* **75** 124023

A novel approach to autoparallels for the theories of symmetric teleparallel gravity

Caglar Pala[1], **Muzaffer Adak**[2]
[1] Erciyes University, 38030 Kayseri, Turkey
[2] Pamukkale University, 20017 Denizli, Turkey

E-mail: [1]`caglar.pala@gmail.com`, [2]`madak@pau.edu.tr`

Abstract. Although the autoparallel curves and the geodesics coincide in the Riemannian geometry in which only the curvature is nonzero among the nonmetricity, the torsion and the curvature, they define different curves in the non-Riemannian ones. We give a novel approach to autoparallel curves and geodesics for theories of the symmetric teleparallel gravity written in the coincident gauge. Then we apply our autoparallel equation to a Schwarzschild-type metric and give remarks about dark matter and orbit equation.

Keywords: Non-Riemannian geometry, Geodesic, Autoparallel curve

Dedicated to Tekin Dereli in honour of his 72nd birthday

1. Introduction

There have been two main ingredients for the theories of gravity since Isaac Newton; field equations and trajectory equations. The field equations are cast for describing the dynamics of a source and the trajectory ones are the for determining path followed by a spinless massive point test particle. For the Newton's theory of gravity they are written respectively as

$$\nabla^2\phi = 4\pi G\rho\,, \tag{1a}$$

$$\frac{d^2\vec{r}}{dt^2} + \vec{\nabla}\phi = 0\,, \tag{1b}$$

where G is the Newton's constant of gravity, $\vec{r}$ is the three-dimensional Euclidean position vector of the test particle, t is the time disjoint from the Euclidean space, ρ is the volume density of mass of the source, $\vec{\nabla}$ is the gradient operator in the three-dimensional Euclidean space and ϕ is the gravitational field from a source at the position of $\vec{r}$. On the other hand, in the Einstein theory of gravity, the general theory of relativity, the effect of the spacetimes curvature on matter fields is interpreted as the gravitational force and one assumes that a massive point test particle has spacetime history that matches to the time-like geodesic associated with the spacetime metric.

$$R_{\mu\nu} - \frac{1}{2}Rg_{\mu\nu} = 4\pi G T_{\mu\nu}\,, \tag{2a}$$

$$\frac{d^2x^\mu}{d\tau^2} + \widetilde{\omega}^\mu_{\ \nu\lambda}\frac{dx^\nu}{d\tau}\frac{dx^\lambda}{d\tau} = 0\,, \tag{2b}$$

where $g_{\mu\nu}$ is the metric tensor of $(1+3)$-dimensional pseudo-Riemannian spacetime, $R_{\mu\nu}$ is its Ricci curvature tensor, R is its curvature scalar, $T_{\mu\nu}$ is the energy-momentum tensor for the source, τ is a affine parameter (proper time in practice), x^μ denotes a chosen coordinate system of the spacetime, $\widetilde{\omega}^{\mu}_{\nu\lambda}$ are the Christoffel symbols (the Levi-Civita connections) of the metric. The first equation is known as the Einstein's equation and the second one is geodesic equation of the pseudo-Riemannian spacetime.

These prescriptions were compatible with the observations for a while. But there are pretty much reasons to go beyond Einstein's theory of gravity coming from the experimental side such as dark matter and dark energy and also from theoretical side such as vain efforts of obtaining a quantum gravity, see for example [1]-[3]. The correlation between the general relativity and the geometry of the spacetime makes it natural to explore modifications to the general relativity based on extending its geometrical framework. In these cases, the geodesic hypothesis based on autoparallels of the Levi-Civita connection may need to be extended in alternative theories of gravity formulated in the non-Riemannian geometries. Shortly the geodesic equation (2b) is replaced with the autoparallel equation

$$\frac{d^2x^\mu}{d\tau^2} + \omega^{\mu}_{\nu\lambda}\frac{dx^\nu}{d\tau}\frac{dx^\lambda}{d\tau} = 0 \tag{3}$$

where $\omega^{\mu}_{\nu\lambda}$ are the non-Riemannian connections containing torsion and/or nonmetricity apart from the Levi-Civita connections. For example, in the reference [4] the authors investigated the geodesic hypothesis based on autoparallels of the Levi-Civita connection in the Brans-Dicke theory in the Riemann-Cartan spacetime and thereby found that if the Brans-Dicke scalar field couples significantly to matter and test particle moves on such worldline, the commenly accepted methods based on the conventional geodesic hypothesis may need refinement. Those results were improved in the work [5] in which it was of interest to compare the autoparallel orbits derived from different metric compatible spacetime connections for a spinless test particle in the metric of a spinning source. In a recent work [6], autoparallels in alternative gravity have been treated again. The authors explicitly derived autoparallels as effective post-Riemannian geometric constructs, and at the same time they argued against postulating autoparallels as fundamental equations of motion for test bodies in alternative gravity theories.

There is not an agreement on which equation, (geodesic of the Levi-Civita connection or autoparallel of the full connection) should be used as the worldline of a spinless test particle at vicinity of a gravitating source in non-Riemennian spacetime geometries. Experiments will guide us whether particles follow metric geodesic paths or their trajectories are in turn autoparallel curves for the full affine connection [7]. However, we can discuss which one seems more natural from a theoretical perspective. At this context the reference [8] states that metric geodesic trajectories seem better aligned with our current understanding of physics.

One of the alternative methods for modifying Einstein's theory of general relativity written in the Riemannian geometry is to propose new theories in a non-Riemannian geometry. Together with our collaborators we have published a series of researches on the modified gravity theories formulated in terms of only nonmetricity without torsion and curvature, the so-called symmetric teleparallel gravity [9]-[14] just after the pioneering work [15]. These works have attracted noteworthy interest in the scientific community [16]-[24]. All the works above investigated new gravity models by starting from a Lagrangian. Then the field equations have been obtained via variation. Finally some possible solutions have been obtained and analyzed. But the trajectory of a test particle has not been discussed so much in the theories of symmetric teleparallel gravity [25],[26]. Therefore this complementary work was performed. In a very recent work in this direction [27], the authors consider the geodesic deviation equation, describing the relative accelerations of nearby particles, and the Raychaudhuri equation, giving the evolution of the kinematical quantities associated with deformations in the Weyl-type $f(Q,T)$ gravity, in which the nonmetricity Q is represented in the standard Weyl form, fully determined by the Weyl vector, while T represents the trace of the matter energy–momentum tensor.

2. Mathematical preliminaries

Spacetime, in general, is denoted by $\{M, g, \nabla\}$ where M is orientable and differentiable manifold, $g = g_{\alpha\beta}e^\alpha \otimes e^\beta$ is the metric tensor written in terms of 1-forms (or coframe) e^α and ∇ is connection associated with connection 1-forms $\omega^\alpha{}_\beta$. In the language of the exterior algebra, the Cartan structure equations define nonmetricity 1-forms, torsion 2-forms and curvature 2-forms, respectively

$$Q_{\alpha\beta} := -\frac{1}{2}Dg_{\alpha\beta} = \frac{1}{2}(-dg_{\alpha\beta} + \omega_{\alpha\beta} + \omega_{\alpha\beta})\,, \tag{4a}$$

$$T^\alpha := De^\alpha = de^\alpha + \omega^\alpha{}_\beta \wedge e^\beta\,, \tag{4b}$$

$$R^\alpha{}_\beta := D\omega^\alpha{}_\beta := d\omega^\alpha{}_\beta + \omega^\alpha{}_\gamma \wedge \omega^\gamma{}_\beta\,, \tag{4c}$$

where $\wedge$ denotes the exterior product, d the exterior derivative and D the covariant exterior derivative [28]. Geometrically, the nonmetricity tensor measures the deformation of length and angle standards during parallel transport. For example, after parallel transportation of a vector along a closed curve, the length of the final vector may be different from that of the initial vector. On the other hand, torsion relates to the translational group. Moreover it is sometimes said that closed parallelograms do not exist in spacetime with torsion. It can be restated in terms of the picture of manifold and bundle as that after a vector is parallel carried around a closed path on a manifold, if torsion is not vanishing, there will be an interval between initial and final points on the fiber of the bundle over the manifold. Finally, curvature is related to the linear group. That is, when a vector is parallel transported along a closed loop, the vector undergoes a rotation due to curvature. These three tensors satisfy the Bianchi identities:

$$DQ_{\alpha\beta} = \frac{1}{2}(R_{\alpha\beta} + R_{\beta\alpha})\,, \tag{5a}$$

$$DT^\alpha = R^\alpha{}_\beta \wedge e^\beta\,, \tag{5b}$$

$$DR^\alpha{}_\beta = 0\,. \tag{5c}$$

The full connection 1-forms can be decomposed uniquely as follows [1]-[3], [29]:

$$\omega^\alpha{}_\beta = \underbrace{(g^{\alpha\gamma}dg_{\gamma\beta} + p^\alpha{}_\beta)/2 + \widetilde{\omega}^\alpha{}_\beta}_{Metric} + \underbrace{K^\alpha{}_\beta}_{Torsion} + \underbrace{q^\alpha{}_\beta + Q^\alpha{}_\beta}_{Nonmetricity} \tag{6}$$

where $\widetilde{\omega}^\alpha{}_\beta$ Levi-Civita connection 1-forms

$$\widetilde{\omega}^\alpha{}_\beta \wedge e^\beta = -de^\alpha\,, \tag{7}$$

$K^\alpha{}_\beta$ contortion tensor 1-forms,

$$K^\alpha{}_\beta \wedge e^\beta = T^\alpha\,, \tag{8}$$

and anti-symmetric 1-forms

$$q_{\alpha\beta} = -(\iota_\alpha Q_{\beta\gamma})e^\gamma + (\iota_\beta Q_{\alpha\gamma})e^\gamma\,, \tag{9}$$

$$p_{\alpha\beta} = -(\iota_\alpha dg_{\beta\gamma})e^\gamma + (\iota_\beta dg_{\alpha\gamma})e^\gamma\,, \tag{10}$$

where ι_α denotes the interior product with respect to the basis vector ∂_α dual to e^α; $e^\alpha(\partial_\beta) = \delta^\alpha_\beta$. This decomposition is self-consistent. To see that it is enough to multiply (6) from right by e^β

and to use definitions above. As moving indices vertically in front of both d and D, a special attention is needed because in general it may be $dg_{\alpha\beta} \neq 0$ and $Dg_{\alpha\beta} \neq 0$. Symmetric part of the full connection comes from (4a)

$$\omega_{(\alpha\beta)} = Q_{\alpha\beta} + \frac{1}{2} dg_{\alpha\beta} \tag{11}$$

and the remainder is anti-symmetric part

$$\omega_{[\alpha\beta]} = \frac{1}{2} p_{\alpha\beta} + \widetilde{\omega}_{\alpha\beta} + K_{\alpha\beta} + q_{\alpha\beta} \ . \tag{12}$$

In a holonomic or coordinate basis in which $e^\alpha = dx^\alpha$, then $de^\alpha = 0$ via the Poincare lemma, this decomposition becomes

$$\omega^\alpha{}_\beta = \frac{1}{2} g^{\alpha\sigma} (\partial_\gamma g_{\sigma\beta} + \partial_\beta g_{\sigma\gamma} - \partial_\sigma g_{\beta\gamma}) e^\gamma + K^\alpha{}_\beta + q^\alpha{}_\beta + Q^\alpha{}_\beta \tag{13}$$

where we used $\iota_\alpha d \equiv \partial_\alpha$. We notice that the first group on the right hand side coincide with the Christoffel symbols. From now on we always work with the coordinate frame.

3. The coincident gauge in the symmetric teleparallel geometry

If only $Q_{\alpha\beta} = 0$, the connection is said to be metric compatible; Einstein-Cartan geometry. If both $Q_{\alpha\beta} = 0$ and $T^\alpha = 0$, the connection is called Levi-Civita; pseudo-Riemannian geometry. If both $R^\alpha{}_\beta = 0$ and $Q_{\alpha\beta} = 0$, it is called teleparallel (Weitzenböck) geometry. If both $R^\alpha{}_\beta = 0$ and $T^\alpha = 0$, it is called symmetric teleparallel (or Minkowski-Weyl) geometry [7]. That is, in the symmetric teleparallel geometry only the nonmetricity tensor is nonzero: $Q_{\alpha\beta} \neq 0 \, , \quad T^\alpha = 0 \, , \quad R^\alpha{}_\beta = 0$. Thus the Cartan structure equations (4) give a set of nonlinear dynamical equation for the full connection, $\omega^\alpha{}_\beta$, in coordinate (holonomic) frame in which $e^\alpha = dx^\alpha$ and then $de^\alpha = 0$,

$$\frac{1}{2}(-dg_{\alpha\beta} + \omega_{\alpha\beta} + \omega_{\alpha\beta}) \neq 0 \, , \tag{14a}$$

$$\omega^\alpha{}_\beta \wedge dx^\beta = 0 \, , \tag{14b}$$

$$d\omega^\alpha{}_\beta + \omega^\alpha{}_\gamma \wedge \omega^\gamma{}_\beta = 0 \, . \tag{14c}$$

Here the full connection 1-form can not be solved analytically. Meanwhile, it is worthy to remember that we could have done it in the Riemannian geometry. However, one special solution is to choose $\omega^\alpha{}_\beta = 0$ yielding $Q_{\alpha\beta} = -\frac{1}{2} dg_{\alpha\beta} \neq 0 \, , \quad T^\alpha = 0 \, , \quad R^\alpha{}_\beta = 0$. This choice is called the *coincident gauge*[1].

3.1. Autoparallels

In the Riemannian geometry one requires that the tangent vector, $\mathcal{T}^\mu$, to the autoparallel curve, $x^\mu(\tau)$, points in the same direction as itself when parallel propagated, and demands that it maintains the same length [30]

$$D\mathcal{T}^\mu = 0 \, . \tag{15}$$

On the other hand, since intuitively autoparallel curves are "those as straight as possible" we do not demand the vector to keep the same length during parallel propagation in STPG. It is

[1] When we firstly realized this trick, we called it as *a gauge fixing* in our paper [13], later we read the nomenclature *coincident gauge* in the literature.

known that nonmetricity is related to length and angle standards at parallel transportation. Therefore we prescribe the parallel propagation of the tangent vector as

$$D\mathcal{T}^\mu = (aQ^\mu{}_\nu + bq^\mu{}_\nu)\mathcal{T}^\nu + cQ\mathcal{T}^\mu \tag{16}$$

where $\mathcal{T}^\mu = dx^\mu(\tau)/d\tau$ is the tangent vector to the curve $x^\mu(\tau)$ with an affine parameter τ, $Q^\mu{}_\nu$ is the nonmetricity 1-form, $Q = Q^\mu{}_\mu$ is the Weyl 1-form, $q^\mu{}_\nu$ is defined by (9) in terms of the nonmetricity and a, b, c are arbitrary constants. This prescription of parallel transport is our novel approach. Here since we are in the coincident gauge, $\omega^\alpha{}_\beta = 0$, we obtain $D\mathcal{T}^\mu = d\mathcal{T}^\mu = (\partial_\alpha \mathcal{T}^\mu)dx^\alpha$. Moreover, we write all one-forms in their components $Q^\mu{}_\nu = Q^\mu{}_{\nu\alpha}dx^\alpha$, $q^\mu{}_\nu = q^\mu{}_{\nu\alpha}dx^\alpha$ and $Q = Q^\nu{}_{\nu\alpha}dx^\alpha$. Then the equation (16) yields

$$\partial_\alpha \mathcal{T}^\mu = (aQ^\mu{}_{\nu\alpha} + bq^\mu{}_{\nu\alpha})\mathcal{T}^\nu + cQ^\nu{}_{\nu\alpha}\mathcal{T}^\mu \,. \tag{17}$$

Now after multiplying this with $\mathcal{T}^\alpha$ and then writing $\mathcal{T}^\alpha\partial_\alpha := \frac{d}{d\tau}$ on the left hand side and $\mathcal{T}^\mu = dx^\mu/d\tau$ on the both sides, we arrive at

$$\frac{d^2x^\mu}{d\tau^2} = (aQ^\mu{}_{\nu\alpha} + bq^\mu{}_{\nu\alpha})\frac{dx^\nu}{d\tau}\frac{dx^\alpha}{d\tau} + cQ^\nu{}_{\nu\alpha}\frac{dx^\mu}{d\tau}\frac{dx^\alpha}{d\tau} \,. \tag{18}$$

Here by using $Q_{\alpha\beta} = -\frac{1}{2}dg_{\alpha\beta}$, the result of the coincident gauge, we get $Q^\mu{}_{\nu\alpha} = -\frac{1}{2}g^{\mu\beta}(\partial_\alpha g_{\beta\nu})$ and accordingly $q^\mu{}_{\nu\alpha} = -\frac{1}{2}g^{\mu\beta}(\partial_\nu g_{\alpha\beta} - \partial_\beta g_{\alpha\nu})$. Consequently, we express the autoparallel curve equation in terms of the metric as follows

$$\frac{d^2x^\mu}{d\tau^2} = g^{\mu\beta}\left[-\frac{a+b}{4}(\partial_\alpha g_{\beta\nu} + \partial_\nu g_{\beta\alpha}) + \frac{b}{2}(\partial_\beta g_{\alpha\nu})\right]\frac{dx^\nu}{d\tau}\frac{dx^\alpha}{d\tau} - \frac{c}{2}g^{\nu\beta}(\partial_\alpha g_{\beta\nu})\frac{dx^\mu}{d\tau}\frac{dx^\alpha}{d\tau} \tag{19}$$

where we symmetrized the first term of the square parentheses $(\nu\alpha) = \frac{1}{2}(\nu\alpha + \alpha\nu)$. We notice that if we fix $a = b = 1$ and $c = 0$, then this equation becomes the same as the autoparallel curve of the Riemannian geometry.

3.2. Geodesics

Intuitively, geodesics are "curves as short as possible". An interval between two infinitesimal points is given by the metric

$$ds^2 = g_{\mu\nu}dx^\mu dx^\nu \tag{20}$$

we can parameterize the curve between endpoints as $x^\mu = x^\mu(\tau)$, then we obtain (in a metric with signature $(-,+,+,+)$)

$$s = \int_{\tau_1}^{\tau_2} (-g_{\mu\nu}\dot{x}^\mu\dot{x}^\nu)^{1/2}\, d\tau \tag{21}$$

where dot denotes a derivative with respect to the affine parameter, τ. We inserted a minus sign because of the Lorentz signature. We wish now to derive the condition on a curve which makes it extremize the length between its endpoints, i.e., wish to find those curves whose length does not change to first order under an arbitrary smooth deformation which keeps the endpoints fixed. This condition gives rise to the Euler-Lagrange equations

$$\frac{d}{d\tau}\left(\frac{\partial L}{\partial \dot{x}^\alpha}\right) - \frac{\partial L}{\partial x^\alpha} = 0 \tag{22}$$

of the action integral $I = \int_{\tau_1}^{\tau_2} L(x^\mu, \dot{x}^\mu, \tau)$. Thus in our case the Lagrangian is

$$L = [-g_{\mu\nu}(x)\dot{x}^\mu\dot{x}^\nu]^{1/2} \tag{23}$$

where x stands for coordinate functions x^μ. Now the Euler-Lagrange equations yield

$$\ddot{x}^\beta + \frac{1}{2} g^{\alpha\beta}(\partial_\mu g_{\alpha\nu} + \partial_\nu g_{\alpha\mu} + \partial_\alpha g_{\mu\nu})\dot{x}^\mu \dot{x}^\nu = \frac{\dot{x}^\beta}{2(g_{\mu\nu}\dot{x}^\mu\dot{x}^\nu)} \frac{d(g_{\mu\nu}\dot{x}^\mu\dot{x}^\nu)}{d\tau} \tag{24}$$

Here we pay a special attention to the last term since the length of a vector needs not to be stay the same during the parallel transportation because of the existence of the nonmetricity. Let us evaluate it. First, we write it as $g_{\mu\nu}\dot{x}^\mu\dot{x}^\nu = g_{\mu\nu}\mathcal{T}^\mu\mathcal{T}^\nu = \mathcal{T}_\mu\mathcal{T}^\mu$. Now,

$$d(\mathcal{T}_\mu\mathcal{T}^\mu) = (Dg_{\mu\nu})\mathcal{T}^\mu\mathcal{T}^\nu + 2g_{\mu\nu}\mathcal{T}^\mu(D\mathcal{T}^\nu) = -2Q_{\mu\nu}\mathcal{T}^\mu\mathcal{T}^\nu + 2g_{\mu\nu}\mathcal{T}^\mu(D\mathcal{T}^\nu) \tag{25}$$

Here the usage of the equation (16) gives

$$d(\mathcal{T}_\mu\mathcal{T}^\mu) = 2(a-1)Q_{\mu\nu}\mathcal{T}^\mu\mathcal{T}^\nu + 2cQ\mathcal{T}_\mu\mathcal{T}^\mu \tag{26}$$

This means that if we choose $a = 1$ and $c = 0$, the geodesic equation of STPG (24) turns out to be the same as the geodesic equation of the Riemannian geometry.

4. An application to a stationary, spherically axisymmetric metric

Geodesics of (time-like and null) of Schwarzschild geometry are in very good agreement with solar experiments. Furthermore, shadow of the M87* super massive black hole is also in very good agreement with Kerr null geodesics [31]. On the other hand, the observed baryonic matter in extragalactic systems does not match gravitational potential of general relativity. This discrepancy is called as dark matter. A classic example is that the rotation curves of disk galaxies become approximately flat ($v \approx const$) when they should be decreasing in a Keplerian way ($v \propto r^{-1/2}$). Accordingly we try to see possible novel effects of our results concretely by dealing with some astrophysical events. Thus we consider a stationary, spherically axisymmetric metric which is written approximately (up to the first order) in the spherical coordinates $x^\mu = (t, r, \theta, \phi)$ and in natural gravitational units ($c = G = 1$) [7]

$$g = -\left(1 + \frac{\mathcal{H}m}{r}\right)dt^2 + \left(1 + \frac{\mathcal{F}m}{r}\right)dr^2 + r^2\left(d\theta^2 + \sin^2\theta d\phi^2\right) + \frac{2\mathcal{G}m\ell}{r}\sin^2\theta dt d\phi \tag{27}$$

where m and $J = m\ell$ are mass and rotational angular momentum of gravitating object, $\mathcal{H}, \mathcal{F}, \mathcal{G}$ are dimensionless constants. We remark that in general relativity, the Kerr metric at large distances yields these constants as $\mathcal{H} = -\mathcal{F} = \mathcal{G} = -2$. One can readily read the components

$$g_{tt} = -\left(1 + \frac{\mathcal{H}m}{r}\right), \qquad g_{rr} = 1 + \frac{\mathcal{F}m}{r}, \qquad g_{\theta\theta} = r^2,$$
$$g_{\phi\phi} = r^2\sin^2\theta, \qquad g_{t\phi} = \frac{\mathcal{G}m\ell}{r}\sin^2\theta, \qquad \text{other } g_{\alpha\beta} = 0, \tag{28}$$

and compute the inverses up to the first order (practically by omitting the terms containing $\mathcal{H}^2, \mathcal{F}^2, \mathcal{G}^2, \mathcal{H}\mathcal{G}, \mathcal{H}\mathcal{F}, \mathcal{F}\mathcal{G}$)

$$g^{tt} = -\left(1 - \frac{\mathcal{H}m}{r}\right), \qquad g^{rr} = 1 - \frac{\mathcal{F}m}{r}, \qquad g^{\theta\theta} = \frac{1}{r^2},$$
$$g^{\phi\phi} = \frac{1}{r^2\sin^2\theta}, \qquad g^{t\phi} = \frac{\mathcal{G}m\ell}{r^3}, \qquad \text{other } g^{\alpha\beta} = 0. \tag{29}$$

Now we calculate the equation (19) explicitly. With the notation $x^t := t$, $x^r := r$, $x^\theta := \theta$, $x^\phi := \phi$, for $\mu = \theta, t, \phi, r$, it reads respectively

$$\frac{d^2\theta}{d\tau^2} = \left\{ -\frac{a+b+2c}{r} + \frac{c(\mathcal{H}+\mathcal{F})m}{2r^2} \right\} \frac{dr}{d\tau}\frac{d\theta}{d\tau} + \frac{2b\mathcal{G}m\ell \sin\theta\cos\theta}{r^3}\frac{dt}{d\tau}\frac{d\phi}{d\tau} + b\sin\theta\cos\theta \frac{d\phi}{d\tau}\frac{d\phi}{d\tau} - c\cot\theta \frac{d\theta}{d\tau}\frac{d\theta}{d\tau}, \tag{30a}$$

$$\frac{d^2t}{d\tau^2} = \left\{ -\frac{2c}{r} + \frac{m\left[(a+b)\mathcal{H} + c(\mathcal{H}+\mathcal{F})\right]}{2r^2} \right\} \frac{dt}{d\tau}\frac{dr}{d\tau} - c\cot\theta \frac{dt}{d\tau}\frac{d\theta}{d\tau} - (a+b)\frac{3\mathcal{G}m\ell\sin^2\theta}{2r^2}\frac{dr}{d\tau}\frac{d\phi}{d\tau}, \tag{30b}$$

$$\frac{d^2\phi}{d\tau^2} = \frac{(a+b)\mathcal{G}m\ell}{2r^4}\frac{dt}{d\tau}\frac{dr}{d\tau} - \frac{(a+b)\mathcal{G}m\ell\cot\theta}{r^3}\frac{dt}{d\tau}\frac{d\theta}{d\tau} + \left\{ -\frac{a+b+2c}{r} + \frac{c(\mathcal{H}+\mathcal{F})m}{2r^2} \right\} \frac{dr}{d\tau}\frac{d\phi}{d\tau} - (a+b+c)\cot\theta\frac{d\theta}{d\tau}\frac{d\phi}{d\tau}, \tag{30c}$$

$$\frac{d^2r}{d\tau^2} = \frac{b\mathcal{H}m}{2r^2}\frac{dt}{d\tau}\frac{dt}{d\tau} + \left\{ -\frac{2c}{r} + \frac{m\left[c\mathcal{H} + (a+c)\mathcal{F}\right]}{2r^2} \right\} \frac{dr}{d\tau}\frac{dr}{d\tau} - c\cot\theta\frac{dr}{d\tau}\frac{d\theta}{d\tau} - b\,(m\mathcal{F} - r)\frac{d\theta}{d\tau}\frac{d\theta}{d\tau} - b\sin^2\theta\,(\mathcal{F}m - r)\frac{d\phi}{d\tau}\frac{d\phi}{d\tau} - \frac{b\mathcal{G}m\ell\sin^2\theta}{r^2}\frac{dt}{d\tau}\frac{d\phi}{d\tau}. \tag{30d}$$

Now we consider a body moving instantly at the equatorial plane. Formally we set $\theta = \pi/2$ and $d\theta/d\tau = 0$ the equation (30a) and obtain $d^2\theta/d\tau^2 = 0$ which means that the body always stays at the equatorial plane. We substitute this result into (30b), (30c) and (30d),

$$\frac{d^2t}{d\tau^2} + \left\{ \frac{2c}{r} - \frac{m\left[(a+b)\mathcal{H} + c(\mathcal{H}+\mathcal{F})\right]}{2r^2} \right\} \frac{dt}{d\tau}\frac{dr}{d\tau} + \frac{3(a+b)\mathcal{G}m\ell}{2r^2}\frac{d\phi}{d\tau}\frac{dr}{d\tau} = 0, \tag{31a}$$

$$\frac{d^2\phi}{d\tau^2} - \frac{(a+b)\mathcal{G}m\ell}{2r^4}\frac{dt}{d\tau}\frac{dr}{d\tau} + \left\{ \frac{a+b+2c}{r} - \frac{c(\mathcal{H}+\mathcal{F})m}{2r^2} \right\} \frac{d\phi}{d\tau}\frac{dr}{d\tau} = 0, \tag{31b}$$

$$\frac{d^2r}{d\tau^2} - \frac{b\mathcal{H}m}{2r^2}\left(\frac{dt}{d\tau}\right)^2 + \left\{ \frac{2c}{r} - \frac{m\left[c\mathcal{H} + (a+c)\mathcal{F}\right]}{2r^2} \right\} \left(\frac{dr}{d\tau}\right)^2 + b(\mathcal{F}m - r)\left(\frac{d\phi}{d\tau}\right)^2 + \frac{b\mathcal{G}m\ell}{r^2}\frac{dt}{d\tau}\frac{d\phi}{d\tau} = 0. \tag{31c}$$

In order to determine the constants of motion we rewrite the equation (31a) in terms of unknown functions $A(r)$ and $B(r)$

$$\frac{d}{d\tau}\left[A(r)\frac{dt}{d\tau} + B(r)\frac{d\phi}{d\tau} \right] = 0 \tag{32}$$

which means that $A(r)\frac{dt}{d\tau} + B(r)\frac{d\phi}{d\tau} = const$. We deliver the derivative and arrange the result.

$$\frac{d^2t}{d\tau^2} + \frac{A'}{A}\frac{dr}{d\tau}\frac{dt}{d\tau} + \frac{B}{A}\frac{d^2\phi}{d\tau^2} + \frac{B'}{A}\frac{dr}{d\tau}\frac{d\phi}{d\tau} = 0 \tag{33}$$

where prime denotes a derivative with respect to r. Now we insert the equation (31b) into the third term.

$$\frac{d^2t}{d\tau^2} + \left[\frac{A'}{A} + \frac{B(a+b)\mathcal{G}m\ell}{2Ar^4} \right] \frac{dr}{d\tau}\frac{dt}{d\tau} + \left\{ \frac{B'}{A} - \frac{B}{A}\left[\frac{a+b+2c}{r} - \frac{cm(\mathcal{H}+\mathcal{F})}{2r^2} \right] \right\} \frac{dr}{d\tau}\frac{d\phi}{d\tau} = 0 \tag{34}$$

By comparing this result with (31a), we arrive at two equations which can be used for determining the functions $A(r)$ and $B(r)$,

$$\frac{A'}{A}+\frac{B(a+b)\mathcal{G}m\ell}{2Ar^4}=\frac{2c}{r}-\frac{m\left[(a+b)\mathcal{H}+c(\mathcal{H}+\mathcal{F})\right]}{2r^2}, \tag{35a}$$

$$\frac{B'}{A}-\frac{B}{A}\left[\frac{a+b+2c}{r}-\frac{cm(\mathcal{H}+\mathcal{F})}{2r^2}\right]=\frac{3(a+b)\mathcal{G}m\ell}{2r^2}. \tag{35b}$$

Here since we obtain a very complicated set of solutions containing the Heun function via the software of Maple, for simplicity from this point we ignore the rotation by setting $\ell=0$. Then we rewrite the equations (31)

$$\frac{d^2t}{d\tau^2}+\left\{\frac{2c}{r}-\frac{m\left[(a+b)\mathcal{H}+c(\mathcal{H}+\mathcal{F})\right]}{2r^2}\right\}\frac{dt}{d\tau}\frac{dr}{d\tau}=0, \tag{36a}$$

$$\frac{d^2\phi}{d\tau^2}+\left\{\frac{a+b+2c}{r}-\frac{c(\mathcal{H}+\mathcal{F})m}{2r^2}\right\}\frac{d\phi}{d\tau}\frac{dr}{d\tau}=0, \tag{36b}$$

$$\frac{d^2r}{d\tau^2}+\left\{\frac{2c}{r}-\frac{m\left[c\mathcal{H}+(a+c)\mathcal{F}\right]}{2r^2}\right\}\left(\frac{dr}{d\tau}\right)^2-\frac{b\mathcal{H}m}{2r^2}\left(\frac{dt}{d\tau}\right)^2+b(\mathcal{F}m-r)\left(\frac{d\phi}{d\tau}\right)^2=0. \tag{36c}$$

Thus we arrive at two constants, E_0 and L_0, of motion through the equations (36a) and (36b) which represent energy and angular momentum conversations,

$$\frac{dt}{d\tau}\simeq\frac{E_0}{r^{2c}}\left\{1-\frac{m\left[(a+b)\mathcal{H}+c(\mathcal{H}+\mathcal{F})\right]}{2r}\right\}, \tag{37a}$$

$$\frac{d\phi}{d\tau}\simeq\frac{L_0}{r^{a+b+2c}}\left\{1-\frac{mc(\mathcal{H}+\mathcal{F})}{2r}\right\}. \tag{37b}$$

We keep the terms only up to the first order. We now substitute the equations (37a) and (37b) to the radial equation (36c)

$$\begin{aligned}\frac{d^2r}{d\tau^2}&+\left\{\frac{2c}{r}-\frac{m\left[c\mathcal{H}+(a+c)\mathcal{F}\right]}{2r^2}\right\}\left(\frac{dr}{d\tau}\right)^2-\frac{b\mathcal{H}mE_0^2}{2r^{2+4c}}\left\{1-\frac{m\left[(a+b)\mathcal{H}+c(\mathcal{H}+\mathcal{F})\right]}{r}\right\}\\&+\frac{bL_0^2(\mathcal{F}m-r)}{r^{2(a+b+2c)}}\left\{1-\frac{mc(\mathcal{H}+\mathcal{F})}{r}\right\}=0\end{aligned} \tag{38}$$

Integration of this second order differential equation yields a first order differential equation

$$\begin{aligned}&r^{4c}\left(1+\frac{m\left[c\mathcal{H}+(a+c)\mathcal{F}\right]}{r}\right)\left(\frac{dr}{d\tau}\right)^2+\frac{bm\mathcal{H}E_0^2}{r}\\&+\frac{2bL_0^2}{r^{2(a+b)-2}}\left\{\frac{(1-a)m\mathcal{F}}{[1-2(a+b)]r}-\frac{1}{2(1-a-b)}\right\}-E_0^2+\varepsilon=0\end{aligned} \tag{39}$$

where ε is a constant of integration. When we look at the linearized Schwarzschild limit of general relativity by setting $a=b=1$, $c=0$, $\mathcal{H}=-\mathcal{F}=-2$, we observe that the case of $\varepsilon=1$ corresponds to a time-like geodesic for massive object and $\varepsilon=0$ to a null geodesic for photon.

Furthermore, in order to gain some insights on dark matter problem we compute speed of a star at a distance r from the center of gravitating source via the equations (37b) and (39), $v^2=(dr/d\tau)^2+r^2(d\phi/d\tau)^2$, as follows

$$\begin{aligned}v^2=&\frac{E_0^2-\varepsilon}{r^{4c}}-\frac{m\{bE_0^2\mathcal{H}+(E_0^2-\varepsilon)[c\mathcal{H}+(a+c)\mathcal{F}]\}}{r^{4c+1}}+\frac{(a-1)L_0^2}{r^{2(a+b+2c)-2}}\\&-\frac{mL_0^2}{r^{2(a+b+2c)-1}}\left\{c(\mathcal{F}+\mathcal{H})+\frac{2b(1-a)\mathcal{F}}{1-2a-2b}+\frac{b[c\mathcal{H}+(a+c)\mathcal{F}]}{1-a-b}\right\}\end{aligned} \tag{40}$$

It is seen that for some values of our peculiar parameters, as the distance of a body to the source increases, its speed approaches a constant, as the observations reveal. For example, in a case of $c = 0$ and $a + b \geq 1$, as $r \to \infty$, $v \to \sqrt{E_0^2 - \varepsilon}$. In a different case, $c > 0$ and $a + b + 2c = 1$, $\lim_{r\to\infty} v \to L_0\sqrt{(a-1)/(a+b-1)}$.

In addition to brief analysis of dark matter we now would like to investigate the orbit equation. In the equation (39) we write $dr/d\tau = (d\phi/d\tau)(dr/d\phi)$ and there use the equation (37b). Thus

$$\left(\frac{dr}{d\phi}\right)^2 + \frac{b}{a+b-1}r^2 = \frac{E_0^2 - \varepsilon}{L_0^2} r^{2(a+b)} - \frac{m[b\mathcal{H}E_0^2 + a\mathcal{F}(E_0^2 - \varepsilon)]}{L_0^2} r^{2(a+b)-1} + \frac{b(3a+2b-2)m\mathcal{F}}{1-3(a+b)+2(a+b)^2} r \,. \tag{41}$$

By changing the dependent variable as $r = 1/u$ and rearranging terms we find

$$\left(\frac{du}{d\phi}\right)^2 + \frac{b}{a+b-1}u^2 = \frac{E_0^2 - \varepsilon}{L_0^2} u^{4-2(a+b)} - \frac{m[b\mathcal{H}E_0^2 + a\mathcal{F}(E_0^2 - \varepsilon)]}{L_0^2} u^{5-2(a+b)} + \frac{b(3a+2b-2)m\mathcal{F}}{1-3(a+b)+2(a+b)^2} u^3 \,. \tag{42}$$

In order to obtain an equation in the form of Newton's orbit equation we derive this with respect to ϕ

$$\frac{d^2u}{d\phi^2} + \frac{b}{a+b-1}u = \frac{[2-(a+b)](E_0^2 - \varepsilon)}{L_0^2} u^{3-2(a+b)} + \frac{3b(3a+2b-2)m\mathcal{F}}{2[1-3(a+b)+2(a+b)^2]} u^2 - \frac{m[5-2(a+b)][b\mathcal{H}E_0^2 + a\mathcal{F}(E_0^2 - \varepsilon)]}{2L_0^2} u^{4-2(a+b)} \,. \tag{43}$$

In this step we look at the Schwarzschild limit by setting $a = b = 1$, $\mathcal{H} = -\mathcal{F} = -2$ and find the result

$$\frac{d^2u}{d\phi^2} + u = \frac{\varepsilon m}{L_0^2} + 3mu^2 \,. \tag{44}$$

This is exactly the orbit equation of planets for $\varepsilon = 1$ and of light for $\varepsilon = 0$ in the Schwarzschild metric of general relativity. Of course, one can obtain new forms of orbit equation for new settings different from above. However, we notice that as long as we stay within the Newtonian limit, which we must observe in the solar system, all possible choices yield the same form of orbit equations. By Newtonian limit we mean that the term u^0 must appear on the right hand side. Correspondingly, there are only two options; (i) $a + b = 2$ and (ii) $a + b = 3/2$. One more remark is that while our free parameter c appears in our galactic velocity analysis, here it disappears automatically. Exact solutions and more detailed analysis is left for our future research projects.

5. Concluding remarks

In general relativity it is commonly assumed that idealised massive spinless test particles have spacetime histories that coincide with time-like geodesic (autoparallel) curves associated with the spacetime metric when acted on by gravitation alone. On the other hand, in modified theories of gravity formulated in a non-Riemannain spacetimes with torsion and/or nonmetricity, there is not a consensus on which equation, geodesic or autoparallel, describes the correct worldlines of spinless test particles under the influence of a gravitating source. In Einstein's pseudo-Riemannian description of gravitation the affine parameter, τ, physically is taken as proper time which is measured by a standard clock that is modeled by any time-like curve $x^\mu(\tau)$. Thus in spite of the fact that proper time passed between two events connected by $x^\mu(\tau)$ is path dependent, a standard Einsteinian clock accepts a proper time parameterization independent of its trajectory. On the other hand, in a non-Riemannian geometry containing nonmetricity the identification of a clock as a device for measuring proper time requires more care. In this context in a recent preprint [32] it is argued that the symmetric teleparallel theories of gravity do not represent phenomenologically viable descriptions of nature due to the second clock effect. Nevertheless, in [33],[34] the definitions of standard clocks in theories of gravitation formulated in the non-Riemannian spacetime geometries are discussed and it seems possible to define a new standard clock that will be able to measure the affine parameter by considering the invariance of action and the full connection under gauge symmetries. Additionally, in lemma 2 of reference [35] it is shown that in symmetric teleparallel spacetimes, the second clock effect does not arise. Also the aurhors of [36] conclude that the definition of proper time for Weyl invariant spacetimes naturally extends to spaces with arbitrary nonmetricity. In the very recent papers [37], [38] it is argued that the second clock effect does not occur in Weyl gauge theories of gravity, which are invariant both under local Poincare transformations and local changes of scale.

Thus, in this paper we discussed the bounded orbital motion of a massive spinless test particle in the background of a symmetric teleparallel geometry in the coincident gauge in terms of worldlines that are autoparallels and geodesics of nonmetric compatible spacetime connections. With a novel suggestion, we waived keeping the length of a tangent vector the same during parallel transport along the curve by the equation (16) containing three free parameters, a, b, c. Then we obtained the autoparallel equations and geodesic equations explicitly in any coordinate chart $\{x^\mu\}$. Finally we observed that they both can match to the geodesic equation of the Riemannian geometry for some certain values of the parameters ($a = b = 1,\ c = 0$) in (16). We started applying our equation to a stationary, spherically axisymmetric metric up to the first order. After a while we pass to a spherically symmetric static metric in order to avoid complexity. Finally we remarked on dark matter and orbit equation. We left the investigation of exact solutions to our autoparallel equations to our future research projects. Of course, new generation experiments with very high sensitivity will decide which curve governs correctly the motion of a spinless test particle [39].

Acknowledgement
The authors thank to emeritus Prof.Dr. Tekin Dereli for stimulating conversations on the non-Riemannian geometries, and the geodesic and autoparallel curves, also to the anonymous referee for instructive criticisms and to Prof.Dr. Bayram Tekin for discussion on the Schwarzschild geodesics and to Prof.Dr. Ozcan Sert for his help on usage of Maple.

References
[1] Hehl F W *et al.* 1995 *Phys. Rep.* **258** 1, DOI: 10.1016/0370-1573(94)00111-F, arXiv: gr-qc/9402012
[2] Tucker R W and Wang C 1995 *Class. Quant. Grav.* **12** 2587, DOI: 10.1088/0264-9381/12/10/016, arXiv: gr-qc/9509011
[3] Adak M 2010 *Non-Riemannian Theories of Gravitation and the Dirac Equation*, Lambert Academic Pub., ISBN: 978-3-8383-5139-1
[4] Dereli T and Tucker R W 2002 *Mod. Phys. Lett. A* **17** 421, DOI: 10.1142/S021773230200662X, arXiv: gr-qc/0104050
[5] Cebeci H, Dereli T and Tucker R W 2004 *Int. J. Mod. Phys. D* **13** 137, DOI: 10.1142/S0218271804004487, arXiv: gr-qc/0311080
[6] Obukhov Y N and Puetzfeld D 2021 *Phys. Rev. D* **104** 044031, DOI: 10.1103/PhysRevD.104.044031, arXiv: 2105.08428
[7] Mao Y *et al.* 2007 *Phys. Rev. D* **76** 104029, DOI: doi.org/10.1103/PhysRevD.76.104029, arXiv: gr-qc/0608121

[8] Jimenez J B and Delhom A 2020 *Eur. Phys. J. C* **80** 585, DOI: 10.1140/epjc/s10052-020-8143-z, arXiv: 2004.11357
[9] Adak M and Sert O 2005 *Turk. J. Phys.* **29** 1, arXiv: gr-qc/0412007
[10] Adak M, Kalay M and Sert O 2006 *Int. J. Mod. Phys. D* **15** 619, DOI: 10.1142/S0218271806008474, arXiv: gr-qc/0505025
[11] Adak M and Dereli T 2008 *EPL* **82** 30008, DOI: 10.1209/0295-5075/82/30008, arXiv: hep-th/0607058
[12] Adak M 2006 *Turk. J. Phys.* **30** 379, arXiv: gr-qc/0611077
[13] Adak M, Sert O, Kalay M and Sari M 2013 *Int. J. Mod. Phys. A* **28** 1350167, DOI: 10.1142/S0217751X13501674, arXiv: 0810.2388
[14] Adak M 2018 *Int. J. Geom. Meth. Modern Phys.* **15** 1850198, DOI: 10.1142/S0219887818501980, arXiv: gr-qc/9809049
[15] Nester J M and Yo H J 1999 *Chinese J. Phys.* **37** 113, arXiv: gr-qc/9809049
[16] Mol I 2017 *Adv. Appl. Clifford Algebras* **27** 2607, DOI: 10.1007/s00006-016-0749-8, arXiv: 1406.0737
[17] Koivisto T 2018 *Int. J. Geom. Meth. Modern Phys.* **15**, 1840006, DOI: 10.1142/S0219887818400066, arXiv: 1802.00650
[18] Golovnev A, Koivisto T and Sandstad M 2017 *Class. Quantum Grav.* **34** 145013, DOI: 10.1088/1361-6382/aa7830 arXiv: 1701.06271
[19] Ong Y C, Izumi K, Nester J M and Chen P 2013 *Phys.Rev. D* **88** 024019, DOI: 10.1103/PhysRevD.88.024019, arXiv: 1303.0993
[20] Daouda M H, Rodrigues M E and Houndjo M J S 2011 *ISRN Astron.Astrophys.* **2011** 341919, DOI: 10.5402/2011/341919, arXiv: 1109.2928
[21] Goordeeva I A, Pan'zhenskii V I and Stepanov S E 2010 *J. Math. Sci.* **169** 342, DOI: 10.1007/s10958-010-0052-5
[22] Baekler P and Hehl F W 2006 *Int. J. Mod. Phys. D* **15** 635, DOI: 10.1142/S0218271806008589, arXiv: gr-qc/0601063
[23] Pradhan A, Maurya D C and Dixit A 2021 *Int. J. Geom. Meth. Mod. Phys.* **18** 2150124, DOI: 10.1142/S0219887821501243
[24] Hohmann M 2021 *Phys. Rev. D* **104** 124077, DOI: 10.1103/PhysRevD.104.124077, arXiv: 2109.01525
[25] Miranda M 2019 *Non-Metricity in Theories of Gravity and Einstein Equivalence Principle*, Ph D Thesis (Universita Degli Studi Di Napoli " Fredreica II", 2018-2019)
[26] Damianos I 2019 *Metric Affine Gravity and Cosmology / Aspects of Torsion and Non-Metricity in Gravity Theories*, PhD Thesis (University of Thessaloniki, 2019), arXiv: 1902.09643
[27] Yang J Z *et al.* 2021 *Eur. Phys. J. C* **81** 111, DOI: 10.1140/epjc/s10052-021-08910-6, arXiv: 2101.09956v2
[28] Cartan E 1923 *On manifolds with an affine connection and the theory of general relativity*, (Bibliopolis, 1986)
[29] Pala C *et al.*, 2022 *Int. J. Geomet. Meth. Mod. Phys.*, **19** 2250045, DOI: 10.1142/S0219887822500451, arXiv: 2011.10982
[30] Wald R M 1984 *General Relativity*, The Universty of Chicago Press, ISBN: 0-226-87033-2
[31] Wielgus M *et al.* 2020 *Ap J* **901** 67, DOI: doi.org/10.3847/1538-4357/abac0d, arXiv: 2009.11842
[32] Quiros I, *Symmetric teleparallel theories in the firing line of gauge symmetry*, arXiv: 2111.05490
[33] Teyssandier P and Tucker R W 1996 *Class. Quantum Grav.* **13** 145, DOI: 10.1088/0264-9381/13/1/013, arXiv: gr-qc/9510045
[34] Teyssandier P, Tucker R W and Wang C 1998 *Acta Phys. Pol. B* **29** 987, https://www.actaphys.uj.edu.pl/R/29/4/987/pdf
[35] Jimenez J B, Heisenberg L and Koivisto T 2020 *Class. Quant. Grav.* **37** 195013, DOI: 10.1088/1361-6382/aba31b, arXiv: 2004.04606
[36] Delhom A et al. 2020 *Eur. Phys. J. C* **80** 415, DOI: 10.1140/epjc/s10052-020-7974-y, arXiv: 2001.10633
[37] Hobson M P and Lasenby A N 2020 *Phys. Rev. D* **102** 084040, DOI: https://doi.org/10.1103/PhysRevD.102.084040, arXiv:2009.06407
[38] Hobson M P and Lasenby A N, *Note on the absence of the second clock effect in Weyl gauge theories of gravity*, arXiv:2112.09967
[39] Belenchia A *et al.*, *Quantum Physics in Space*, arXiv: 2108.01435

DERELI-FS-2021 IOP Publishing
Journal of Physics: Conference Series **2191** (2022) 012018 doi:10.1088/1742-6596/2191/1/012018

Quantum Correlations in Jahn-Teller Molecular Systems Simulated with Superconducting Circuits

Ali Pedram, Onur Pusuluk, and Özgür E. Müstecaplıoğlu

Department of Physics, Koç University, Sarıyer, Istanbul, 34450, Turkey

E-mail: apedram19@ku.edu.tr, opusuluk@ku.edu.tr, omustecap@ku.edu.tr

Abstract. We explore quantum correlations, in particular, quantum entanglement, among vibrational phonon modes as well as between electronic and vibrational degrees of freedom in molecular systems, described by Jahn-Teller mechanism. Specifically, to isolate and simplify the phonon- electron interactions in a complex molecular system, the basis of our discussions is taken to be the proposal of simulating two-frequency Jahn- Teller systems using superconducting circuit quantum electrodynamics systems (circuit QED) by Tekin Dereli and co-workers in 2012. We evaluate the quantum correlations, in particular entanglement between the vibrational phonon modes, and present analytical explanations using a single privileged Jahn-Teller mode picture. Furthermore, spin-orbit entanglement or quantum correlations between electronic and vibrational degrees of freedom are examined. We conclude by discussing experimental feasibility to detect such quantum correlations, considering the dephasing and decoherence in state-of-the-art superconducting two-level systems (qubits).

1. Introduction

Quantum entanglement in natural systems, in particular molecules, has attracted much attention for its fundamental and practical significance [1–15]. Due to their complex structure and many interactions within, it is difficult to isolate the quantum entanglement route in molecules. One such route is the case where electronic and vibrational degrees of freedom are coupled via so-called Jahn-Teller interaction [6, 8, 9, 12]. Dereli and coworkers have proposed that quantum simulation of molecules using superconducting circuits can serve to capture the essential physics of multi-mode Jahn-Teller effect [16]. Here, we take advantage of Dereli and co-workers' methods and explore quantum entanglement in two-mode Jahn-Teller molecules simulated by superconducting circuits .

The contribution by Dereli and coworkers is not only serving the purpose of illuminating complex quantum processes in molecules but also provides an unprecedented application direction for the ultrastrong coupling regime of superconducting circuit QED. Jahn-Teller coupling can be comparable to involved frequencies of the vibrational phonons and hence it requires an artificial system where ultrastrong coupling can be realized [6, 17, 18]. Realization of such strong interaction regimes is challenging in the usual cavity QED but it is possible with the advent of superconducting circuit QED [19–21]. Flux qubits and transmission line resonators are ideal platforms to realize such strong photon-qubit interactions, where photons can play the role of vibrational phonons. Vibrational modes in Jahn-Teller molecules can be optical and acoustic phonons and hence an ideal platform to simulate such a system would need tunable and wide range of frequencies in addition to the strong coupling. In the case of superconducting circuits even optomechanical-like interactions can be simulated by using transmission line resonators [22–25].

Typical focus on Jahn-Teller molecules in chemistry is limited to energetically low-lying states and their spectral properties, such as optical absorption. We explore quantum correlations and look for potential quantum entanglement in ground state of the molecular system with two-mode Jahn-Teller coupling. Due to routine methods of measuring quantum correlations in superconducting circuit QED, our results can be significant for practical entanglement generation in addition to illuminating the nature of entanglement in natural molecules. In addition, our quantum simulation and effective privileged mode approach can be further extended to explore the interplay of quantum correlations, chaos, and phase transitions [26].

The rest of the paper is organized is as follows. We first give a short review of the two-mode Jahn-Teller model and the method of privileged mode as described by Dereli and coworkers in Sec. II. We present our quantum entanglement measures and their evaluation results in Sec. III. Sec. IV is the conclusion of our analysis of the quantum entanglement results in different parameter regimes from the perspective of effective privileged mode approach.

2. Privileged mode approach to two-mode Jahn-Teller model: A short review

2.1. two-mode Jahn-Teller model and circuit QED simulation

Jahn-Teller transition happens in some molecules that can end up in a preferred geometrical configuration among a set of degenerate ones due to the symmetry breaking by vibrational phonon and localized electron interaction. Hamiltonian describing this phenomenon is known as Jahn-Teller model [27]. Jahn-Teller coupling belongs to the general class of spin-boson interactions. Vibrational degrees of freedom of molecules are quantized in terms of phonons with boson statistics. Optical and acoustic phonons have high and low frequencies and can be described by an approximate two-mode model, a particular Jahn-Teller model, also known as the Herberg-Teller model, which will be our focus here. Dereli and coworkers have shown that $\mathrm{E}\times(\beta_1 + \beta_2)$ type Herzberg-Teller model can be simulated by using the superconducting circuit (sc) Hamiltonian (we take $\hbar = 1$)

$$\hat{H}_{\text{sc}} = \hat{H}_0 + \hat{H}_{\text{int}} \tag{1}$$

where the non-interacting and interacting terms respectively are given by

$$\hat{H}_0 = \frac{\omega_q}{2}\hat{\sigma}_z + \sum_{i=1}^{2}\omega_i\hat{a}_i^\dagger\hat{a}_i, \tag{2}$$

$$\hat{H}_{\text{int}} = \sum_{i=1}^{2} g_i(\hat{a}_i + \hat{a}_i^\dagger)\hat{\sigma}_x + J(\hat{a}_1^\dagger\hat{a}_2 + \hat{a}_2^\dagger\hat{a}_1). \tag{3}$$

Here, ω_q and ω_i denote the transition frequency of the qubit and the frequencies of the modes of two lumped element LC resonators ($i = 1, 2$). Creation ($\hat{a}_i^\dagger$) and annihilation ($\hat{a}_i$) operators of the modes obey the Weyl-Heisenberg (bosonic) algebra. The two-qubit states simulate the diabatic electronic states in actual molecules. Two energy levels of the qubit, $|1\rangle, |2\rangle$ are represented by the Pauli spin-$1/2$ operators $\hat{\sigma}_z = |2\rangle\langle 2| - |1\rangle\langle 1|, \hat{\sigma}_x = |1\rangle\langle 2| + |2\rangle\langle 1|$. The resonators are coupled to a flux qubit with the interaction coefficients g_i. Their difference can simulate the case of spatial anisotropy of molecules. We scale the interaction coefficients with the corresponding mode frequencies and write $g_i = \omega_i k_i$. The displacement-spin coupling form of the interaction is the signature of the Jahn-Teller model. The last term, however, makes the circuit QED model different than the original Jahn-Teller Hamiltonian. It arises due to the inductive coupling of the two resonators. This term can be made relatively smaller than the ultrastrong coupling coefficient by optimizing the system geometry by extending the distance between the LC resonators using a sufficiently large flux qubit. In the privileged mode description, this term will not play a significant role.

2.2. *Privileged Mode Approach to Two-Mode Jahn-Teller Model*

Privileged mode is an energetically favored superposition of the two resonator modes which can be expressed in terms of the annihilation operators as

$$\hat{b}_1 = \frac{1}{k_p}(k_1\hat{a}_1 + k_2\hat{a}_2), \tag{4}$$

$$\hat{b}_2 = \frac{1}{k_p}(k_2\hat{a}_1 - k_1\hat{a}_2), \tag{5}$$

where $k_p^2 = k_1^2 + k_2^2$. Next to the privileged mode, there is also an energetically disadvantaged mode, whose annihilation operator is denoted by $\hat{b}_2$, and the transformation from the resonator modes to the privileged and disfavored modes is one-to-one. The orthogonal transformation coefficients are determined by maximizing $k_p^2\omega_p$ and using the orthogonality condition of the transformation [28].

The terms of the transformed Hamiltonian for the privileged mode is the same as the single mode Jahn-Teller model

$$H_{\text{JT}} = \frac{\omega_q}{2}\hat{\sigma}_z + \omega_p\hat{b}_1^\dagger\hat{b}_1 + g_p(\hat{b}_1 + \hat{b}_1^\dagger)\hat{\sigma}_x, \tag{6}$$

where the privileged mode frequency and its coupling rate to the flux qubit are denoted by $\omega_p = (\omega_1 k_1^2 + \omega_2 k_2^2)/k_p^2$ and $g_p = \omega_p k_p$, respectively. The rest of the total transformed Hamiltonian terms include the hopping between the privileged and disadvantaged modes. The full hamiltonian for the superconducting circuit after the transformation becomes

$$\hat{H}_{\text{sc}} = \frac{\omega_q}{2}\hat{\sigma}_z + (\omega_p + \frac{Jk_1k_2}{k_p^2})\hat{b}_1^\dagger\hat{b}_1 + (\tilde{\omega}_p - \frac{Jk_1k_2}{k_p^2})\hat{b}_2^\dagger\hat{b}_2 + [c + J\frac{k_2^2 - k_1^2}{k_p^2}](\hat{b}_1^\dagger\hat{b}_2 + \hat{b}_2^\dagger\hat{b}_1) \\ + k_p[\omega_p(\hat{b}_1^\dagger + \hat{b}_1) + c(\hat{b}_2^\dagger + \hat{b}_2)]\hat{\sigma}_x. \tag{7}$$

Here we have $\tilde{\omega}_p = (\omega_1 k_2^2 + \omega_2 k_1^2)/k_p^2$. To ensure the terms for the disadvantaged mode are negligible, perturbation theory requires that the parameters J and $c = \Delta k_1 k_2/k_p^2$ must be sufficiently small where $\Delta = \omega_1 - \omega_2$. The latter condition can be realized if one resonator mode interacts with the flux qubit in the ultrastrong coupling regime (without loss of generality, we label it as $i = 1$). The other mode, $i = 2$, can be coupled strongly to the flux qubit, and we demand the condition $g_1 > g_2$ to be satisfied. The hopping rate J between the modes must be smaller or comparable to g_2 [16].

3. Results: Quantum Entanglement in the Jahn-Teller Ground State and Privileged Mode Interpretation

Some entanglement measures quantify the resource value of entanglement for certain tasks. Distillable entanglement, entanglement cost, and entanglement of formation are such operational measures. However, certain tasks that can be achieved by means of molecular correlations are outside our scope. Abstract measures such as logarithmic negativity and relative entropy of entanglement are sufficient for capturing the correlations in our systems. The latter requires an optimization that increases the computational cost of the simulations. Therefore, we quantify the quantum entanglement in the ground state of the Jahn-Teller model using the logarithmic negativity [29] as the measure of entanglement. The logarithmic negativity, which is an upper bound to the distillable entanglement, is defined as

$$E_N(A|\bar{A}) = log_2\left\|\rho_{A\bar{A}}^{T_A}\right\|, \tag{8}$$

where A is a subsystem of a total quantum system in which its complement is $\bar{A}$, ρ^{T_A} signifies partial transpose with respect to the subsystem A, and $\|\rho\|$ is the trace norm that equals to $\text{tr}[\sqrt{\rho\rho^\dagger}]$. Here,

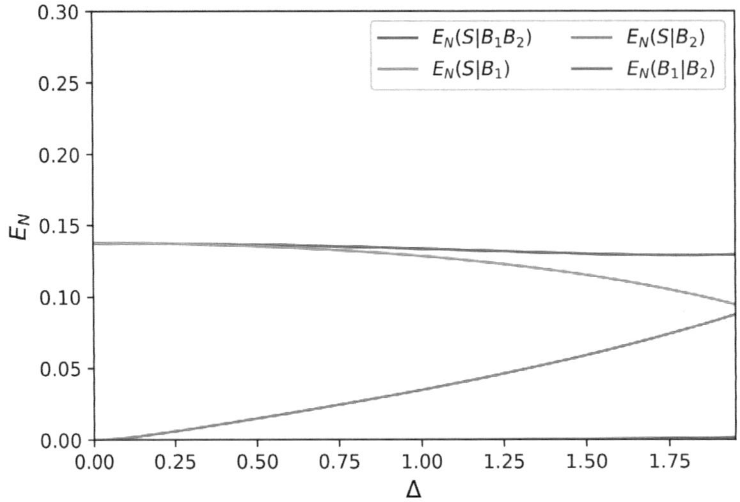

Figure 1. Ground state entanglement in Jahn-Teller model in terms of the logarithmic negativity E_N. $\Delta = \omega_1 - \omega_2$ is the frequency difference between the vibration modes. The parameters are fixed to explore the strong coupling regime such that $J = 0$ and $k_1 = k_2 = k = 0.1/\sqrt{2}$.

$E_N(A|\bar{A})$ quantifies the entanglement shared between the subsystems A and $\bar{A}$ when their joint state is described by the density matrix $\rho_{A\bar{A}}$.

For the case of our model, correlations might arise between the flux qubit (S), the privileged mode (B_1), and the disadvantaged mode (B_2). To capture an adequate description of the correlations in the tripartite ground state $\rho_{SB_1B_2}$ due to the Jahn-Teller Hamiltonian, we consider the bipartitions $S|B_1B_2$, $S|B_1$, $S|B_2$, and $B_1|B_2$ in what follows. To measure the entanglement shared between two of the three subsystems, we first compute the reduced density matrix of the subsystems under consideration by tracing out the degrees of the freedom of the third one.

According to our calculations of $E_N(S|B_1B_2)$, $E_N(S|B_1)$, $E_N(S|B_2)$, and $E_N(B_1|B_2)$, it is sufficient to set the Fock space dimension N of the resonators to 10 to analyze the behaviour of ground-state quantum entanglement with respect to the Hamiltonian parameters. We have done the same calculations for higher-dimensional Hilbert spaces up until $N = 20$, and they gave similar results for the strong and ultra-strong coupling regimes.

We first examine the effect of the frequency difference between the vibration modes on the entanglement shared in the Jahn-Teller ground state. To this end, and for the sake of simplicity, we take $k_1 = k_2 = k$, $J = 0$ and $\omega_{1/2} = 1 \pm \Delta/2$ by rescaling all the parameters with respect to the qubit transition frequency ω_q. It is clear from Eq. 7 that as Δ is increased, both the coupling between the two resonators and the coupling of the flux qubit to the disadvantaged mode becomes stronger. Therefore, the flux qubit and the disadvantaged mode correlations are expected to constitute a significant portion of the total correlations as we increase Δ. Fig. 1 justifies this expectation in the strong coupling regime where $k = 0.1/\sqrt{2}$. The effective single privileged mode description of the system is valid for $|\Delta| < 0.1$. The entanglement between the flux qubit and the disadvantaged mode remains negligible until this limit. In the two-mode regime, $E_N(S|B_2)$ rises slowly with further increases of Δ. When $\Delta = 2$, the flux qubit becomes coupled to both modes with the same strength, as a result of which $E_N(S|B_1)$ and $E_N(S|B_2)$ take the same value. Although the hopping constant J is set to zero, $E_N(B_1|B_2)$ takes a small but nonzero value at this point.

For the case of ultra-strong coupling ($k = 1/\sqrt{2}$) shown in Fig. 2, the general trend of ground-

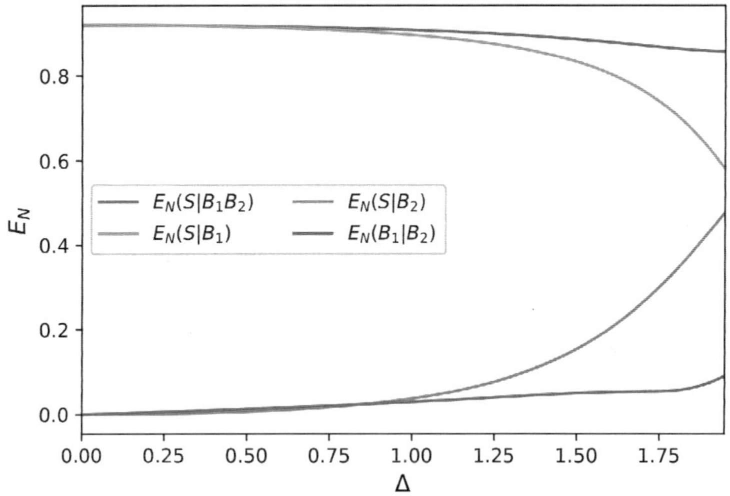

Figure 2. Ground state entanglement in Jahn-Teller model in terms of the logarithmic negativity E_N. $\Delta = \omega_1 - \omega_2$ is the frequency difference between the vibration modes. The parameters are fixed to explore the ultra-strong coupling regime such that $J = 0$ and $k_1 = k_2 = k = 1/\sqrt{2}$.

state entanglement does not appear different from the strong coupling case. However, the stronger coupling produces stronger entanglement between the flux qubit and the two resonators in this case. The observation of the steeper increase in the entanglement between the flux qubit and the disadvantaged mode is delayed until $|\Delta| = 1$, only after which the privileged mode approximation fails.

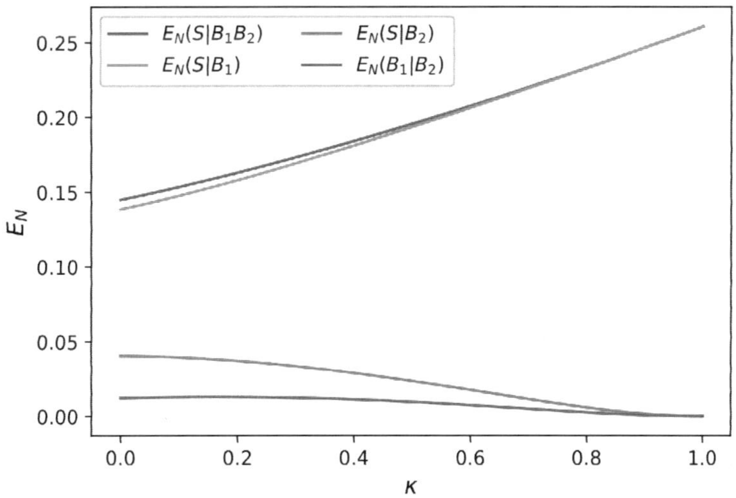

Figure 3. Ground state entanglement in Jahn-Teller model in terms of the logarithmic negativity E_N. $\kappa = k_1 - k_2$ is the difference between the factors that rescale the coupling constants g_i with respect to frequencies ω_i. The parameters are fixed such that $J = 0$ and $\omega_1 = 2\omega_2 = 0.1$.

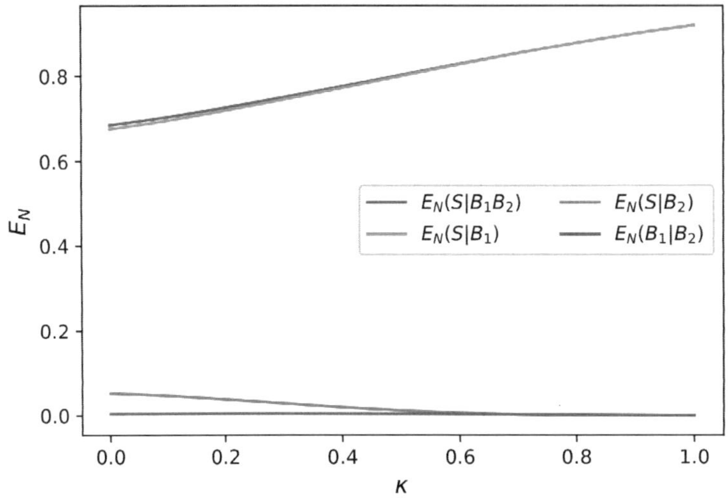

Figure 4. Ground state entanglement in Jahn-Teller model in terms of the logarithmic negativity E_N. $\kappa = k_1 - k_2$ is the difference between the factors that rescale the coupling constants g_i with respect to frequencies ω_i. The parameters are fixed such that $J = 0$ and $\omega_1 = 2\omega_2 = 1$.

We also explore the ground state entanglement by changing the difference between the scaling factors appearing in the Jahn-Teller couplings $g_i = k_i\omega_i$. To do so, we set $J = 0$, $\omega_1 = 2\omega_2$, and $k_{1/2} = (1 \pm \kappa)/2$. We use $k_1 - k_2 = \kappa$ as our control parameter. The results for $\omega_1 = 0.1$ and $\omega_1 = 1$ are respectively given in Figs. 3 and 4. When the flux qubit and the privileged mode are resonant, i.e., $\omega_q = \omega_1 = 1$, we end up with higher values for $E_N(S|B_1)$. The coupling of the privileged mode to the flux qubit is one order of magnitude larger than both the coupling of the disadvantaged mode and the hopping between the modes for $\kappa \geq 2/3$ in both cases. So, $E_N(S|B_1)$ approximately equals $E_N(S|B_1B_2)$ in both cases as long as the privileged mode approximation works well.

So far, a significant amount of entanglement has not appeared between the privileged and disadvantaged modes in the two-mode Jahn-Teller regime. This led us to change both ω_i and k_i at the same time. For the sake of simplicity, we set $J = 0$, $\omega_{1/2} = 1 \pm \Delta/2$, and $k_1 = k_2 = \Delta$ in Fig. 5. Here, the privileged mode approximation fails when $\Delta \geq 0.2$. However, $E_N(S|B_1B_2)$ and $E_N(S|B_1)$ seem to overlap until Δ reaches 1. $E_N(B_1|B_2)$ rises rapidly after this point with further increases of Δ. As of $\Delta = 1.5$, $E_N(S|B_1)$ shows a sharp decline, whereas $E_N(S|B_1B_2)$ remains fixed at unit negativity. When the flux qubit becomes coupled to both modes with the same strength at $\Delta = 2$, a significant amount of entanglement arises between the flux qubit and the two-mode subsystem and between the two modes themselves. However, neither the privileged nor the disadvantaged are strongly entangled with the flux qubit in this limit.

Finally, we consider the role of hopping constant J on the ground state entanglement. In Fig. 6, it is allowed to be comparable to g_2 but bounded from above by g_1 to faithfully simulate the Jahn-Teller interaction by the circuit QED model. The remaining parameters are fixed such that $k_1 = k_2 = k = 1/\sqrt{2}$ and $\omega_1 = 2\omega_2 = 0.2$. For this parametric regime the privileged mode approximation works well, i.e., the privileged mode is coupled to the flux qubit with a strength one order of magnitude larger than both its coupling strength to the disadvantaged mode and the flux qubit-disadvantaged mode coupling strength. However, $E_N(B_1|B_2)$ and $E_N(S|B_2)$ are always comparable to $E_N(S|B_1B_2)$ and $E_N(S|B_1)$. If we doubled k, $E_N(S|B_1B_2)$ would also double but $E_N(B_1|B_2)$ would increase more than six times in this figure. Thus, a non-vanishing hopping between the two modes results in nontrivial changes in the distribution of entanglement between different subsystems in the Jahn-Teller model.

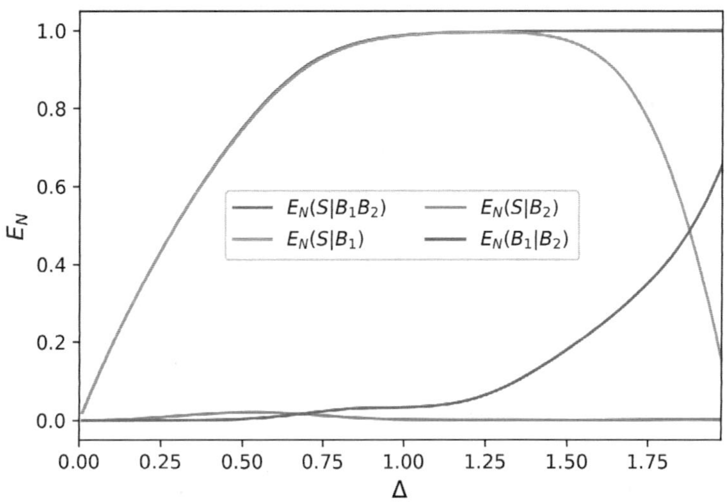

Figure 5. Ground state entanglement in Jahn-Teller model in terms of the logarithmic negativity E_N when $J = 0$ and $\omega_1 - \omega_2 = \Delta = k_1 = k_2$.

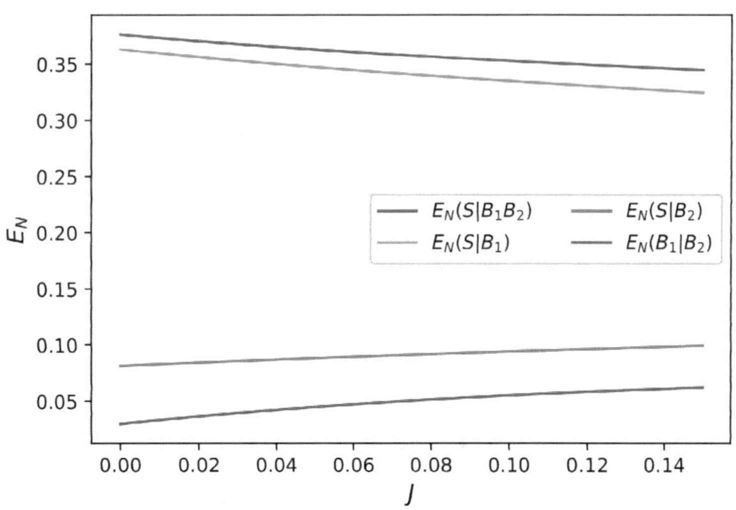

Figure 6. Ground state entanglement in Jahn-Teller model in terms of the logarithmic negativity E_N when $J \neq 0$. Here, $k_1 = k_2 = k = 1/\sqrt{2}$ and $\omega_1 = 2\omega_2 = 0.2$.

4. Conclusion

We examined the quantum entanglement in the two-mode Jahn-Teller ground state simulated by a superconducting circuit using privileged mode description following the approach by Dereli and co-workers [16].

We find that the two-mode Jahn-Teller ground state exhibits bipartite entanglement between the qubit and both modes in the strong coupling regime of the superconducting circuit quantum simulator. This entanglement is weakly dependent on the frequency difference (detuning) between the modes. The disadvantaged mode is approximately disentangled from the qubit at slight frequency differences, and it is disentangled from the privileged mode at all detunings. As the detuning increases towards the regime where the privileged mode approximation fails, both modes have comparable entanglement to the qubit. Similar behavior with stronger entanglement is found in the ultrastrong regime, too. The entanglement between the privileged mode and the qubit increases significantly, while the disadvantaged mode entanglement to the qubit is slowly decreasing, with the difference of the Jahn-Teller coupling constants, simulated by the flux qubit and the transmission line resonator interaction. We identified a parameter regime to entangle the privileged mode to the disadvantaged mode where both modes are equally strongly coupled to the qubit, and their detuning is significant so that the privileged mode approximation fails. In this case, the qubit is disentangled from the modes, and we only have entanglement between the bosonic modes. These conclusions are made when the modes are not directly coupled. We have also investigated the effect of direct coupling between the modes, which can be possible technically in superconducting circuits. Direct interaction of the modes can further increase the qubit-mode and the mode-mode entanglements. The enhancement is, however, more pronounced for mode-mode entanglement.

Our results can be significant to simulate molecular systems with strong and ultrastrong coupled superconducting circuits to illuminate their quantum correlations and utilize them as large-scale quantum resources for quantum technologies.

4.1. Acknowledgments

O. P. acknowledges support by the Scientific and Technological Research Council of Turkey (TÜBİTAK), Grant No. (120F089).

References

[1] Boguslawski K, Tecmer P, Barcza G, Legeza Ö and Reiher M 2013 *Journal of Chemical Theory and Computation* **9** 2959–2973 ISSN 1549-9618 publisher: American Chemical Society URL https://doi.org/10.1021/ct400247p

[2] Pusuluk O, Yesiller M H, Torun G, Müstecaplıoğlu Ö E, Yurtsever E and Vedral V 2021 *arXiv:2107.13992 [physics, physics:quant-ph]* ArXiv: 2107.13992 URL http://arxiv.org/abs/2107.13992

[3] Ding L, Mardazad S, Das S, Szalay S, Schollwöck U, Zimborás Z and Schilling C 2021 *Journal of Chemical Theory and Computation* **17** 79–95 ISSN 1549-9618 publisher: American Chemical Society URL https://doi.org/10.1021/acs.jctc.0c00559

[4] Kurashige Y, Chan G K L and Yanai T 2013 *Nature Chemistry* **5** 660–666 ISSN 1755-4349 bandiera_abtest: a Cg_type: Nature Research Journals Number: 8 Primary_atype: Research Publisher: Nature Publishing Group Subject_term: Biophysical chemistry;Quantum chemistry Subject_term_id: biophysical-chemistry;quantum-chemistry URL https://www.nature.com/articles/nchem.1677

[5] Hou X W, Chen J H and Hu B 2005 *Physical Review A* **71** 034302 publisher: American Physical Society URL https://link.aps.org/doi/10.1103/PhysRevA.71.034302

[6] Hines A P, Dawson C M, McKenzie R H and Milburn G J 2004 *Physical Review A* **70** 022303 publisher: American Physical Society URL https://link.aps.org/doi/10.1103/PhysRevA.70.022303

[7] Hou X W, Wan M F and Ma Z Q 2006 *Chemical Physics Letters* **426** 469–473 ISSN 0009-2614 URL https://www.sciencedirect.com/science/article/pii/S0009261406008566

[8] Liberti G, Zaffino R L, Piperno F and Plastina F 2007 *Physical Review A* **76** 042332 publisher: American Physical Society URL https://link.aps.org/doi/10.1103/PhysRevA.76.042332

[9] Streltsov S V and Khomskii D I 2020 *Physical Review X* **10** 031043 publisher: American Physical Society URL https://link.aps.org/doi/10.1103/PhysRevX.10.031043

[10] Boguslawski K and Tecmer P 2015 *International Journal of Quantum Chemistry* **115** 1289–1295 ISSN 1097-461X _eprint: https://onlinelibrary.wiley.com/doi/pdf/10.1002/qua.24832 URL https://onlinelibrary.wiley.com/doi/abs/10.1002/qua.24832

[11] Lin Y, Leibrandt D R, Leibfried D and Chou C w 2020 *Nature* **581** 273–277 ISSN 1476-4687 bandiera_abtest: a Cg_type: Nature Research Journals Number: 7808 Primary_atype: Research Publisher: Nature Publishing Group Subject_term: Atomic and molecular physics;Quantum information;Qubits Subject_term_id: atomic-and-molecular-physics;quantum-information;qubits URL https://www.nature.com/articles/s41586-020-2257-1

[12] McKemmish L K, McKenzie R H, Hush N S and Reimers J R 2011 *The Journal of Chemical Physics* **135** 244110 ISSN

0021-9606 publisher: American Institute of Physics URL https://aip.scitation.org/doi/10.1063/1.3671386

[13] Mottet M, Tecmer P, Boguslawski K, Legeza Ö and Reiher M 2014 *Physical Chemistry Chemical Physics* **16** 8872–8880 ISSN 1463-9084 publisher: The Royal Society of Chemistry URL https://pubs.rsc.org/en/content/articlelanding/2014/cp/c4cp00277f

[14] Barcza G, Legeza Ö, Marti K H and Reiher M 2011 *Physical Review A* **83** 012508 publisher: American Physical Society URL https://link.aps.org/doi/10.1103/PhysRevA.83.012508

[15] Szalay S, Barcza G, Szilvási T, Veis L and Legeza Ö 2017 *Scientific Reports* **7** 2237 ISSN 2045-2322 bandiera_abtest: a Cc_license_type: cc_by Cg_type: Nature Research Journals Number: 1 Primary_atype: Research Publisher: Nature Publishing Group Subject_term: Electronic structure of atoms and molecules;Quantum information;Theoretical physics Subject_term_id: electronic-structure-of-atoms-and-molecules;quantum-information;theoretical-physics URL https://www.nature.com/articles/s41598-017-02447-z

[16] Dereli T, Gül Y, Forn-Díaz P and Müstecaplıoğlu Ö E 2012 *Physical Review A* **85** 053841 publisher: American Physical Society URL https://link.aps.org/doi/10.1103/PhysRevA.85.053841

[17] Larson J 2008 *Physical Review A* **78** 033833 publisher: American Physical Society URL https://link.aps.org/doi/10.1103/PhysRevA.78.033833

[18] Meaney C P, Duty T, McKenzie R H and Milburn G J 2010 *Physical Review A* **81** 043805 publisher: American Physical Society URL https://link.aps.org/doi/10.1103/PhysRevA.81.043805

[19] Yoshihara F, Fuse T, Ashhab S, Kakuyanagi K, Saito S and Semba K 2017 *Nature Physics* **13** 44–47 ISSN 1745-2481 bandiera_abtest: a Cg_type: Nature Research Journals Number: 1 Primary_atype: Research Publisher: Nature Publishing Group Subject_term: Quantum information;Qubits;Single photons and quantum effects Subject_term_id: quantum-information;qubits;single-photons-and-quantum-effects URL https://www.nature.com/articles/nphys3906

[20] Forn-Díaz P, Lamata L, Rico E, Kono J and Solano E 2019 *Reviews of Modern Physics* **91** 025005 publisher: American Physical Society URL https://link.aps.org/doi/10.1103/RevModPhys.91.025005

[21] Frisk Kockum A, Miranowicz A, De Liberato S, Savasta S and Nori F 2019 *Nature Reviews Physics* **1** 19–40 ISSN 2522-5820 bandiera_abtest: a Cg_type: Nature Research Journals Number: 1 Primary_atype: Reviews Publisher: Nature Publishing Group Subject_term: Atomic and molecular interactions with photons;Atomic and molecular physics;Optics and photonics;Quantum physics Subject_term_id: atomic-and-molecular-interactions-with-photons;atomic-and-molecular-physics;optics-and-photonics;quantum-physics URL https://www.nature.com/articles/s42254-018-0006-2

[22] Johansson J R, Johansson G and Nori F 2014 *Physical Review A* **90** 053833 publisher: American Physical Society URL https://link.aps.org/doi/10.1103/PhysRevA.90.053833

[23] Eichler C and Petta J 2018 *Physical Review Letters* **120** 227702 publisher: American Physical Society URL https://link.aps.org/doi/10.1103/PhysRevLett.120.227702

[24] Rimberg A J, Blencowe M P, Armour A D and Nation P D 2014 *New Journal of Physics* **16** 055008 ISSN 1367-2630 publisher: IOP Publishing URL https://doi.org/10.1088/1367-2630/16/5/055008

[25] Kim E j, Johansson J R and Nori F 2015 *Physical Review A* **91** 033835 publisher: American Physical Society URL https://link.aps.org/doi/10.1103/PhysRevA.91.033835

[26] Majerníková E and Shpyrko S 2011 *Journal of Physics A: Mathematical and Theoretical* **44** 065101 ISSN 1751-8121 publisher: IOP Publishing URL https://doi.org/10.1088/1751-8113/44/6/065101

[27] Bersuker I 2006 *The Jahn-Teller Effect* (Cambridge) iSBN: 9780521822121 9780511524769 9780521169363 Publisher: Cambridge University Press URL https://www.cambridge.org/core/books/jahnteller-effect/05766FA392314EE9022AA86A53AC52E5

[28] Barentzen H and Polansky O E 1978 *The Journal of Chemical Physics* **68** 4398–4405 ISSN 0021-9606 publisher: American Institute of Physics URL https://aip.scitation.org/doi/10.1063/1.435519

[29] Plenio M B 2005 *Phys. Rev. Lett.* **95**(9) 090503 URL https://link.aps.org/doi/10.1103/PhysRevLett.95.090503

Entangled neutrinos in a supernova

Y. Pehlivan
Mimar Sinan Fine Arts University, Sisli, Istanbul, 34380, Turkey
E-mail: yamac.pehlivan@msgsu.edu.tr

Savas Birol
Mimar Sinan Fine Arts University, Sisli, Istanbul, 34380, Turkey
E-mail: savas.birol@gmail.com

Abstract. We consider the entanglement of neutrinos evolving adiabatically under the effect of vacuum oscillations and self interactions through decreasing neutrino density similar to a core collapse supernova. For an initial state which consists only of electron type neutrinos, we analytically calculate the asymptotic value of the entropy of entanglement between the lowest energy neutrino and the rest of the system as a function of the number of neutrinos. We find that, as we increase neutrino number the entanglement entropy grows at first. But after going through a maximum, it approaches to zero in the limit of infinite number of neutrinos. We find that the number of neutrinos for which the entropy maximizes depends on the mixing angle.

1. Introduction

Since neutrinos interact with each other via neutral current weak force, they can form a self interacting many-body system. Flavor evolution of neutrinos in such a system is different from the simple oscillations that each would display in vacuum. The emergent many-body effects due to the neutrino self interactions is usually referred to as the *collective flavor oscillations*. For the collective effects to be sizable, the mean free path of the neutrinos under self interactions should be comparable to or smaller than their vacuum oscillation wavelengths. It is believed that this condition is satisfied in some astrophysical environments such as the core collapse supernova [1, 2, 3, 4, 5], black hole accretion disks [6, 7, 8, 9, 10], neutron star mergers [11], and the Early Universe [12, 13, 14]. Collective flavor oscillations of neutrinos in such environments is extensively studied in the literature. For reviews see Refs. [15, 16].

Like most other many-body systems in physics, the system of self interacting neutrinos is usually studied by adopting the *mean field approximation*. This approximation replaces the picture of mutually interacting neutrinos with a collection of *independent* particles, each moving in an external *mean field*. This mean field is collectively formed by all neutrinos and averages the effect of the system on each neutrino. For self consistency, the mean field is recalculated at every time step as the neutrinos themselves evolve. For a detailed derivation and discussion in the context of neutrinos, see Ref. [17]. For a general introduction to mean field approximation, see Chapter II of Ref. [18].

Mean field approximation is a standard tool in various areas of physics where many-body interactions arise, such as condense matter systems, cold atoms, or atomic nuclei. The general wisdom in those fields is that the mean field approximation becomes more and more accurate as the number of particles increase. But these systems differ from self interacting neutrinos in at least two curial aspects: First, they are usually found close to their ground states where the mean field changes slowly, allowing the scheme of averaging over the particles to work smoothly. Second, one usually has experimental access to these systems which makes it possible to compare mean field calculations with measurements. Neither of these is true in the case of self interacting neutrinos in astrophysical environments. The suitable set of initial states is far from equilibrium and fast oscillations in mean field is expected [19]. For recent studies on these fast oscillations see, for example, Refs. [20, 21].

These considerations led to a number of studies which addressed the adequacy of the mean field approximation in describing self interacting neutrinos [22, 23, 24, 17, 25, 26, 27, 28, 29]. The research on this subject revolves around the importance of entangled neutrino states in the dynamics. This is because the mean field approximation naturally eliminates all entangled states from the Hilbert space. After all, neutrinos cannot develop entanglement while they are evolving independently in an external field[1]. As a result, a large built up of entanglement in the original many-body system indicates that the mean field approximation is not applicable. In particular Refs. [27, 28, 29] calculate the von Neumann bi-partite entanglement entropy and conclude that this entropy increases with the number of neutrinos.

However, these studies are not conclusive: For n neutrinos, each with d flavor degrees of freedom, the Hilbert space dimension of the full many-body problem is d^n. As a result, solutions are feasible either for small numbers of neutrinos or with additional simplifying assumptions, For example, Ref. [27] employs relatively realistic conditions but includes only $n = 9$ neutrinos while Refs. [28, 29] can go up to $n = 128$ neutrinos in a simplified setup. But both cases have the potential to lead to artificial results. And in this paper, we present a specific example in which this is the case.

For this, we adopt a setup which is similar to the one considered in Ref. [27]: An effective two flavor mixing scheme, a decreasing neutrino density representing emission from an astrophysical source, and initial state which consists only of electron neutrinos in a box-like energy distribution. For this particular initial state, Ref. [27] observes that the entanglement entropy approaches to an asymptotic value as the neutrino density drops, and that this asymptotic value increases with the number of neutrinos up to $n = 9$. In this paper, by imposing the additional assumption that the evolution is adiabatic in this setup, we calculate the asymptotic value of the entanglement entropy analytically for any number of neutrinos. We find that, although the asymptotic entropy increases with n at first, it maximizes after a while and then starts to decrease. The particular number at which the asymptotic entropy maximizes depends on the adopted mixing angle. But we find that for all mixing angles, it eventually approaches to zero at large n.

It is important to note that Ref. [27] considers a variety of initial states and finds that the results depend strongly on the initial conditions. What allows us to analytically calculate the asymptotic entropy as a function of n is the fact that the initial state that we work with is symmetrical under the exchange of any two particles. This is a high degree of symmetry which makes it difficult to draw generic conclusions from our results.

But we think that this example still provides a useful point of view. In particular, we find that the number of neutrinos at which the entanglement entropy maximizes decreases with the mixing angle. For large mixing angles maximization happens at a few tens of neutrinos while for small mixing angles it happens after a few hundreds. This observation can be turned into a practical tool: Even if an analytical calculation is not possible for a different initial state, a numerical study can be carried out with a finite number of neutrinos in which one may try different mixing angles to see if this maximization occurs at a relatively small n. If it does, the

[1] However, the flavor evolutions of the neutrinos are still correlated because they are interacting with each other indirectly through their contributions to the mean field.

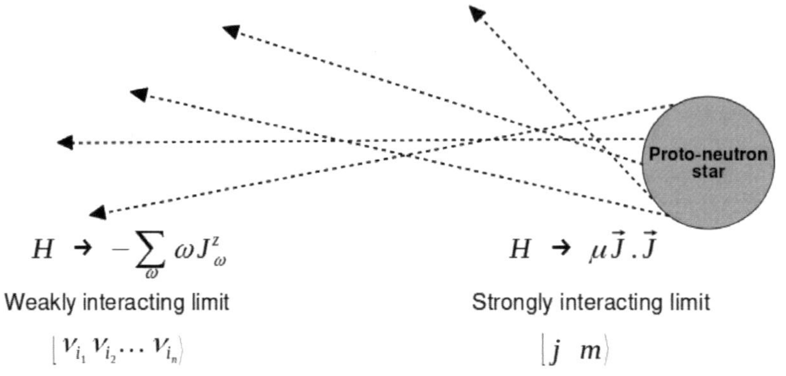

Figure 1: Schematic description of the neutrinos emitted from a proto-neutron star. Close to the center, the neutrino density is so large that neutrino-neutrino interactions dominate the dynamics. Away from the center, we can ignore the neutrino-neutrino interactions.

same effect that we observe here may be at play.

In what follows we introduce the model Hamiltonian describing the supernova neutrinos in Section II. In Section III, we discuss its spectrum and the adiabatic evolution of the particular initial state that we are interested in. In Section III we calculate the asymptotic value of the entanglement entropy as an analytical function of the number of neutrinos. In Section IV, we give our conclusions.

2. Hamiltonian

In this paper, we consider neutrinos emitted by a newly formed proto-neutron star at the center of a core collapse supernova as shown in Fig. 1. But this is only for the sake of providing a backdrop for our discussion. The actual model that we work with does not capture the rich physics of supernova neutrinos. Basically, we consider the proto-neutron star as a spherically symmetric source which emits only electron neutrinos, and assume that there are no other background particles. Although neutrino-neutrino interaction cross section depends on the angle between them [30], we take this angle dependence into account in an effective way using the *neutrino bulb model* which was developed in Ref. [31]. Also we assume that there are only two mixing neutrino flavors.

The mathematical formalism we use is the same as in Ref. [26]. It is built upon the neutrino isospin operator $\vec{J}_\omega$ which is a fictitious *spin operator* analogous to nuclear isospin. It is defined with respect to the doublet

$$\begin{pmatrix} |\nu_1(\omega)\rangle \\ |\nu_2(\omega)\rangle \end{pmatrix}. \tag{1}$$

Here, ν_i for $i = 1, 2$ denotes the neutrino eigenstate with the definite mass m_i. We use the vacuum oscillation frequency ω to label the neutrinos. It is related to the neutrino's energy E by

$$\omega = \frac{m_2^2 - m_1^2}{2E}. \tag{2}$$

For simplicity we assume that the oscillation modes are evenly spaced as

$$\omega = \omega_0, 2\omega_0, 3\omega_0, \ldots, n\omega_0, \tag{3}$$

and that each mode is occupied by only one neutrino. When we change the number of neutrinos, we do it in such a way that the highest frequency $n\omega_0$ is fixed.

Journal of Physics: Conference Series **2191** (2022) 012019 doi:10.1088/1742-6596/2191/1/012019

By definition the components of the isospin operator obey the relations

$$J^z_\omega \, |\nu_1(\omega)\rangle = \frac{1}{2} \, |\nu_1(\omega)\rangle, \qquad J^z_\omega \, |\nu_2(\omega)\rangle = -\frac{1}{2} \, |\nu_2(\omega)\rangle, \qquad \vec{J}_\omega \cdot \vec{J}_\omega \, |\nu_i(\omega)\rangle = \frac{3}{4} \, |\nu_i(\omega)\rangle, \tag{4}$$

and satisfy $SU(2)$ commutators among themselves:

$$[J^+_\omega, J^-_{\omega'}] = 2\delta_{\omega\omega'} J^0_\omega, \qquad [J^0_\omega, J^\pm_{\omega'}] = \pm\delta_{\omega\omega'} J^\pm_\omega. \tag{5}$$

For a system of n neutrinos, we denote the total isospin by $\vec{J}$:

$$\vec{J} = \sum_\omega \vec{J}_\omega. \tag{6}$$

In terms of the isospin operators, the Hamiltonian describing their flavor evolution is given by

$$H(\mu) = \sum_\omega -\omega J^z_\omega + \mu(r) \vec{J} \cdot \vec{J}. \tag{7}$$

The first term of this Hamiltonian describes the vacuum oscillations and the second term describes the neutrino-neutrino interactions. Here r represents the distance from the center of the proto-neutron star. The effective interaction strength $\mu(r)$ is given by [30, 31]

$$\mu(r) = \frac{\sqrt{2} G_F}{V} (1 - \sqrt{1 - (R_\nu/r)^2})^2. \tag{8}$$

Here, G_F is the Fermi coupling constant, and V is the normalization volume used for box quantization of the neutrinos. By definition it is the volume in which the total probability of finding each neutrino is 1. Although the normalization volume appears explicitly in the Hamiltonian, what emerges from the calculations is always the neutrino density n/V. As the neutrinos expand from the source, V increases and the neutrino-neutrino interaction rate drops. The term in the parenthesis in Eq. (8) is the geometrical factor of the neutrino bulb model [31] with R_ν denoting the radius of the proto-neutron star. This factor serves two purposes: First, neutrino-neutrino interaction cross section depends on the angle between them, which, on the average, becomes smaller as the neutrinos expand from a source. Second, although the term $\vec{J} \cdot \vec{J}$ involves interactions between every pair of neutrinos, in reality not every neutrino can meet and interact with every other neutrino due to the geometry of the problem. Each neutrino can interact with only those emitted from a restricted part of the surface of the proto-neutron star. Also, while a given neutrino can meet and interact with many other neutrinos when it is near the surface, this is not the case when it is far away. The geometrical factor takes these into account by suitably scaling down the interaction rate with distance.

The Hamiltonian in Eq. (7) involves the neutrino mass difference but not the vacuum mixing angle because it is written in the mass basis. The mixing angle enters the problem as an initial condition when the neutrinos start their life in weak interaction eigenstates. These eigenstates are given by

$$\begin{aligned} |\nu_e(\omega)\rangle &= \cos\theta |\nu_1(\omega)\rangle + \sin\theta |\nu_2(\omega)\rangle, \\ |\nu_x(\omega)\rangle &= -\sin\theta |\nu_1(\omega)\rangle + \cos\theta |\nu_2(\omega)\rangle. \end{aligned} \tag{9}$$

Here ν_e denotes the electron flavor whereas ν_x denotes a general superposition of muon and tau flavors. We treat the mixing angle θ as a free parameter for illustration purposes. We also assume that the mass hierarchy is normal, i.e., $m_2 > m_1$.

3. Eigenstates and the adiabatic evolution

Generic evolution of the system requires knowledge of the evolution operator but it is difficult to calculate. However, adiabatic evolution requires only an understanding of the energy spectrum which is examined in detail in Refs. [32, 26]. In what follows we summarize the relevant points from these references.

The Hamiltonian in Eq. (7) conserves the total number of neutrinos, as well as the z component of the total isospin operator:

$$[H(\mu), J^z] = 0, \quad [H, N] = 0, \quad [N, J^z] = 0. \tag{10}$$

Here $\hat{N}$ is the total neutrino number operator. Its conservation is due to the fact that there are no terms creating or annihilating neutrinos in our formulation. We assume that neutrinos simply free stream outside the proto-neutron star. The commutation of J^z with the Hamiltonian is also evident: It commutes with the vacuum oscillation term because they are both isospin z components. It also commutes with the interaction term because the latter is the Casimir operator of the total isospin algebra. We denote the simultaneous eigenstates and corresponding eigenvalues of these operators as

$$\begin{aligned} H(\mu)\,|E(\mu), n, m\rangle &= E(\mu)\,|E(\mu), n, m\rangle, \\ \hat{N}\,|E(\mu), n, m\rangle &= n\,|E(\mu), n, m\rangle, \\ J^z\,|E(\mu), n, m\rangle &= m\,|E(\mu), n, m\rangle. \end{aligned} \tag{11}$$

Clearly the eigenvalue of the number operator, denoted by n, can be written as

$$n = n_1 + n_2, \tag{12}$$

where n_i denotes the total number of neutrinos of ν_i type for $i = 1, 2$. As can be seen from Eq. (4), the eigenvalue of J_z is then equal to

$$m = \frac{n_1 - n_2}{2}. \tag{13}$$

This tells us that each many-body eigenstate comes with a definite value for n_1 and n_2.

It is easy to find the many-body eigenstates and the corresponding eigenvalues of the Hamiltonian in the strongly and weakly interacting regimes. When the neutrinos are close to the surface of the proto-neutron star, the neutrino density so high that that we can take $\mu \to \infty$ limit and ignore the vacuum oscillation term in the Hamiltonian. (See Fig. 1.) In this limit, we have

$$\lim_{\mu\to\infty} H(\mu) = \mu \vec{J} \cdot \vec{J}. \tag{14}$$

This tells us that the total isospin states denoted by $|j\ m\rangle$ are the many-body eigenstates of the system, i.e.,

$$\lim_{\mu\to\infty} |E(\mu), n, m\rangle = |j\ m\rangle. \tag{15}$$

Here j denotes the total isospin quantum number which can take values starting from $j = 0$ or $1/2$ up to $j = n/2$, and m is given by Eq. (13). It may look like we lost a quantum number in Eq. (15), but this is due to the fact that total isospin quantum number j comes with many degeneracies when we add more than two isospins. The energy eigenvalues in this limit are given by

$$\lim_{\mu\to\infty} E(\mu) = \mu j(j+1). \tag{16}$$

Conversely, when neutrinos are far from the source, the neutrino density decreases so much that we can take $\mu \to 0$ limit in the Hamiltonian and ignore the interactions. In this limit, we have

$$\lim_{\mu\to 0} H(\mu) = -\sum_{\omega} \omega J^z_\omega \tag{17}$$

and the many-body eigenstates reduce to the products of individual mass eigenstates. In other words, they have the form

$$\lim_{\mu\to 0} |E(\mu), n, m\rangle = |\nu_{i_1}\ \nu_{i_2}\ \dots\ \nu_{i_n}\rangle, \tag{18}$$

where $i_1, i_2, \dots, i_n = 1, 2$. Note that when we write the single particle states is this form, we always arrange them in the order of increasing ω. Therefore, the first neutrino in Eq. (18) has the lowest oscillation frequency ω_0 whereas the last one has the highest oscillation frequency $n\omega_0$. The energy eigenvalues corresponding to the eigenstates given in Eq. (18) are

$$\lim_{\mu\to 0} E(\mu) = \frac{\omega_0}{2} \sum_k (-1)^{i_k} k, \tag{19}$$

where we used Eq. (3) and Eq. (4).

Under adiabatic conditions, the dynamical evolution of the eigenstates follow their own slow change with the external conditions. Here, the external condition is the interaction strength μ which decreases with neutrino density. Specifically, adiabaticity requires the flavor oscillation wavelength of the neutrinos to be much smaller than the distance scale at which the density changes. It is believed that this condition is satisfied in the core collapse supernovae. As a result, as we go from strongly to weakly interacting regime, the eigenstates in Eq. (15) should dynamically evolve into those given in Eq. (18). However, it is not immediately obvious which eigenstate evolves into which one because usually there are many crossings between energy eigenvalues. Therefore, one must also diagonalize the Hamiltonian in the intermediate regime in order to analytically connect the eigenstates of the strongly and weakly interacting regimes.

In the intermediate regime, the eigenstates and eigenvalues should be calculated using a numerical scheme. Fig. 2 shows the eigenvalues for $n = 10$ neutrinos as a function of the interaction strength μ. This figure is taken from Ref [26] and shows the eigenstates grouped according to the value of the conserved quantity J^z given in Eq. (13). Calculations are carried out using the Bethe ansatz technique which is based on the dynamical symmetries of the Hamiltonian in Eq. (7) examined in detail in Refs. Note that in Fig. 2 the eigenvalues are plotted in the direction of increasing μ. As the neutrinos go from the center of supernova to the outer regions as in Fig. 1, the eigenvalues should be traced from right to left [32, 26].

The insets in Fig. 2 show that these eigenvalues cross each other several times in the low μ region. However, for each m value, the highest energy eigenvalue never undergoes a crossing with the others. Our calculation of the entanglement entropy of the final state after an adiabatic evolution strongly depends on this observation. But before we move into the details, we must emphasize two points.

First, although our calculations depend on the assumption that it is true, there is no algebraic proof in the literature that the highest energy eigenstates of the Hamiltonian in Eq. (7) for each m are always non-degenerate. However, various proofs for special cases, and even *experimental evidence* exists in that direction. The neutrino Hamiltonian given in Eq. (7) is the same as the Hamiltonian describing pairing interactions between Cooper electrons in superconductors and cold atoms systems [32]. One just has to replace the neutrino isospin operators with pair quasispin operators, which also obey SU(2) commutation relations, and multiply the Hamiltonian with an overall minus sign. As a result, the highest energy eigenstates of the

 2191 (2022) 012019 doi:10.1088/1742-6596/2191/1/012019

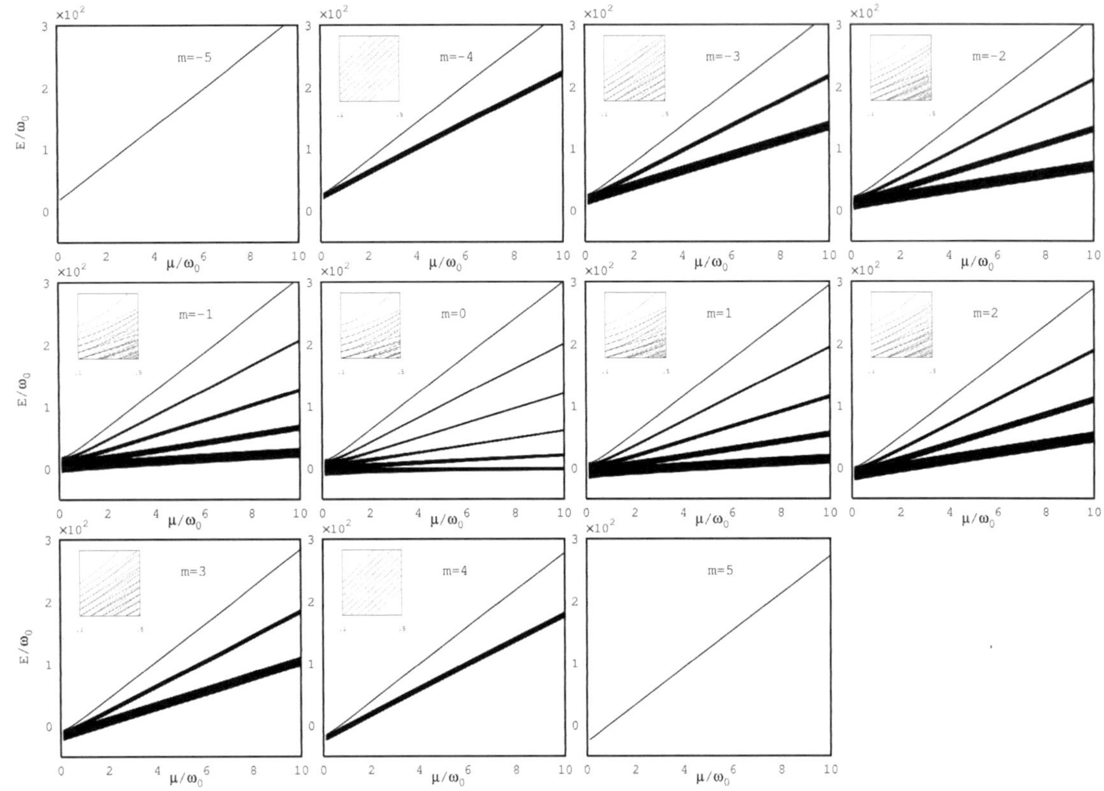

Figure 2: Eigenvalues for $n = 10$ neutrinos grouped according to the value of the conserved quantity J^z. Taken from Ref. [26].

neutrino Hamiltonian are the ground states of the pairing Hamiltonian. The non-degeneracy of these states is a well known phenomenon which is observed in numerical solutions with up to $n = 1600$ and $\kappa = 800$ in Ref. [33] for the case of equally spaced oscillation modes similar to the one introduced in Eq. (3). An analytical proof for large values of n and m is originally given by Gaudin [34], and later elaborated by Richardson [35].

One can also discuss this non-degeneracy from the point of view of experiments involving cold atomic systems. In such experiments one can control the strength of the pairing interaction and take the system adiabatically from weakly interacting regime to strongly interacting regime. This results in the famous Bardeen-Cooper-Schrieffer to Bose-Einstein condensation crossover [36]. The fact that it is a crossover, rather than a phase transition, indicates that the ground state of the system never undergoes a level crossing with one of the excited levels. A level crossing would cause abrupt changes in the measurable quantities which is not experimentally observed. For details see, for example, Refs. [37, 38, 39].

The second point to be emphasized is that the non-crossing is not required for our adiabaticity assumption. Although adiabatic evolution is generally identified with the presence of energy gaps, it is in fact possible for a system to evolve adiabatically even when energy eigenvalues cross each other which is called the gapless adiabaticity [40]. This is discussed in the context of self interacting neutrinos in Ref. [27]. In our case the non-crossing of the highest energy eigenstate is useful in identifying which eigenstate in the highly interacting limit is adiabatically connected to which eigenstate in the weakly adiabatic limit.

At the strongly interacting limit, Eq. (16) tells us that the highest energy eigenstate for a given m is the one with the highest possible total isospin quantum number, which is $j = n/2$. At the weakly interacting limit, Eq. (19) tells us that the highest energy eigenstate is the one in which there are $n_1 = n/2 + m$ neutrinos of ν_1 type occupying the lowest oscillation frequencies, and $n_2 = n/2 - m$ neutrinos of ν_2 type occupying the highest oscillation frequencies. As a result, the states

$$|\frac{n}{2},\ m\rangle \longrightarrow |\underbrace{\nu_1,\ldots,\nu_1}_{\frac{n}{2}+m},\underbrace{\nu_2,\ldots,\nu_2}_{\frac{n}{2}-m}\rangle \tag{20}$$

are analytically connected through the decreasing interaction strength so that, under adiabatic conditions, the one on the left would evolve into the one on the right.

Theory of angular momentum addition tells us that the states with the highest total isospin quantum number are maximally symmetric. In other words, $|\frac{n}{2},\ m\rangle$ is invariant under the exchange of any two neutrinos. Those states with $j < n/2$ do not have this property. Considering that the state in which all neutrinos are ν_e is also symmetric under this exchange, one expects that this particular initial state can be decomposed entirely in terms of the $|\frac{n}{2},\ m\rangle$ states. That this is indeed the case can be shown by repeatedly applying Eq. (9), and using the relevant Clebsh-Gordon coefficients [26]:

$$|\psi\rangle_{\text{initial}} = |\nu_e, \nu_e, \ldots, \nu_e\rangle = \sum_{m=-n/2}^{n/2} (\cos\theta)^{(\frac{n}{2}+m)}(\sin\theta)^{(\frac{n}{2}-m)}\sqrt{\binom{n}{\frac{n}{2}+m}}|n/2, m\rangle. \tag{21}$$

Since this decomposition includes only the highest energy eigenstates in the $\mu \to \infty$ limit, its adiabatic evolution leads to

$$|\psi\rangle_{\text{final}} = \sum_{m=-n/2}^{n/2} (\cos\theta)^{(\frac{n}{2}+m)}(\sin\theta)^{(\frac{n}{2}-m)}\sqrt{\binom{n}{\frac{n}{2}+m}}e^{i\varphi_m}|\underbrace{\nu_1,\ldots,\nu_1}_{\frac{n}{2}+m},\underbrace{\nu_2,\ldots,\nu_2}_{\frac{n}{2}-m}\rangle \tag{22}$$

as per Eq. (20). Here φ_m are some phases (both dynamical and geometrical) whose exact forms are not needed here. Eq. (22) is an analytical expression for the asymptotic final state of the adiabatic evolution in $\mu \to 0$ limit. This is clearly an entangled many-body state whose entropy of entanglement is calculated in the next section.

4. Entanglement entropy

Entanglement entropy is a measure of entanglement between two parts of a system. We use von Neumann entanglement entropy which is defined as the regular thermodynamic entropy of either one of these parts. The idea is that, even when the full system is in a pure state, each part lacks a definite quantum state due to its entanglement to the other. Therefore individual parts would be in mixed states and have thermodynamic entropies.

Let us denote the two parts of the system by $\mathcal{A}$ and $\mathcal{B}$. The reduced density matrix $\rho_{\mathcal{A}}$ of part $\mathcal{A}$ is obtained by tracing the density matrix of the full system ρ over those degrees of freedom belonging to part $\mathcal{B}$, and vice versa:

$$\rho_{\mathcal{A}} = \text{Tr}_{\mathcal{B}}(\rho), \qquad \rho_{\mathcal{B}} = \text{Tr}_{\mathcal{A}}(\rho). \tag{23}$$

The entropy of the entanglement between parts $\mathcal{A}$ and $\mathcal{B}$ is defined as

$$S = -\text{Tr}\left(\rho_{\mathcal{A}} \ln \rho_{\mathcal{A}}\right) = -\text{Tr}\left(\rho_{\mathcal{B}} \ln \rho_{\mathcal{B}}\right). \tag{24}$$

The entropy of entanglement is a conjoint quantity. Whether we trace over part $\mathcal{A}$ and calculate the entropy of part $\mathcal{B}$, or do the opposite, the result is the same as indicated by Eq. (24).

However, the entropy depends on how we partition the system. In particular, the zero value for entanglement entropy indicates that the quantum state $|\psi\rangle$ of the full system is a direct multiplication of pure states $|\psi\rangle_{\mathcal{A}}$ and $|\psi\rangle_{\mathcal{B}}$ of the parts i.e.,

$$|\psi\rangle = |\psi\rangle_{\mathcal{A}}|\psi\rangle_{\mathcal{B}}. \tag{25}$$

While this may be the case for one way of partitioning the system, it may not be so for another possible partition.

In our case, the full system is n neutrinos. We take part $\mathcal{A}$ to be the first $n-1$ neutrino and part B to be the last neutrino, which is the neutrino with the highest oscillation frequency. Initially entropy of entanglement between these parts is zero because the initial wave function of the full system given in Eq. (21) is factorizable. But this is clearly not the case in the final state given in Eq. (22). Tracing over all possible states of the first $n-1$ neutrino, one show that the reduced density matrix of the last neutrino is given by

$$\rho_{\mathcal{B}} = (\cos\theta)^{2n}|\nu_1\rangle\langle\nu_1| + (\cos\theta)^{2n-1}(\sin\theta)\sqrt{\binom{n}{n-1}}\left(e^{i(\varphi_{n-1}-\varphi_n)}|\nu_2\rangle\langle\nu_1| + e^{-i(\varphi_{n-1}-\varphi_n)}|\nu_1\rangle\langle\nu_2|\right)$$
$$+\sum_{k=0}^{n-1}(\cos\theta)^{2k}(\sin\theta)^{2n-2k}\binom{n}{k}|\nu_2\rangle\langle\nu_2|. \tag{26}$$

The trace of this density operator is equal to 1 as can be easily confirmed by noting that the sum of the coefficients of $|\nu_1\rangle\langle\nu_1|$ and $|\nu_2\rangle\langle\nu_2|$ is equal to $(\cos^2\theta + \sin^2\theta)^n$.

In general, the von Neumann entropy of a two-state density operator such as the one in Eq. (26), can be more easily calculated in terms of its polarization (or Bloch) vector [27]. This vector is defined by expanding the density operator in terms of the Pauli spin operators

$$\rho_{\mathcal{B}} = \frac{1}{2}\left(I + \vec{P}\cdot\vec{\sigma}\right), \tag{27}$$

where

$$\sigma_1 = |\nu_1\rangle\langle\nu_2| + |\nu_2\rangle\langle\nu_1|, \quad \sigma_2 = -i|\nu_1\rangle\langle\nu_2| + i|\nu_2\rangle\langle\nu_1|, \quad \sigma_3 = |\nu_1\rangle\langle\nu_1| - |\nu_2\rangle\langle\nu_2|. \tag{28}$$

With this expansion, the eigenvalues of the density operator are easily identified as $(1 \pm P)/2$. This allows us to easily apply the formula in Eq. (24) in the basis where the density operator is diagonal. The result is

$$S = -\frac{1+P}{2}\ln\frac{(1+P)}{2} - \frac{1-P}{2}\ln\frac{(1-P)}{2}. \tag{29}$$

Note that since $0 \le P \le 1$, the entanglement entropy given in Eq. (29) is always positive. $P = 1$ limit corresponds to the case in which both $\mathcal{A}$ and $\mathcal{B}$ are in pure states because the eigenvalues of $\rho_{\mathcal{B}}$ are 1 and 0. In this limit Eq. (29) gives $S = 0$. In the opposite limit where $P = 0$, both eigenvalues of $\rho_{\mathcal{B}}$ are 1/2. This corresponds to the case that $\mathcal{A}$ and $\mathcal{B}$ are maximally entangled. In this limit Eq. (29) gives $S = \ln 2$ which is the maximum possible entropy for this particular partitioning of the system.

The value of the polarization vector can be directly calculated by substituting Eq. (26) into Eq. (27) and using the identity $x^n = e^{n\ln x}$. The result is

$$P = \sqrt{1 - 4e^{-2\beta n} + 4(1 + n\tan^2\theta)e^{-4\beta n}} \tag{30}$$

for $n \geq 2$, where $\beta = -\ln(\cos\theta)$. As can be seen from this formula, for finite n we always have $P < 1$ which indicates that the last neutrino is entangled to the rest of the system. This can be shown by noting that $0 \leq (1+n\tan^2\theta)e^{-2\beta n} \leq 1$. In fact, as we increase the number of neutrinos, P goes through a minimum at which point the entanglement entropy becomes maximum. This can be clearly seen in Fig. 3 where we plot the entanglement entropy S as a function of n obtained by substituting Eq. (30) in Eq. (29). The figure shows results for various mixing angles indicated next to the lines. For clarity, we plot large mixing angles on the left and small mixing angles on the right. We see that the points at which entropy becomes maximum depends on the adopted value of the mixing angle. For large mixing angles, maximization occurs at a few tens of neutrinos while for small mixing angles it occurs after a few hundreds of neutrinos.

However, Eq. (30) shows that in $n \to \infty$ limit we have $P \to 1$ and consequently $S \to 0$ for any value of the mixing angle, indicating that the last neutrino is not entangled to the rest of the system. For large mixing angles entropy approaches to zero quickly while for small mixing angles it decreases more slowly. This can be easily understood by looking at the final state in Eq. (22). The squared coefficients of the single particle states in that equation can be approximated by a Gaussian distribution as

$$(\cos\theta)^{(n+2m)}(\sin\theta)^{(n-2m)}\binom{n}{\frac{n}{2}+m} \approx \frac{1}{\sqrt{2\sigma^2\pi}}\exp\left(-\frac{(m-\bar{m})^2}{2\sigma^2}\right), \tag{31}$$

where

$$\bar{m} = n(\cos^2\theta - \frac{1}{2}), \qquad \sigma = \sqrt{n}\cos\theta\sin\theta. \tag{32}$$

As n increases, the fractional width of this Gaussian becomes smaller. This width represents the ratio of those terms which are considerably different from zero to the total number of terms. Since the former is proportional to $\sigma \sim \sqrt{n}$ whereas the latter is proportional to n, this ratio approaches to zero in $n \to \infty$ limit, indicating that

$$\lim_{n\to\infty} |\psi\rangle_{\text{final}} = e^{i\varphi_{\bar{m}}}|\underbrace{\nu_1,\ldots,\nu_1}_{\frac{n}{2}+\bar{m}},\underbrace{\nu_2,\ldots,\nu_2}_{\frac{n}{2}-\bar{m}}\rangle. \tag{33}$$

As a result, the final state is unentangled in the limit of a large number of particles.

5. Conclusions

In this study, we considered the entanglement of self interacting neutrinos in a setting similar to that of a core collapse supernova. Without taking matter effects into account, we focused on the adiabatic evolution of neutrinos under the vacuum oscillations and self interactions. For an initial condition in which all neutrinos are emitted as electron neutrinos from the surface of the proto-neutron star, we showed that the asymptotic value of the entanglement entropy far from the center can be calculated analytically. This calculation was possible due to the symmetry of the initial state under the exchange of any two neutrinos. This symmetry allows the initial state to project only on those states which undergo no level crossings during the evolution, and makes it possible to analytically trace its adiabatic evolution from the center of the supernova to the outer regions.

We find that the entropy of entanglement between the neutrino with the highest oscillation frequency and the rest of the system initially grows as we increase the number of particles in the system. However, after maximizing at a certain value, it approaches to zero in the $n \to \infty$ limit. We conclude that, for large numbers of neutrinos the final state is not entangled. We also find that the value at which the entanglement is maximum depends on the mixing angle. But the entanglement vanishes at large numbers for any value of the mixing angle.

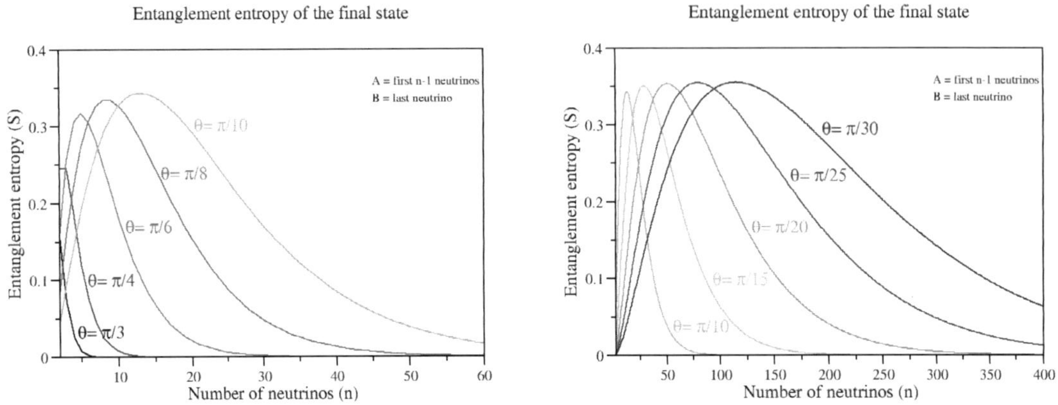

Figure 3: Asymptotic value of the entanglement entropy in $\mu \to 0$ limit as a function of the number of neutrinos n in the system obtained by substituting Eq. (30) in Eq. (29). For clarity, we plot the curves corresponding to larger mixing angles on the left and smaller mixing angles on the right. For large mixing angles, the entropy maximizes when we have a few tens of neutrinos. For small mixing angles it maximizes when we have a few hundreds of neutrinos.

It is not possible to draw generic conclusions from our study because our calculations depend significantly on the symmetry of the initial state. Ref. [27] finds that entanglement builds up in the system for a variety of initial states, including those which do not have the above-mentioned symmetry. In particular, those initial states which are not symmetric under the exchange of neutrinos would undergo level crossings. Whether the system evolves adiabatically or non-adiabatically through these crossings, the growth of the entanglement can be very different than the results obtained here.

Still, our results shine a new light on the problem. We believe that the dependence of the entanglement entropy on the mixing angle is our most important result. For those initial states for which analytical calculations are not possible, it may be best to try both large and small mixing angles. If one finds that the entropy increases in one case but decreases in the other, it is possible that a similar effect observed here is in effect. For example, Ref. [27] adopts the mixing angle $\theta \approx \pi/20$ and finds that the entanglement entropy increases until $n = 9$ neutrinos. Our results indicate that, for this mixing angle it would continue to increase until $n \approx 60$ before starting to decrease. However, for a mixing angle of $\theta = \pi/3$ it would start decreasing with n right away as can be seen from our Fig. 3. A similar comment may also be given on Refs. [28, 29] which adopt zero mixing angle and find that the entropy of entanglement still increases after 128 neutrinos. Although the setting and the initial state of Ref. [28, 29] are very different from our case (they adopt a constant neutrino density, only two energy modes, and a less symmetric initial state), it would be worthwhile to repeat this study for a finite mixing angle.

Acknowledgments
We thank B. Balantekin, A. Pathwardhan, S. Shashank and L. Jones for useful discussion.

[1] Burrows A 1990 *Ann.Rev.Nucl.Part.Sci.* **40** 181–212
[2] Raffelt G G 1996 *Stars as Laboratories for Fundamental Physics* (University of Chicago Press)
[3] Kotake K, Sato K and Takahashi K 2006 *Rept.Prog.Phys.* **69** 971–1144 (*Preprint* `astro-ph/0509456`)
[4] Beacom J F 2010 *Ann.Rev.Nucl.Part.Sci.* **60** 439–462 (*Preprint* `1004.3311`)
[5] Fuller G M, Mayle R W, Wilson J R and Schramm D N 1987 **322** 795–803
[6] Ruffert M and Janka H 1999 *Astron.Astrophys.* **344** 573–606 (*Preprint* `astro-ph/9809280`)

[7] Popham R, Woosley S and Fryer C 1999 *Astrophys.J.* **518** 356–374 (*Preprint* astro-ph/9807028)
[8] Narayan R, Piran T and Kumar P 2001 *Astrophys. J.* **557** 949 (*Preprint* astro-ph/0103360)
[9] Matteo T D, Perna R and Narayan R 2002 *Astrophys.J.* **579** 706–715 (*Preprint* astro-ph/0207319)
[10] Chen W X and Beloborodov A M 2007 *Astrophys.J.* **657** 383–399 (*Preprint* astro-ph/0607145)
[11] Faber J A and Rasio F A 2012 *Living Rev. Rel.* **15** 8 (*Preprint* 1204.3858)
[12] Dolgov A 2002 *Phys.Rept.* **370** 333–535 (*Preprint* hep-ph/0202122)
[13] Hannestad S 2006 *Ann.Rev.Nucl.Part.Sci.* **56** 137–161 (*Preprint* hep-ph/0602058)
[14] Savage M J, Malaney R A and Fuller G M 1991 **368** 1–11
[15] Duan H, Fuller G M and Qian Y Z 2010 *Ann.Rev.Nucl.Part.Sci.* **60** 569–594 (*Preprint* 1001.2799)
[16] Chakraborty S, Hansen R, Izaguirre I and Raffelt G 2016 *Nucl. Phys.* **B908** 366–381 (*Preprint* 1602.02766)
[17] Balantekin A and Pehlivan Y 2007 *J.Phys.* **G34** 47–66 (*Preprint* astro-ph/0607527)
[18] Bonitz M 1998 *Quantum Kinetic Theory* (Teubner-texte physik)
[19] Sawyer R 2005 *Phys.Rev.* **D72** 045003 (*Preprint* hep-ph/0503013)
[20] Chakraborty S, Hansen R S, Izaguirre I and Raffelt G 2016 *JCAP* **03** 042 (*Preprint* 1602.00698)
[21] Dasgupta B, Mirizzi A and Sen M 2017 *JCAP* **1702** 019 (*Preprint* 1609.00528)
[22] Friedland A and Lunardini C 2003 *Phys.Rev.* **D68** 013007 (*Preprint* hep-ph/0304055)
[23] Bell N F, Rawlinson A A and Sawyer R 2003 *Phys.Lett.* **B573** 86–93 (*Preprint* hep-ph/0304082)
[24] Friedland A and Lunardini C 2003 *JHEP* **0310** 043 (*Preprint* hep-ph/0307140)
[25] Volpe C, Väänänen D and Espinoza C 2013 *Phys.Rev.* **D87** 113010 (*Preprint* 1302.2374)
[26] Birol S, Pehlivan Y, Balantekin A B and Kajino T 2018 (*Preprint* 1805.11767)
[27] Cervia M J, Patwardhan A V, Balantekin A, Coppersmith d S and Johnson C W 2019 *Phys. Rev. D* **100** 083001 (*Preprint* 1908.03511)
[28] Roggero A 2021 (*Preprint* 2102.10188)
[29] Roggero A 2021 (*Preprint* 2103.11497)
[30] Sigl G and Raffelt G 1993 *Nucl.Phys.* **B406** 423–451
[31] Duan H, Fuller G M, Carlson J and Qian Y Z 2006 *Phys.Rev.* **D74** 105014 (*Preprint* astro-ph/0606616)
[32] Pehlivan Y, Balantekin A, Kajino T and Yoshida T 2011 *Phys.Rev.* **D84** 065008 (*Preprint* 1105.1182)
[33] Roman J, Sierra G and Dukelsky J 2002 *Nucl.Phys.* **B634** 483–510 (*Preprint* cond-mat/0202070)
[34] Gaudin M 1968 *États propres et valeurs propres de l'Hamiltonien d'appariement,* unpublished. Saclay preprint. Included in: Travaux de M. Gaudin, Modèles Exactement Résolus (Les Èditions de Physique), 1995
[35] Richardson R W 1977 *Journal of Mathematical Physics* **18** 1802–1811
[36] Pehlivan Y, Subaşı A L, Ghazanfari N, Birol S and Yüksel H 2017 *Phys. Rev.* **D95** 063022 (*Preprint* 1603.06360)
[37] Bloch I, Dalibard J and Zwerger W 2008 *Rev. Mod. Phys.* **80**(3) 885–964 URL http://link.aps.org/doi/10.1103/RevModPhys.80.885
[38] Giorgini S, Pitaevskii L P and Stringari S 2008 *Rev. Mod. Phys.* **80**(4) 1215–1274 URL http://link.aps.org/doi/10.1103/RevModPhys.80.1215
[39] Randeria M and Taylor E 2014 *Ann. Rev. Condensed Matter Phys.* **5** 209–232 (*Preprint* 1306.5785)
[40] Avron J and Elgart A 1999 *Commun. Math. Phys.* **203** 445–463 (*Preprint* math-ph/9805022)

Journal of Physics: Conference Series **2191** (2022) 012020 doi:10.1088/1742-6596/2191/1/012020

Scalar Pre-potentials for Spinor and Tensor Fields on Spacetime

R. W. Tucker[1] **and T. J. Walton**[2]

[1] Department of Physics, University of Lancaster and Cockcroft Institute, Daresbury Laboratory, Warrington, UK

[2] Jeremiah Horrocks Institute, School of Natural Sciences, University of Central Lancashire, Preston, UK

E-mail: `r.tucker@lancaster.ac.uk` and `twalton@uclan.ac.uk`

Abstract. We review a technique for solving a class of classical linear partial differential systems of relevance to physics in Minkowski spacetime. All the equations are amenable to analysis in terms of complex solutions in the kernel of the scalar Laplacian and a complexified Hertz potential. The complexification prescription ensures the existence of regular physical solutions with chirality and propagating, non-singular, pulse-like characteristics that are bounded in all three spatial dimensions. The technique is applied to the source-free Maxwell, Bopp-Landé-Podolsky and linearised Einstein field systems, and particular solutions are used for constructing classical models describing single-cycle laser pulses and a mechanism is discussed for initiating astrophysical jets. Our article concludes with a brief introduction to spacetime Clifford algebra ideals that we use to represent spinor fields. We employ these to demonstrate how the same technique used for tensor fields enables one to construct new propagating, chiral, non-singular, pulse-like spinor solutions to the massless Dirac equation in Minkowski spacetime.

1. Introduction

In this article we discuss a class of solutions to a set of linear field equations on Minkowski spacetime. The class of solutions will all be constructed from a scalar field α satisfying $\Box\,\alpha = 0$, where $\Box$ is the Lorentz-invariant *D'Alembertian* (or *Laplacian operator*), and a set of covariantly-constant, antisymmetric tensor fields (2-forms) $\Pi^{\nu,\chi}$. We shall refer to these elements as "pre-potentials" since, in general, (like most "potentials" in physics) they define gauge equivalence classes of solutions to "gauge-invariant" equations. The "gauge-invariant" equations under consideration here include the source-free Maxwell system for the Minkowski-Maxwell field tensor, one of its modifications proposed by Bopp-Landé-Podolsky [1, 2, 3], the tensor equation for perturbations of Einstein's gravitational field equation in a matter-free background Minkowski spacetime and the electromagnetically-neutral massless Dirac equation. It will be shown how all these linear partial differential systems can be formulated in terms of certain differential tensor and spinor operators that render them amenable to analysis using the pre-potentials $\{\alpha, \Pi^{\nu,\chi}\}$. The class of particular solutions derived from these offers useful models of interest in laser physics and astrophysics. In particular, we indicate how to construct both Maxwell free-space models describing propagating finite-energy *multi-chiral* laser pulses that are bounded in all three spatial dimensions, and propagating gravitational pulses with similar characteristics. We argue that the former may offer new channels for quantum encryption and that the latter play a role in the formation of astrophysical jets observed in X-ray spectroscopy. The derivations of these models are discussed briefly since full details can be found in the indicated references.

Our formulation is given in terms of the language of differential geometry and section 2 establishes essential definitions and our notational conventions. Section 3 describes the common technique used to solve the field systems above and draws upon ideas that seem to have originated in J. L. Synge's (1956) efforts to construct a classical model of the photon [4]. Section 4 sketches how a similar technique enables one to solve the linearised vacuum Einstein system for multi-chiral focussed gravitational pulses. In section 5 we indicate how Minkowski spacetime spinor fields can be formulated in terms of sections of a $(3,1)$ Clifford algebra bundle and how the above pre-potential technique provides the necessary tools for deriving spinor solutions for the neutral, massless Dirac equation. Finally, in section 6, we summarise our conclusions.

2. Definitions and Notational Conventions

The natural mathematical language to discuss the differential properties of tensor fields on spacetime is in terms of differential forms and their associated exterior calculus [5]. In this section we give a brief summary of relevant notation used throughout the article.

Let $\mathcal{M}$ denote spacetime modelled as a four-dimensional orientable Lorentzian manifold with *metric tensor field* g of signature $(3,1) \equiv (-,+,+,+)$. The metric tensor field can always be written in a local cobasis $\{e^a\}$ of differential 1-forms as

$$\mathsf{g} = \eta_{ab}\, e^a \otimes e^b = -e^0 \otimes e^0 + e^1 \otimes e^1 + e^2 \otimes e^2 + e^3 \otimes e^3$$

with $\eta_{ab} = \mathsf{g}(X_a, X_b) = \mathrm{diag}(-1,1,1,1)$ in terms of the dual local basis $\{X_a\}$ of vector fields, defined so that $e^a(X_b) = \delta^a_b$ $(a, b = 0, 1, 2, 3)$. This induces the *inverse metric tensor field*

$$\mathsf{g}^{-1} = \eta^{ab}\, X_a \otimes X_b = -X_0 \otimes X_0 + X_1 \otimes X_1 + X_2 \otimes X_2 + X_3 \otimes X_3$$

where $\eta^{ab} = \mathsf{g}^{-1}(e^a, e^b)$ with $\eta^{ab}\eta_{bc} = \delta^a_c$. In our discussion of spinor fields below, based on a Clifford algebra, it will prove important to introduce the physically dimensionless tensor fields:

$$\breve{\mathsf{g}} = \Upsilon \mathsf{g} \qquad \text{and} \qquad \breve{\mathsf{g}}^{-1} = \frac{\mathsf{g}^{-1}}{\Upsilon} \tag{1}$$

where Υ is some chosen positive real number.

The tensor fields g, g^{-1} establish an isomorphism between the space of *vector fields* on $\mathcal{M}$ – sections, denoted $\Gamma T\mathcal{M}$, of the *tangent bundle* $T\mathcal{M}$ – and the space of *differential 1-forms* on $\mathcal{M}$ – sections, denoted $\Gamma T^*\mathcal{M}$, of the *cotangent bundle* $T^*\mathcal{M}$. We represent this isomorphism with a 'tilde':

$$\begin{aligned} \text{vector field}, X \quad &\longrightarrow \quad \text{differential 1-form}, \widetilde{X} \equiv \mathsf{g}(X, -) \\ \text{differential 1-form}, \alpha \quad &\longrightarrow \quad \text{vector field}, \widetilde{\alpha} \equiv \mathsf{g}^{-1}(\alpha, -) \end{aligned}$$

where, for typographical economy, we use the same 'tilde' symbol for both mappings. These are referred to as *metric duals* and the tilde map satisfies:

$$\widetilde{\widetilde{X}} = X \qquad \text{and} \qquad \widetilde{\widetilde{\alpha}} = \alpha$$

for any vector field X and differential 1-form α. Differential forms of arbitrary degree ($0 \leq p \leq 4$) are identified with sections $\Gamma\Lambda\mathcal{M}$ of an *exterior bundle* $\Lambda\mathcal{M}$: a *differential p-form* α is associated with a section of the pth exterior bundle $\Lambda^p\mathcal{M}$ whose elements are the totally antisymmetric sections of the type $(p, 0)$ tensor bundle on $\mathcal{M}$ (hence $\alpha \in \Gamma\Lambda^p\mathcal{M}$). The bundle of differential 0-forms is defined so that $\Gamma\Lambda^0\mathcal{M} \equiv \mathcal{F}(\mathcal{M})$, the smooth scalar functions on $\mathcal{M}$ and $\Gamma\Lambda^1\mathcal{M} \equiv \Gamma T^*\mathcal{M}$. Given a metric tensor field on $\mathcal{M}$, it is possible to relate the exterior algebra bundle $\Lambda\mathcal{M}$ with the bundle of Clifford algebras $C_{3,1}\mathcal{M}$. These algebras constitute the fibres of a *Clifford bundle* whose local sections define *Clifford elements* on $\mathcal{M}$. *Spinor fields* can be identified with local sections lying in certain minimal left ideals of this Clifford bundle. A Clifford calculus based on this Clifford bundle structure will be discussed in the last section on spinors.

Contraction of a differential p-form α with respect to a vector field X is denoted $i_X\alpha$. The *interior operator* i_X is a graded derivation defined by:

$$i_X(\alpha \wedge \beta) = i_X\alpha \wedge \beta + \eta\,\alpha \wedge i_X\beta \qquad \text{for any} \quad \alpha \in \Gamma\Lambda^p\mathcal{M}, \beta \in \Gamma\Lambda^q\mathcal{M}$$

in terms of the *involution*:

$$\eta\,\alpha \equiv (-1)^p\,\alpha \qquad \text{for any} \quad \alpha \in \Gamma\Lambda^p\mathcal{M},$$

and satisfying $i_X i_X\alpha = 0$. If $p = 1$ then one defines $i_X\alpha = \alpha(X)$ and if $p = 0$: $i_X\alpha = 0$. For an arbitrary p-form α, one has the useful identity:

$$p\,\alpha = e^a \wedge i_{X_a}\alpha \tag{2}$$

in terms of any arbitrary dual bases $\{e^a\}$ and $\{X_a\}$.

In the g-orthonormal basis $\{e^a\}$ one has a canonical local 4-form denoted:

$$\mathcal{V} \equiv e^0 \wedge e^1 \wedge e^2 \wedge e^3.$$

The *Hodge map* $\star$ is a linear isomorphism mapping p-forms to $(4-p)$-forms on $\mathcal{M}$. It is defined by the relations:

$$\alpha \in \Gamma\Lambda^p\mathcal{M}, X \in \Gamma T\mathcal{M}, f \in \mathcal{F}(\mathcal{M}) : \left\{ \begin{array}{rcl} \star(\alpha \wedge \widetilde{X}) & = & i_X \star \alpha \\ \star(\,f\alpha\,) & = & f \star \alpha \\ \star 1 & = & \mathcal{V}. \end{array} \right. .$$

In spacetime with signature $(3, 1)$, the Hodge map and interior operator satisfy the following useful relations:

$$\alpha \in \Gamma\Lambda^p\mathcal{M}, X \in \Gamma T\mathcal{M} : \left\{ \begin{array}{rcl} \star\star\,\alpha & = & -\eta\,\alpha \\ \star i_X\alpha & = & -\star\alpha \wedge \widetilde{X}. \end{array} \right. \tag{3}$$

Since the Hodge map is an isomorphism it has an inverse denoted $\star^{-1}$:

$$\star^{-1}\alpha = -\eta \star \alpha \qquad \text{for any} \quad \alpha \in \Gamma\Lambda^p\mathcal{M}.$$

Journal of Physics: Conference Series **2191** (2022) 012020 doi:10.1088/1742-6596/2191/1/012020

This implies $\star\star\alpha = -\eta\,\alpha$. The metric tensor field uniquely defines the *Levi-Civita covariant derivative operator* ∇_X with respect to any vector field X satisfying:

$$\begin{aligned}
\nabla_X(f_1Y + f_2Z) &= (\nabla_X f_1)Y + f_1\nabla_X Y \\
&\quad + (\nabla_X f_2)Z + f_2\nabla_X Z && \text{(linearity)}\\
\nabla_{f_1X+f_2Y}Z &= f_1\nabla_X Z + f_2\nabla_Y Z && (\mathcal{F}\text{-linearity})\\
\nabla_X\,[\alpha(Y)] &= (\nabla_X\alpha)(Y) + \alpha(\nabla_X Y) && \text{(commutes with contractions)}\\
\nabla_X \mathsf{g} &= 0, && \text{(metric compatibility)}\\
\nabla_X Y - \nabla_Y X &= [X,Y] && \text{(torsion-free)}\\
\nabla_X\star &= \star\nabla_X && \text{(commutes with the Hodge star)}
\end{aligned}$$

for any $X, Y, Z \in \Gamma T\mathcal{M}$, $f_1, f_2 \in \mathcal{F}(\mathcal{M})$ and $\alpha \in \Gamma\Lambda^1\mathcal{M}$ where $[X,Y] \equiv XY - YX$ denotes the *Lie commutator bracket* (or simply *Lie bracket*) between X and Y. In terms of the arbitrary local dual bases $\{e^a\}$ and $\{X_a\}$, any covariant derivative can be used to define a set of local *connection* 1*-forms* $\{\omega^a{}_b\}$ associated with each basis $\{X_a\}$ for $\Gamma T\mathcal{M}$:

$$\nabla_{X_a} X_b = \omega^c{}_b(X_a)\,X_c.$$

Since ∇_{X_a} commutes with contractions:

$$\nabla_{X_a} e^c = -\omega^c{}_b(X_a)\,e^b. \tag{4}$$

The *Christoffel symbols* $\{\Gamma_{ab}{}^c\}$ are components of $\nabla_{X_a}X_b$ in this basis:

$$\nabla_{X_a} X_b = \Gamma_{ab}{}^c\,X_c$$

and hence

$$\omega^a{}_b = \Gamma_{cb}{}^a e^c.$$

Since a Levi-Civita covariant derivative is torsion-free one has, in any g-*orthonormal* basis

$$\omega_{ab} = -\omega_{ba}$$

where $\omega_{ab} = \mathsf{g}_{ac}\,\omega^c{}_b$. We define the *Levi-Civita covariant differential operator* on tensor fields:

$$\nabla \equiv e^a \otimes \nabla_{X_a}. \tag{5}$$

The covariant derivative operator associated with g is used to define the *curvature operator of* ∇:

$$\boldsymbol{R}_{X,Y} \equiv [\nabla_X, \nabla_Y] - \nabla_{[X,Y]} \qquad \text{for all} \quad X, Y \in \Gamma T\mathcal{M}.$$

This operator defines the $(3,1)$ *curvature tensor field* `Riem` of ∇:

$$\mathtt{Riem}(X,Y,Z,\beta) = \beta(\boldsymbol{R}_{X,Y}Z), \qquad \text{for all} \quad X,Y,Z \in \Gamma T\mathcal{M}, \beta \in \Gamma\Lambda^1\mathcal{M}.$$

Contracting the curvature tensor field yields the $(2,0)$ *Ricci tensor field* `Ric` of ∇:

$$\mathtt{Ric}(X,Y) = \mathtt{Riem}(X_a, X, Y, e^a), \qquad \text{for all} \quad X, Y \in \Gamma T\mathcal{M}$$

where the arbitrary bases $\{X_a\}$, $\{e^a\}$ are dual. The Ricci tensor can be *contracted* to construct the *Ricci curvature scalar* $\mathcal{R}$:

$$\mathcal{R} \equiv \mathtt{Ric}(X_a, X^a)$$

where $X^a = \mathrm{g}^{ab} X_b$.

While ∇_X has a type-preserving action on any tensor field, the *exterior derivative* d is defined only to act on antisymmetric tensor fields (differential forms) and is nilpotent: $d \circ d = 0$. Provided $p + q < 4$, the exterior derivative satisfies

$$d(\alpha \wedge \beta) = d\alpha \wedge \beta + \eta\, \alpha \wedge d\beta, \qquad \text{for any} \quad \alpha \in \Gamma\Lambda^p\mathcal{M}, \beta \in \Gamma\Lambda^q\mathcal{M},$$

otherwise $d(\alpha \wedge \beta) = 0$. Given an arbitrary local cobasis $\{e^a\}$ with dual basis $\{X_a\}$ one may verify the useful identity:

$$d = e^a \wedge \nabla_{X_a}. \tag{6}$$

From these relations and the nilpotency of d, it follows immediately that the *co-derivative* δ defined by

$$\delta\alpha \equiv \star^{-1} d \star \eta\, \alpha \qquad \text{for any} \quad \alpha \in \Gamma\Lambda^p\mathcal{M}. \tag{7}$$

is also nilpotent: $\delta \circ \delta = 0$. Unlike d, however, δ is not a graded derivation on differential forms. On spacetime one has $\delta\alpha = \star d \star \alpha$ for any p-form α. Using (6) and the properties above, one may verify the relation:

$$\delta = -i_{X^a} \nabla_{X_a} \tag{8}$$

where $X^a = \widetilde{e^a} = \mathrm{g}^{-1}(e^a, -) = \eta^{ab} X_b$.

The Lie bracket gives rise to an $\mathbb{R}$-*linear derivation* $\mathcal{L}_X$ on functions $f, g \in \mathcal{F}(\mathcal{M})$:

$$[X, Y] \equiv \mathcal{L}_X Y, \qquad \mathcal{L}_X f = Xf, \qquad \mathcal{L}_{cX} Y = c\mathcal{L}_X Y, \qquad \mathcal{L}_X(fg) = (\mathcal{L}_X f)g + f(\mathcal{L}_X g)$$

for $c \in \mathbb{R}$. The operator $\mathcal{L}_X$ is called the *Lie derivative with respect to* X and can be generalised to act on arbitrary tensor fields:

$$\mathcal{L}_X(\mathtt{S} \otimes \mathtt{T}) = \mathcal{L}_X \mathtt{S} \otimes \mathtt{T} + \mathtt{S} \otimes \mathcal{L}_X \mathtt{T} \tag{9}$$

for any $X \in \Gamma T\mathcal{M}$ and arbitrary tensor fields $\mathtt{S}, \mathtt{T}$. In particular, it commutes with contractions and satisfies:

$$\mathcal{L}_X(f\mathtt{T}) = (Xf)\mathtt{T} + f\mathcal{L}_X \mathtt{T}.$$

From (9), it is also a derivation on the exterior algebra of differential forms:

$$\mathcal{L}_X(\alpha \wedge \beta) = \mathcal{L}_X \alpha \wedge \beta + \alpha \wedge \mathcal{L}_X \beta$$

for arbitrary $X \in \Gamma T\mathcal{M}$, $\alpha \in \Gamma\Lambda^p\mathcal{M}$ and $\beta \in \Gamma\Lambda^q\mathcal{M}$. From the definitions above, one has the identities [5]:

$$d\mathcal{L}_X = \mathcal{L}_X d \qquad \text{(commutes with exterior derivative)} \tag{10}$$

$$\mathcal{L}_X[\alpha(Y)] = (\mathcal{L}_X \alpha)(Y) + \alpha(\mathcal{L}_X Y) \qquad \text{(commutes with contraction)}$$

$$\mathcal{L}_X = i_X d + d i_X \qquad \text{(Cartan's magic formula)} \tag{11}$$

for any $X, Y \in \Gamma T\mathcal{M}$, arbitrary tensor fields $\mathtt{T}, \mathtt{S}$, $f \in \mathcal{F}(\mathcal{M})$ and $\alpha \in \Gamma\Lambda^1\mathcal{M}$.

A spacetime $\mathcal{M}$ is said to admit a *conformal Killing vector field* $\mathcal{K}$ if

$$\mathcal{L}_{\mathcal{K}}\mathrm{g} = 2\nu\,\mathrm{g}, \qquad \nu \in \mathcal{F}(\mathcal{M}),\ \nu > 0.$$

If $\nu = 0$, the vector field $\mathcal{K}$ is a *Killing vector field* K with

$$\mathcal{L}_K\mathrm{g} = 0. \tag{12}$$

It follows from (12) that the Lie derivative with respect to any Killing vector field K commutes with the Hodge star map:

$$\mathcal{L}_K\star = \star\mathcal{L}_K. \tag{13}$$

Furthermore, if $\nu \neq 0$ for any metric tensor field on a manifold with dimension $2m$, one has for any m-form F the identity:

$$\mathcal{L}_{\mathcal{K}}(\star F) = \star(\mathcal{L}_{\mathcal{K}}F).$$

This relation has relevance to the conformal symmetry of source-free Maxwell equations for the Maxwell 2-form F in spacetime where $m = 2$.

3. Electromagnetic Pulses in Spacetime Vacua

The method of pre-potentials can be used to construct solutions to the source-free linear vacuum Maxwell system in Minkowski spacetime:

$$dF = 0 \qquad \text{and} \qquad \delta F = 0$$

where F denotes the electromagnetic 2-form. The first equation is satisfied locally on some open spacetime domain by $F = dA$ for some *electromagnetic potential* 1-form A. The second equation then becomes:

$$\delta dA = \star d \star dA = 0 \tag{14}$$

using (7). We seek particular local *complex solutions* to (14) with the ansätz:

$$A = \star d\,(\alpha\Pi) \tag{15}$$

for the electromagnetic potential in terms of a *complex scalar pre-potential* α and a *complex* 2-form[1] Π. The 2-form Π is chosen so that, in general, it is both locally *closed* and *co-closed*:

$$d\Pi = 0 \qquad \text{and} \qquad \delta\Pi = 0 \tag{16}$$

respectively. It follows from (6) and (8) that if $\nabla_{X_a}\Pi = 0$ for all basis elements in $\{X_a\}$ the form Π is both closed and co-closed (such a Π is said to be locally *parallel* or *covariantly constant*). Using (3), (6), (8) and (16), the ansätz (15) yields:

$$\star dA = \delta d\alpha \wedge \Pi + \nabla_{X_a} d\alpha \wedge i_{X^a}\Pi. \tag{17}$$

[1] When the 2-form $\alpha\Pi$ is *real* it reduces to the Hodge dual of a *real Hertz potential* for a Maxwell system.

Using (2) and (4), one has the identity:

$$\nabla_{X_a}\beta = i_{X_a}d\beta + di_{X_a}\beta - \omega^c{}_a \wedge i_c\beta \qquad \text{for any} \quad \beta \in \Gamma\Lambda^p\mathcal{M}.$$

This can be applied to simplify the last term in (17) yielding:

$$\star dA = \delta d\alpha \wedge \Pi + d\,(i_{X_a}d\alpha \wedge i_{X^a}\Pi)$$

where (16) is also used. From the nilpotency of d and (16) one has:

$$d\star F = d\star dA = d\delta d\alpha \wedge \Pi.$$

Hence a particular sub-class of solutions to (14) can be found when α satisfies the spacetime scalar wave equation:

$$\Box\alpha \equiv \delta d\alpha = 0. \tag{18}$$

Thus, *any* solution to (18) for the pre-potential α can, given a covariantly constant 2-form Π, be used to construct a solution to the source-free Maxwell system using (15) and $F = dA$. Of course, (18) has many solutions, however with the advent of the laser *non-singular spatially bounded solutions in all three dimensions* are of particular interest as models for the electromagnetic fields in modern single-cycle laser devices.

In a Cartesian chart with local *dimensionless* coordinates $\{x^\mu\} \equiv \{t, x, y, z\}$ the *dimensionless* Minkowski spacetime metric tensor field g is given by

$$\mathsf{g}|_{\text{Cart}} = -dt \otimes dt + dx \otimes dx + dy \otimes dy + dz \otimes dz.$$

A particularly simple class of such solutions can be generated from the complex axially-symmetric scalar field [4, 6, 7, 8, 9]

$$\alpha(t,x,y,z) = \frac{\ell_0^2}{x^2 + y^2 + [\,a + i(z-t)\,]\,[\,b - i(z+t)\,]} \tag{19}$$

where ℓ_0, a, b are real constants. When a, b are *strictly positive* constants, (19) defines a non-singular solution to $\Box\alpha = 0$. This scalar field is a non-separable solution to (18) with the relative sizes of a and b determining both the direction of propagation along the z-axis of the dominant peak of the real or imaginary part of the scalar field and the number of oscillations in its peak. When $a \gg b$, the dominant peak propagates along the z-axis to the right. The parameter ℓ_0 determines the magnitude of such a peak.

To construct a source-free vacuum Minkowski Maxwell solution, we must also define a *covariantly constant* 2-form. One may define a *complex six-dimensional chiral eigen-basis of covariantly constant 2-forms* $\Pi^{s,\chi}$ satisfying

$$\frac{1}{i}\,\mathcal{L}_{\partial/\partial\phi}\,\Pi^{s,\chi} = \chi\,\Pi^{s,\chi} \tag{20}$$

with $s \in \{\text{CE}, \text{CM}\}$, $\chi \in \{1, 0, -1\}$ where $x = \rho\cos(\phi)$, $y = \rho\sin(\phi)$. Such a basis takes the form:

$$\begin{cases} \Pi^{\text{CE},\pm 1} = d(x \pm iy) \wedge dt = e^{\pm i\phi}\,(d\rho \pm i\rho\, d\phi) \wedge dt, \\ \Pi^{\text{CE},0} = dz \wedge dt, \\ \Pi^{\text{CM},\chi} = \star\,\Pi^{\text{CE},\chi} \end{cases} \tag{21}$$

The index s indicates that the CE (CM) chiral family contain electric (magnetic) fields that are orthogonal to the z-axis when $\chi = 0$. Furthermore, the rationale for the χ-labelling of the 2-form solution

$$F^{s,\chi} = d\left[\star d(\alpha \Pi^{s,\chi})\right] \tag{22}$$

in this chiral basis follows from an exploration of the properties of spacetime dimensionless proper-time τ parameterised curves $x^\mu = C^\mu(\tau)$ of massive test particles interacting with such electromagnetic fields according to the covariant Lorentz force equations of motion:

$$\begin{cases} \nabla_{\dot{C}(\tau)}\dot{C}(\tau)\Big|_{C(\tau)} = \dfrac{q}{m_0}\,\widetilde{i_{\dot{C}(\tau)}F}\Big|_{C(\tau)} \\ \mathsf{g}(\dot{C}(\tau),\dot{C}(\tau))\Big|_{C(\tau)} = -1 \end{cases} \tag{23}$$

where $\dot{C}(\tau) \equiv C_\star(\partial/\partial\tau)|_{C(\tau)}$ denotes the tangent vector of $C(\tau)$ and q, m_0 are dimensionless charge and mass parameters respectively.

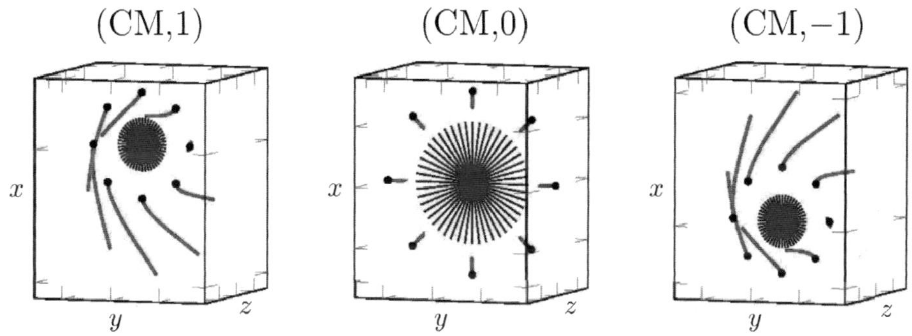

Figure 1. Three-dimensional space-curves for particles subject to an incident (CM, 1) laser pulse (left), (CM, 0) laser pulse (centre) and (CM, −1) laser pulse (right). Each particle has initial velocity in the z-direction. The shaded circular disc region indicates the initial laser spot size relative to the black markers on the space-curves that denote the initial positions of the charged test particles.

In terms of dimensionless Minkowski cylindrical polar coordinates $\{t, \rho, \phi, z\}$ the metric tensor field g takes the form:

$$\mathsf{g}|_{\text{polar}} = -dt \otimes dt + d\rho \otimes d\rho + \rho^2\, d\phi \otimes d\phi + dz \otimes dz.$$

A choice of values for the dimensionless parameters $\{\ell_0, a, b, q, m_0\}$ can be used to solve (23) numerically for a collection of trajectories for charged particles, each arranged initially around the circumference of a circle in a plane orthogonal to the propagation axis of incident CM type electromagnetic pulses with different chirality. The resulting space-curves in 3-dimensions, displayed in figure 1 clearly exhibit the different responses of charged matter to CM pulses with distinct chirality values [10]. Similar space-curves arise from charged particles interacting with chiral CE type modes.

The *dimensionless* energy, linear and angular momentum parameters for the pulse in vacuo can be calculated from *the Maxwell drive* 3-*forms* τ_K [5, 11, 12]:

$$\tau_K = \frac{1}{2}(i_K \mathcal{F} \wedge \star \mathcal{F} - \mathcal{F} \wedge i_K \star \mathcal{F})$$

where $\mathcal{F} = \mathrm{Re}(F)$ and K represents a Killing vector field generating time, space and axially symmetric rotations respectively. With $K = \partial/\partial t$ a timelike Killing vector field on Minkowski spacetime, the total *physical* pulse electromagnetic energy associated with a spatially bounded pulse in all three dimensions can be calculated using:

$$\mathcal{E}[\mathcal{S}] = E_0 \int_{-\infty}^{\infty} dt \int_{\mathcal{S}} \mathcal{J}_K(t, \rho, \phi, z)$$

where E_0 is a real positive constant with the physical dimensions of energy, $\mathcal{S}$ is any plane with constant $z = z_0 \geq 0$ and $\mathcal{J}_K \equiv -i_K \tau_K$ is the dimensionless *energy current* 2-*form* [11], related to the *Poynting vector field*. We can then write

$$\mathcal{E}[\mathcal{S}] = E_0 \int_{-\infty}^{\infty} dt \int_0^{\infty} d\rho \int_0^{2\pi} d\phi \, P(t, \rho, \phi)$$

in terms of a *total power density profile* 0-form P defined by

$$P(t, \rho, \phi) \equiv i_{\partial/\partial\phi} \, i_{\partial/\partial\rho} \, \mathcal{J}_K(t, \rho, \phi, z)\big|_{z=z_0}.$$

Figure 2 displays a clearly pronounced principle maximum in P as a function of $x = \rho\cos(\phi)$, $y = \rho\sin(\phi)$ at $z = 0$, $t = 0$ for the specific choice of parameters used to produce the three-dimensional space-curves in figure 1.

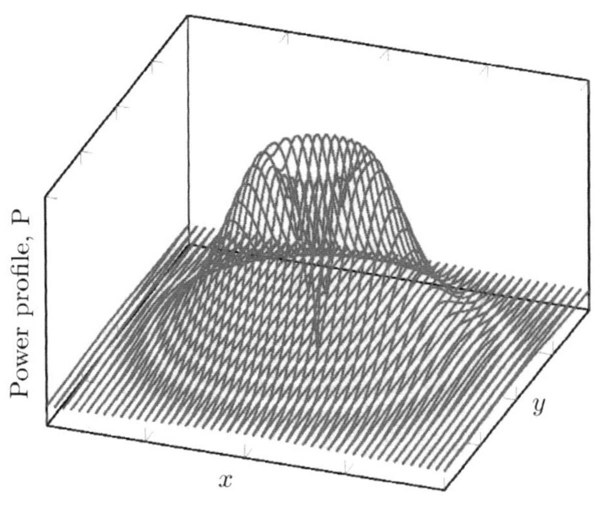

Figure 2. Dimensionless power profile for a (CM,1) laser pulse indicating the spatial boundedness of the pulse in space.

A similar analysis using spacelike translational or rotational Killing vectors can be used to determine the total integrated flux of linear or angular momentum current (total Newtonian force or torque) respectively. We deduce that the pulse momentum and angular momentum in the propagation direction can transfer an impulsive force and torque respectively to charges lying in an orthogonal plane. More generally, the classical configurations of a high energy pulse labelled CE or CM could be distinguished experimentally by its interaction with different arrangements of charged matter. Particular choices of pulse parameters $\{\ell_0, a, b\}$ can be made to model the physical characteristics of existing laser pulses and have been used, for electromagnetic micropulses of duration t_0 with $\mathcal{E}t_0 \lesssim \hbar$, to model an *effective* description of general non-stationary quantum states of a laser pulse [10]. It is suggested that such a description may have utility for simulating a novel transfer of quantum information and may be worthy of further investigation for technological applications such as quantum computing and encryption.

Minkowski spacetime admits Killing vector fields that generate the group of spatial and temporal translations, as well as spatial rotations and Lorentzian boosts (hyperbolic rotations). For example, in the Cartesian chart the Killing vector field $K_z = \partial/\partial z$ generates unit translations in the direction $\partial/\partial z$ while the Killing vector field:

$$K_\phi = x\frac{\partial}{\partial y} - y\frac{\partial}{\partial x} = \frac{\partial}{\partial \phi} \tag{24}$$

generates $O(2)$ rotations about that direction and the Killing vector field $K_t = \partial/\partial t$ generates unit timelike translations.

From the definition (3), if K is any local Killing vector field it commutes with the Hodge map and the exterior derivative d on arbitrary differential forms. Hence it also commutes with the coderivative δ on forms. So, for any Minkowski spacetime Maxwell solution F with source 1-form j satisfying $\delta F = j$ and $dF = 0$, one has:

$$\delta(\mathcal{L}_K F) = \mathcal{L}_K j \qquad \text{and} \qquad d(\mathcal{L}_K F) = 0$$

i.e. $\mathcal{L}_K F$ is also a Maxwell solution with source 1-form $\mathcal{L}_K j$. In any *source-free domain* $\mathcal{U} \in \mathcal{M}$, this implies that in a Cartesian chart one can generate new source-free Maxwell solutions in $\mathcal{U}$:

$$F^{(p,q,r)} = \mathcal{L}^p_{\partial/\partial x}\mathcal{L}^q_{\partial/\partial y}\mathcal{L}^r_{\partial/\partial z}F, \qquad p,q,r = 0,1,2,\ldots$$

in terms of any 2-form F satisfying $dF = 0$, $\delta F = 0$ in $\mathcal{U}$ and where

$$\mathcal{L}^p_K F \equiv \underbrace{\mathcal{L}_K\mathcal{L}_K\cdots\mathcal{L}_K}_{p\text{ times}} F.$$

Such solutions are proportional to electrostatic Cartesian electric point multipoles if F is static ($\mathcal{L}_{\partial/\partial t}F = 0$) and describes the electric monopole (Coulomb) solution at some point located in some region of spacetime not belonging to $\mathcal{U}$. Similarly, it follows from (10), (11) and (13) that the operator $\mathcal{L}^r_{K_\pm}$, where

$$K_\pm = \mathsf{g}^{-1}\Big(d(\rho e^{\pm i\phi}), -\Big) = e^{\pm i\phi}\left(\frac{\partial}{\partial\rho} \pm \frac{i}{\rho}\frac{\partial}{\partial\phi}\right) \tag{25}$$

is a complex Killing vector ($\mathcal{L}_{K_\pm}\mathsf{g} = 0$) generating translations in the transverse (x, y)-plane, defines the 2-form

$$F^{s,\chi\pm r} \equiv \mathcal{L}^r_{K_\pm}F^{s,\chi} \qquad \text{satisfying} \qquad \frac{1}{i}\mathcal{L}_{\partial/\partial\phi}F^{s,\chi\pm r} = (\chi \pm r)\,F^{s,\chi\pm r} \qquad (r = 0,1,2,\ldots)$$

describing a family of *multi-chiral* source-free Maxwell solutions including those that exhibit a non-singular, propagating, non-stationary pulse-like nature, spatially bounded in all three dimensions.

As a further application of pre-potentials to the solution of a linear system of differential equations, we briefly shift attention to generalizations of Maxwell's theory. Motivated by the desire to ameliorate the divergences in perturbative QED a number of generalized theories of electromagnetism have been proposed. To date, there is little experimental evidence for testing their predicted departures from Maxwell's theory. However, with the increase in laser technology one may now be entering regimes that can discriminate between such theories. In particular, *Bopp-Landé-Podolsky electrodynamics* are *linear* in the electromagnetic fields, contain a fundamental real non-zero constant λ_0 and approach Maxwell's theory when this constant tends to zero. Unlike Maxwell's theory, they contain solutions describing static point charges with *finite* electromagnetic energy.

The classical source-free Bopp-Landé-Podolsky field equation in vacuo for a complex electromagnetic field tensor $F = dA$ is [13]:

$$d \star F - \lambda_0^2 \, d\delta d \star F \; = \; 0. \tag{26}$$

Clearly, all classical source-free vacuum Maxwell solutions satisfy this equation. However, there exist additional solutions to (26) that are *not* source-free vacuum Maxwell solutions. For example, such additional solutions of (26) follow from (15) provided:

$$\Box \alpha + \lambda_0^2 \Box^2 \alpha \; = \; 0 \tag{27}$$

when supplemented by a closed and co-closed choice of Π. Furthermore, it follows that as $\lambda_0 \to 0$, one recovers the class of Maxwell solutions discussed above. For non-zero λ_0, there exist particular non-Maxwellian solutions to (27) satisfying

$$\Box \alpha + \frac{1}{\lambda_0^2} \alpha \; = \; 0, \qquad \lambda_0 \neq 0.$$

Thus *any* solution to this equation for the pre-potential α can, given *any* covariantly constant 2-form Π, be used to construct a solution to the source-free Bopp-Landé-Podolsky system using (15) and $F = dA$.

In dimensionless Minkowski cylindrical coordinates $\{t, \rho, \phi, z\}$, a particular non-separable axially symmetric spatially bounded solution in three dimensions is given for arbitrary non-zero real dimensionless Λ_0 by:

$$\alpha(t, \rho, z) \; = \; \frac{\Lambda_0^2}{\zeta(t, \rho, z)} \, K_1\!\left(\frac{\zeta(\rho, t, z)}{\lambda_0} \right)$$

where

$$\zeta(t, \rho, z) \; \equiv \; \sqrt{\rho^2 + [a + i(z - t)]\,[b - i(z + t)]}$$

in terms of strictly positive (real) parameters a, b that determine pulse propagation along the positive z-axis and where $K_1(z)$ denotes the first order modified Bessel function of the second kind. The function $\zeta(t, \rho, z)$ is non-zero for all real t, ρ, z and hence both the real and imaginary parts of $\alpha(t, \rho, z)$ are bounded. To our knowledge, this constitutes a new source-free finite-energy solution for F to Bopp-Landé-Podolsky electrodynamics that does not arise in classical Maxwell electrodynamics [14].

The interaction of the Bopp-Landé-Podolsky pulse with classical charged test particles follows from the divergence of the Bopp-Landé-Podolsky stress-energy-momentum tensor [13], and may offer an experimental means to discriminate between the Maxwell and Bopp-Landé-Podolsky descriptions of source-free electromagnetic fields in vacuo.

4. Gravitational Pulses in Vacua

The pre-potential approach can also be used to analyse gravitational field equations. In particular, it can be used to generate solutions to the *linearised Einstein vacuum system.* In any matter-free domain of spacetime $\mathcal{U} \subset \mathcal{M}$, an Einsteinian gravitational field is described by a real symmetric covariant rank-two metric tensor field $\widehat{\mathtt{g}}$ with Lorentzian signature that satisfies the vacuum Einstein equation

$$\mathtt{Ein}^{(\widehat{\mathtt{g}})} = 0 \tag{28}$$

where

$$\mathtt{Ein}^{(\widehat{\mathtt{g}})} \equiv \mathtt{Ric}^{(\widehat{\mathtt{g}})} - \frac{1}{2}\,\mathrm{Tr}^{(\widehat{\mathtt{g}})}(\mathtt{Ric}^{(\widehat{\mathtt{g}})})\,\widehat{\mathtt{g}} \tag{29}$$

and $\mathtt{Ric}^{(\widehat{\mathtt{g}})}$ is the Levi-Civita Ricci tensor associated with the torsion-free, metric-compatible Levi-Civita connection $\nabla^{(\widehat{\mathtt{g}})}$. By taking the $\widehat{\mathtt{g}}$-trace of (28), one also has $\mathtt{Ric}^{(\widehat{\mathtt{g}})} = 0$. A coordinate independent linearisation of (29) about an arbitrary Lorentzian metric can be found in [15, 16]. In particular, a linearisation about a real flat Minkowski spacetime metric η on $\mathcal{U}$ determines the linearised metric

$$\mathtt{g} = \eta + \mathtt{h}$$

and to first order one writes $\widehat{\mathtt{g}} = \mathtt{g} + O(\kappa^2)$. The real variable κ is a dimensionless parameter in $\mathtt{h}$ used to keep track of the expansion order and

$$\eta = -\,e^0 \otimes e^0 + \sum_{k=1}^{3} e^k \otimes e^k \;=\; \eta_{ab}\, e^a \otimes e^b \tag{30}$$

in *any* real η-orthonormal coframe[2] on $\mathcal{U}$. Since we explore the source-free Einstein equation (relevant to the motion of test-matter far from any sources) the scale associated with any linearised solutions must be fixed by the solutions themselves rather than any coupling to self-gravitating matter. Furthermore since only dimensionless relative scales have any significance we define the real tensor field $\mathtt{h}$ to be a *perturbation* of η on $\mathcal{U}$ relative to *any* local η-orthonormal coframe $\{e^a\}$ provided

$$|\,\mathtt{h}(X_a, X_b)\,| < 1 \qquad \text{on } \mathcal{U} \qquad \text{for all } a,b = 0,1,2,3 \tag{31}$$

where $e^a \in \Gamma T^*\mathcal{U}$, $X_b \in \Gamma T\mathcal{U}$, $e^a(X_b) = \delta^a_b$. It should be noted that the perturbation order of any component of the η-covariant *derivative* of a tensor and its η-trace relative to such a coframe is not necessarily of the same order as that assigned to the tensor. Thus perturbation order is

[2] An arbitrary coframe on $\mathcal{U}$ is a set of 1-forms $\{e^a\}$ satisfying $e^0 \wedge e^1 \wedge e^2 \wedge e^3 \neq 0$. If $\beta = \beta_{ab} e^a \otimes e^b$ in *any* coframe $\{e^a\}$ on $\mathcal{U}$, $\mathrm{Tr}^{(\eta)}(\beta) \equiv \beta_{ab}\eta^{ab}$ with $\eta^{ab}\eta_{bc} = \delta^a_c$.

not synonymous with "scale" in this context. We use the conditions (31) to *define* perturbative Lorentzian spacetime to be sub-domains $\mathcal{P}_{\mathcal{U}} \subset \mathcal{U}$ where

$$\max_{a,b} |\mathtt{h}(X_a, X_b)| < 1. \tag{32}$$

The tensor $\mathtt{h} \equiv \mu^{(\eta)}(\xi')$ with $\xi' \equiv \mathrm{Re}(\xi)$ may be constructed [16] from any complex covariant symmetric rank two tensor ξ satisfying the *linearised source-free Einstein equation on the perturbative domain* $\mathcal{P}_{\mathcal{U}}$:

$$\mathrm{Lap}^{(\eta)}(\xi) - 2\mu^{(\eta)}\left(\mathrm{Sym}\nabla^{(\eta)}(\,\mathrm{Div}^{(\eta)}(\xi)\,)\right) = 0. \tag{33}$$

Here and below, $\nabla^{(\eta)}$ denotes the operator of Levi-Civita covariant differentiation associated with η, $X^a \equiv \eta^{ab}X_b$, $Y \equiv \nabla^{(\eta)}_{X_a} X^a$, for any tensor $\mathtt{T}$ on $\mathcal{U}$:

$$\mathrm{Lap}^{(\eta)}(\mathtt{T}) \equiv \nabla^{(\eta)}_Y \mathtt{T} - \nabla^{(\eta)}_{X_a}\nabla^{(\eta)}_{X^a}\mathtt{T} \tag{34}$$

$$\mu^{(\eta)}(\mathtt{T}) \equiv \mathtt{T} - \frac{1}{2}\mathrm{Tr}^{(\eta)}(\mathtt{T})\,\eta$$

and for all complex covariant symmetric rank-two tensors $\mathtt{T}$ on $\mathcal{U}$:

$$\mathrm{Div}^{(\eta)}(\mathtt{T}) \equiv (\nabla^{(\eta)}_{X_a}\mathtt{T})(X^a, -). \tag{35}$$

It follows from (34) that for any scalar field α:

$$\mathrm{Lap}^{(\eta)}(\alpha) = -\delta d\alpha. \tag{36}$$

Since for any $\mathtt{g}$ the trace-reverse map $\mu^{(\mathtt{g})}$ satisfies $\mu^{(\mathtt{g})} \circ \mu^{(\mathtt{g})} = \mathrm{Id}$, if ξ' is trace-free with respect to η, then $\mathtt{h} = \xi'$. If ξ is also divergence-free with respect to η, then $\mathrm{Lap}^{(\eta)}(\xi) = 0$. Thus, divergence-free, trace-free solutions ξ satisfy:

$$\mathrm{Lap}^{(\eta)}(\xi) = 0 \qquad \text{and} \qquad \mathrm{Div}^{(\eta)}(\xi) = 0. \tag{37}$$

We now construct complex solutions ξ to (37) in terms of a complex 0-form pre-potential α. These can then be used to generate a real linearised metric $\mathtt{g}$. The properties of the scalar field α needed to generate chiral, wave-like and pulse-like solutions to the linearised source-free Einstein equations have been developed in [17]. To emulate the methodology used above for the electromagnetic pulse solutions we note that a key role is played by the commutativity of certain exterior operators with covariantly constant *antisymmetric* tensors and the Hodge de-Rham Laplacian δd on scalar fields in Minkowski spacetime. In Einsteinian gravitation one seeks similar properties involving the tensor Laplacian operator $\mathrm{Lap}^{(\mathtt{g})}$, $\nabla^{(\mathtt{g})}$ and *symmetric* tensors in spacetimes with a metric tensor $\mathtt{g}$. The key general identity is the relation between $\mathrm{Lap}^{(\mathtt{g})}$, $\nabla^{(\mathtt{g})}$ and the curvature operator $\boldsymbol{R}^{(\mathtt{g})}_{X,Y}$ of the torsion-free, metric compatible connection $\nabla^{(\mathtt{g})}$:

$$\mathrm{Lap}^{(\mathtt{g})}\left(\nabla^{(\mathtt{g})}\alpha\right) = \nabla^{(\mathtt{g})}\mathrm{Lap}^{(\mathtt{g})}(\alpha) + \boldsymbol{R}^{(\mathtt{g})}_{X_j,\widetilde{d\alpha}}\,e^j$$

valid for any scalar field α. On flat Minkowski spacetime with metric η, this yields the commutation relation[3] :

$$[\mathrm{Lap}^{(\eta)}, \nabla^{(\eta)}]\alpha = [\mathrm{Lap}^{(\eta)}, d]\alpha = 0 \qquad \text{on scalar fields } \alpha. \tag{38}$$

[3] For any scalar field α and metric tensor $\mathtt{g}$, $\nabla^{(\mathtt{g})}\alpha = d\alpha$ and is independent of $\mathtt{g}$.

Journal of Physics: Conference Series **2191** (2022) 012020 doi:10.1088/1742-6596/2191/1/012020

Thus, if α is *any* complex (four times differentiable) scalar field on $\mathcal{U}$ then

$$\nabla^{(\eta)}\nabla^{(\eta)}\,\mathrm{Lap}^{(\eta)}(\alpha) \;=\; \mathrm{Lap}^{(\eta)}\left(\nabla^{(\eta)}\nabla^{(\eta)}\alpha\right) \;=\; \mathrm{Lap}^{(\eta)}\left(\nabla^{(\eta)}d\alpha\right) \tag{39}$$

since $\nabla^{(\eta)}\alpha = d\alpha$ for all 0-forms α. Thus, if we define

$$\boldsymbol{\xi} \;\equiv\; \nabla^{(\eta)}d\alpha \tag{40}$$

then from (34) and (39):

$$\mathrm{Lap}^{(\eta)}(\alpha) \;=\; -\delta d\alpha \;=\; 0 \qquad \Longrightarrow \qquad \mathrm{Lap}^{(\eta)}\left(\nabla^{(\eta)}d\alpha\right) \;=\; \mathrm{Lap}^{(\eta)}(\boldsymbol{\xi}) \;=\; 0.$$

Furthermore, directly from (35):

$$\mathrm{Div}^{(\eta)}(\nabla^{(\eta)}\beta) \;=\; \mathrm{Lap}^{(\eta)}(\beta) \qquad \text{for all} \quad \beta\in\Gamma\Lambda^1\mathcal{M} \tag{41}$$

and so with $\beta = d\alpha$, it follows from (40), (41), (38) and (36) respectively:

$$\mathrm{Div}^{(\eta)}(\boldsymbol{\xi}) \;=\; \mathrm{Div}^{(\eta)}(\nabla^{(\eta)}d\alpha) \;=\; \mathrm{Lap}^{(\eta)}(d\alpha) \;=\; d(\mathrm{Lap}^{(\eta)}(\alpha)) \;=\; -d(\delta d\alpha),$$

which vanishes when $\delta d\alpha = 0$. Hence $\mathrm{Lap}^{(\eta)}(\boldsymbol{\xi}) = 0$ and $\mathrm{Div}^{(\eta)}(\boldsymbol{\xi}) = 0$ are both satisfied for *any* suitably differentiable complex scalar field α satisfying $\delta d\alpha = 0$. This should be compared with (18), the equation determining the class of pulsed electromagnetic field solutions discussed in the previous section.

From the Levi-Civita differential operator (5), for any scalar field α:

$$\boldsymbol{\xi} \;=\; \nabla^{(\eta)}d\alpha \;=\; e^a\otimes e^b\left[\nabla_{X_a}\nabla_{X_c}\alpha - \nabla_{\nabla_{X_a}X_c}\alpha\right]$$

which describes a symmetric, second-rank covariant tensor. By definition:

$$\mathrm{Tr}^{(\eta)}(\boldsymbol{\xi}) \;=\; \mathrm{Tr}^{(\eta)}\left(\nabla^{(\eta)}d\alpha\right) \;=\; (\nabla^{(\eta)}d\alpha)(X_i, X^i) \;=\; -\mathrm{Lap}^{(\eta)}(\alpha) \;=\; \delta d\alpha$$

using (34) and hence $\boldsymbol{\xi} = \nabla^{(\eta)}d\alpha$ is traceless if $\delta d\alpha = 0$.

Thus, we conclude that any complex scalar field α satisfying $\delta d\alpha = 0$ will give rise to the real metric:

$$\mathsf{g} \;=\; \boldsymbol{\eta} + \boldsymbol{\xi}' \;=\; \boldsymbol{\eta} + \mathrm{Re}(\boldsymbol{\xi}) \;=\; \boldsymbol{\eta} + \mathrm{Re}(\,\nabla^{(\eta)}d\alpha\,)$$

generating a solution to the *linearised source-free Einstein equation on the perturbative domain* $\mathcal{P}_\mathcal{U}$ (33).

In a local chart $\Phi_\mathcal{U}$ possessing dimensionless coordinates $\{t,\rho,\phi,z\}$ with $t\geq 0$, $\rho>0$, $\phi\in[0,2\pi)$ and $|z|\geq 0$ on a spacetime domain $\mathcal{U}\subset\mathcal{M}$, a local coframe $\mathcal{C}$ adapted to these co-ordinates is $\{e^0 = dt,\, e^1 = d\rho,\, e^2 = \rho\, d\phi,\, e^3 = dz\}$. With $\boldsymbol{\eta}$ given by (30), this coframe is $\boldsymbol{\eta}$-orthonormal but not in general g-orthonormal. Such a chart facilitates the coordination of a series of massive test particles initially arranged in a series of concentric rings with different values of ρ lying in spatial planes with different values of z at $t = 0$. Furthermore, we define for any metric $\mathsf{g} = \boldsymbol{\eta} + \mathtt{h}$ on $\mathcal{U}$:

$$\mathcal{H}(t,\rho,\phi,z) \;\equiv\; \max_{0\leq a,b\leq 3}\left|\mathtt{h}(X_a, X_b)\right|.$$

The particular complex scalar α of relevance here satisfying $\mathrm{Lap}^{(\boldsymbol{\eta})}(\alpha) = 0$ is given in the (t, ρ, ϕ, z) chart $\Phi_{\mathcal{U}}$ above as[4]

$$\alpha(t,\rho,z) = \frac{\ell_0^2}{\rho^2 + [\,a + i(z-t)\,]\,[\,b - i(z+t)\,]} \tag{42}$$

with strictly positive real dimensionless constants. The scalar $\alpha(t, \rho, z)$ is then singularity-free in t, ρ and z and clearly axially-symmetric with respect to rotations about the z-axis. It also gives rise to an *axially-symmetric* complex tensor $\boldsymbol{\xi}^{(0)}$ satisfying[5] $\mathcal{L}_{\partial/\partial\phi}\boldsymbol{\xi}^{(0)} = 0$. In $\mathcal{U}$, this yields the real axially-symmetric metric tensor $\mathbf{g}^{(0)}$ (satisfying $\mathcal{L}_{\partial/\partial\phi}\,\mathbf{g}^{(0)} = 0$) with

$$\begin{cases} g^{(0)}_{00} = -1 + \dfrac{\partial^2\alpha'}{\partial t^2}, & g^{(0)}_{01} = g^{(0)}_{10} = \dfrac{\partial^2\alpha'}{\partial t\,\partial\rho}, \quad g^{(0)}_{03} = g^{(0)}_{30} = \dfrac{\partial^2\alpha'}{\partial t\,\partial z} \\ g^{(0)}_{11} = 1 + \dfrac{\partial^2\alpha'}{\partial\rho^2}, & g^{(0)}_{13} = g^{(0)}_{31} = \dfrac{\partial^2\alpha'}{\partial\rho\,\partial z} \\ g^{(0)}_{22} = 1 + \dfrac{1}{\rho}\dfrac{\partial\alpha'}{\partial\rho}, & g^{(0)}_{33} = 1 + \dfrac{\partial^2\alpha'}{\partial z^2} \end{cases}$$

where $\alpha' \equiv \mathrm{Re}(\alpha)$.

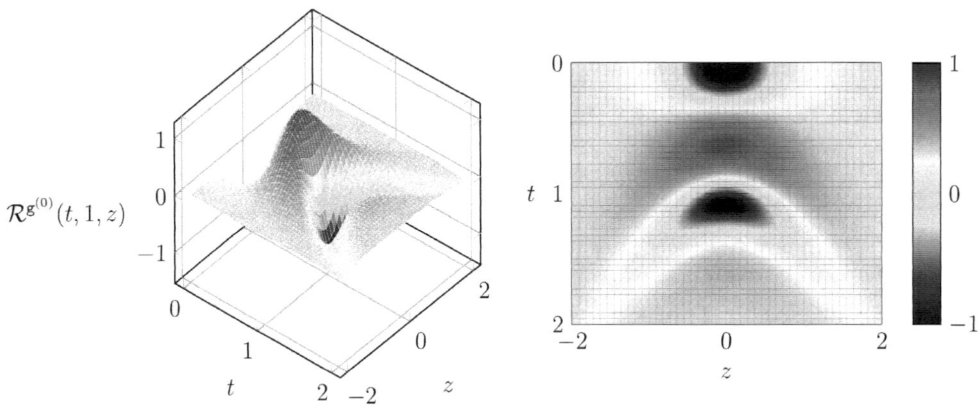

Figure 3. This figure illustrates the nature of the spacetime geometry determined by g_0 on $\mathcal{M}$ with associated Ricci curvature scalar $\mathcal{R}^{g^{(0)}}(t, \rho, z)$. Regions where $\mathcal{R}^{g^{(0)}}(t, 1, z)$ change sign are clearly visible in the right side where a 2-dimensional density plot shows a pair of prominent loci that separately approach the future ($t \geq 0$) light-cone of the event at $\{\rho = 1,\ t = 0,\ z = 0\}$. A more detailed graphical description of $\mathcal{R}^{g^{(0)}}(t, 1, z)$ is given in the left hand 3-dimensional plot where an initial pulse-like maximum around $t \simeq 0$ evolves into a pair of enhanced loci with peaks at values of z with *opposite signs* when $t \geq 1$. This Ricci curvature scalar is generated from a metric perturbation pulse with parameters $a = b = 1$ and $\ell_0 = 1/4$.

Complex symmetric tensors $\boldsymbol{\xi}^{(m)}$ with integer chirality $m > 0$ satisfying $\mathrm{Lap}^{(\boldsymbol{\eta})}(\boldsymbol{\xi}^{(m)}) = 0$, $\mathrm{Div}^{(\boldsymbol{\eta})}(\boldsymbol{\xi}^{(m)}) = 0$, $\mathrm{Tr}^{(\boldsymbol{\eta})}(\boldsymbol{\xi}^{(m)}) = 0$ and $\frac{1}{i}\mathcal{L}_{\partial/\partial\phi}\boldsymbol{\xi}^{(m)} = m\boldsymbol{\xi}^{(m)}$ may be generated from $\boldsymbol{\xi}^{(0)}$ by

[4] This should be compared with (19).

[5] Since $\nabla^{(\boldsymbol{\eta})}$ is a flat connection, if K is an $\boldsymbol{\eta}$-Killing vector then the operator $\nabla^{(\boldsymbol{\eta})}\mathcal{L}_K = \mathcal{L}_K\nabla^{(\boldsymbol{\eta})}$ on all tensors.

repeated covariant differentiation with respect to a particular η-null, η-Killing complex vector field K_+:

$$\xi^{(m)} = \underbrace{\nabla^{(\eta)}_{K_+} \cdots\cdots \nabla^{(\eta)}_{K_+}}_{m \text{ times}} \xi^{(0)}$$

where

$$K_+ \equiv \eta^{-1}\Big(d(\rho e^{i\phi}), - \Big) = e^{i\phi}\left(\frac{\partial}{\partial\rho} + \frac{i}{\rho}\frac{\partial}{\partial\phi}\right).$$

Solutions with negative integer chirality can be obtained by complex conjugation of the positive chirality complex eigen-solutions. Each $\xi^{(m)}$ defines a real spacetime metric $g^{(m)} = \eta + \mathrm{Re}(\xi^{(m)})$ on $\mathcal{U}$ which, for $m \neq 0$, is not axially symmetric: $\mathcal{L}_{\partial/\partial\phi} g^{(m)} \neq 0$.

A longstanding problem in astrophysics has been to understand the origin of the extensive jets that have been observed in X-ray spectroscopy emanating from a number of galaxies. These are considered to be the result of matter falling onto accretion disks that initiate complex magnetohydrodynamic processes. However, it is unclear whether such processes are sufficient to account for the energetics involved. Another puzzle is why some galaxies are observed with bi-directional jets whereas others are not. It has been suggested that gravitation may also play a role in initiating such processes [18]. We have suggested in [19, 17] that an analysis of the timelike geodesics associated with a propagating, gravitational chiral pulse may provide some insight into the relevance of gravitational chirality in initiating the magnetohydrodynamic mechanism. As we shall show, chiral pulses can be shown to account for both uni-directional and bi-directional jet structures.

Given ξ and hence $g = \eta + h$, all dimensionless proper-time parametrised timelike spacetime geodesics C on $\mathcal{U}$ (with tangent vector $\dot{C}$) associated with g must satisfy a differential-algebraic system. For any worldline C with components $C^\mu(\tau)$ in any local chart on $\mathcal{U}$ with dimensionless coordinates $\{x^\mu\}$ and $\dot{C}^\mu(\tau) = \partial C^\mu(\tau)/\partial\tau$:

$$\begin{aligned} \nabla^{(g)}_{\dot{C}(\tau)} \dot{C}(\tau)\Big|_{C(\tau)} &= 0 \\ g(\dot{C}(\tau), \dot{C}(\tau))\Big|_{C(\tau)} &= -1. \end{aligned} \tag{43}$$

i.e.:

$$\nabla^{(g)}_{\dot{C}(\tau)} \dot{C}(\tau)\Big|_{C(\tau)} \equiv \frac{D\dot{C}^\mu(\tau)}{d\tau}\frac{\partial}{\partial x^\mu}\bigg|_{C(\tau)} = \left(\frac{d\dot{C}^\mu(\tau)}{d\tau} + (\Gamma^\mu_{\alpha\beta}\circ C)(\tau)\,\dot{C}^\alpha(\tau)\,\dot{C}^\beta(\tau)\right)\frac{\partial}{\partial x^\mu}\bigg|_{C(\tau)}$$

where $\Gamma^\mu_{\alpha\beta}$ denotes a Christoffel symbol associated with $\nabla^{(g)}$.

In the following only solutions to (43) that lie in the perturbative domains $\mathcal{P}_{\mathcal{U}}$, as determined by (32), are displayed.

An indication of the nature of the spacetime geometry determined by $g^{(m)}$ on $\mathcal{M}$ is given by the structure of the associated Ricci curvature scalar $\mathcal{R}^{g^{(m)}}(t, \rho, z)$. Unlike gravitational wave spacetimes this scalar is not identically zero. For $m = 0$ it is axially symmetric and in the chart

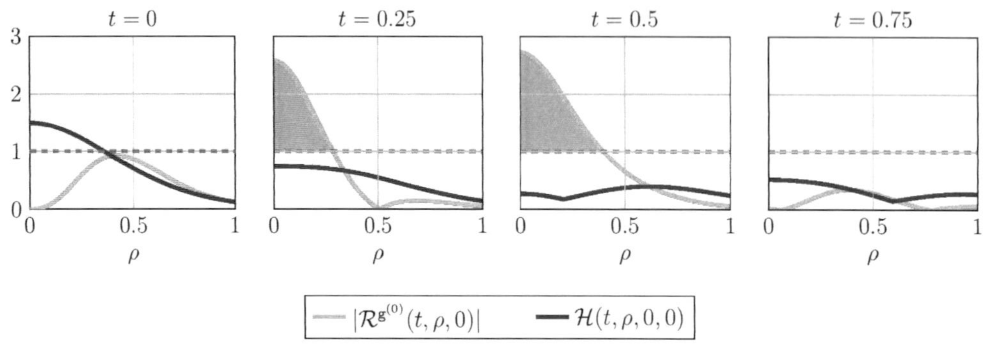

Figure 4. The axially-symmetric expressions $|\mathcal{R}^{\mathbf{g}^{(0)}}(t,\rho,0)|$ and $\mathcal{H}(t,\rho,0,0)$ are plotted as functions of ρ for $t = 0,\ 0.25,\ 0.5,\ 0.75$ and parameters $a = b = 1$, $\ell_0 = 1/4$. Regions where the blue curves lie under the red dotted line denote perturbative regions $\mathcal{P}_{\mathcal{U}}$. The light blue shaded regions clearly indicate curvature scalars that are greater in magnitude than unity despite lying within $\mathcal{P}_{\mathcal{U}}$ regions.

$\Phi_{\mathcal{U}}$ its independence of ϕ means that for values of fixed radius ρ_0 its structure can be displayed for a range of t and z values given a choice of parameters (a, b, ℓ_0). Regions where $\mathcal{R}^{\mathbf{g}^{(0)}}(t, 1, z)$ change sign are clearly visible in the right side of figure 3 where a 2-dimensional density plot shows a pair of prominent loci that separately approach the future $(t \geq 0)$ light-cone of the event at $\{\rho = 1, t = 0, z = 0\}$. A more detailed graphical description of $\mathcal{R}^{\mathbf{g}^{(0)}}(t, 1, z)$ is given in the left hand 3-dimensional plot in figure 3 where an initial pulse-like maximum around $t \simeq 0$ evolves into a pair of enhanced loci with peaks at values of z with *opposite signs* when $t \geq 1$. In this presentation the maximum pulse height has been normalised to unity. This characteristic behaviour is similar to that possessed by $\mathrm{Re}(\,\alpha(t, 1, z)\,)$. It suggests that "tidal forces" (responsible for the geodesic deviation of neighbouring geodesics [20, 21, 22]) are concentrated in spacetime regions where components of the Riemann tensor of $\mathbf{g}^{(0)}$ have pulse-like behaviour in domains similar to those possessed by $\mathcal{R}^{\mathbf{g}^{(0)}}(t, \rho, z)$.

Explicit formulae for $\mathcal{R}^{\mathbf{g}^{(0)}}(t, \rho, z)$ and $\mathcal{H}(t, \rho, \phi, z)$ are not particularly illuminating[6]. However, for fixed values of the parameters (a, b, ℓ_0), their values can be plotted numerically in order to gain some insight into their relative magnitudes in any perturbative domain $\mathcal{P}_{\mathcal{U}}$. With z fixed at zero, figure 4 displays such plots as functions of ρ and a set of t values. It is clear that in perturbative domains the curvature scalar may exceed unity. Since in general:

$$\mathcal{R}^{\mathbf{g}^{(0)}}(t, \rho, z) \;=\; \mathcal{Q}(t, \rho, z)\kappa^2 + O(\kappa^3)$$

where $\mathcal{Q}$ is a non-singular rational function of its arguments and the tensor $\mathtt{h}$ is, by definition, of order κ, figure 4 demonstrates that relative tensor κ-orders are not, in general, indicators of their corresponding relative magnitudes.

By modelling thick accretion disks by a finite number of massive point particles occupying a number of planar rings with varying ρ and varying z the system (43) has been explored numerically. Sets of space-curves in spacelike sections of the perturbative spacetimes defined above are displayed in figure 5 and figure 6 for various choices of the dimensionless parameters a, b, ℓ_0 and dimensionless evolution proper-time. The resulting jet-like structures of these space-curves with dimensionless proper-time parameter $\tau \in [0, \tau_{\max}]$ are quantified in terms of aspect

[6] Since $\mathbf{g}^{(0)}$ is axially-symmetric, the function $\mathcal{H}$ is independent of ϕ.

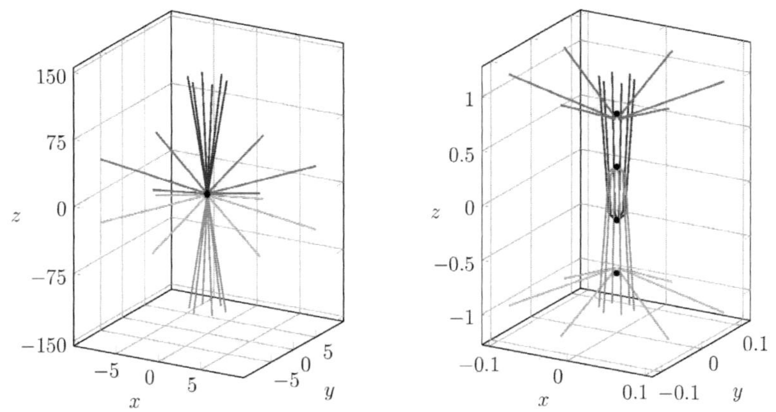

Figure 5. On the left six geodesics are shown emanating from six locations with ϕ values $0, \pi/3, 2\pi/3, \pi, 4\pi/3, 5\pi/3$ on a ring with radius 10^{-4} in the plane $z = 0.735$ and six from similarly arranged points on rings of the same radius at $z = 0.245$, $z = -0.245$ and $z = -0.735$. The initial locations are not resolved in these figures. The 24 geodesics each evolve from $\tau = 0$ to $\tau = 10^4$ and clearly display an axially symmetric bi-directional jet structure from the rings in conformity with the expectations based on the spacetime structure of $\mathcal{R}^{g^{(0)}}(t, 1, z)$ in figure 3. The figure on the right resolves the structure of this jet array for $0 \leq \tau \leq 100$. All geodesics are generated with the additional initial conditions $\dot{\rho}(0) = 0$, $\dot{z}(0) = 0$, $\dot{\phi}(0) = 0.4$ and the background perturbation pulse has parameters $a = b = 1$, $\ell_0 = 1/6$. A single uni-directional jet-array arises when only one ring is populated with matter. This figure demonstrates that the jets from the sources at $z = \pm 0.245$ have a dimensionless aspect ratio $\mathcal{A}(10^4) = 64.7$ much greater than those produced from the sources at $z = \pm 0.735$ where $\mathcal{A}(10^4) = 3.22$.

ratios defined by:

$$\mathcal{A}(\tau_{\max}) \equiv \left| \frac{\widehat{z}(\tau_{\max}) - \widehat{z}(0)}{\widehat{\rho}(\tau_{\max}) - \widehat{\rho}(0)} \right| .$$

5. Spinor Pre-potentials

In the above sections we have outlined how a particular 0-form singularity-free, dispersive pulse-like solution α on spacetime for the scalar equation (18) in a Minkowski background can be used to generate solutions to *linear* partial differential equations of relevance to mathematical physics, namely the Minkowski source-free vacuum Maxwell system for electromagnetic fields, the linear system arising from the Minkowski Bopp-Landé-Podolsky source-free vacuum system and the linearised equation of source-free Einstein theory of gravitation perturbed about Minkowski spacetime. A notable feature of all these solutions is that they include non-stationary, non-separable functions when expressed in local coordinates. Furthermore they can all be classified in terms of chirality eigenvalues of certain differential operators with respect to Minkowski Killing vector fields.

This raises the question of whether there exist classical linear spinor equations that exhibit similar spinor pulse-like solutions that can be generated from the particular complex pulse-like solutions to (18). To show that this is indeed the case it proves expedient to exploit the close relation of spinor representations of the covering of the g-orthonormal group $\mathrm{SO}(3,1)$ with the spacetime Clifford algebra $\mathrm{C}_{3,1}(V, \breve{g})$ where $V = \Gamma\Lambda^1\mathcal{M}$ is the underlying vector space with cobasis $\{e^0, e^1, e^2, e^3\}$ and $\breve{g}$ has signature $(-,+,+,+)$.

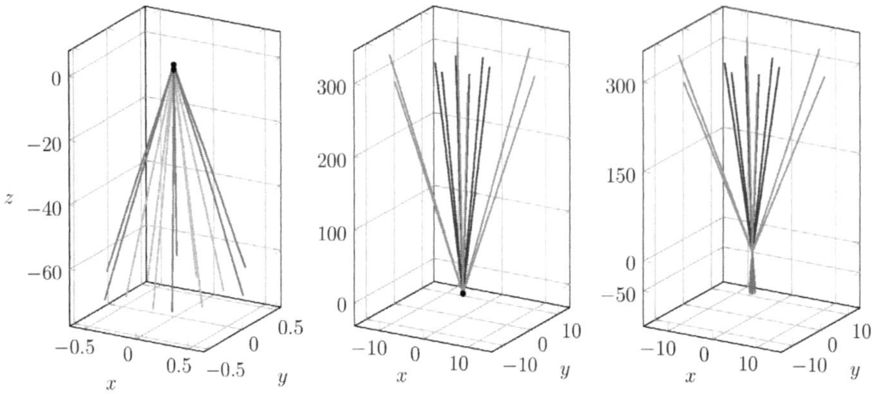

Figure 6. On the left, six geodesics are shown emanating from six locations with ϕ values $0, \pi/3, 2\pi/3, \pi, 4\pi/3, 5\pi/3$ on a ring with radius 10^{-4} in the plane $z = 0.735$ and six from similarly arranged points on rings of the same radius at $z = -0.735$. The initial locations are not resolved in these figures. The 12 geodesics each evolve from $\tau = 0$ to $\tau = 10^4$ and clearly display an axially symmetric uni-directional jet structure from the rings. All geodesics are generated with the additional initial conditions $\dot{\rho}(0) = 0$, $\dot{z}(0) = 0$, $\dot{\phi}(0) = 0.4$ and the background perturbation pulse has parameters $a = 1$, $b = 3$, $\ell_0 = 1/6$. The figure in the centre shows an oppositely directed jet evolving from similar initial conditions, but with initial $z = 0.245$ and $z = -0.245$. The figure on the right displays the "z-asymmetric" jet structure obtained by merging both pairs of sources with *dominant* component belonging to the jet in the left-hand figure having aspect ratio $\mathcal{A}(10^4) = 178.8$.

In order to maintain the dimensional coherence of equations that relate physical observables, it is convenient to take all the elements in V to be physically dimensionless so that all elements in the Clifford algebra $C_{3,1}(V, \breve{g})$ are also physically dimensionless. This can be readily achieved if all local charts for Minkowski spacetime are defined with dimensionless coordinate labels. In which case, $\breve{g} \equiv g$ is dimensionless as in the sections above. In scenarios where one chooses to assign any other physical dimension to the metric tensor g which takes the value Λ_0^2 in some system of physical units (i.e. $[g] = \Lambda_0^2$) then $\breve{g} \equiv g/\Lambda_0^2$ ensures that $\breve{g}$ is physically dimensionless with signature $(3, 1)$ and each element in V is dimensionless. In this section, the Levi-Civita connection associated with the metric tensor $\breve{g}$ will be denoted ∇, as elsewhere.

We denote by $\Gamma\Lambda_{3,1}\mathcal{M}$ the 16-dimensional vector space with algebra product $\wedge$ where an arbitrary element of $\Gamma\Lambda_{3,1}\mathcal{M}$ is a superposition of all p-forms $(p = 0, 1, 2, 3)$. If the coefficients of each p-form in this superposition belong to $\mathbb{R}$ (or $\mathbb{C}$) the elements are said to be *over the field* $\mathbb{R}$ (or $\mathbb{C}$). The space $\Gamma\Lambda_{3,1}\mathcal{M}$ is a space of local sections of the exterior algebra bundle over Minkowski spacetime $\mathcal{M}$ and we call its elements *inhomogeneous forms* when projected to $\mathcal{M}$ by pullback. Homogeneous elements of $\Gamma\Lambda_{3,1}\mathcal{M}$ are then eigen-forms of the involution η where

$$\eta\,\Phi = (-1)^p\,\Phi \qquad \Phi \in \Gamma\Lambda^p\mathcal{M} \quad (p = 0, 1, 2, 3).$$

Suppose σ is a 1-form in $\Gamma\Lambda_{3,1}\mathcal{M}$ and Ω an arbitrary element in $\Gamma\Lambda_{3,1}\mathcal{M}$, then one defines a new algebra $C_{3,1}(V, \breve{g})$ with underlying vector space V and algebra product $\vee$ by the rules:

$$\sigma \vee \Omega = \sigma \wedge \Omega + i_{\tilde{\sigma}}\Omega \tag{44}$$

$$\Omega \vee \sigma = \sigma \wedge \eta\,\Omega - i_{\tilde{\sigma}}(\eta\,\Omega) \tag{45}$$

Journal of Physics: Conference Series **2191** (2022) 012020 doi:10.1088/1742-6596/2191/1/012020

where the tilde map is now taken with respect to $\breve{g}$. The associativity of this Clifford product completely determines the action of $\vee$ between arbitrary forms and the vector space of exterior forms is thereby associated with a Clifford algebra. Thus, if $\{e^a\}$ and $\{X_b\}$ are any dual bases of $\Gamma T^*\mathcal{M}$ and $\Gamma T\mathcal{M}$ respectively, and Ω_1, Ω_2 any exterior forms on $\mathcal{M}$, the relations:

$$\Omega_1 \vee \Omega_2 = \sum_{p=0}^{4} \frac{(-1)^{\lfloor p/2 \rfloor}}{p!} (\eta^p\, i_{X_{a_1}} i_{X_{a_2}} \cdots i_{X_{a_p}} \Omega_1) \wedge (i_{\widetilde{e^{a_1}}} i_{\widetilde{e^{a_2}}} \cdots i_{\widetilde{e^{a_p}}} \Omega_2) \tag{46}$$

$$\Omega_1 \wedge \Omega_2 = \sum_{p=0}^{4} \frac{(-1)^{\lfloor p/2 \rfloor}}{p!} (i_{X_{a_1}} i_{X_{a_2}} \cdots i_{X_{a_p}}\, \eta^p\, \Omega_1) \vee (i_{\widetilde{e^{a_1}}} i_{\widetilde{e^{a_2}}} \cdots i_{\widetilde{e^{a_p}}} \Omega_2) \tag{47}$$

follow recursively from (44) and (45). In these relations the notation $\lfloor r \rfloor$ denotes the integer part of r. If one takes $\Omega \equiv e^a$ and $\sigma \equiv e^b$ in (44), (45) one has:

$$e^b \vee e^a = e^b \wedge e^a + \breve{g}^{-1}(e^a, e^b)$$
$$e^a \vee e^b = -e^b \wedge e^a + \breve{g}^{-1}(e^a, e^b)$$

or

$$e^b \vee e^a + e^a \vee e^b = 2\breve{g}^{-1}(e^a, e^b)$$

in terms of the metric tensor $\breve{g}^{-1}$. We denote sections of the Clifford bundle over $\mathcal{M}$ by $\Gamma C_{3,1}(V, \breve{g})\mathcal{M}$. *In particular*, if $\{e^a\}$ is a *local* $\breve{g}$-*orthonormal basis* (44) and (45) yield:

$$[e^b \vee e^c,\, e^a] = 2(\eta^{ac}\, e^b - \eta^{ab}\, e^c)$$

where $\eta^{ab} \equiv \breve{g}^{-1}(e^a, e^b)$ and

$$[\Omega_1,\, \Omega_2] \equiv \Omega_1 \vee \Omega_2 - \Omega_2 \vee \Omega_1 \tag{48}$$

denotes the *Clifford commutator* between arbitrary elements in the Clifford algebra. If the Levi-Civita connection 1-forms are denoted $\{\omega_{bc}\}$ in this cobasis, one defines for any local vector field $Y \in \Gamma T\mathcal{M}$:

$$\Sigma_Y \equiv \frac{1}{4}\,\omega_{bc}(Y)\, e^b \vee e^c \tag{49}$$

so that the relation

$$\nabla_{X_a} e^c = -\omega^c{}_b(X_a)\, e^b$$

yields the identity

$$\nabla_Y e^a \equiv [\Sigma_Y, e^a], \qquad a = 0, 1, 2, 3 \quad \text{and} \quad Y \in \Gamma T\mathcal{M} \tag{50}$$

in terms of the Clifford commutator. Since the exterior product of *mutually* $\breve{g}$-*orthogonal* 1-forms is the same as their Clifford product then (50) extends to:

$$\nabla_Y e^I = [\Sigma_Y, e^I]$$

where $\{e^I\}$ denotes a $\breve{g}$-orthonormal Clifford product multi-basis for $\Gamma C_{3,1}(V, \breve{g})\mathcal{M}$.

The Clifford algebra $\mathrm{C}_{3,1}(V, \breve{g})$ can be decomposed into four *minimal left ideals* characterised by a complete set of four minimal rank (primitive) idempotents $\{\mathrm{P}_1, \mathrm{P}_2, \mathrm{P}_3, \mathrm{P}_4\}$ that are:

$$\begin{aligned} \text{"pairwise orthogonal"}: \quad \mathrm{P}_i \vee \mathrm{P}_j &= 0 \qquad (\text{for } i \neq j) \\ \text{and idempotent}: \quad \mathrm{P}_i^2 &= \mathrm{P}_i \qquad (i = 1, 2, 3, 4) \end{aligned}$$

with

$$\sum_{j=1}^{4} \mathrm{P}_j = 1.$$

The set is said to be *maximal* when it is not possible to find a set of more idempotents having these properties. The *primitivity* of each P_j implies

$$\mathrm{P}_j \vee \Omega \vee \mathrm{P}_j = 4\, S_0(\Omega \vee \mathrm{P}_j)\, \mathrm{P}_j \qquad \text{for all } \mathrm{P}_j, \Omega \in \Gamma \mathrm{C}_{3,1}(V, \breve{g})\mathcal{M}$$

where the map S_0 extracts the 0-form part of the Clifford product $\Omega \vee \mathrm{P}_j$ using (46).

For example, a particular set of idempotents for $\mathrm{C}_{3,1}(V, \breve{g})$ is given by:

$$\begin{cases} \mathrm{P}_1 \equiv \frac{1}{4}(1 + e^1) \vee (1 + e^0 \vee e^2) \\ \mathrm{P}_2 \equiv \frac{1}{4}(1 + e^1) \vee (1 - e^0 \vee e^2) \\ \mathrm{P}_3 \equiv \frac{1}{4}(1 - e^1) \vee (1 + e^0 \vee e^2) \\ \mathrm{P}_4 \equiv \frac{1}{4}(1 - e^1) \vee (1 - e^0 \vee e^2). \end{cases} \tag{51}$$

Since e^0, e^3 and $e^0 \vee e^3$ all have Clifford squares equal to ± 1, they all have Clifford inverses $(e^0)^{-1}$, $(e^3)^{-1}$ and $(e^0 \vee e^3)^{-1}$ respectively. Thus the above four primitive idempotents are all *similar* under the transformations:

$$\begin{cases} e^3 \vee \mathrm{P}_1 \vee (e^3)^{-1} = \mathrm{P}_3 \\ e^0 \vee \mathrm{P}_1 \vee (e^0)^{-1} = \mathrm{P}_4 \\ (e^0 \vee e^3) \vee \mathrm{P}_1 \vee (e^0 \vee e^3)^{-1} = \mathrm{P}_2. \end{cases} \tag{52}$$

Since $\mathrm{C}_{3,1}(V, \breve{g})$ over $\mathbb{R}$ or $\mathbb{C}$ is isomorphic to a total matrix algebra, it is always possible to construct a basis $\boldsymbol{m}_{ij} \in \Gamma \mathrm{C}_{3,1}(V, \breve{g})\mathcal{M}$ $(i, j = 1, 2, 3, 4)$ for it, satisfying:

$$\begin{aligned} \boldsymbol{m}_{ij} \vee \boldsymbol{m}_{jk} &= \boldsymbol{m}_{ik} \qquad (\text{no sum}) \\ \boldsymbol{m}_{ij} \vee \boldsymbol{m}_{pk} &= 0 \qquad j \neq p,\ k = 1, 2, 3, 4. \end{aligned}$$

There exists a whole class of such bases related by the *inner automorphism*:

$$\boldsymbol{m}_{ij} \longmapsto s \vee \boldsymbol{m}_{ij} \vee s^{-1}$$

where s is any invertible element in $\Gamma C_{3,1}(V, \breve{g})\mathcal{M}$. Elements of this class constitute a matrix basis for $\Gamma C_{3,1}(V, \breve{g})\mathcal{M}$ since if we write $\phi = \sum_{i,j} \phi_{ij}\, \boldsymbol{m}_{ij}$, $\psi = \sum_{i,j} \psi_{ij}\, \boldsymbol{m}_{ij}$ where ϕ_{ij}, ψ_{ij} are complex functions on $\mathcal{M}$ then:

$$\phi \vee \psi = \sum_{i,j} \rho_{ij}\, \boldsymbol{m}_{ij}$$

where $\rho_{ij} \equiv \sum_{k=1}^{4} \phi_{ik}\psi_{kj}$. In terms of $\{\mathrm{P}_j\}$, the basis elements $\{\boldsymbol{m}_{ij}\}$ satisfy:

$$\boldsymbol{m}_{ij} = \mathrm{P}_i \vee c_{ij} \vee \mathrm{P}_j = \mathrm{P}_i \vee c_{ij} = c_{ij} \vee \mathrm{P}_j \qquad \text{(no sum)}$$

for some Clifford sections $c_{ij} \in \Gamma C_{3,1}(V, \breve{g})\mathcal{M}$.

Furthermore, the relations (52) enable one to readily construct a real (4×4)-matrix (Majorana) representation for any real element in $\Gamma C_{3,1}(V, \breve{g})\mathcal{M}$ [5, p. 17].

Any local section $\Psi \in \Gamma C_{3,1}(V, \breve{g})\mathcal{M}$ belongs to a *minimal left ideal* denoted $\Gamma\mathcal{I}_L \mathrm{P}\mathcal{M}$ represented by a local primitive idempotent $\mathrm{P} \in \Gamma C_{3,1}(V, \breve{g})\mathcal{M}$ if $\Psi = \Psi \vee \mathrm{P}$. Such a vector space carries a representation of the groups $\mathrm{PIN}(3,1)$ and $\mathrm{SPIN}(3,1)$. As such, the elements Ψ are identified as local *spinor field representations*. In Minkowski spacetime $\mathcal{M}$ they can be matched across overlapping charts on $\mathcal{M}$ to generate global spinor representations of these spin groups.

Recall that the (torsion-free) Levi-Civita covariant derivative is defined to be a type-preserving operator on sections of the bundle of tensors over spacetime $\mathcal{M}$ satisfying the Leibniz rule (over $\otimes$ and hence $\wedge$):

$$\nabla_Y(\alpha \otimes \beta) = \nabla_Y\alpha \otimes \beta + \alpha \otimes \nabla_Y\beta \qquad \text{for all } Y \in \Gamma T\mathcal{M},\ \alpha \in \Gamma T^{r_1}_{s_1}\mathcal{M},\ \beta \in \Gamma T^{r_2}_{s_2}\mathcal{M}$$

that is also compatible with the spacetime metric-tensor field g:

$$Y\Big(\mathsf{g}(Z_1, Z_2)\Big) = \mathsf{g}(\nabla_Y Z_1, Z_2) + \mathsf{g}(Z_1, \nabla_Y Z_2) \qquad \text{for all } Y, Z_1, Z_2 \in \Gamma T\mathcal{M}.$$

The construction of covariant derivatives of spinor fields differs fundamentally from the construction of the Levi-Civita covariant derivative of tensor fields in a number of ways. For the bundle $\Gamma C_{3,1}(V, \breve{g})\mathcal{M}$ spinor covariant derivatives may be classified in terms of the *involution operators* $\{\xi, \xi\eta, \xi^*, \xi\eta^*\}$ on elements of the Clifford bundle. For any such Clifford sections Ω_1, Ω_2:

$$\begin{aligned}
\eta\Big(\Omega_1 \vee \Omega_2\Big) &= \eta\,\Omega_1 \vee \eta\,\Omega_2\\
\xi\Big(\Omega_1 \vee \Omega_2\Big) &= \xi\,\Omega_2 \vee \xi\,\Omega_1\\
\Big(\Omega_1 \vee \Omega_2\Big)^* &= \Omega_1^* \vee \Omega_2^*.
\end{aligned}$$

In these rules, $*$ denotes complex conjugation and:

$$\xi\Omega = \xi\sum_{r=0}^{4}\Omega_r = \sum_{r=0}^{4}\xi\Omega_r = \sum_{r=0}^{4}(-1)^{\lfloor r/2\rfloor}\Omega_r,$$

where $\Omega_r \in \Gamma\Lambda^r\mathcal{M}$ when Ω is expressed as a superposition of degree r-forms in[7] $\Gamma\Lambda\mathcal{M}$. An alternative notation for $\xi\Omega$ that we shall use below is Ω^ξ. Furthermore, this notation will be

[7] Equivalently, ξ maps any scalar or element of V to itself but reverses the order of elements in any Clifford product.

adopted for any of the involutions given above. If J denotes any one of these four involutions, the *adjoint of any spinor field* $\Psi \in \Gamma\mathcal{I}_L\mathrm{P}\mathcal{M}$ is defined as:

$$\Psi^\dagger \equiv W \vee \Psi^J$$

for some chosen element $W \in \mathrm{P} \vee \Gamma\mathrm{C}_{3,1}(V, \breve{g})\mathcal{M} \vee \mathrm{P}^J$ such that

$$(\Omega \vee \Psi)^\dagger = \Psi^\dagger \vee \Omega^J \qquad \text{for all } \Omega \in \Gamma\mathrm{C}_{3,1}(V, \breve{g})\mathcal{M}. \tag{53}$$

The equation (53) in turn defines an "inner product" or *spinor bilinear form* (,) on the space of spinor sections generated by P and W:

$$(\phi, \psi) \qquad \text{with} \qquad \phi, \psi \in \Gamma\mathcal{I}_L\mathrm{P}\mathcal{M}$$

such that

$$(\phi, \psi)\mathrm{P} = \phi^\dagger \vee \psi.$$

Note that this inner product depends upon both P, the element W and the choice of involution J.

One may now define a spinor-type-preserving *spinor covariant derivative* S_Y with respect to any vector field Y on spacetime (i.e. if $\Psi = \Psi \vee \mathrm{P}$, $S_Y\Psi = S_Y\Psi \vee \mathrm{P}$ for all Y), satisfies a Leibniz rule over Clifford products:

$$S_Y(\,\Omega \vee \Psi\,) = \nabla_Y\Omega \vee \Psi + \Omega \vee S_Y\Psi \qquad \text{for all } \Omega \in \Gamma\mathrm{C}_{3,1}(V, \breve{g})\mathcal{M},\ \Psi \in \Gamma\mathcal{I}_L\mathrm{P}\mathcal{M}$$

and is compatible with the chosen spinor adjoint:

$$Y\,(\Phi, \Psi) = (S_Y\Phi, \Psi) + (\Phi, S_Y\Psi).$$

In general, these last two conditions do not completely fix S_Y for any given P, W, J. However, for our purpose here, we shall choose $J = \xi^*$ in which case the definition

$$\begin{aligned} S_Y : \Gamma\mathcal{I}_L\mathrm{P}\mathcal{M} &\longrightarrow \Gamma\mathcal{I}_L\mathrm{P}\mathcal{M} \\ \Psi &\longmapsto S_Y\Psi = \nabla_Y\Psi \vee \mathrm{P} + (i_Y\lambda)\Psi \end{aligned} \tag{54}$$

uniquely defines S_Y when the 1-form λ satisfies:

$$(i_Y\lambda) \vee W = \frac{1}{2}\mathrm{P} \vee \nabla_Y W \vee \mathrm{P}^J.$$

If follows that S_Y also depends upon P, W and $J \equiv \xi^*$. For other choices of involution, the reader may consult [5, 23, 24].

In Minkowski spacetime there exists a global chart with dimensionless coordinates $\{x^0, x^1, x^2, x^3\}$ and a globally defined ∇-*parallel* $\breve{g}$-orthonormal cobasis of 1-forms $\{e^a = dx^a\}$ $(a = 0, 1, 2, 3)$ satisfying $\nabla e^a = 0$. This implies that the idempotent projectors (51) are all ∇-parallel. Using (49) and (50) the spinor covariant derivative (54) can, in this case, be expressed in terms of Σ_Y as:

$$S_Y\Psi = \nabla_Y\Psi + \Psi \vee \Sigma_Y \qquad \text{for any } \Psi \in \Gamma\mathcal{I}_L\mathrm{P}_j\mathcal{M} \quad (j = 1, 2, 3, 4).$$

In this ∇-parallel cobasis, all connection 1-forms vanish: $\omega^a{}_b = 0$ for all $a, b = 0, 1, 2, 3$, and hence $S_Y \Psi = \nabla_Y \Psi$ for all $Y \in \Gamma T\mathcal{M}$. Therefore, the projectors P_j are *S-parallel with respect to the involution* $J = \xi^*$: $S_Y \mathrm{P}_j = 0$ for all $Y \in \Gamma T\mathcal{M}$. However, in a non-$\nabla$-parallel cobasis $\omega^a{}_b \neq 0$ for some a, b and one may exploit (54) for some alternative choice of P, W or J. For example, if one selects a local cylindrically polar chart with dimensionless coordinates $\{t, \rho, \phi, z\}$ in which a local $\breve{\mathrm{g}}$-orthonormal cobasis is:

$$e^0 = dt, \quad e^1 = d\rho, \quad e^2 = \rho\, d\phi, \quad e^3 = dz,$$

one finds an adapted projector set $\{\widehat{\mathcal{P}}_i\}$ $(i = 1, 2, 3, 4)$ in this chart in which *not all projectors are* ∇, and hence S, parallel with respect to $J = \xi^*$:

$$\begin{cases} \widehat{\mathcal{P}}_1 \equiv \dfrac{1}{4}(1 - e^2) \vee (1 + ie^1 \vee e^3) \\ \widehat{\mathcal{P}}_2 \equiv \dfrac{1}{4}(1 + e^2) \vee (1 + ie^1 \vee e^3) \\ \widehat{\mathcal{P}}_3 \equiv \dfrac{1}{4}(1 - e^2) \vee (1 - ie^1 \vee e^3) \\ \widehat{\mathcal{P}}_4 \equiv \dfrac{1}{4}(1 + e^2) \vee (1 - ie^1 \vee e^3). \end{cases}$$

However, one may find an alternative adapted projector set $\{\mathcal{P}_i\}$ $(i = 1, 2, 3, 4)$ in which all projectors *are* ∇ and S-parallel with respect to $J = \xi^*$:

$$\begin{cases} \mathcal{P}_1 \equiv \dfrac{1}{4}(1 + ie^0) \vee (1 + ie^1 \vee e^2) \\ \mathcal{P}_2 \equiv \dfrac{1}{4}(1 + ie^0) \vee (1 - ie^1 \vee e^2) \\ \mathcal{P}_3 \equiv \dfrac{1}{4}(1 - ie^0) \vee (1 - ie^1 \vee e^2) \\ \mathcal{P}_4 \equiv \dfrac{1}{4}(1 - ie^0) \vee (1 + ie^1 \vee e^2). \end{cases} \tag{55}$$

If $\{e^a\}$ and $\{X_b\}$ are naturally dual bases (not necessarily $\breve{\mathrm{g}}$-orthonormal) we define, in general, the *Dirac differential operator* on spinor sections in $\Gamma\mathcal{I}_L\mathrm{P}\mathcal{M}$:

$$\not{S} \equiv e^a \vee S_{X_a}$$

and for all $\Psi = \Psi \vee \mathrm{P} \in \Gamma\mathcal{I}_L\mathrm{P}\mathcal{M}$ we call

$$\not{S}\Psi = m_0 \Psi, \qquad m_0 \geq 0 \tag{56}$$

a *classical generalised Dirac equation* for an electromagnetically neutral spinor field Ψ with dimensionless mass parameter m_0. Dirac's discovery (1928) of the electron-positron matrix field equation was based on a $U(1)$-coupling to the real Maxwell potential 1-form A on spacetime. In the context of fields in $\Gamma\mathcal{I}_L\mathrm{P}\mathcal{M}$ this employs the *complex* $U(1)$ spinor covariant derivative:

$$S'_Y \Psi = S_Y \Psi + i\, q\, A(Y)\, \Psi$$

and his wave equation is a matrix representation of the Clifford bundle equation:

$$\not{S}\Psi + i\, q\, e^a A(X_a)\, \Psi - m_e \Psi = 0 \qquad \text{for all } \Psi \in \Gamma\mathcal{I}_L\mathrm{P}\mathcal{M},\ q, m_e \in \mathbb{R}^+. \tag{57}$$

In the quantum interpretation of this equation in Minkowski spacetime, stationary time-periodic amplitudes are promoted to operators in a Fock space that create and annihilate "fundamental" electron or positron particles in states with properties obtained from a classical analysis of *separable solutions* to (57). Since the recent discovery of *massive* neutrinos, there is little room left for the physical interpretation of solutions to (56) with $m_0 = 0$ in the Standard Model of fundamental particle physics. Notwithstanding this observation we assert that the same pre-potential complex pulse solution α that generates Maxwell, Bopp-Landé-Podolsky and linearised vacuum Einstein solutions does indeed generate exact classical non-stationary, non-separable solutions to the massless electrically neutral Dirac equation (56) in Minkowski spacetime.

To illustrate this assertion we first choose the minimal left ideal generated by the idempotent P_1. By analogy with (21), we next introduce a complex 2-form $\Pi^{\nu,\chi}$ and construct the 1-form:

$$A^{\nu,\chi} = \star d(\,\alpha\,\Pi^{\nu,\chi}\,)$$

with $\nu \in \{1,2\}$, $\chi \in \{-1,0,1\}$,

$$\begin{cases} \Pi^{1,\pm 1} &= d(x \pm iy) \wedge dt = e^{\pm i\phi}\,(d\rho \pm i\rho\, d\phi) \wedge dt, \\ \Pi^{1,0} &= dz \wedge dt, \\ \Pi^{2,\chi} &= \star\,\Pi^{1,\chi} \end{cases} \tag{58}$$

and define the 2-form:

$$F^{\nu,\chi} = dA^{\nu,\chi}. \tag{59}$$

In a dimensionless Minkowski chart $\{t,x,y,z\}$ with $\breve{g}$-orthonormal cobasis $\{e^0 = dt, e^1 = dx, e^2 = dy, e^3 = dz\}$ consider first the chirality-zero $\Pi^{1,0} = dz \wedge dt \equiv dz \vee dt$. The associated complex 2-form for this choice of $\Pi^{\nu,\chi}$ is then:

$$\begin{aligned} F^{1,0} &= dA^{1,0} = d \star d(\,\alpha\, dz \wedge dt\,) \\ &= \frac{\partial^2\alpha}{\partial t\,\partial x}\, e^0 \vee e^2 - \frac{\partial^2\alpha}{\partial t\,\partial y}\, e^0 \vee e^3 + \frac{\partial^2\alpha}{\partial x\,\partial z}\, e^1 \vee e^2 - \frac{\partial^2\alpha}{\partial x\,\partial y}\, e^1 \vee e^3 - \left(\frac{\partial^2\alpha}{\partial y^2} + \frac{\partial^2\alpha}{\partial z^2}\right) e^2 \vee e^3 \end{aligned}$$

for any complex 0-form $\alpha \in \Gamma\Lambda^0\mathcal{M}$. A direct computation now verifies that if α is the complex pulse solution (19) satisfying

$$\Box\alpha = \delta d\alpha = \frac{\partial^2\alpha}{\partial t^2} - \frac{\partial^2\alpha}{\partial x^2} - \frac{\partial^2\alpha}{\partial y^2} - \frac{\partial^2\alpha}{\partial z^2} = 0$$

then the spinor field $\Psi^{1,0} = \Psi^{1,0} \vee \mathrm{P}_1 \equiv F^{1,0} \vee \mathrm{P}_1 \in \Gamma\mathcal{I}_L\mathrm{P}\mathcal{M}$ satisfies the massless spinor field equation:

$$\not{\$}\Psi^{1,0} = 0. \tag{60}$$

Similarly, in the cylindrical polar chart we write $\Phi^{1,0} \equiv = F^{1,0} \vee \mathcal{P}_1 = \Phi^{1,0} \vee \mathcal{P}_1 \in \Gamma\mathcal{I}_L\mathcal{P}_1\mathcal{M}$ and also find

$$\not{\$}\Phi^{1,0} = 0. \tag{61}$$

This follows since $\Phi^{1,0}$ and $\Psi^{1,0}$ are (differentiable) spinor fields equivalent under the action of elements of the Clifford group that relate the generators of the ideals in different coordinate charts. To appreciate the inner consistency of the above definitions and the significance of these results, we have for any spinor field Ξ:

$$\not{S}\,\Xi \equiv e^a \vee S_{X_a}\Xi.$$

Then for *any* primitive idempotent P and 2-form $\boldsymbol{F}$, with $\Xi = \boldsymbol{F} \vee \mathrm{P}$:

$$\begin{aligned} e^a \vee S_{X_a}(\boldsymbol{F} \vee \mathrm{P}) &= e^a \vee (S_{X_a}\boldsymbol{F} \vee \mathrm{P} + \boldsymbol{F} \vee S_{X_a}\mathrm{P}) \\ &= e^a \vee \nabla_{X_a}\boldsymbol{F} \vee \mathrm{P} + e^a \vee \boldsymbol{F} \vee S_{X_a}\mathrm{P} \\ &= e^a \wedge \nabla_{X_a}\boldsymbol{F} \vee \mathrm{P} + i_{\widetilde{e^a}}\nabla_{X_a}\boldsymbol{F} \vee \mathrm{P} + e^a \vee \boldsymbol{F} \vee S_{X_a}\mathrm{P} \end{aligned}$$

using the spinor Leibniz rule and (44). Thus

$$\not{S}(\boldsymbol{F} \vee \mathrm{P}) = d\boldsymbol{F} \vee \mathrm{P} - \delta\boldsymbol{F} \vee \mathrm{P} + e^a \vee \boldsymbol{F} \vee S_{X_a}\mathrm{P}. \tag{62}$$

Hence if $\boldsymbol{F}$ is *any* closed and co-closed 2-form, the first two terms on the right-hand side of (62) vanish. Furthermore, in the adapted global ∇-parallel Cartesian chart, $S_{X_a}\mathrm{P} = 0$ $(a = 0, 1, 2, 3)$ and one has the result (60) with $\Xi \equiv \Psi^{1,0}$. The covariance of the spinor covariant derivative under ideal transformations induced by coordinate transformations is manifest by the vanishing of all terms on the right-hand side of (62) in the above cylindrically polar Minkowski chart with $\mathrm{P} = \mathcal{P}$. This again yields the result (61) with $\Xi \equiv \Psi^{1,0}$. However, in general not all polar primitives $\mathcal{P}'$ may generate a solution of the form $\boldsymbol{F} \vee \mathcal{P}'$ for closed and co-closed $\boldsymbol{F}$ since not all polar primitives $\mathcal{P}'$ satisfy $S_{X_a}\mathcal{P}' = 0$ with $J = \xi^*$ for all basis elements in the span of $\{X_a\}$. Nevertheless, among the solutions to (61) we may construct propagating, non-singular, spinor pulse-like chiral solutions based upon the pre-potential (42):

$$\alpha(t, \rho, z) = \frac{\ell_0^2}{\rho^2 + [\,a + i(z - t)\,]\,[\,b - i(z + t)\,]} \tag{63}$$

and any complex 2-form $\Pi^{\nu,\chi}$ given by (58).

In sections 3 and 4 we have indicated how to generate new vacuum propagating multi-chiral pulse solutions to the source-free Maxwell, Bopp-Podolsky-Landé and linearised Einstein gravitational field equations in a Minkowski spacetime background by taking suitable derivatives of the *tensor fields* constructed from solutions based upon certain pre-potentials, with respect to Killing vector fields that generate the rotation groups $O(2)$. Although the operator $\mathcal{L}_X$ is well-defined for any vector field $X \in \Gamma T\mathcal{M}$, it will only act as a derivation on a Clifford product of forms if X is a (conformal) Killing vector field K of $\breve{\mathrm{g}}$. In that case, we have for the Levi-Civita covariant derivative ∇ the identity:

$$\mathcal{L}_K\Omega = \nabla_K\Omega + \left[\frac{1}{4}d\widetilde{K}, \Omega\right] \qquad \text{for all } \Omega \in \Gamma\mathrm{C}_{3,1}(V, \breve{\mathrm{g}})\mathcal{M}$$

in terms of the Clifford commutator (48). If K_1, K_2 are any (conformal) Killing vector fields with *Lie bracket* $[K_1, K_2] \equiv \mathcal{L}_{K_1}K_2$ one has the identity:

$$[\mathcal{L}_{K_1}, \mathcal{L}_{K_2}] = \mathcal{L}_{[K_1, K_2]}$$

acting on *tensor* fields. This motivates the definition of a *spinor Lie derivative* operator $\widehat{\mathcal{L}}_K$ acting on any spinor field $\Psi \in \Gamma\mathcal{I}_L\mathrm{P}\mathcal{M}$ where K is any (conformal) Killing vector field on $\mathcal{M}$:

$$\widehat{\mathcal{L}}_K\Psi \equiv S_K\Psi + \frac{1}{4}d\widetilde{K} \vee \Psi,$$

since the right-hand side stays within the ideal generated by P and in terms of Clifford commutators:

$$\left[\widehat{\mathcal{L}}_{K_1}, \widehat{\mathcal{L}}_{K_2}\right] = \widehat{\mathcal{L}}_{[K_1,K_2]}$$

for all (conformal) Killing vector fields K_1, K_2.

As noted above, the *chiral* properties of tensor fields constructed from the pre-potential α and chiral 2-form $\Pi^{\nu,\chi}$ are defined with respect to the Killing vector field $\partial/\partial\phi$ and the tensor operators $\nabla_{\partial/\partial\phi}$ or $\mathcal{L}_{\partial/\partial\phi}$. The spinor fields $\Psi^{1,0}$ and $\Phi^{1,0}$ above are defined in terms of projector idempotents adapted to either the Cartesian or cylindrical polar chart in Minkowski spacetime. These projector sets are special since they are both associated with the differential operator $\mathcal{L}_{\partial/\partial\phi}$ as well as $\not{S}$ (when $\Box\,\alpha = 0$):

$$\mathcal{L}_{\partial/\partial\phi}\mathrm{P}_i = \mathcal{L}_{\partial/\partial\phi}\mathcal{P}_i = \mathcal{L}_{\partial/\partial\phi}\widehat{\mathcal{P}}_i = 0 \qquad (i = 1, 2, 3, 4).$$

This then implies that any spinor solution generated by a primitive P of the form:

$$\Xi^{\nu,\chi} = F^{\nu,\chi} \vee \mathrm{P} \tag{64}$$

with $\mathcal{L}_{\partial/\partial\phi}\mathrm{P} = 0$ satisfies

$$\not{S}\,\Xi^{\nu,\chi} = 0. \tag{65}$$

It is notable that using projectors adapted to different coordinate systems the spinor Lie derivative plays no role in this construction.

Furthermore, by analogy with the construction of multi-chiral Maxwell solutions, it follows that for all chiral solutions $F^{\nu,\chi}$ (59) we can construct a family of *multi-chiral spinor fields* defined by:

$$\Xi^{\nu,r\pm\chi} \equiv F^{\nu,r\pm\chi} \vee \mathrm{P} = \left(\mathcal{L}^{r}_{K_\pm} F^{\nu,\chi}\right) \vee \mathrm{P} \quad \text{for all P such that } \mathcal{L}_{\partial/\partial\phi}\mathrm{P} = 0$$

where $K_\pm$ is the complex Killing vector (25) generating translations in the transverse (x, y)-plane. Then

$$\frac{1}{i}\,\mathcal{L}_{\partial/\partial\phi}\,\Xi^{\nu,r\pm\chi} = (r \pm \chi)\,\Xi^{\nu,r\pm\chi} \qquad (r = 0, 1, 2, \ldots) \tag{66}$$

(i.e. the spinor $\Xi^{\nu,r\pm\chi}$ has chirality $r \pm \chi$). For projectors that also satisfy:

$$e^a \vee F^{\nu,r\pm\chi} \vee S_{X_a}\mathrm{P} = 0 \qquad (r = 0, 1, 2, \ldots) \tag{67}$$

it follows from (62) that such fields are zero-mass Dirac solutions:

$$\not{S}\,\Xi^{\nu,r\pm\chi} = 0. \tag{68}$$

For P belonging to $\{\mathrm{P}_i\}$ and $\{\mathcal{P}_i\}$ $(i = 1, 2, 3, 4)$ we have therefore shown solutions to (67), and hence (68), do exist. Amongst such solutions are those describing a 3-parameter family of propagating, non-stationary, non-singular, pulse-like spinor fields, spatially bounded in all three dimensions.

Journal of Physics: Conference Series **2191** (2022) 012020 doi:10.1088/1742-6596/2191/1/012020

In summary, if $\mathrm{P} \in \Gamma\mathcal{I}_L\mathrm{P}\mathcal{M}$ is an S-parallel Clifford idempotent satisfying $\mathcal{L}_{\partial/\partial\phi}\mathrm{P} = 0$ where $\partial/\partial\phi$ is the real Minkowski Killing vector field given in (24), α any complex scalar field satisfying $\Box\,\alpha = 0$ and $\Pi^{\nu,\chi}$ a ∇-parallel 2-form in Minkowski spacetime satisfying (58), then the spinor field $\Xi^{\nu,\chi} \equiv F^{\nu,\chi} \vee \mathrm{P}$ with $F^{\nu,\chi} \equiv dA^{\nu,\chi}$, $A^{\nu,\chi} \equiv \star d(\,\alpha\Pi^{\nu,\chi}\,)$ will satisfy the Dirac spinor equation (65). All such solutions have chirality $\chi = \{-1, 0, 1\}$. Particular solutions with α given by (63) generate a 3-parameter family of non-singular propagating spinor pulses with chirality χ. Furthermore, in general, spinor fields $\Xi^{\nu,\chi\pm r}$ with chirality $\chi \pm r$ $(r = 0, 1, 2, \ldots)$ can be generated from $F^{\nu,\chi}$ as indicated in (66) by Lie differentiation with respect to the complex Killing vector field $K_\pm$ (25).

6. Concluding Remarks

In this article we have illustrated a technique for analysing a class of linear partial differential equations in terms of non-separable solutions α to the complex scalar Lorentz-covariant Laplace equation $\Box\,\alpha = 0$ and a set of complex, anti-symmetric, rank-two eigen-tensors of a chiral operator in Minkowski spacetime. The technique has been applied to the Maxwell, Bopp-Landé-Podolsky, linearised Einstein and massless electrically-neutral Dirac field systems in source-free spacetime domains. Particular solution families have been constructed for each system that exhibit axially symmetric, dispersive, propagating solutions bounded in all three spatial dimensions of these domains.

Since neutrinos are no longer considered to be massless fundamental particles, the massless Dirac spinor pulses derived here lie outside the Standard Model of particle physics. In principle, such chiral spinor beams could affect the time-like geodesic structure of spacetime via their coupling to gravitation through their stress-energy-momentum tensor and hence influence the geodesic motion of massive test particles. This is analogous to our illustration of how the chiral Maxwell pulses influence the motion of electrically-charged test particles via their coupling through the covariant Lorentz force and to the chiral gravitational pulses that initiate astrophysical jets. The generation of fields with chiral characteristics is a common feature of all solutions constructed with the use of our pre-potentials. As we have shown, these features can be used to model phenomenology on vastly different physical scales. In particular, they may also offer scope for exploring observable properties of pulsars and the formation of spiral galaxies, as well as astrophysical jets.

A further direction for applications is to extend the technique to linear spinor-tensor field equations (for example, the Rarita-Schwinger field equation for spin-3/2 particles). This may offer phenomenological models for systems that exhibit linearised supergravity or supersymmetry (in any spacetime dimension). In this context, we comment that both tensor and spinor field systems arise naturally as different sections of a Clifford bundle over spacetime. Furthermore, the even sub-algebra of the Clifford algebra admits spinor representations (Weyl spinors) and spinor field equations that are also amenable to analysis in terms of chiral pre-potentials.

Finally we note that the chirality of classical free-space propagating solutions should not be confused with the notion of "helicity". In particular, the helicity of free-space photons, regarded as the quanta associated with free-space plane waves, is restricted to ± 1 and is defined in terms of source-free Maxwell separable-mode solutions. Such modes have no longitudinal polarisation in the direction of propagation unlike propagating free-space chiral modes. Whether it is useful to regard any of the particular propagating chiral solutions described in this paper as a Fourier-superposition of quantised fundamental particle (separable) modes, we shall leave to the reader.

Acknowledgments

It is with great pleasure for the authors to pay their respects to Tekin Dereli on his 70th birthday Festschrift celebration. He and RWT have for many years been close research colleagues and family friends. One of the immeasurable benefits arising from scientific collaborations abroad can be discovering the history and cultural heritage of other countries. Besides their mutual enthusiasm for physics, RWT will be forever grateful to Tekin for revealing to him both the history and geography of his beautiful country.

The authors RWT and TJW are also grateful to STFC (ST/G008248/1) and EPSRC (EP/J018171/1) for support, and acknowledge useful discussions with colleagues in the ALPHA-X project and Cockcroft Institute.

References

[1] Bopp F 1940 *Annalen der Physik* **430** 345–384
[2] Landé A 1941 *Physical Review* **60** 121
[3] Podolsky B 1942 *Physical Review* **62** 68
[4] Synge J L 1956 *Relativity: the Special Theory* (North-Holland Publishing Company, Amsterdam)
[5] Benn I M and Tucker R W 1988 *An Introduction to Spinors and Geometry with Applications in Physics* (Adam Hilger: IoP Publishing, Bristol)
[6] Brittingham J N 1983 *Journal of Applied Physics* **54** 1179–1189
[7] Visser M 2003 *Physics Letters A* **315** 219–224
[8] Ziolkowski R W 1985 *Journal of Mathematical Physics* **26** 861–863
[9] Ziolkowski R W 1989 *Phys. Rev. A* **39** 2005
[10] Goto S, Tucker R W and Walton T J 2016 *Journal of Physics A: Mathematical and Theoretical* **49** 265203:1–12
[11] Tucker R W and Walton T J 2009 *Il Nuovo Cimento C* **32** 205–229
[12] Goto S, Tucker R W and Walton T J 2011 *Proceedings of the Royal Society A: Mathematical, Physical and Engineering Sciences* **467** 59–78
[13] Gratus J, Perlick V and Tucker R W 2015 *Journal of Physics A: Mathematical and Theoretical* **48** 435401:1–28
[14] Goto S, Tucker R W and Walton T J 2016 *Nuclear Instruments and Methods in Physics Research Section B: Beam Interactions with Materials and Atoms* **369** 40–44
[15] Stewart J M 1979 *Astronomy and Astrophysics* **367** 527–538
[16] Clark S J and Tucker R W 2000 *Classical and Quantum Gravity* **17** 4125–4157
[17] Tucker R W and Walton T J 2017 *Classical and Quantum Gravity* **34** 035005:1–18
[18] Chicone C, Mashhoon B and Rosquist K 2011 *Physics Letters A* **375** 1427–1430
[19] Tucker R W and Walton T J 2018 *International Journal of Geometric Methods in Modern Physics* **15** 1840004:1–14
[20] Schutz B F 1985 On generalised equations of geodesic deviation *Galaxies, Axisymmetric Systems, and Relativity* (Cambridge University Press, Cambridge) p 237
[21] Philipp D, Puetzfeld D and Lämmerzahl C 2019 On the applicability of the geodesic deviation equation in general relativity *Relativistic Geodesy* (Springer, Cham) pp 419–451
[22] Perlick V 2008 *General Relativity and Gravitation* **40** 1029–1045
[23] Tucker R W 1986 A Clifford calculus for physical field theories *Clifford Algebras and their Applications in Mathematical Physics* (Springer, Dordrecht) pp 177–199
[24] Schray J, Tucker R W and Wang C H T 1996 Lucy: a Clifford algebra approach to spinor calculus *Clifford algebras with numeric and symbolic computations* (Birkhäuser, Boston) pp 121–143

Journal of Physics: Conference Series **2191** (2022) 012021 doi:10.1088/1742-6596/2191/1/012021

A model of non-minimally coupled gravitation and electromagnetism in (1+2) dimensions

Kıvanç İ. Ünlütürk[1] **and Cem Yetişmişoğlu**[2]

Department of Physics, Koç University, 34450, Sarıyer, İstanbul, Turkey

E-mail: [1]`kunluturk17@ku.edu.tr` and [2]`cyetismisoglu@ku.edu.tr`

Abstract. Following earlier works of Dereli and collaborators, we study a three dimensional toy model where we extend the topologically massive gravity with electrodynamics by the most general RF^2-type non-minimal coupling terms. Here R denotes the possible curvature terms and F denotes the electromagnetic 2-form. We derive the variational field equations and look for exact solutions on constant negative curvature space-times with a constant, self-dual electromagnetic field. The notion of self-dual electromagnetic fields in three dimensions is introduced by Dereli and collaborators in the study of exact solutions of models with gravity-electromagnetism couplings. We note the conditions that the parameters of the model have to satisfy for these self-dual solutions to exist.

1. Introduction

Three dimensions has the virtue of providing many interesting toy models. One of the main reasons these toy models are investigated is that these toy models help us gain insight to their more complicated four dimensional analogues. For instance, there is a plethora of three dimensional gravitational models which are studied to better understand the quantum gravity problem in four dimensions, see for instance [1–10]. Although three dimensional general relativity with cosmological constant has no local degrees of freedom [11, 12], one can obtain intriguing models that possess propagating degrees of freedom by adding a Lorentz Chern-Simons term to the action, one of the earliest examples being the Topologically Massive Gravity (TMG) theory [1, 2]. On the one hand its properties have been studied extensively: it admits a celebrated black hole solution [13] and a unitary quantum theory of gravity for the wrong sign of the Einstein-Hilbert term [14] which is also renormalizable [15]. On the other hand, different generalizations that contain additional fields have also been studied, e.g. in [16–19].

In this manuscript we are going to study a generalisation of topologically massive gravity with electrodynamics that contains RF^2-type non-minimal couplings between electrodynamics and gravity. Similar toy models in four dimensions has been considered in [20–24] where in general a single RF^2-type interaction is added to the Einstein-Maxwell theory. Although four dimensional Einstein-Maxwell theory, where the electromagnetic field is minimally coupled to gravity, is classically a well-established theory, non-minimal coupling terms between curvature and electrodynamics may be introduced, motivated by perturbative QED on curved backgrounds, in particular coming from vacuum polarization. On the tree level, photons propagate along null geodesics, whereas at the 1-loop level, due to vacuum polarization, a photon may exist as a virtual electron-positron pair. Consequently, photons attain a size at the order of magnitude of the Compton wavelength of the electron. Therefore, photon propagation can be influenced by the space-time curvature where a curved space-time acts as an optically active medium. Such effects can be conveniently described by non-minimal couplings of the electromagnetic field to gravity [25]. Indeed, the effective action density that arises in QED at the 1-loop level is given by [25]

$$\frac{\gamma_1}{2}R_{ab}\wedge F^{ab}\star F+\frac{\gamma_2}{2}Ric_a\wedge\iota^a F\wedge\star F+\frac{\gamma_3}{2}RF\wedge\star F+\frac{\gamma_4}{2}\mathrm{d}\star F\wedge\star(\mathrm{d}\star F)\,,$$

where the coupling constants $\{\gamma_i\}$ depend on the fine structure constant and mass of the electron. The notation used here will be explained in detail shortly.

Motivated by these interactions, in [20–24] the authors consider various extensions of Einstein-Maxwell theory with RF^2-type interaction terms and look for solutions. Particularly authors, consider an $R_{ab}\wedge F^{ab}\star F$ coupling term and show that a class of exact non-trivial pp-wave solutions exist in [20]; add coupling terms of the form $Y(R)F\wedge\star F$ and show that static, spherically symmetric solutions exist; look at exact solutions of anisotropic inflation for the same coupling in [23]. Finally in [24] authors consider $RF\wedge\star F$ and $RF\wedge F$ couplings and show that there exists a charge screening solution (with vanishing total charge) for this model. That is, a spherically symmetric, asymptotically de Sitter solution with a constant electromagnetic field 2-form exists, provided that the coupling constant for the non-minimal coupling term is related to the radius of space-time in a certain way. Here, we extend three dimensional topologically massive gravity with electrodynamics to a model that contains all of the non-minimal coupling terms coming from QED. It is noteworthy to see similar effects coming from the non-minimal coupling terms of this toy model. We derive the field equations using first order constrained variational principle. We then study an exact class of AdS solutions with a constant self-dual electromagnetic field. The notion of a self-dual electromagnetic 2-form in three dimensions is introduced for the study of exact solutions of Einstein-Maxwell-Chern-Simons [26] and TMG with electrodynamics [18, 19]. Here, what we mean by self-duality[1] is that the Faraday 2-form contains a single electric field component and a magnetic field whose magnitudes are equal. We show that similar to the result in [24], our model admits exact constant self-dual solutions on AdS space-time given that the coupling constants of the model are constrained by two algebraic relations.

The organization of the paper is as follows. We start by giving our notations and conventions. Then, in the second section, we give the action formulation of our model and derive the variational field equations. In the third section we look for the exact solutions to the variational field equations and in the final section we conclude with some final remarks.

Notations and Conventions: We will be using the language of differential forms on three dimensional Riemann-Cartan manifolds. The metric of space-time is expressed as $g=\eta_{ab}e^a\otimes e^b$ where $\eta_{ab}=\mathrm{diag}(-,+,+)$ and $\{e^a\}$ denotes the set-of g-orthonormal co-frames which are dual to the g-orthonormal frames $\{X_a\}$. The frame indices are raised and lowered by the use of η_{ab}. The operations $\star:\Omega^pM\to\Omega^{3-p}M$, $\mathrm{d}:\Omega^pM\to\Omega^{p+1}M$ and $\iota_a:\Omega^pM\to\Omega^{p-1}M$ denote the Hodge duality with respect to the metric g, exterior differentiation and interior product with respect to a frame X_a operations, respectively. We use the shorthand notation $e^{ab\cdots}=e^a\wedge e^b\wedge\cdots$ and $\iota_{ab\cdots}=\iota_a\iota_b\cdots$.

[1] In three dimensions, electromagnetic 2-form cannot be self-dual in the sense that it cannot be in an eigenspace of the Hodge duality operator.

Journal of Physics: Conference Series **2191** (2022) 012021 doi:10.1088/1742-6596/2191/1/012021

A linear metric connection is specified by a set of 1-forms $\{\omega^a{}_b\}$ taking values in the three dimensional Lorentz algebra. The torsion and curvature 2-forms are defined via the first and second structure equations as

$$T^a = \mathrm{d}e^a + \omega^a{}_b \wedge e^b, \tag{1}$$
$$R^a{}_b = \mathrm{d}\omega^a{}_b + \omega^a{}_c \wedge \omega^c{}_b, \tag{2}$$

respectively. They satisfy the Bianchi identities

$$DT^a = R^a{}_b \wedge e^b, \tag{3}$$
$$DR^a{}_b = 0, \tag{4}$$

where $D : \Omega^p M \to \Omega^{p+1} M$ denotes the exterior covariant derivative operator with respect to the linear connection. The Ricci 1-forms $\{Ric_a\}$ and the curvature scalar R are obtained as contractions of the curvature 2-forms, that is

$$Ric_a = \iota_b R^b{}_a, \qquad R = \iota^a Ric_a. \tag{5}$$

In three dimensions, the curvature 2-forms, Ricci 1-forms and the curvature scalar are related to each other by the identity

$$R_{ab} = \epsilon^c{}_{ab} \star Ric_c + \frac{R}{2} e_a \wedge e_b, \tag{6}$$

where ϵ_{abc} is the totally anti-symmetric Levi-Civita symbol with $\epsilon_{012} = 1$.

2. Action and variational field equations

The action functional for our model

$$S[A, e^a, \omega^a{}_b, \lambda^a] = \int_M \mathcal{L} \tag{7}$$

depends on four independent variables: the electromagnetic potential 1-form A with Faraday 2-form $F = \mathrm{d}A$, co-frame 1-forms $\{e^a\}$, connection 1-forms $\{\omega^a{}_b\}$ and Lagrange multiplier 1-forms $\{\lambda^a\}$. The Lagrangian 3-form can be decomposed into four parts as

$$\mathcal{L} = \mathcal{L}_{\text{grav}} + \mathcal{L}_{\text{EM}} + \mathcal{L}_{\text{coup}} + \mathcal{L}_{\text{constraint}}, \tag{8}$$

where $\mathcal{L}_{\text{grav}}$ and $\mathcal{L}_{\text{EM}}$ denote the action densities involving solely gravitational and electrodynamical terms, $\mathcal{L}_{\text{coup}}$ denotes the non-minimal coupling terms and $\mathcal{L}_{\text{constraint}}$ is the constraint term which will set the torsion 2-forms to zero.

The gravitational action density

$$\mathcal{L}_{\text{grav}} = \frac{1}{\mu}\left(\omega^a{}_b \wedge \mathrm{d}\omega^b{}_a + \frac{2}{3}\omega^a{}_b \wedge \omega^b{}_c \wedge \omega^c{}_a\right) + \frac{1}{2} R_{ab} \wedge \star e^{ab} - \Lambda \star 1 \tag{9}$$

contains the usual Einstein-Hilbert term with the cosmological constant Λ. Furthermore, we also couple the Lorentz Chern-Simons term with the coupling constant μ. The electromagnetic action density

$$\mathcal{L}_{\text{EM}} = -\frac{1}{2} F \wedge \star F - \frac{m}{2} A \wedge F, \tag{10}$$

contains the Maxwell density and Maxwell Chern-Simons term with coupling constant m. For the couplings between gravitational and electromagnetic fields, we consider the non-minimal coupling terms of type RF^2, i.e.

$$\begin{aligned}\mathcal{L}_{\text{coup}} &= \frac{\gamma_1}{2}F^{ab}R_{ab}\wedge\star F + \frac{\gamma_2}{2}Ric_a\wedge\iota^a F\wedge\star F + \frac{\gamma_3}{2}RF\wedge\star F\\ &= \left(\frac{\gamma_1}{2}+\frac{\gamma_2}{4}\right)F^{ab}R_{ab}\wedge\star F + \left(\frac{\gamma_2}{4}+\frac{\gamma_3}{2}\right)RF\wedge\star F, \end{aligned}\tag{11}$$

where in the last equality we make use of the curvature identity (6) to get rid of the term proportional to Ricci 1-forms for technical ease. Finally, the constraint term reads

$$\mathcal{L}_{\text{constraint}} = \lambda_a\wedge T^a. \tag{12}$$

Upon the variation of the Lagrange multiplier 1-forms $\{\lambda_a\}$, we obtain that torsion 2-forms vanish identically, that is $T^a = 0$. This ensures that our connection 1-forms are the unique Levi-Civita connections 1-forms. Therefore in what follows, it is understood that all curvatures and covariant derivatives are to be calculated using the Levi-Civita connection 1-forms. The variation with respect to the electromagnetic potential 1-form yield modified Maxwell's equations

$$\begin{aligned}0 = &-\mathrm{d}\star F - mF + \left(\frac{\gamma_1}{2}+\frac{\gamma_2}{4}\right)\mathrm{d}\Big[R\star F - 2\left(\iota^a\star F\right)Ric_a + F_{ab}\star R^{ab}\Big]\\ &+\left(\frac{\gamma_2}{4}+\frac{\gamma_3}{2}\right)\mathrm{d}\left(2R\star F\right). \end{aligned}\tag{13}$$

The variation with respect to the connection 1-forms $\{\omega^{ab}\}$ yield

$$\begin{aligned}\frac{1}{2}\left(e_a\wedge\lambda_b - e_b\wedge\lambda_a\right) = &-\frac{2}{\mu}R_{ab} + \frac{1}{2}D\star e_{ab} + \left(\frac{\gamma_1}{2}+\frac{\gamma_2}{4}\right)D\left(F_{ab}\star F\right)\\ &+\left(\frac{\gamma_2}{4}+\frac{\gamma_3}{2}\right)D\iota_{ba}\left(F\wedge\star F\right). \end{aligned}\tag{14}$$

This equation may be solved algebraically for the Lagrange multiplier 1-forms. The solution reads

$$\lambda_a = -\frac{4}{\mu}Y_a + \left(\gamma_1+\frac{\gamma_2}{2}\right)\left[\iota^b D\left(F_{ba}\star F\right) - \frac{1}{4}\iota^{cb}D\left(F_{bc}\star F\right)e_a\right] \tag{15}$$

$$+\left(\frac{\gamma_2}{2}+\gamma_3\right)\left[\iota^b D\iota_{ab}\left(F\wedge\star F\right) - \frac{1}{4}\iota^{cb}D\iota_{cb}\left(F\wedge\star F\right)e_a\right], \tag{16}$$

where $Y_a = Ric_a - (1/4)Re_a$ are the Schouten 1-forms. Finally, the co-frame variations produce the Einstein field equations

$$\begin{aligned}0 = &\frac{1}{2}\epsilon_{abc}R^{bc} - \frac{\Lambda}{2}\epsilon_{abc}e^{bc} - \frac{1}{2}\iota_a F\wedge\star F + \frac{1}{2}\left(\iota_a\star F\right)F + D\lambda_a\\ &+\left(\frac{\gamma_1}{2}+\frac{\gamma_2}{4}\right)\Big[\frac{1}{2}F^{bc}\left(\iota_a\star R_{bc}\right)F - 2F_a{}^b Ric_b\wedge\star F\\ &\qquad + 2F_a{}^b\left(\iota^c\star F\right)R_{bc} + \frac{1}{2}F^{bc}\left(\iota_a\star F\right)R_{bc}\Big]\\ &+\left(\frac{\gamma_2}{4}+\frac{\gamma_3}{2}\right)\Big[-\epsilon_{abc}\star\left(F\wedge\star F\right)R^{bc} - 2Ric_a F\wedge\star F\Big], \end{aligned}\tag{17}$$

where the Lagrange multiplier 1-forms are given by the expression (16). We note that when the non-minimal coupling constants are taken to be zero, the Einstein field equations (17) and modified Maxwell's equations (13) become that of TMG with electrodynamics as expected. Now, we will be looking for exact solutions of these equations.

Journal of Physics: Conference Series **2191** (2022) 012021 doi:10.1088/1742-6596/2191/1/012021

3. AdS$_3$ solution with constant electromagnetic field

We will be working on constant negative curvature background, that is AdS$_3$ space-time. We will be working with the left-invariant 1-forms on AdS$_3 \cong$ SO(2, 2)/SO(2, 1), which satisfy

$$\mathrm{d}e^a = -\frac{1}{\rho}\epsilon^a{}_{bc}e^b \wedge e^c, \tag{18}$$

where $\rho = (-\Lambda)^{-1/2}$ is the AdS radius.[2] In this coordinate system Levi-Civita connection 1-forms read

$$\omega^a{}_b = -\frac{1}{\rho}\epsilon^a{}_{bc}e^c. \tag{19}$$

Finally, the curvature 2-forms, Ricci 1-forms and curvature scalar are given by

$$R^a{}_b = -\frac{1}{\rho^2}e^a \wedge e_b, \quad Ric_a = -\frac{2}{\rho^2}e_a, \quad R = -\frac{6}{\rho^2}, \tag{20}$$

respectively. For the electromagnetic field, we take the "self-dual" ansatz

$$F = -Ee^{01} + Be^{12}, \tag{21}$$

with $E = kB$ where $k^2 = 1$. Furthermore we assume that the electric and magnetic fields are constant. These assumptions give, in particular, $F \wedge \star F = 0$ which leads to great simplifications. After a tedious computation, Einstein field equations (17) and modified Maxwell's equation (13) become algebraic relations in terms of the AdS radius and coupling constants:

$$2\rho^2 + 14\gamma_1 + \gamma_2 - 12\gamma_3 = 0, \tag{22}$$

$$m\rho^3 + 2\rho^2 + 4\gamma_1 + 8\gamma_2 + 12\gamma_3 = 0. \tag{23}$$

These equations effectively carve out a three dimensional subspace of the parameter space $\{\rho, m, \gamma_1, \gamma_2, \gamma_3\}$. To be concrete, we can take the coupling constants $\gamma_1, \gamma_2, \gamma_3$ for the non-minimal couplings to be arbitrary and they completely determine the AdS radius and electromagnetic Chern-Simons coupling constant. Although the standard values of the parameters γ_i are calculated in four dimensions, to get a rough idea one may calculate the orders of magnitude of ρ and m using these values. This yields

$$\rho \sim 10^{-2}\lambda\!\!\!^{-}_{\mathrm{e}} \sim 10^{-14}\ \mathrm{m},$$
$$m \sim 10 \cdot m_{\mathrm{e}} \sim 10^{-29}\ \mathrm{kg},$$

where $\lambda\!\!\!^{-}_{\mathrm{e}} = \hbar/m_{\mathrm{e}}c$ is the reduced Compton wavelength of the electron.

4. Conclusion

Starting from the action formulation of topologically massive gravity with electrodynamics extended by the non-minimal coupling terms, we derived variational field equations. We showed that there exists exact constant negative curvature background with constant self-dual electromagnetic fields provided that the parameters of the model satisfy the constraints (22)-(23). Although the solution given in this work is a highly restrictive one, the existence of a solution is theoretically stimulating since this shows, in particular, that the model is not inconsistent. Moreover, this simple solution may serve as a starting point for looking for more general solutions. The solution we have found is similar to the work of Dereli and Şenikoğlu [24], where the existence of a spherically symmetric, asymptotically de Sitter solution in four dimensions with a constant electromagnetic field 2-form was shown, for a simple RF^2-type non-minimal interaction.

[2] A detailed account of this coordinate system may be found in the Appendix of [27].

Acknowledgments
The authors would like to thank Prof. Tekin Dereli, in particular for stimulating discussions on three dimensional gravity models and in general for his everlasting guidance. A similar model, which is a subcase of the model considered here, was suggested by Tekin Dereli to one of us (CY) as an independent study problem. Although the independent study was completed, the results were never published.

References
[1] Deser S, Jackiw R and Templeton S 1982 *Phys. Rev. Lett.* **48** 975
[2] Deser S and Jackiw R 1988 *Ann. Phys.* **185** 406
[3] Witten E 1988 *Nucl. Phys. B* **311** 46
[4] Witten E 2007 Three-dimensional gravity revisited (*Preprint* 0706.3359)
[5] Horne J H and Witten E 1989 *Phys. Rev. Lett.* **62** 501–504
[6] Li W, Song W and Strominger A 2008 *JHEP* **04** 082 (*Preprint* 0801.4566)
[7] Bergshoeff E A, Hohm O and Townsend P K 2009 *12th Marcel Grossmann Meeting on General Relativity* pp 2329–2331
[8] Bergshoeff E, Hohm O, Merbis W, Routh A J and Townsend P K 2014 *Class. Quant. Grav.* **31** 145008 (*Preprint* 1404.2867)
[9] Tekin B 2015 *Phys. Rev. D* **92** 024008 (*Preprint* 1503.07488)
[10] Carlip S 1998 *Quantum Gravity in 2+1 Dimensions* (Cambridge University Press)
[11] Deser S and Jackiw R 1984 *Ann. Phys.* **153** 405–416
[12] Deser S, Jackiw R and 't Hooft G 1984 *Ann. Phys.* **152** 220
[13] Bañados M, Teitelboim C and Zanelli J 1992 *Phys. Rev. Lett.* **69** 1849 (*Preprint* hep-th/9204099)
[14] Brown J D and Henneaux M 1986 *Comm. Math. Phys.* **104** 207–226
[15] Oda I 2009 Renormalizability of topologically massive gravity (*Preprint* 0905.1536)
[16] Dengiz S, Kilicarslan E and Tekin B 2012 *Phys. Rev. D* **86** 104014 (*Preprint* 1209.1251)
[17] Deser S and Kay J H 1983 *Phys. Lett. B* **120** 97–100
[18] Dereli T and Sarıoğlu Ö 2000 *Phys. Lett. B* **492** 339–343 (*Preprint* gr-qc/0009090)
[19] Dereli T and Sarıoğlu Ö 2001 *Phys. Rev. D* **64** 027501 (*Preprint* gr-qc/0009082)
[20] Dereli T and Sert O 2011 *Phys. Rev. D* **83** 065005 (*Preprint* 1101.1177)
[21] Dereli T and Sert O 2011 *Eur. Phys. J. C* **71** 1589 (*Preprint* 1102.3863)
[22] Dereli T and Sert O 2011 *Mod. Phys. Lett. A* **26** 1487–1494 (*Preprint* 1105.4579)
[23] Adak M, Akarsu O, Dereli T and Sert O 2017 *JCAP* **11** 026 (*Preprint* 1611.03393)
[24] Dereli T and Şenikoğlu Y 2020 *EPL* **131** 10006 (*Preprint* 2004.07505)
[25] Drummond I T and Hathrell S J 1980 *Phys. Rev. D* **22** 343
[26] Dereli T and Obukhov Yu N 2000 *Phys. Rev. D* **62** 024013 (*Preprint* gr-qc/0001017)
[27] Dereli T and Yetişmişoğlu C 2016 *Phys. Rev. D* **94** 064067 (*Preprint* 1604.00463)

AUTHOR INDEX